CW01370331

PHYSIOLOGICALLY-BASED PHARMACOKINETIC (PBPK) MODELING AND SIMULATIONS

PHYSIOLOGICALLY-BASED PHARMACOKINETIC (PBPK) MODELING AND SIMULATIONS

Principles, Methods, and Applications in the Pharmaceutical Industry

Sheila Annie Peters
AstraZeneca R&D Mölndal
Mölndal, Sweden

WILEY

JOHN WILEY & SONS, INC.

Copyright © 2012 by John Wiley & Sons. All rights reserved.

Published by John Wiley & Sons, Inc., Hoboken, New Jersey
Published simultaneously in Canada

No part of this publication may be reproduced, stored in a retrieval system, or transmitted in any form or by any means, electronic, mechanical, photocopying, recording, scanning, or otherwise, except as permitted under Section 107 or 108 of the 1976 United States Copyright Act, without either the prior written permission of the Publisher, or authorization through payment of the appropriate per-copy fee to the Copyright Clearance Center, Inc., 222 Rosewood Drive, Danvers, MA 01923, 978-750-8400, fax 978-750-4470, or on the web at www.copyright.com. Requests to the Publisher for permission should be addressed to the Permissions Department, John Wiley & Sons, Inc., 111 River Street, Hoboken, NJ 07030, 201-748-6011, fax 201-748-6008, or online at http://www.wiley.com/go/permission.

Limit of Liability/Disclaimer of Warranty: While the publisher and author have used their best efforts in preparing this book, they make no representations or warranties with respect to the accuracy or completeness of the contents of this book and specifically disclaim any implied warranties of merchantability or fitness for a particular purpose. No warranty may be created or extended by sales representatives or written sales materials. The advice and strategies contained herein may not be suitable for your situation. You should consult with a professional where appropriate. Neither the publisher nor author shall be liable for any loss of profit or any other commercial damages, including but not limited to special, incidental, consequential, or other damages.

For general information on our other products and services or for technical support, please contact our Customer Care Department within the United States at 877-762-2974, outside the United States at 317-572-3993 or fax 317- 572-4002.

Wiley also publishes its books in a variety of electronic formats. Some content that appears in print may not be available in electronic formats. For more information about Wiley products, visit our web site at www.wiley.com.

Library of Congress Cataloging-in-Publication Data:

Peters, Sheila Annie.
 Physiologically-based pharmacokinetic (PBPK) modeling and simulations : principles, methods, and applications in the pharmaceutical industry / Sheila Annie Peters.
   p. ; cm.
 Includes bibliographical references and index.
 ISBN 978-0-470-48406-7 (cloth)
  1. Pharmacokinetics—Mathematical models. 2. Drugs—Design—Mathematical models. I. Title.
 [DNLM: 1. Pharmacokinetics. 2. Drug Design. 3. Models, Biological. QV 38]
 RM301.5.P48 2011
 615′.7—dc23

2011019939

Printed in the United States of America

10 9 8 7 6 5 4 3 2 1

This book is dedicated to my parents, friends, and Alfred and Christina who have always believed in me.

CONTENTS

Preface	xv
Acknowledgments	xvii

SECTION I. PRINCIPLES AND METHODS — 1

1 MODELING IN THE PHARMACEUTICAL INDUSTRY — 3

1.1	Introduction	3
1.2	Modeling Approaches	4
1.3	Steps Needed to Maximize Effective Integration of Models into R&D Workflow	7
1.4	Scope of the Book	8
	Keywords	10
	References	12

2 PHYSIOLOGICALLY-BASED MODELING — 13

2.1	Introduction	13
2.2	Examples of Physiological Modeling	14
2.3	Need for Physiological Models in the Pharmaceutical Industry	15
2.4	Organs as Compartments	15
2.5	Bottom-Up vs. Top-Down Modeling in Pharmacokinetics	16
	References	16

3 REVIEW OF PHARMACOKINETIC PRINCIPLES — 17

3.1	Introduction	18
3.2	Routes of Administration	18
3.3	Drug Disposition	18
	3.3.1 Absorption	18

vii

	3.3.2	Plasma Protein Binding, Blood–Plasma Ratio	20
	3.3.3	Distribution, Elimination, Half-Life, and Clearance	23
	3.3.4	Role of Transporters in ADME	29
3.4	Linear and Nonlinear Pharmacokinetics		34
3.5	Steady-State Pharmacokinetics		34
3.6	Dose Estimations		37
3.7	Successful PK Optimization in Drug Discovery		40
Keywords			40
References			41

4 PHYSIOLOGICAL MODEL FOR ABSORPTION 43

4.1	Introduction		44
4.2	Drug Absorption and Gut Bioavailability		44
	4.2.1	Solubility and Dissolution Rate	44
	4.2.2	Permeability: Transcelluar, Paracellular, and Carrier-Mediated Pathways	51
	4.2.3	Barriers to Membrane Transport—Luminal Degradation, Efflux, and Gut Metabolism	53
4.3	Factors Affecting Drug Absorption and Gut Bioavailability		56
	4.3.1	Physiological Factors Affecting Oral Drug Absorption and Species Differences in Physiology	56
	4.3.2	Compound-Dependent Factors	62
	4.3.3	Formulation-Dependent Factors	63
4.4	*In Silico* Predictions of Passive Permeability and Solubility		66
	4.4.1	*In Silico* Models for Permeability	66
	4.4.2	*In Silico* Models for Solubility	67
4.5	Measurement of Permeability, Solubility, Luminal Stability, Efflux, and Intestinal Metabolism		67
	4.5.1	*In Vitro*, *in Situ* and *in Vivo* Assays for Permeability	67
	4.5.2	Measurement of Thermodynamic or Equilibrium Solubility	72
	4.5.3	Luminal Stability	74
	4.5.4	Efflux	74
	4.5.5	*In Vitro* Models for Estimating Extent of Gut Metabolism	76
4.6	Absorption Modeling		76
Keywords			83
References			84

5 PHYSIOLOGICAL MODEL FOR DISTRIBUTION 89

    5.1 Introduction   90
    5.2 Factors Affecting Tissue Distribution of Xenobiotics   91
        5.2.1 Physiological Factors and Species Differences in Physiology   91
        5.2.2 Compound-Dependent Factors   98
    5.3 *In Silico* Models of Tissue Partition Coefficients   98
    5.4 Measurement of Parameters Representing Rate and Extent of Tissue Distribution   105
        5.4.1 Assessment of Rate and Extent of Brain Penetration   105
    5.5 Physiological Model for Drug Distribution   110
    5.6 Drug Concentrations at Site of Action   111
    Keywords   114
    References   115

6 PHYSIOLOGICAL MODELS FOR DRUG METABOLISM AND EXCRETION 119

    6.1 Introduction   119
    6.2 Factors Affecting Drug Metabolism and Excretion of Xenobiotics   120
    6.3 Models for Hepatobiliary Elimination and Renal Excretion   124
        6.3.1 *In Silico* Models   124
        6.3.2 *In Vitro* Models for Hepatic Metabolism   125
        6.3.3 *In Vitro* Models for Transporters   127
    6.4 Physiological Models   136
        6.4.1 Hepatobiliary Elimination of Parent Drug and Metabolites   136
        6.4.2 Renal Excretion   141
    References   144

7 GENERIC WHOLE-BODY PHYSIOLOGICALLY-BASED PHARMACOKINETIC MODELING 153

    7.1 Introduction   153
    7.2 Structure of a Generic Whole Body PBPK Model   154
    7.3 Model Assumptions   157
    7.4 Commercial PBPK Software   158
    References   159

8 VARIABILITY, UNCERTAINTY, AND SENSITIVITY ANALYSIS 161

8.1 Introduction 161
8.2 Need for Uncertainty Analysis 162
8.3 Sources of Physiological, Anatomical, Enzymatic, and Transporter Variability 163
8.4 Modeling Uncertainty and Population Variability with Monte Carlo Simulations 169
8.5 Sensitivity Analysis 172
8.6 Conclusions 174
Keywords 174
References 175

9 EVALUATION OF DRUG–DRUG INTERACTION RISK WITH PBPK MODELS 183

9.1 Introduction 184
9.2 Factors Affecting Drug–Drug Interactions 186
9.3 *In Vitro* Methods to Evaluate Drug–Drug Interactions 190
    9.3.1 Candidate Drug as a Potential Inhibitor 190
    9.3.2 Candidate Drug as a Potential Victim of Inhibition 192
9.4 Static Models to Evaluate Drug–Drug Interactions 193
9.5 PBPK Models to Evaluate Drug–Drug Interactions 195
    9.5.1 Intrinsic Clearance of Victim (V) in the Absence of Inhibitor or Inducer 195
    9.5.2 Intrinsic Clearance of Victim (V) in the Presence of Inhibitor 196
    9.5.3 Time-Dependent Changes in the Abundance of an Enzyme Isoform Inhibited by an MBI 197
    9.5.4 Intrinsic Clearance of Victim (V) in the Presence of Inducer 197
9.6 Comparison of PBPK Models and Static Models for the Evaluation of Drug–Drug Interactions 198
Keywords 201
References 202

10 PHYSIOLOGICALLY-BASED PHARMACOKINETICS OF BIOTHERAPEUTICS 209

10.1 Introduction 210
10.2 Therapeutic Proteins 210

	10.2.1 Peptides and Proteins	210
	10.2.2 Monoclonal Antibodies	212
10.3	Pharmacokinetics of Therapeutic Proteins	214
	10.3.1 Peptides and Proteins	215
	10.3.2 Monoclonal Antibodies	224
10.4	PBPK/PD Modeling for Therapeutic Proteins	230
	10.4.1 Need for PBPK Modeling for Therapeutic Proteins	230
	10.4.2 PBPK Modeling for Therapeutic Proteins	231
	10.4.3 Pharmacokinetic Scaling	239
	10.4.4 Applications of PBPK Models of Therapeutic Proteins	242
	10.4.5 PBPK Integration with Pharmacodynamics	244
10.5	Antisense Oligonucletides and RNA Interferance	245
	10.5.1 Antisense Oligonucletides (ASOs)	245
	10.5.2 Ribonucleic Acid Interference (RNAi)	245
	10.5.3 Pharmacokinetics of ASOs[50] and Double-Stranded RNAs	247
	10.5.4 Design and Modifications of ASOs to Improve Target Affinity and PD: the First, Second, and Third Generation ASOs	249
	10.5.5 Integration of PK/PBPK and PD Modeling	253
Keywords		254
References		256

SECTION II. APPLICATIONS IN THE PHARMACEUTICAL INDUSTRY   261

11 DATA INTEGRATION AND SENSITIVITY ANALYSIS   263

11.1 Introduction	263
11.2 Examples of Data Integration with PBPK Modeling	264
11.3 Examples of Sensitivity Analysis with PBPK Modeling	267
References	271

12 HYPOTHESIS GENERATION AND PHARMACOKINETIC PREDICTIONS   273

12.1 Introduction	274
12.2 PBPK Simulations of Pharmacokinetic Profiles for Hypothesis Generation and Testing	274

	12.2.1 Methodology	274
	12.2.2 *In Vivo* Solubility	278
	12.2.3 Delayed Gastric Emptying	280
	12.2.4 Regional Variation in Intestinal Loss: Gut Wall Metabolism, Intestinal Efflux, and Luminal Degradation	282
	12.2.5 Enterohepatic Recirculation	284
	12.2.6 Inhibition of Drug-Metabolizing Enzymes	286
	12.2.7 Inhibition of Hepatic Uptake	286
	12.2.8 Inhibition of Hepatobiliary Efflux	290
12.3	Pharmacokinetic Predictions	293
	12.3.1 Human Predictions from Preclinical Data	293
	12.3.2 Pharmacokinetic Predictions in Clinical Development	293
References		294

13 INTEGRATION OF PBPK AND PHARMACODYNAMICS — 299

13.1	Introduction	300
13.2	Pharmacodynamic Principles	300
	13.2.1 Pharmacological Targets and Drug Action	300
	13.2.2 Functional Adaptation Processes: Tolerance, Sensitization, and Rebound (Fig 13.2)	301
13.3	Pharmacodynamic Modeling	307
	13.3.1 Concentration–Effect, Dose–Response Curves, and Sigmoid E_{max} Models	307
	13.3.2 Mechanism-Based PD Modeling	315
	13.3.3 Simple Direct Effects	315
	13.3.4 Models Accommodating Delayed Pharmacological Response	321
	13.3.5 Models Accommodating Nonlinearity in Pharmacological Response with Respect to Time	332
13.4	Pharmacokinetic Modeling: Compartmental PK and PBPK	335
13.5	Integration of PK or PBPK with PD Modeling	335
13.6	Reasons for Poor PK/PD Correlation	339
13.7	Applications of PK or PBPK/PD Modeling in Drug Discovery and Development	340
	13.7.1 Need for a Mechanistic PBPK/PD Integration	341

	13.7.2 Applications of PK or PBPK/PD in Drug Discovery	342
	13.7.3 Applications of PK or PBPK/PD in Drug Development	360
13.8	Regulatory Perspective	370
13.9	Conclusions	371
	Keywords	372
	References	376

14 PHYSIOLOGICALLY-BASED PHARMACOKINETIC MODELING OF POPULATIONS   383

14.1	Introduction	383
14.2	Population Modeling with PBPK	384
14.3	Healthy to Target Patient Population: Impact of Disease on Pharmacokinetics	386
14.4	Modeling Subpopulations: Impact of Age, Gender, Co-morbidities, and Genetics on Pharmacokinetics	389
14.5	Personalized Medicine with PBPK/PD	392
	Keyword	395
	References	396

15 PBPK MODELS ALONG THE DRUG DISCOVERY AND DEVELOPMENT VALUE CHAIN   401

15.1	Summary of Applications of PBPK Models along Value Chain	401
15.2	Obstacles and Future Directions for PBPK Modeling	403
	Keyword	405
	References	405

Appendices	**407**
Index	**423**

PREFACE

Physiologically-based pharmacokinetic (PBPK) modeling has made rapid strides in the pharmaceutical industry in the last decade or so, thanks to an increasing awareness of the potential applications of this powerful tool. As pharmaceutical companies are working to integrate PBPK modeling into their lead selection cycle and clinical development, the availability of commercial software has played a key role in enabling even those without modeling expertise to come on board. However, this entails the risk of misuse, misinterpretation, or overinterpretation of modeling results, if the principles and underlying assumptions of PBPK modeling are not clearly understood by the users. Today, the challenge facing pharmaceutical companies is educating and training their staff to achieve an effective application of PBPK/ pharmacodynamics (PD) in projects across the value chain. In the future, providers of education should take on the responsibility of making available, modelers with appropriate skills. Given the complexity of PBPK modeling, it is certainly not an easy task for a beginner with little or no background to understand the model structure and to be aware of its limitations. The lack of a textbook on PBPK has been a further deterrent. It is hoped that this book will serve as a primary source of information on the principles, methods, and applications of PBPK modeling, exposing the power of a largely hidden and unexplored tool. Applications in the pharma sector will be the main focus, as applications in environmental toxicology and human health risk assessment have already been the subject of a previous publication.

Target audiences for the book include students and researchers in academia, apart from scientists and modelers in the pharmaceutical industry. The book can also be a resource for R&D managers in the pharmaceutical industry, seeking a quick overview of the benefits of applying PBPK modeling along the drug discovery and development value chain. An understanding of the principles of PBPK modeling by R&D management would enhance their acceptance and appreciation, which in turn can translate to effective managerial support for PBPK modeling. This book is intended to serve the interests of both the general reader, who may only want an overview of the applications of PBPK modeling without wanting an in-depth understanding of the underlying methods, and the specialist reader, who may be interested to build new models. For the general reader, keywords appear in boldface and are explained at the end of the chapters. No particular expertise is assumed in order to keep the book accessible to a diverse audience. An extensive list of bibliographic references will help the specialist reader to build on the concepts developed in the book. A generous use of figures to illustrate concepts will help the reader gain valuable insights into this fascinating subject.

The book comprises two parts. The first part provides a detailed and systematic treatment of the principles behind physiological modeling of pharmacokinetic processes, interindividual variability, and drug interactions for small-molecule drugs and biologics. The second part exposes the reader to the powerful applications of **PBPK** modeling along the value chain in drug discovery and development.

SHEILA ANNIE PETERS

ACKNOWLEDGMENTS

I would like to thank Bernd Meibohm and several of my colleagues at AstraZeneca—Balaji Agoram, Ulf Bredberg, James Bird, Hugues Dolgos, Ulf Ericsson, Marcus Friden, Rasmus Jansson Löfmark, Martin Hayes, Sarah Kelly, Maria Learoyd, James Tucker, Pete Webborn, and Anna-Lena Ungell, who helped review the chapters. I would like to record my deep appreciation for the meticulous work of Tony Johansson, whose positive attitude and hard work has resulted in the excellent figures in this book. This work would not have been possible without the consistent support extended by my friends and family.

S. A. P.

SECTION I

PRINCIPLES AND METHODS

1

MODELING IN THE PHARMACEUTICAL INDUSTRY

CONTENTS

1.1 Introduction   3
1.2 Modeling Approaches   4
1.3 Steps Needed to Maximize Effective Integration of Models
     into R&D Workflow 7
1.4 Scope of the Book 8
     Keywords 10
     References. 12

1.1 INTRODUCTION

In an effort to reduce the attrition rates of drugs, pharmaceutical companies are constantly looking to improve and understand compound behavior through the use of novel tools. Modeling is one such tool that has gradually gained recognition in the pharmaceutical industry, over the last couple of decades, as a means of achieving quality, efficiency, and significant cost savings. Modeling

Physiologically-Based Pharmacokinetic (PBPK) Modeling and Simulations: Principles, Methods, and Applications in the Pharmaceutical Industry, First Edition. Sheila Annie Peters.
© 2012 John Wiley & Sons, Inc. Published 2012 by John Wiley & Sons, Inc.

and simulation methods have played a crucial role in the pharmaceutical industry in identifying and validating target, predicting the efficacy, absorption, distribution, metabolism, excretion, toxicity (ADMET), and safety of drug candidates, aiding a better understanding of data through effective integration and extraction of knowledge, predicting the human dose, developing new formulations, designing safety and efficacy trials, and guiding regulatory decisions. Most models are used in a build–validate–learn–refine cycle in which all available knowledge that can aid prediction of a property of interest is initially captured during model building. It is then used for predicting observations (validation phase), and any discrepancies of the predicted from observed is then understood on a scientific basis (learning phase) and appropriately incorporated in the model (refine phase). Once a model has been tested to provide satisfactory results, it can be used on a routine basis, reserving animal studies and other resource-intensive experiments for confirmation only. The use of *in silico* technologies can reduce the cost of drug development by up to 50% according to some analysts.[1] The impact of integrating modeling into the research and development (R&D) workflow has been so encouraging that many companies have increased their investments in this sector. The Food and Drug Administration (FDA) "critical path" document[2] recommends **model-based drug development** for improved knowledge management and decision making. The key elements of such a model-based drug development and how they fit together to aid strategy and decision making in drug development is outlined by Lalonde et al.[3]

1.2 MODELING APPROACHES

From understanding a disease to bringing a safe and effective new treatment to patients, it takes about 10–15 years for a pharmaceutical company to discover a potential drug (drug discovery) and to develop it as a final product (drug development). A schematic of a drug discovery and development pipeline is shown in Figure 1.1. Advances in genomics and proteomics and an increase in computational power have contributed to increasing our knowledge of disease at the level of genes, proteins, and cells. This understanding leads to the identification of proteins, which are involved in a disease of interest. A single protein/gene that has been validated to be relevant in a disease and to be **druggable** is chosen as the target. Hits to this target are identified through **virtual screening** and high-throughput screening (HTS) assays. Compounds that can best modulate the target are chosen as hits. Hits are classified into a small set of lead series (**lead generation**). The most promising series showing potential drug activity, reduced off-target toxicity, and with physicochemical and metabolic profiles that are compatible with acceptable *in vivo* bioavailability progress into the **lead optimiztion** stage. The objective at this stage is to select a candidate drug that meets predefined criteria with respect to efficacy,

1.2 MODELING APPROACHES

pharmacokinetics, and safety. The candidate drug is then developed to a final drug product, after sufficient testing in animals (preclinical development) and humans (clinical development) to confirm the efficacy and safety of the drug. The modeling methods along the drug discovery and development value chain are indicated in Figure 1.1.

Models allow us to understand how complex interactions and processes work. Sometimes, modeling provides a unique way to understanding a system. Figure 1.2 summarizes the reasons for employing models in the pharmaceutical industry. It is important to be aware that all models are only approximations of the system they represent. Underlying assumptions should be carefully weighed to get the best benefits from a model. In addition, different modeling approaches differ in their strengths and limitations. Quantitative structure–activity relationships (**QSAR**) and quantitative structure property relationships (**QSPR**) **models** rely on combining appropriate descriptors for compounds in a training set. Their key strengths are simplicity and ease of use. However, the predictive power of these models is restricted to compounds within the same chemical space as that of the training set. Empirical or data-driven models are built and refined only after the experimental data is collected (cannot be prespecified) and its parameters lack physical/physiological/biochemical interpretation. They are best employed for exploratory data analysis. On the other hand, mechanistic models are prespecified and capture the underlying mechanisms of the system they represent to the extent known, with parameters corresponding to some physical entities of the system. These models can, therefore, be used to predict the next set of data. An example of this is physiologically-based models.

Pharmacokinetic (PK) modeling provides information about processes that affect the kinetics of a compound in a species, such as absorption, distribution, metabolism, and excretion using the concentration–time profile. Traditionally, this has been done with **compartmental PK modeling**, a data-driven approach in which the model structure is defined by the data (therefore empirical). The fall in concentration with respect to time is fitted to a series of exponential terms whose decay constants and pre-exponents are related to the rates of absorption, distribution, metabolism, and elimination. The model that best fits the *in vivo* PK data, according to some defined statistical criterion, is chosen as the final model. Empirical models are case specific and the potential for credible extrapolations using these models is limited. Since the pharmacodynamic response of a drug need not necessarily parallel its pharmacokinetics, PK models are combined with pharmacodynamic (PD) effect–concentration profiles at different doses. This data-driven, exploratory **PK/PD modeling**[4] has long been used in drug development, for getting a continuous description of the effect–time course resulting directly from the administration of a certain dose. Physiologically-based pharmacokinetic (PBPK) modeling offers a mechanistic approach to predicting the disposition of a drug, which can then be combined with a PD model (**PBPK/PD**). Although the principles behind a PBPK approach has long been known through the work of Teorell[5] in 1937, the mathematical complexity of the model and the lack of physiological data needed for the

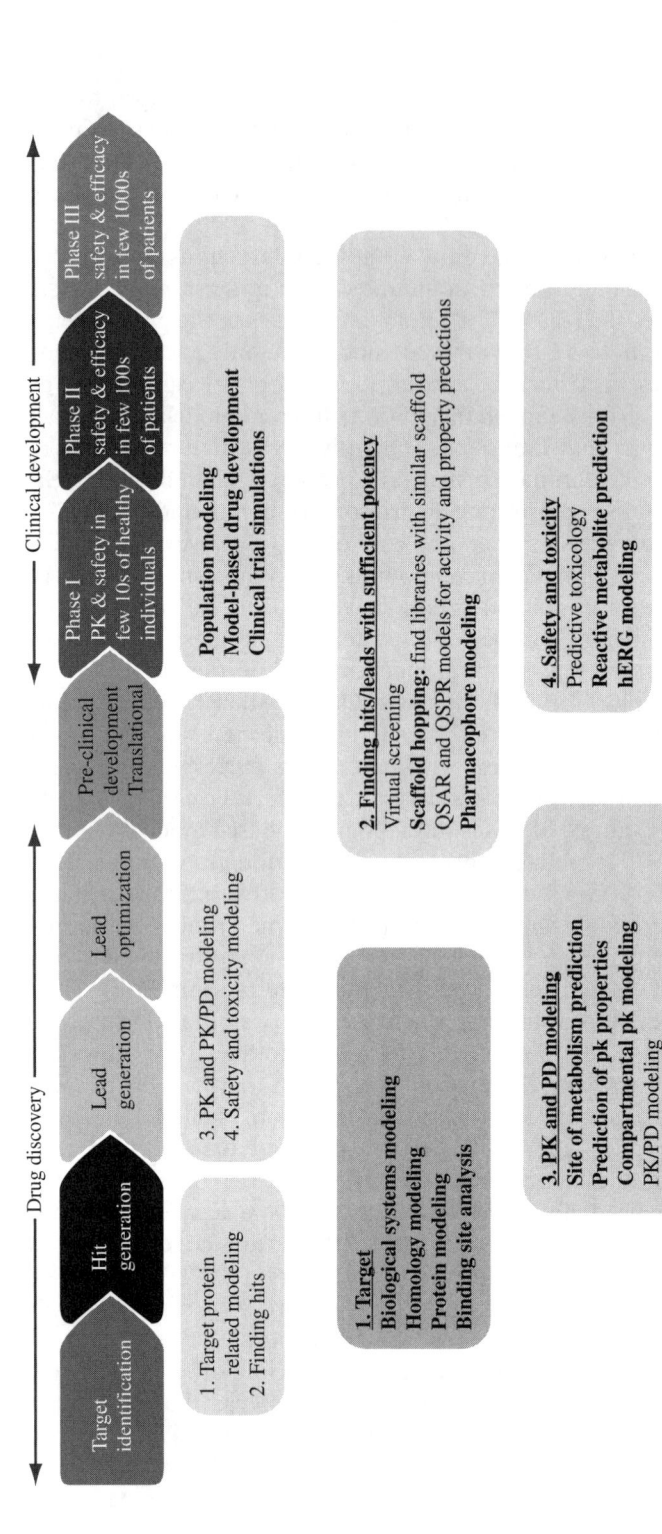

Figure 1.1. Modeling at various stages of drug discovery and development.

1.3 STEPS NEEDED TO MAXIMIZE EFFECTIVE INTEGRATION OF MODELS   7

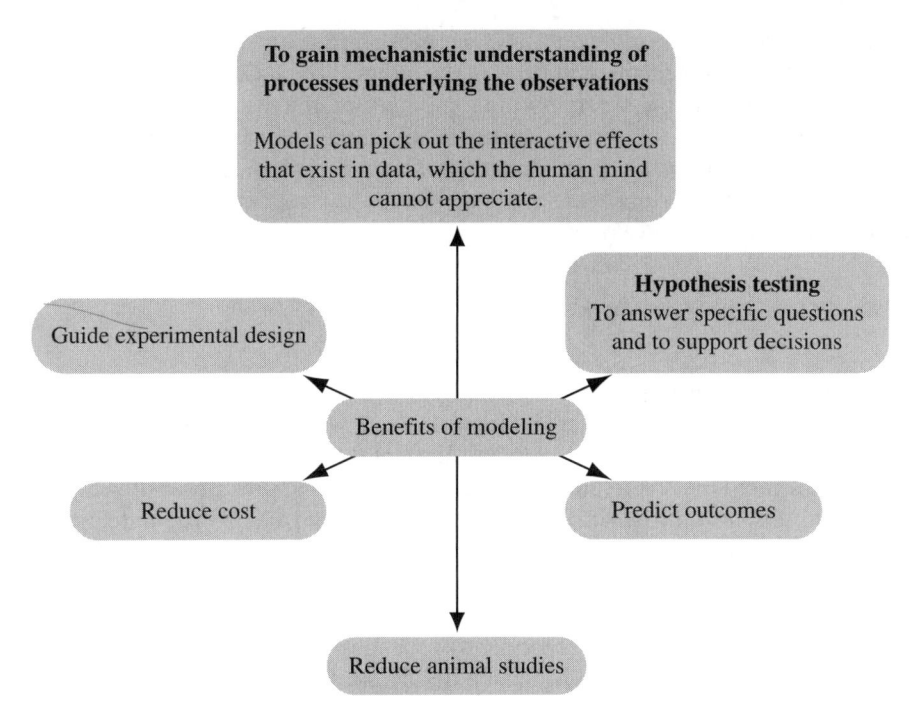

Figure 1.2. Need for modeling.

model meant that the idea remained dormant until the 1960s.[6] The tremendous increase in computational power at relatively low cost paved the way for complex PBPK models to be built. PBPK models help simulate the concentration–time profile of a drug in a species by integrating the physicochemical properties of the compound with the physiology of the species. Being mechanistic, PBPK models can be used to simulate and to predict the next set of data and to plan the next experiment.

1.3 STEPS NEEDED TO MAXIMIZE EFFECTIVE INTEGRATION OF MODELS INTO R&D WORKFLOW

Although PBPK models were developed for cancer drugs even during the 1960s and 1970s by Bischoff et al.[7] and Bischoff and Dedrick,[8] the pharma industry has been slow to exploit the power of PBPK. While the importance of integrating modeling, simulation, and other *in silico* technologies in the R&D workflow is clearly acknowledged by leaders in the industry and by regulatory authorities, practical implementation has been slow especially in some areas of modeling. A number of reasons have been identified. The lack of trained/skilled scientists, sceptical attitude from project teams, and lack of commitment on the part of leadership to implement are the most important among them.[9] In all this, the role of management in driving the integration is seen as key to bringing about a change in the workflow and mindset of the scientists as well as to allocate resources for training scientists. Gaining acceptance among project teams is vital to ensure that modeling results are seriously considered and incorporated in decisions, thus paving the way for cost-effective and efficient drug discovery and development.

1.4 SCOPE OF THE BOOK

Physiologically-based pharmacokinetic modeling for the discovery and development of small-molecule and biological drugs will be the main focus of the book, as applications of PBPK in environmental toxicology and human health risk assessment have already been the subject of a previous publication.[10] The chapters in the first section will cover the basics of PBPK modeling and simulation, while the second section will deal with its applications in drug discovery and development.

Chapters 2–6 will elaborate on the principles essential for integrating species physiology with compound-dependent properties. Chapter 7 will put together all of the absorption, distribution, metabolism, and excretion (ADME) physiological models for small-molecule drugs.

Physiologically-based PK modeling involves the use of a number of compound-dependent and physiology-dependent parameters. Being a parameter-intensive model, the predicted outcome could be associated with a high level of uncertainty. It is, therefore, important to consider the propagation of error arising from the uncertainties in input parameters. These uncertainties can be modeled using the Monte Carlo approach, which forms the subject of Chapter 8.

As late failures in the drug development process become more costly, the desire to evaluate the potential for risks earlier in the drug discovery process has become a growing industry trend. An early assessment of the potential for drug–drug interactions (DDI) with co-medications mediated by inhibition/induction of cytochrome P450 (CYP) enzymes or from transporters is, therefore, seen as imperative even in the lead optimization stage. PBPK models provide a mechanistic approach to integrating relevant information on a potential inhibitor and a substrate for the prediction of DDI risk. Chapter 9 details the differential equations that describe the mutually dependent kinetics of coadministered drugs and wraps up with a discussion on the advantages of physiological models over static models in the evaluation of drug–drug interactions.

Biologicals (or biologics) are fast emerging as alternative therapeutics to small molecules. Biologicals are proteins such as monoclonal antibodies, cytokines, growth factors, enzymes, and thrombolytics that can treat a variety of diseases. Since the launch of Eli Lilly's recombinant human insulin in 1982, more than 100 biologicals have received marketing approval in the United States, highlighting their importance as a source of new drugs and new revenues. With an increasing fraction of pharmaceutical R&D devoted to

1.4 SCOPE OF THE BOOK

biologicals, it is expected to have a significant role in drug development in the future. Chapter 10 is devoted to examining the differences between biologicals and small molecules with respect to PK behavior and how these differences can be accommodated within PBPK models.

Section II of the book will cover applications of PBPK modeling in drug discovery and development with examples. Applications in the pharmaceutical sector will be the main focus. PBPK modeling can be used as a prediction, simulation, or as an extrapolation tool. PK properties such as absorption, distribution, and elimination of compounds are influenced not only by compound properties but also by the physiology of the species in which they are observed. PBPK modeling attempts to integrate available structural, *in silico* or *in vitro* physicochemical, and human-specific biochemical compound data in a physiological context for the predictions of PK parameters such as absorption and distribution or time profiles of plasma concentrations of drugs. Chapter 11 describes how PBPK models provide an excellent framework for enabling data integration and human PK predictions. Chapter 11 also describes the applications of parameter sensitivity analysis for optimizing lead compounds during drug discovery. In the lead optimization stage, understanding the effects of modulating key ADME-determining compound-dependent properties on a desired PK outcome is often needed in order to optimize the physicochemical space. The PK outcome could be metabolic liability, absorption, distribution, or bioavailability of compounds. The effects of modulation depend very much on the physicochemical space chosen initially.

The value of a PBPK model as a prediction tool is sometimes limited by the lack of reliable input parameters especially for clearance, where the *in vitro* measurements for intrinsic clearance rarely match up to the *in vivo*. The mechanistic structure of PBPK models can be better exploited when it is used as a simulation tool. In a simulation, the focus is not on quantitative predictions. Instead, the emphasis is on gaining valuable insights into processes driving the pharmacokinetics of a compound, through hypothesis generation and testing. This neglected area, holding the promise of improving the quality of selected leads, reducing animal studies and cost, is the subject of Chapter 12. The mechanistic basis of PBPK models makes them ideal for extrapolation.

The structure of PBPK models allows the prediction of tissue concentrations, which can be valuable in human health risk assessment or for linking with pharmacodynamic models. PBPK models when combined with PD models can be powerful in predicting the time-course of drug effects under physiological and pathological conditions. The integration of PBPK models with PD models aid a robust design of clinical trials and is covered in Chapter 13. PBPK–based predictions aid the optimal use of all available compound information within a physiological context, making experiments confirmatory rather than exploratory. These have a tremendous impact in reducing preclinical and clinical studies thereby reducing costs.

Applications of PBPK in population modeling form the subject of Chapter 14. Drug failures can sometimes result from considering only an average person and neglecting physiological and genomic variability that can lead to a spread in both plasma drug concentrations and drug response. Chapter 14 describes how targeted therapy and personalized medicine can be achieved with PBPK/PD modeling.

Chapter 15 aims to seamlessly integrate all the applications of PBPK along the drug discovery and development value chain.

KEYWORDS

Binding Site Analysis: Use of computational tools for the prediction of potential ligand-binding active sites in a target protein, given its three-dimensional structure. This is achieved through searching for surface features of the protein (geometry and functional groups) that provide the best shape complementarity and interactions with a set of known ligands.

Biological Systems Modeling: Involves computer simulations of biological systems to analyze and visualize the complex connections of cellular processes such as the networks of metabolites and enzymes that comprise metabolism, signal transduction pathways, and gene regulatory networks.

Clinical Trial Simulation: Combining structural and stochastic elements of pharmacokinetic and pharmacodynamic models to produce a data set that will resemble the results of an actual trial.

Compartmental PK Modeling: Uses kinetic models to describe the concentration–time profile. The compartments do not relate to meaningful physiologic spaces.

Druggable: A druggable target is a protein whose activity can be modulated by a small molecule drug. A druggable target is crucial in determining the progression of a drug discovery project to the lead generation stage.

hERG Modeling: The human ether-à-go-go related gene (hERG) codes for the potassium ion channel *Kv11.1*, a protein that mediates the repolarizing I_{Kr} current in the cardiac action potential. Drugs inhibiting the channel can cause a potentially fatal QT prolongation with a concomitant risk of sudden death. In computational drug design, there are 2 main approaches to hERG modeling. Pharmacophoric or ligand-based modeling relies on determining the physicochemical features associated with the channel block to predict the hERG blocking potential of compounds. Target-based partial homology models of the hERG channel have also been built to interpret electrophysiological and mutagenesis studies.

Homology Modeling: Involves taking a known sequence with an unknown structure and mapping it against a known structure of one or more homologous proteins in an effort to gain insights into three-dimensional structure of the protein.

Lead Generation: A phase in drug discovery in which the objective is to identify one or more chemical series with potential drug activity, reduced off-target toxicity, and with physicochemical and metabolic profile that are compatible with acceptable *in vivo* bioavailability.

Lead Optimization: Phase in drug discovery, following lead generation, in which the objectives are to optimize the PK and PD (efficacy, selectivity and safety) in the screening stage and to select for drug development, a high-quality candidate drug that satisfies a preset target profile in the drug candidate selection stage.

Model-Based Drug Development: Statistical and mathematical modeling that allows for quantitative and effective use of prior information (preclinical efficacy and safety models) and clinical data (information across drugs, end points, trials and doses) for improved data analysis, clinical study design, knowledge management and decision making in clinical drug development.

PBPK/PD Modeling: Linking a physiologically-based pharmacokinetic (PBPK) model, which relates a drug's exposure to its dose with a pharmacodynamic (PD) model, which relates the pharmacological response to exposure.

Pharmacophore Modeling: A ligand-based approach to virtual screening, which makes use of two- or three-dimensional pharmacophores generated from a set of known active compounds to the selected target.

PK/PD Modeling: Linking a pharmacokinetic (PK) model, which relates a drug's exposure to its dose with a pharmacodynamic (PD) model, which relates the pharmacological response to exposure.

Population Modeling: Seeks to identify and quantify the pathophysiological factors that cause changes in the dose–concentration relationship, so that any resulting clinically significant shifts in the therapeutic index can be addressed through appropriate dose adjustments.

Prediction of PK Properties: Generally QSPR models that employ the structure-dependent properties of a compound to arrive at pharmacokinetic properties such as fraction of drug unbound in plasma, volume of distribution, renal elimination, fraction of compound absorbed, or bioavailability. Physiological models are also used to predict PK properties.

Protein Modeling is the prediction of the three-dimensional (secondary, tertiary, and quaternary) structure of a protein from its amino acid sequence.

Reactive Metabolite Prediction: Using the chemical structure of a compound and a database of known reactive metabolites to predict the likelihood that a compound of interest might produce reactive metabolites.

QSAR and QSPR Models: Quantitative structure–activity relationships (QSAR) are mathematical equations relating pharmacological activity to chemical structure for a series of structurally related compounds. Quantitative structure–property relationships (QSPR) relate physicochemical properties of compounds to their structures. QSARs are derived using regression and pattern recognition techniques.

Scaffold Hopping: Computational approaches that use a set of known active compounds to find structurally novel compounds with chemically completely different core structures, and yet binding to the same receptor by modifying the central core structure of the molecule.

Site of Metabolism Prediction: Given the structure of a lead compound, to predict the sites that are prone to metabolic activity. Databases of known reactivity and/or principles of chemical reactivity are employed to predict sites of metabolism. If metabolizing enzyme is identified, then its protein structure is also used for getting poses of a compound of interest at the active site of the enzyme. Machine learning and semiempirical quantum chemical calculation can also be incorporated into prediction models.

Virtual Screening: A computational technique used in early drug discovery for rapid *in silico* assessment of large libraries of chemical structures against three-dimensional structure of a target protein in order to identify structures that are most likely to bind to a target protein.

REFERENCES

1. PricewaterhouseCoopers. *Pharma 2005 Silicon Rally: The Race to e-R&D.* Paraxel's Pharmaceutical R&D statistical sourcebook, 2002/2003.
2. U.S. FDA. Critical path initiative. Available at: http://www.fda.gov/ScienceResearch/SpecialTopics/CriticalPathInitiative/default.htm.
3. Lalonde RL, et al. Model-based drug development. *Clin Pharmacol Ther*. 2007; 82(1):21–32.
4. Dingemanse J, Appel-Dingemanse S. Integrated pharmacokinetics and pharmacodynamics in drug development. *Clin Pharmacokinet*. 2007;46(9):713–737.
5. Teorell T. Kinetics of distribution of substances administered to the body. *Archives Internationales de Pharmacodynamie et de Thérapie*. 1937;57:205–240.
6. Rowland M, Balant L, Peck C. Physiologically-based pharmacokinetics in drug development and regulatory science: A workshop report (Georgetown University, Washington, DC, May 29–30, 2002). *AAPS J*. 2004;6(1):56–67.
7. Bischoff KB, Dedrick RL, Zaharko DS, Longstreth JA. Methotrexate pharmacokinetics. *J Pharm Sci*. 1971;60(8):1128–1133.
8. Bischoff KB, Dedrick RL. Thiopental pharmacokinetics. *J Pharm Sci*. 1968;57(8): 1346–1351.
9. Edginton AN, Theil FP, Schmitt W, Willmann S. Whole body physiologically-based pharmacokinetic models: Their use in clinical drug development. *Expert Opin Drug Metab Toxicol*. 2008;4(9):1143–1152.
10. Reddy MB, Yang RSH, Clewell HJ, Eds. *Physiologically-Based Pharmacokinetic Modeling Science and Applications*. Hoboken, NJ: Wiley, 2005.

2

PHYSIOLOGICALLY-BASED MODELING

CONTENTS

2.1　Introduction . . . .　13
2.2　Examples of Physiological Modeling . . . .　14
2.3　Need for Physiological Models in the Pharmaceutical Industry . . . .　15
2.4　Organs as Compartments . . . .　15
2.5　Bottom-Up vs. Top-Down Modeling in Pharmacokinetics . . . .　16
　　References . . . .　16

2.1 INTRODUCTION

Physiological modeling or physiology-based mathematical modeling aims to integrate knowledge of physiological processes to the extent known with physicochemical attributes or other known/measured information about compounds in order to predict or simulate complex biological properties. The level of detail in a physiological model can vary depending on the nature of the property to be predicted or the process to be simulated, extent of knowledge

---

Physiologically-Based Pharmacokinetic (PBPK) Modeling and Simulations: Principles, Methods, and Applications in the Pharmaceutical Industry, First Edition. Sheila Annie Peters.
© 2012 John Wiley & Sons, Inc. Published 2012 by John Wiley & Sons, Inc.

available, and the level of complexity required to meet an acceptable degree of prediction/simulation accuracy. Physiological models of cells (focusing on cellular processes at molecular level), tissues, a system of organs, and whole organisms aid an increased mechanistic understanding of biological systems, and the potential benefits for a pharmaceutical company are very high. Despite the unique advantages that physiological models can provide, the sheer size and complexity of these models, as well as the problems of parameterization, have been deterrents to their widespread application for a long time. However, the tremendous progress in high-performance computing has removed the constraints in computing power, allowing for a dramatic increase in the comprehensiveness and complexity of physiological models. A steady growth in biological pathway information and genetic data, information on molecular mechanisms, and technological advances in biological measurements have made it possible to obtain the data needed for parameterization of models. Physiological models of today are, therefore, considerably more complex, seeking to answer more demanding questions and addressing the interdependence of component models.

2.2 EXAMPLES OF PHYSIOLOGICAL MODELING

A number of examples of physiological modeling can be found in the field of medicine. Mathematical models of the coupling between membrane ionic currents, energy metabolism [adenosine triphosphate (ATP) regeneration via phosphocreatine buffer effect, glycolysis, and mitochondrial respiration], blood–brain barrier exchanges, and hemodynamics in the brain, physiological model for liver or muscle metabolism, and tumor modeling to understand the signaling pathways leading to angiogenic activities of tumors are some of the examples of physiological organ modeling. Apart from these, physiological models of the cardiovascular system and respiratory system (including respiratory organs such as lung, respiratory muscles such as diaphragm, and peripheral organs such as nasal cavity and oral cavity) are commonly used as a useful way of monitoring conditions of a whole system. Physiological models that simulate human internal thermal physiological systems, including muscle and blood, predict thermoregulatory responses such as metabolic heat generation by computing heat flow by conduction, convection, and mass transport. Thus, physiological models aim to provide meaningful predictions based on a physical and biological understanding of the underlying processes. There is considerable scientific interest in modeling not just the various tissues or systems of organs but the entire human body. Physiological modeling of the human body, and to some extent the modeling of any animal body, provides very precise information that can help improve the well-being or healing time of individuals facing health issues. In addition, the growing complexity and accuracy of physiological modeling brings critical information to the pharma industry. The pathophysiological conditions of diseases can be modeled[1] for the advancement of basic knowledge of a disease and the development of new disease diagnostics. Important whole-body parameters to be considered are volumes, blood flow rates, circulating pools of endogenours or exogenous proteins and nutrients, and other aggregate properties of the body such as cardiac output, body weight, and the like.

2.3 NEED FOR PHYSIOLOGICAL MODELS IN THE PHARMACEUTICAL INDUSTRY

Physiological modeling approaches are important for transitioning biology from a descriptive to a predictive science. Pharmaceutical companies identify molecular interventions that can lead to therapies. Physiological models that integrate an understanding of known biological mechanisms with all available compound information can greatly improve the efficiency of transforming targets into therapies. In the pharma industry, physiological models are used to predict pharmacokinetics, to simulate clinical outcomes and organ-level behavior, and to predict human response to drugs. Patel[2] describes a single physiological model to explain the acute and chronic changes in sodium and water balance in the human body in response to changes in the physiological mechanisms regulating sodium. Noble[3] describes a physiological model of the heart that provides a unified description of organ-level physiology in terms of protein-level biology. The model provides nonintuitive explanations for how antiarrhythmia drugs might work. An extensive knowledge of cell–cell organization, signaling pathways, and the tissue geometry of the heart were necessary to build the model. A similar attempt to integrate proteins to organ-level systems[4] is also described in the literature. Physiological models of type II diabetes have been used for over a decade to understand key parameters, such as insulin sensitivity and β-cell function, pertaining to pathology in patients.[5] Such models provide new insights into important biological processes, thus aiding the development of new medicines and treatment regimes. Physiological models also facilitate the incorporation of interindividual differences in enzymology and receptor densities, making it possible to apply pharmacogenomic principles.

2.4 ORGANS AS COMPARTMENTS

Compartments are areas of the body where muscle, nerves, and blood vessels are confined to relatively inflexible spaces bounded by skin, fascia, and bone. Many physiological models consider each organ in the body as one or more compartments. The contents within a compartment are expected to be homogenous. The complexity of the model is a function of the number of compartments. As such, this number should not exceed more than what is critical for characterizing the system.

2.5 BOTTOM-UP VS. TOP-DOWN MODELING IN PHARMACOKINETICS

In pharmacokinetics, compartmental PK models are widely used in the drug discovery stage. The compartments in these models do not have a physiological relevance. Parameters have no physical or biochemical meaning. These empirical models describe the data but are of little value in understanding the mechanisms underlying the observations and cannot be extended to the next set of data. They do not account for the sequential metabolism of a drug in different organs, do not consider metabolite kinetics, and cannot distinguish the effects of transporter barriers between drug and metabolites.

Population PK models are another class of empirical models that are used during clinical development to understand the sources of observed variability in the drug concentrations among the individuals of a target population through correlations of observed variability with demographical, pathophysiological, and therapeutical variations. These empirical models are useful in evaluating the need for dosage changes to the drug in the event of clinically significant shifts in the therapeutic index. Compartmental and population PK models are conventional approaches seeking to get to the system characteristics starting from the observed data. Contrary to this top-down modeling approach, PBPK models present the opportunity for bottom-up modeling, in which prior information on the system characteristics and other variables that may result in an observation are assembled together to predict an outcome. Any deviations from the predicted outcome can then provide insights into the mechanisms of the underlying processes. PBPK model structure is independent of any observed drug data. PBPK models and systems biology[6] models are good examples of a bottom-up approach. In practice, a combination of these two approaches can be very valuable in understanding drug response in a population. Model parameters are extracted from one set of observations in a system and then used to predict possible outcomes in another.

REFERENCES

1. Butcher EC, Berg EL, Kunkel EJ. Systems biology in drug discovery. *Nat Biotechnol.* 2004;22(10):1253–1259.
2. Patel S. Sodium balance—an integrated physiological model and novel approach. *Saudi J Kidney Dis Transpl.* 2009;20(4):560–569.
3. Noble D. Systems biology and the heart. *BioSystems.* 2006;83(2–3):75–80.
4. Hunter PJ, Borg TK. Integration from proteins to organs: The physiome project. *Nat Rev Mol Cell Biol.* 2003;4(3):237–243.
5. Kansal AR. Modeling approaches to type 2 diabetes. *Diabetes Technol Ther.* 2004; 6(1):39–47.
6. Kohl P, Crampin EJ, Quinn TA, Noble D. Systems biology: An approach. *Clin Pharmacol Ther.* 2010;88(1):25–33.

3

REVIEW OF PHARMACOKINETIC PRINCIPLES

CONTENTS

3.1	Introduction	18
3.2	Routes of Administration	18
3.3	Drug Disposition	18
	3.3.1 Absorption	18
	3.3.2 Plasma Protein Binding, Blood–Plasma Ratio	20
	3.3.3 Distribution, Elimination, Half-Life, and Clearance	23
	3.3.4 Role of Transporters in ADME	29
3.4	Linear and Nonlinear Pharmacokinetics	34
3.5	Steady-State Pharmacokinetics	34
3.6	Dose Estimations	37
3.7	Successful PK Optimization in Drug Discovery	40
	Keywords	40
	References	41

Physiologically-Based Pharmacokinetic (PBPK) Modeling and Simulations: Principles, Methods, and Applications in the Pharmaceutical Industry, First Edition. Sheila Annie Peters.
© 2012 John Wiley & Sons, Inc. Published 2012 by John Wiley & Sons, Inc.

3.1 INTRODUCTION

Pharmacokinetics is the study of the rate and extent of drug transport in the body to the various tissues, right from the time of its administration to its elimination. The rate of drug transport to the target tissue of pharmacological action determines the rate of drug action. It is dependent on the rate of drug absorption from the site of administration into the capillaries and the blood flow rates to the organs of elimination and to target tissue. The extent of drug that reaches the site of drug action at any point in time is dependent on protein binding, extent of metabolism, or biotransformation and elimination. The absorption, distribution, metabolism and elimination (ADME) of a drug should be such that the drug is delivered at the target site at a rate and concentration consistent with a once or twice daily administration. To ensure this, the effective elimination half-life of a drug, determined by the rates and extent of ADME, should ideally be equal to the dosing interval. With successive administrations of a drug, the rate of its elimination tends to equal the rate of administration, at which point an equilibrium steady state is said to have been attained. The dosing frequency of a drug in humans is dictated by the half-life of the drug and by its unbound steady-state concentration, which should equal its pharmacologically effective concentration. The forthcoming chapters will draw heavily upon these concepts, and the reader is encouraged to use this chapter as a reference to basic pharmacokinetic principles.

3.2 ROUTES OF ADMINISTRATION

Common routes of drug administration include per oral (PO), intramuscular (IM), subcutaneous (SC), and intravenous (IV) injections and IV infusion. Other routes include buccal, sublingual, rectal, transdermal, inhalational, and topical. The oral route is the most preferred route, but it is not suitable for drugs that are not stable in the gut, such as for example, peptide and protein drugs.

3.3 DRUG DISPOSITION

3.3.1 Absorption

Other than the IV administrations, all other routes require the drug to be absorbed into the capillaries surrounding the site of administration. The rate of absorption from IM and SC routes depends on the type of tissue at the site of administration—the density, vascularity, and fat content. The IM route, for example, has a higher rate of absorption compared to the SC because of lesser fat and greater vascularity of the dense muscles. Oral drug absorption (Fig. 3.1) refers to the transport of drug molecules across the enterocytes lining the gastrointestinal (GI) tract into the venous capillaries along the gut wall. The

3.3 DRUG DISPOSITION 19

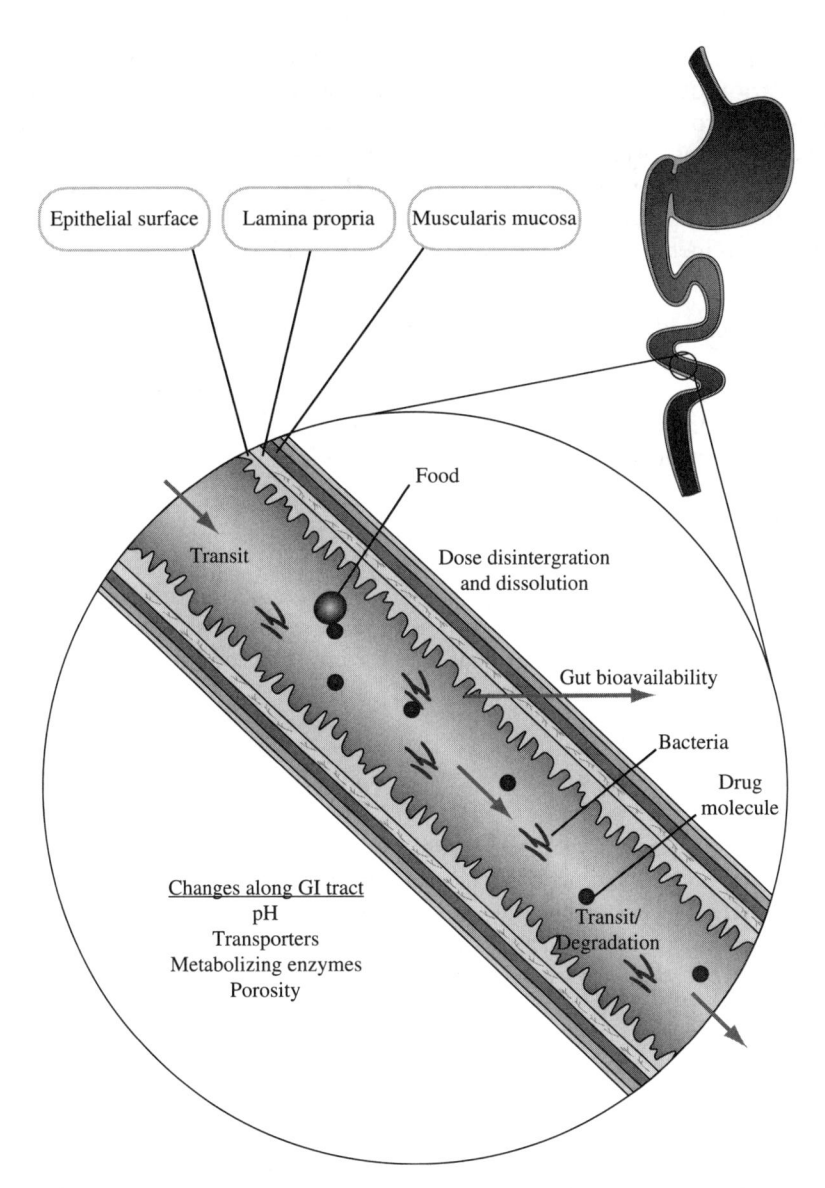

Figure 3.1. Orally administered drug disintegrates, dissolves, transits, and permeates the enterocytes. Along its way down the gastrointestinal tract, the drug may bind to luminal contents—food, bacteria, etc., which prevents its absorption. Transcellular drug absorption can result in transporter-mediated efflux or drug metabolism by intestinal enzymes such as cytochrome P450s (CYPs) and uridine 5'-diphospho-glucuronosyltransferases (UGTs). Small, hydrophilic drugs rely on the paracellular route, while large hydrophilic molecules rely on transcytosis and receptor-mediated endocytosis for absorption. Molecules possessing certain special groups such as peptide linkages are transported across the membrane by carrier transporters. Most drug molecules are sufficiently lipophilic for transcellular passive absorption.

rate of drug absorption is dependent upon a number of physiology-, drug-, and formulation-dependent factors such as the gastric emptying rate, intestinal motility, porosity of tight junctions, luminal and mucosal enzymology, carrier and efflux transporters, small intestinal secretions (bile and digestive enzymes), food interactions, regional differences in pH, permeability, solubility, dissolution rate and particle size, among others. An orally absorbed drug is subjected to first-pass metabolism in the liver before it is available in systemic circulation.

3.3.2 Plasma Protein Binding, Blood–Plasma Ratio

Drugs reversibly bind to plasma proteins depending upon their lipophilicity and ionizability. In general, the greater the lipophilicity of a compound, the greater is its plasma protein binding. The binding equilibrium can be represented as

$$\text{Protein} + \text{drug} \underset{}{\overset{K_A}{\rightleftharpoons}} \text{protein} - \text{drug complex}$$
$$[P] \qquad C_u \qquad\qquad\qquad C_b$$

where [P] is the protein concentration, and C_u and C_b are the unbound and bound concentrations of the drug at equilibrium. The equilibrium constant, K_A, also called the affinity constant, is given by

$$K_A = \frac{C_b}{C_u \times [P] \times n} \tag{3.1}$$

where n is the number of binding sites per mole of the binding protein. Since therapeutic concentrations of most drugs are low, [P] can be assumed to be the total protein concentration $[P]_{\text{Total}}$. The fraction unbound drug in plasma (f_{up}) can be obtained from equation 3.1 in terms of $[P]_{\text{Total}}$ or in terms of the concentrations of the plasma proteins α_1-acidic glycoprotein (AGP), $[P]_{\text{AGP}}$, and albumin $[P]_{\text{albumin}}$:

$$C_u + K_A C_u \times [P]_{\text{Total}} \times n = C_u + C_b$$

$$C_u(1 + K_A \times [P]_{\text{Total}} \times n) = C_u + C_b$$

$$f_{up} = \frac{C_u}{C_u + C_b} = \frac{1}{1 + (K_A \times [P]_{\text{Total}} \times n)} \tag{3.2}$$

$$= \frac{1}{1 + (K_{A,\text{AGP}} \times [P]_{\text{AGP}} \times n_{\text{AGP}}) + (K_{A,\text{albumin}} \times [P]_{\text{albumin}} \times n_{\text{albumin}})}$$

The fraction unbound in plasma (f_{up}) thus depends on the concentration of protein and the affinity of the drug to the protein. Albumin is the principal protein to which many drugs bind, followed by AGP. Other plasma proteins include lipoproteins and the globulins. The concentrations of various plasma

TABLE 3.1. Plasma Proteins

Plasma Proteins	Binding	Molecular Weight (Da)	Concentration (μM)
Albumin	Binds mainly to anionic compounds	67,000	500–700
α_1-Acidic glycoprotein (AGP)	Binds mainly to cationic drugs, e.g., tricyclic antidepressants	42,000	9–23
Lipoproteins		200,000	Variable
α-, β-, and γ-Globulins		53,000	0.6–1.4

proteins are shown in Table 3.1. Albumin is distributed in intravascular (plasma: 43 g/kg organ) and extravascular organs (muscle: 2.3 g/kg, skin: 7.7 g/kg, liver: 1.4 g/kg, gut: 5 g/kg, and other tissues: 3 g/kg). Albumin exists abundantly in the interstitial fluids.

Albumin has six distinct binding sites, two of which specifically bind to long-chain fatty acid, another selectively binds to bilirubin, and the remaining two bind to acidic and lipophilic drugs. One of these drug-binding sites binds to drugs such as warfarin and phenylbutazone, while the other to drugs such as diazepam and ibuprofen. Drugs binding to different binding sites do not compete with one another. When more than 20% of the sites are occupied, concentration dependence of binding begins to get appreciable, ultimately leading to saturation at higher concentrations. Saturation of albumin is rare and restricted to drugs (especially acids) with high therapeutic concentration. However, the binding sites of a few drugs such as tolbutamide and some sulfonamides are saturated even at therapeutic concentrations. AGP concentrations being much lower compared to albumin, saturation of AGP occurs at lower therapeutic concentrations. The concentrations of several plasma proteins can be altered by many factors including stress, surgery, liver or surgery dysfunction, and pregnancy. Most commonly, disease states increase AGP concentration while reducing albumin concentration. Higher levels of AGP have been reported in obese patients with nephrosis. Stress, cancer, and arthritis have been associated with lower AGP levels. Neonates have higher AGP levels. AGP has a higher degree of interindividual variability compared to albumin. Reduced levels of albumin have been reported in myalgia patients. Drugs that are highly bound to plasma proteins are confined to the vascular space and are not readily available for distribution to other tissues and organs. Many carboxylic acid drugs are not easily displaced from plasma proteins and have a low distribution volume. However, this is not true if the affinity of a drug to tissue proteins is higher than that to plasma proteins. Ultrafiltration and equilibrium dialysis are the two commonly employed methods for the determination of plasma protein binding[1]. Albumin is the principal drug-binding protein in tissues followed by ligandin. Measurement of tissue binding is not as straightforward as that in plasma, as the tissue must be disrupted and it is not readily accessible to sampling. Tissue proteins cannot be easily separated into their constituents and cannot easily be quantified.

While the liver can extract high extraction drugs, even if they are highly bound to plasma proteins, the clearance (CL) of many low hepatic extraction drugs are limited by protein binding. Only the unbound drug is available for glomerular filtration and, therefore, for renal elimination. An increase in unbound drug concentration due to a reduced plasma protein binding will enable higher tissue distribution and higher CL. However, since the half-life of a drug is directly proportional to the distribution volume and inversely proportional to CL, there is no net effect on the half-life. Thus, changes in plasma protein binding of a drug are not likely to be clinically relevant[2] except in the following cases:

1. The drug is >98% bound to plasma proteins. In this case even a small shift in plasma protein binding can have a substantial effect on the clearance but less so on the distribution volume, thus temporarily altering the unbound drug concentrations, until new equilibrium is attained.
2. The drug has a high hepatic extraction. The clearance of such drugs will be dependent only on the hepatic blood flow rate and not on the product of $f_{up} \times CL_{int}$, where CL_{int} is the intrinsic clearance of a comound in the liver. Thus an increase in distribution volume is not sufficiently compensated by an increase in CL, leading to a temporary increase in unbound drug concentrations.
3. The drug has a narrow therapeutic window/safety margin such that the differences in unbound drug concentrations discussed in case 1 or 2 will have a greater impact.
4. There is a rapid equilibrium (short pharmacological response time) between drug concentration and pharmacological response (e.g., lidocaine with a PK–PD equilibration time of 2 min) compared to the time required for the body to regain equilibrium (about 30 min). Many antiarrhythmic drugs and anesthetics require only a short time for a change in concentration to cause a change in drug effect. In these cases, the response is sensitive to small transient changes in unbound drug concentrations.
5. The pharmacological response sensitive to changes in drug concentration.

Scaling of PK parameters such as clearance or volume of distribution or pharmacodynamc properties[3] from preclinical species to humans should always be done with the unbound parameters. Any comparisons/correlations of PK parameters should also be done with unbound values.

Some drugs also bind to and distribute into erythrocytes, the main drivers being lipophilicity, pK_a, and active uptake into the erythrocytes. Binding sites within erythrocytes are proteins such as hemoglobin, carbonic anhydrase, as well as plasma membrane. Blood–plasma concentration ratio[4] (R) of a drug is a measure of its binding and distribution to erythrocytes relative to plasma.

3.3 DRUG DISPOSITION

A compound having similar extent of binding to constituents of erythrocytes and plasma has a blood–plasma ratio of 1. Acids tend to have R values around 0.55 and never exceed 1, and bases tend to have higher range of values, often exceeding 1, while neutrals and ampholytes have values around 1.[5] Uchimura et al.[4] describe several methods to determine R. The minimum value of R can be only 0.55, corresponding to no distribution into erythrocytes. However, there is no upper limit. For tacrolimus, R is as high as 55 and exhibits concentration dependence.[6] R is determined by measuring the concentrations of ^{14}C-labeled drug in erythrocytes (C_e) and plasma (C_p) in freshly collected blood. Then, knowing the hematocrit, H (the relative volume of blood occupied by erythrocytes), R is obtained as follows:

$$R = \frac{C_e \times H + C_p \times (1 - H)}{C_p} \tag{3.3}$$

R can also be predicted.[7] Partitioning into erythrocytes can be fast for some drugs, and distribution equilibrium is reached within a few seconds to minutes. However, many drugs with primary amine groups show delayed equilibrium probably due to formation of Schiff bases with membrane fatty acids and aldehydes. While the displacement of the plasma-protein-bound drug to the unbound is rapid except for protein molecules, displacement of erythrocyte-bound drug is relatively slow. For acids with high plasma protein binding, distribution into erythrocytes can significantly affect its distribution volume, as other tissue compartments are not as significant. If blood to plasma ratios exceed 1, as for lipophilic bases, the plasma clearance significantly overestimates blood clearance and could even exceed hepatic blood flow. This is because the concentrations measured in plasma will always be much smaller compared to that measured in whole blood due to the higher distribution into the erythrocytes, when R is > 1. Thus, if C_b is the concentration of the drug in blood, CL_p is the plasma clearance and CL_b is the blood clearance of the drug, the following relation holds:

$$R = \frac{C_b}{C_p} = \frac{CL_p}{CL_b}$$

3.3.3 Distribution, Elimination, Half-Life, and Clearance

The following derivations will aid an appreciation for the relationship between distribution, elimination, and half-life. The rate of change of drug concentrations in the blood (C) can be given by the first-order rate equation

$$\frac{-dC}{dt} = \frac{-dA}{Vdt} = k_{el} \times C$$

$$\frac{-dA}{dt} = V \times k_{el} \times C = CL \times C = k_{el} \times A \tag{3.4}$$

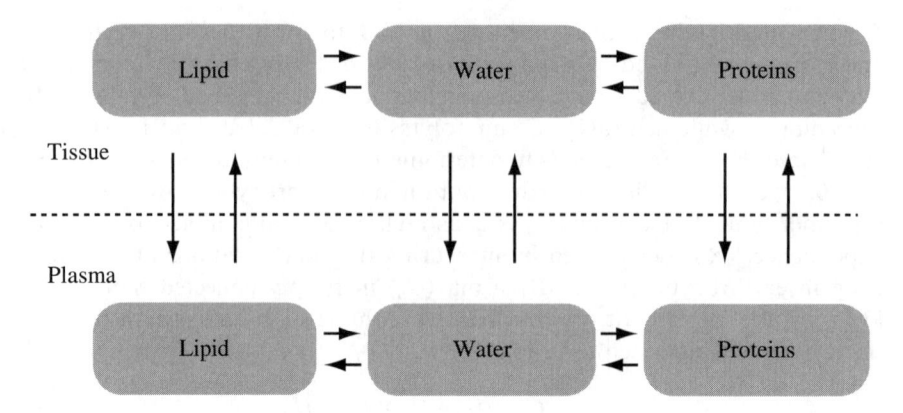

Figure 3.2. Both the blood plasma and the tissues are composed of water, lipid, and proteins. The proportion of the constituents varies in different tissues. A drug equilibrates between the different constituents of plasma and tissues, depending on its lipophilicity.

where A is the amount of drug in the body at any time, t, k_{el} is the first-order elimination rate constant, and V is the volume of distribution of the drug. The higher the lipophilicity and unbound fraction of the drug, the greater is the V. Tissue partition coefficient ($K_{T,P}$) is the ratio of concentration of the drug in a tissue and plasma at steady state and is, therefore, a measure of distribution into that tissue. Tissue partition coefficients are different for different tissues depending on the composition of the tissue and the nature of the compound (Fig. 3.2). The product of k_{el} and V is defined as the total clearance, CL, of the drug from blood. Integrating the equation $-dA/dt = k_{el} \times A$ yields

$$A = A_0 \times e^{-k_{el} t} \tag{3.5}$$

where A_0 is the initial amount of drug in the body. A_0 is simply the dose administered for an IV bolus, but for all other non-IV routes, it is $F \times$ dose, where F is the systemic bioavailability of the drug. The F of an oral dose is

$$F = f_{abs} \times f_{gut} \times f_{hep} \tag{3.6}$$

where f_{abs} is the fraction of dose absorbed, f_{gut} is the fraction escaping gut extraction, and f_{hep} is the fraction escaping hepatic extraction. The oral bioavailability of a drug can be estimated from the areas under the PK curves (AUC) of intravenous and oral doses (AUC_{IV} and AUC_{oral}), assuming that the clearance does not vary between the IV and oral routes:

$$F = \frac{CL_{IV}}{CL_{oral}} = \frac{AUC_{oral}/\text{dose}_{oral}}{AUC_{IV}/\text{dose}_{IV}} \tag{3.7}$$

Bringing A_0 to the left-hand side of equation 3.5, taking the natural logarithms on both sides of the resulting equation, and defining the half-life ($t_{1/2}$) of a drug

3.3 DRUG DISPOSITION

to be the time taken for half of the drug amount to get eliminated from the body (i.e., A becomes $A_0/2$), the half-life is given by

$$t_{1/2} = \frac{\ln 2}{k_{el}} = \frac{\ln 2 \times V}{CL} \tag{3.8}$$

Integrating the equation $-dA = CL \times C\, dt$ yields

$$\int_{t=0}^{t} dA = \int_{t=0}^{t} CL \times C\, dt = CL \times \int_{t=0}^{t} C\, dt = CL \times AUC \tag{3.9}$$

Recognizing that the integral dA over time is the dose, equation 3.9 becomes Dose $= CL \times AUC$, where AUC is the area under the concentration–time profile. The total clearance, CL, is the sum of clearance from every eliminating organ but mainly hepatic (CL_H), renal (CL_R), and biliary (CL_B):

$$CL = CL_H + CL_R + CL$$

$$CL_{organ} = \frac{Q_{organ} \times (C_{ART} - C_{VEN})}{C_{ART}} \tag{3.10}$$

where CL_{organ} is the clearance from an eliminating organ, Q_{organ} is the blood flow rate to that organ, and C_{ART} and C_{VEN} are the arterial and venous concentrations. $Q_{organ} \times C_{ART} - Q_{organ} \times C_{VEN}$ is the rate of elimination from that organ.

The liver is the most important eliminating organ, where phase I and phase II metabolism of the small-molecule drugs convert them into more hydrophilic compounds, which can then be renally eliminated. About 40 human CYP genes have been cloned and classified according to sequence homology. Of these, only 3 CYP families and less than 12 unique enzymes play a substantial role in the hepatic metabolism of drugs in humans. The rate of such an enzyme-driven biotransformation reaction, v, depends on the concentration of the drug according to the Michaelis–Menten equation:

$$v = CL_{int} \times C = \frac{v_{max} \times C}{K_M + C} \tag{3.11}$$

CL_{int} is the intrinsic clearance of the drug, dependent only on the intrinsic chemical nature of the drug; v_{max} is the maximum velocity of the reaction; and K_M is the Michaelis–Menten constant (Fig. 3.3). The therapeutic concentration ranges of most drugs are very low compared to their K_M, and equation 3.11 then becomes

$$v = \frac{v_{max} \times C}{K_M} \tag{3.12}$$

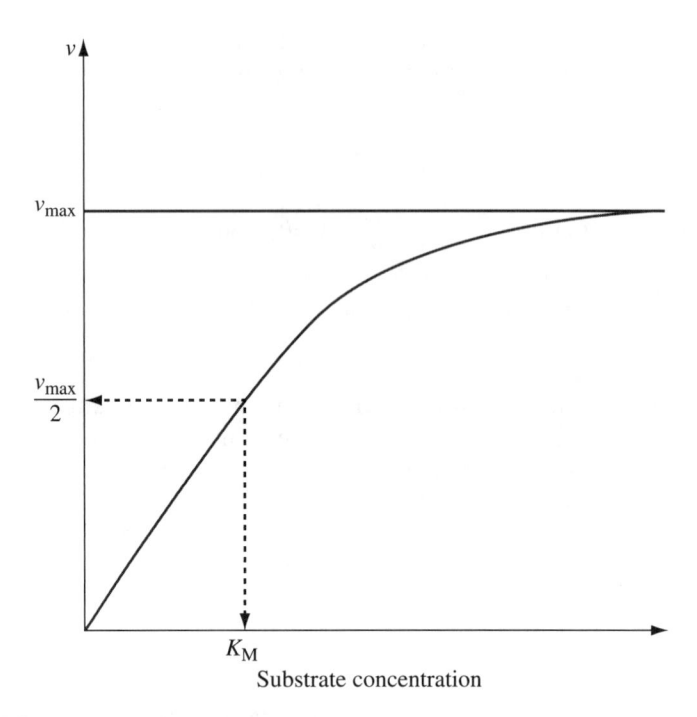

Figure 3.3. Rate of an enzyme-catalyzed reaction as a function of substrate concentration.

Equation 3.12 suggests that the rate of a metabolic reaction varies linearly with the drug concentration at concentrations not exceeding K_M. CL_{int} then equals the ratio of v_{max} to K_M, and it is independent of the drug concentration. However, when drug concentrations are comparable with or exceeds K_M, which can be the case with high doses of drugs, CL_{int} is dependent on C (see equation 3.11). The greater the drug concentration, the greater the extent of enzyme saturation and smaller the CL_{int}. The maximal rate of metabolism is reached. Under these conditions, zero-order kinetics is said to prevail, and a constant amount of drug is eliminated per unit time, independent of the amount of drug in the body. Well-known examples of drugs exhibiting nonlinear clearance include phenytoin, ethanol, methyl salicylate, and theophylline (in some individuals). Apart from CL_{int}, the hepatic clearance CL_H is dependent on how fast the drug is delivered to the enzymes in the hepatocytes, which is basically governed by the total blood flow rate to the liver. A basic tenet of pharmacokinetics is that only the fraction of drug that is not bound to the plasma proteins is available for distribution into tissues and for hepatic, biliary, or renal clearance. These dependencies of CL_H are encapsulated in the well-stirred model in which the liver is considered to be a well-stirred compartment. According to this model, the hepatic clearance from blood, CL_H, is given by

3.3 DRUG DISPOSITION

$$CL_H = \frac{Q_{LI} \times f_{ub} \times CL_{int}}{Q_{LI} + f_{ub} \times CL_{int}} \quad (3.13)$$

where f_{ub} is the fraction unbound in blood. Since the unbound concentrations in plasma and blood are expected to be the same, f_{ub}, f_{up}, and R are related as follows:

$$R = \frac{C_b}{C_p} = \frac{C_{u,b}}{C_{u,p}} \times \frac{f_{up}}{f_{ub}}$$

$$f_{ub} = \frac{f_{up}}{R} \quad (3.14)$$

Equation 3.13 shows that when the product $f_{up} \times CL_{int}$ is high compared to Q_{LI}, then CL_H is simply determined by the blood flow rate to the liver. The drug clearance is limited by the rate at which it is delivered to the drug-metabolizing enzymes. On the other hand, when the product $f_{up} \times CL_{int}$ is low compared to Q_{LI}, then the hepatic clearance linearly varies with the product of fraction unbound in plasma and the intrinsic clearance. For high CL_{int} compounds, the displacement of the drug from plasma proteins is rapid and equilibrium cannot be established between concentrations of the bound and unbound drug in blood (C_b, $C_{u,b}$) and in liver ($C_{u,liver}$). This can be represented as

$$C_b \rightarrow C_{u,b} \rightarrow C_{u,liver}$$

The liver is capable of extracting even the bound drug. For low CL_{int} compounds, there is equilibrium between the bound drug and unbound drug in blood and liver and only the $C_{u,liver}$ is available to the drug-metabolizing enzymes. The equilibrium between the different drug concentrations is shown below:

$$C_b \leftrightarrow C_{u,b} \leftrightarrow C_{u,liver}$$

The hepatic extraction ratio of a drug is obtained from equation 3.13, by taking the Q_{LI} to the left-hand side:

$$E_H = \frac{CL_H}{Q_{LI}} = \frac{f_{ub} \times CL_{int}}{Q_{LI} + f_{ub} \times CL_{int}} \quad (3.15)$$

For hydrophilic drugs, the parent compound can get eliminated in the urine, unchanged. For example, the renal route is the predominant route of elimination for about 60% of anti-infection compounds.[8] The fraction of the dose excreted in the urine unchanged (f_e) is

$$f_e = \frac{A_{e,\text{unchanged}}}{\text{dose}} = \frac{CL_R}{CL} \quad (3.16)$$

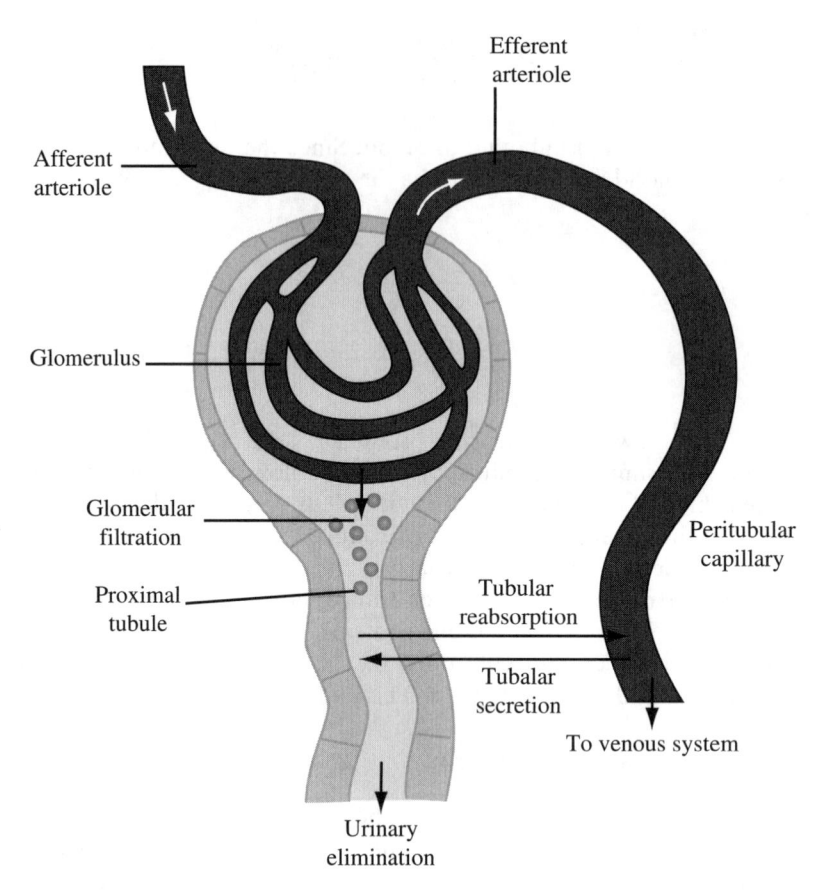

Figure 3.4. Renal elimination of a drug: glomerular filtration, active tubular secretion, and tubular reabsorption. Lipophilic drugs are readily reabsorbed, making renal elimination an important route only for hydrophilic drugs.

where $A_{e,\text{unchanged}}$ is the amount of drug excreted in the urine unchanged. Therefore, CL_R is

$$CL_R = \frac{A_{e,\text{unchanged, IV}}}{AUC_{\text{IV}}} = \frac{A_{e,\text{unchanged, oral}}}{AUC_{\text{oral}}}$$

and $CL_R = f_{\text{ub}} \times \text{GFR}$, where GFR is the **glomerular filtration rate** for a drug for which there is no tubular active secretion or tubular reabsorption (Fig. 3.4). Active tubular secretion is evident if $CL_R > f_{\text{ub}} \times \text{GFR}$ and tubular reabsorption is apparent when $CL_R < f_{\text{ub}} \times \text{GFR}$. Reported values of GFR are measured with endogenous filtration markers such as creatinine, which is freely filtered and secreted (15%) in the proximal tubule. However, since the synthesis and blood concentration of creatinine are influenced by several factors,

3.3 DRUG DISPOSITION

including age, sex, ethnicity, muscle mass, and chronic illness, other markers such as serum cystatin C, ^{51}CrEDTA, or inulin are employed. Since the maximum renal clearance can be about 1.8 mL/min/kg, (the GFR in humans) in the absence of active secretion, drugs cleared exclusively in the kidney (hydrophilic acids and bases with high polar surface area (PSA) and rotatable bond count) generally tend to have low clearances. Apart from low clearances, these compounds are also unaffected by CYP-related issues such as **polymorphisms**, drug-drug interactions (DDI), and reactive metabolites.

Amphiphilic compounds (compounds with both acidic and basic groups) with molecular weights >350 also have the possibility of being actively transported into the bile and excreted via feces. Biliary clearance can be estimated by determining the concentration of a drug in the bile (C_{bile}) collected from a bile-duct-cannulated preclinical species:

$$CL_B = \frac{\text{bile flow} \times C_{bile}}{C} \quad (3.17)$$

The parent drug in bile is emptied into the duodenal section of the small intestinal tract and may be reabsorbed back into the portal vein as it transits down the small intestine. This is called the **enterohepatic recirculation (EHR)** of a parent drug. A metabolite of the drug can also be emptied into bile and into the duodenum (Fig. 3.5). Certain phase II metabolites such as glucuronides are then reconverted to the parent drug by the gut microflora and reabsorbed and recirculated. Although the pharmacokinetic profile is very similar to that associated with parent EHR, the measured C_{bile} for the parent drug in this case would be low, allowing one to distinguish between parent and metabolite EHR.

Absorption, distribution, renal elimination, and biliary elimination are all dependent on physicochemical properties of the drug, particularly lipophilicity and acid/base/neutral characteristics and physiology of the species such as blood flow, organ volumes, transit rates, and the like. The chemical structure of a drug dictates its rate and extent of biotransformation as well as affinity to transporters.

3.3.4 Role of Transporters in ADME

Transporters (Fig. 3.6a) play an important role in enhancing or limiting the disposition of drugs[9] in plasma, target issue, as well as in the blood–brain barrier,[10] as shown in Table 3.2. About 400 genes code for transporters in humans and animals, and these are classified into 2 large superfamilies—the solute carrier, or SLC, superfamily containing 364 proteins in 48 subfamilies[11] and the ATP-binding cassette, or ABC, transporters of 49 proteins in 7 subfamilies. The SLC transporters mediate cellular influx of substrates either by facilitated diffusion or by secondary active transport, driven by co-transport (symport or antiport) of endogenous organic ions.[12–15] ABC transporters

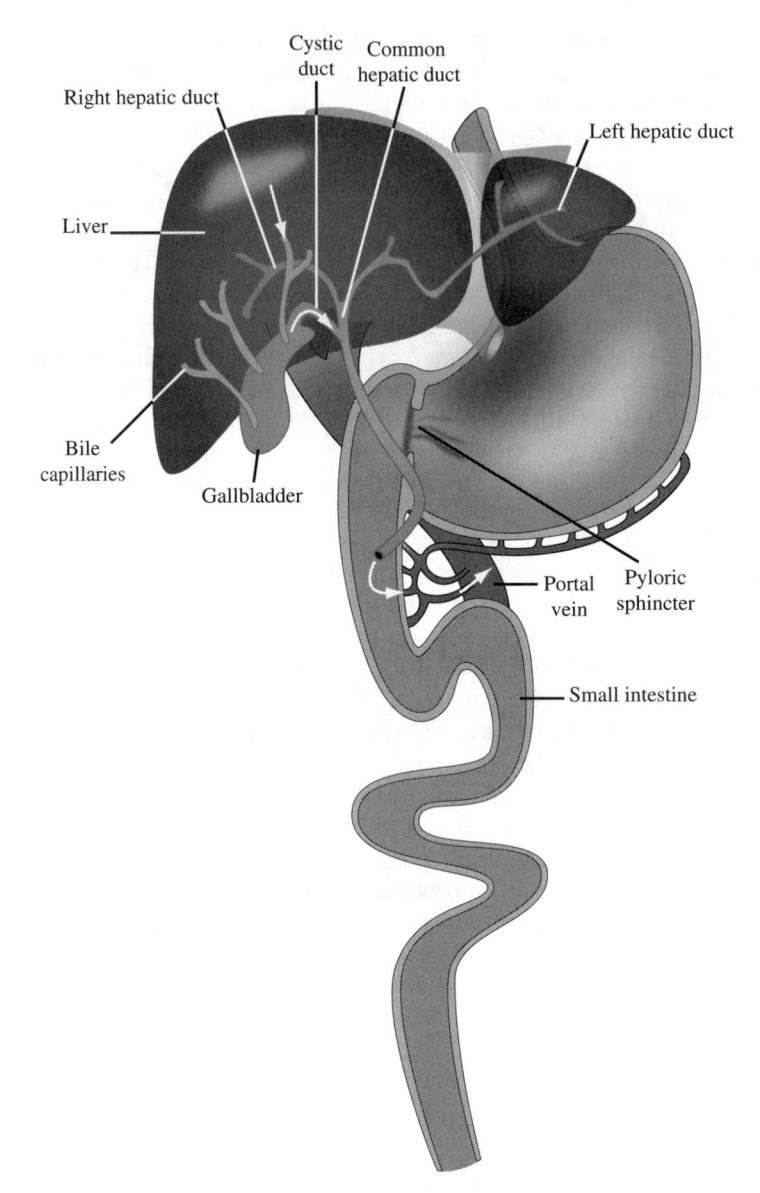

Figure 3.5. Entero-hepatic recirculation of a parent drug or a metabolite.

mediate the primary active transport of unidirectional efflux of drugs often against a steep diffusion gradient, deriving energy from ATP hydrolysis.[16–18] The transporters that play an important role in human drug disposition are the active SLC transporters OATP1B1, 1B3 and 2B1, SLC transporters that mediate facilitated diffusion OCTs 1 and 2, SLC antiport transporters MATE1 and MATE2K, and the ABC efflux transporters P-gp (MDR1), MRP2

3.3 DRUG DISPOSITION

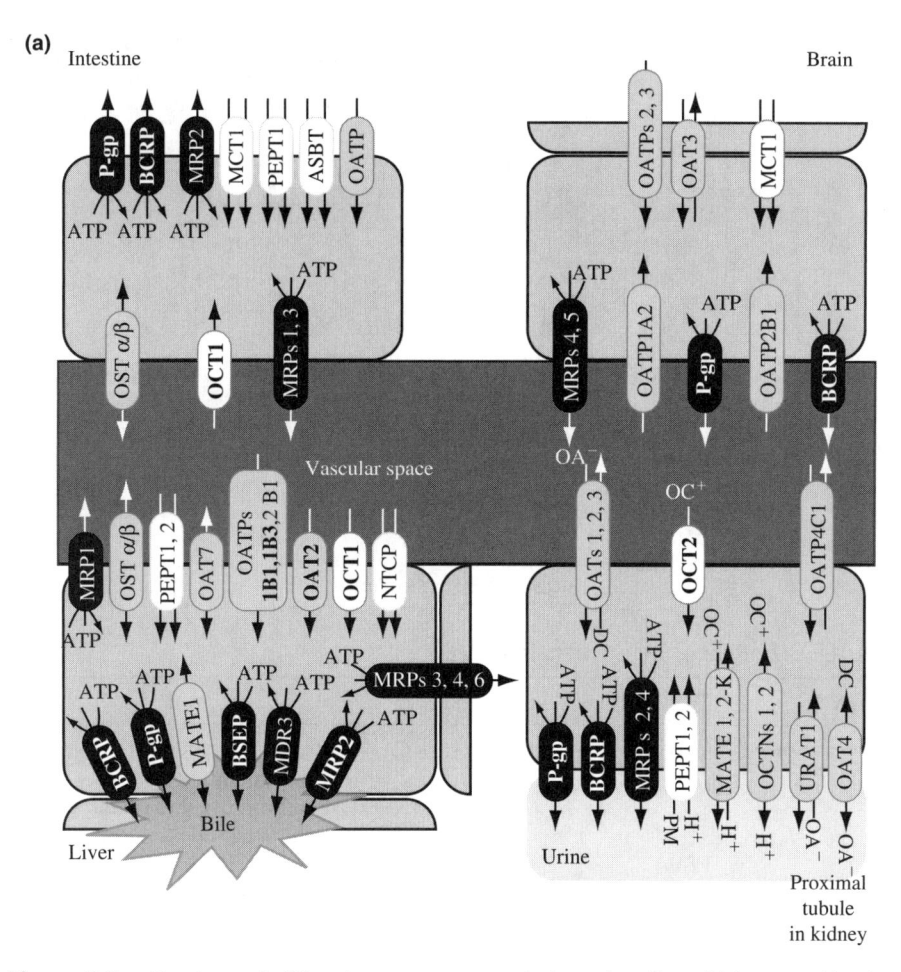

Figure 3.6a. Uptake and efflux drug transporters in intestine, liver, kidney, and brain. OAT, organic anion transporter; OATP, organic anion-transporting polypeptide; NTCP, sodium taurocholate co-transporting peptide; OCT, organic cation transporter; OCTN, novel-type OCT; PEPT, oligopeptide transporter; ASBT, apical sodium-dependent bile acid transporter; MDR, multidrug resistant (or resistance); MRP, multidrug resistance-associated protein; BCRP, breast cancer resistance protein; MATE, multidrug and toxin extrusion; MCT, monocarboxylate transporter; OST, organic solute transport; URAT, urate transporter. Primary active ATP-binding cassette (ABC) efflux transporters derive energy from the hydrolysis of ATP to ADP and are shown in dark shading. The secondary active transporter families or solute carrier (SLC) uptake transporters utilize either ion gradients across membranes or co-transport with intracellular and extracellular ions. These are either symports (unshaded) or antiports (lightly shaded). Multidrug efflux pump, MATE is a cation-coupled family of transporters. OST α/β is a heterodimer that together transports bile acids, conjugated steroids, and structurally-related molecules. Clinically relevant transporters are shown in bold.

(b)

Co-transported ion,
usually H^+, Na^+, HCO_3^-

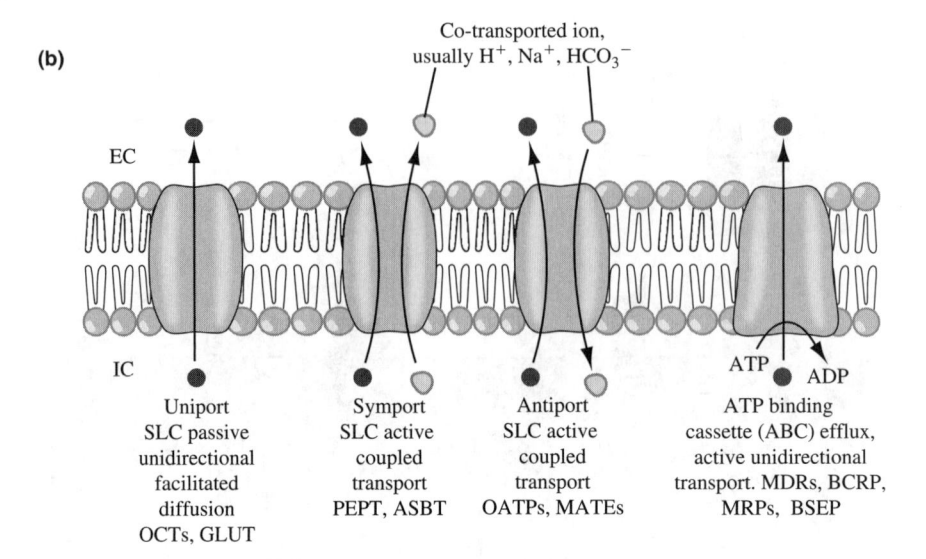

Uniport	Symport	Antiport	ATP binding
SLC passive	SLC active	SLC active	cassette (ABC) efflux,
unidirectional	coupled	coupled	active unidirectional
facilitated	transport	transport	transport. MDRs, BCRP,
diffusion	PEPT, ASBT	OATPs, MATEs	MRPs, BSEP
OCTs, GLUT			

Figure 3.6b. (*Continued*).

TABLE 3.2. Role of Transporters in ADME

ADME	Transporter	Example	Role
Absorption	Carrier	PEPT	Enhances absorption by transporting hydrophilic compounds with specific groups such as peptide linkages.
	Efflux	P-gp, BCRP, MRP2	Limits absorption of large lipophilic molecules. P-gp has broad specificity.
	Uptake	OATP	Uptake of organic ions.
Distribution	Uptake efflux	Various	Increases or decreases tissue distribution.
Metabolism	Uptake	OATP1B1, OATP1B3, OATP2B1	Increases or decreases exposure to drug-metabolizing enzymes thereby increasing or decreasing metabolism.
Biliary elimination	Efflux	P-gp, BCRP	Increases biliary elimination of lipophilic, amphiphilic compounds.
Renal elimination	Uptake	OAT 1, 2, and 3, OCT2, OATP4C1	Increases uptake and elimination of hydrophilic compounds.
	Efflux	P-gp, MATE1, MATE2-K, MRP2, 4	Removal (active secretion) of toxins.

(important in the transport of glucuronide and glutathione conjugates), BCRP, and BSEP. ATP-independent secondary efflux transporters are probably meant to be an additional mechanism for toxin extrusion even if cells are energy deprived. The different types of transporters are shown in Figure 3.6b.

The role of the efflux transporter P-gp in restricting absorption, transporting amphiphilic compounds into bile, and keeping out lipophilic compounds from the brain has long been recognized. It is difficult to predict the disposition of compounds that are transported, as the substrate specificity to most transporters are dependent on the chemical structure of the drug. However, some general principles for identifying transporter substrates are there.[19–23] For example, P-gp substrates are mostly large and lipophillic. Substrates of the uptake transporter OATP1B1 are mostly carboxylic acids and so on. Efflux transporters such as MRP2 and BCRP have similar substrate specificity to uptake transporters, reducing the possibility for toxin accumulation.

Although *in vitro* assays may identify potential substrates of transporters, their relevance *in vivo* depends very much on the expression levels of transporters in the tissue and on the concentration of the drug relative to its K_M. If the affinity of the drug to the protein is high (low K_M) and/or the drug concentrations are high, the transporters are easily saturated. For example, P-gp efflux in the enterocytes is easily saturated by many of its substrates. The rate of P-gp efflux in the gut also depends on the permeability of the drug and the rate of gut metabolism. For high permeability drugs with high gut extraction, the rate of efflux is low. The impact of uptake or efflux transporters in increasing or decreasing intracellular concentrations of their substrates is much more for low-permeability drugs compared to high-permeability drugs. P-gp and CYP3A4 share a common substrate specificity. In the gut, where drug concentrations can be high, this enzyme–transporter interplay[24] increases the probability of contact between the drug and enzyme, by making it possible for CYP3A4 to remain unsaturated due to the dilution effects of P-gp efflux. The mean residence time of drug and enzyme also increases, as the drug is presented to the enzymes repeatedly while it transits down the small intestine. P-gp efflux, however, also offers a competing mechanism to metabolism both in the liver and in the gut. Only a physiologically-based model that simultaneously considers the physiological parameters such as transit rates and blood flow rates along with drug-dependent parameters such as permeability, enzyme and transporter affinity, and rates of metabolism can adequately address the complex balance of enzyme and transporter systems *in vivo*. Renal anionic and cationic transporters have been reviewed.[25,26]

A potent inhibitor/inducer (perpetrator) of a drug-metabolizing enzyme/ transporter/binding protein in the plasma can cause an increase or decrease in the exposure of a coadministered victim drug leading to reduced safety or efficacy. A perpetrator of a DDI has a low measured inhibitor constant (K_i) and relatively high plasma concentrations (high dose). A coadministered victim drug is said to be sensitive to DDI if it relies heavily on the inhibited or induced route for its elimination. Thus, a drug with low fractions of biliary and/or renal elimination (high fraction metabolized, f_m) or a drug that is reliant on a single CYP enzyme or transporter and has high gut and/or hepatic extractions is likely to be a victim. An understanding of the role of transporters in mediating DDI,[27] their impact in drug discovery and development,[28,29] and strategies to address it during drug development[30,31] are emerging.

3.4 LINEAR AND NONLINEAR PHARMACOKINETICS

In Section 3.3.3, it was seen that the saturation of drug-metabolizing enzymes can cause nonlinear clearance. Solubility limitations and saturation of plasma-binding proteins and transporter proteins can also lead to nonlinear pharmacokinetics. Generally, the therapeutic concentrations of most drugs are not high enough to saturate the drug-metabolizing enzymes and transporters. However, for drugs that are administered at high doses, such as antiretroviral drugs, the risk for nonlinear PK is high. An important clinical implication of nonlinear pharmacokinetics[32] is the altered half-life, which leads to a longer or shorter time to achieve a given fraction of steady state. Since dose predictions are often done for the steady state, nonlinearity makes any predictions of disposition highly uncertain.

3.5 STEADY-STATE PHARMACOKINETICS

Steady-state concentrations of a drug in plasma are reached following a constant rate IV infusion or successive administration of the drug at equal intervals. The elimination of an IV bolus dose generally follows an exponential decay starting from an initial concentration C_0. If the initial concentration is assumed to be the steady-state concentration, C_{ss}, the concentration of the drug at any time, t (C_t), after the cessation of infusion is given by

$$C_t = C_{ss} e^{-k_{el} t} \qquad (3.18)$$

When a drug is infused intravenously at a constant rate, the plasma concentration continues to rise until elimination equals the rate of delivery into the body, at which point a steady state is said to have been reached. This is illustrated in Figure 3.7a. Mathematically, the time dependence of the infusion curve is obtained by subtracting the exponential term from 1 and expressed as follows:

$$C_t(\text{inf}) = C_{ss} \times (1 - e^{-k_{el} t}) \qquad (3.19)$$

where $C_t(\text{inf})$ is the concentration of the drug at any time t, following a constant rate infusion of the drug. Equation 3.19 suggests that this

3.5 STEADY-STATE PHARMACOKINETICS

Figure 3.7. Steady state concentrations following (a) constant rate infusion and (b) oral drug administration.

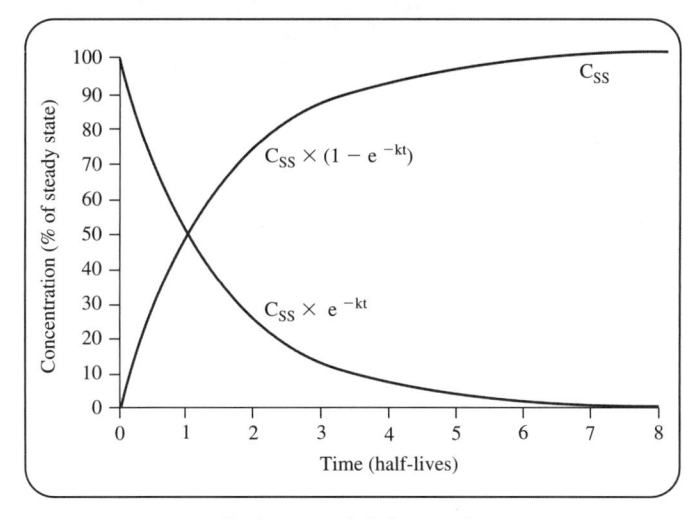

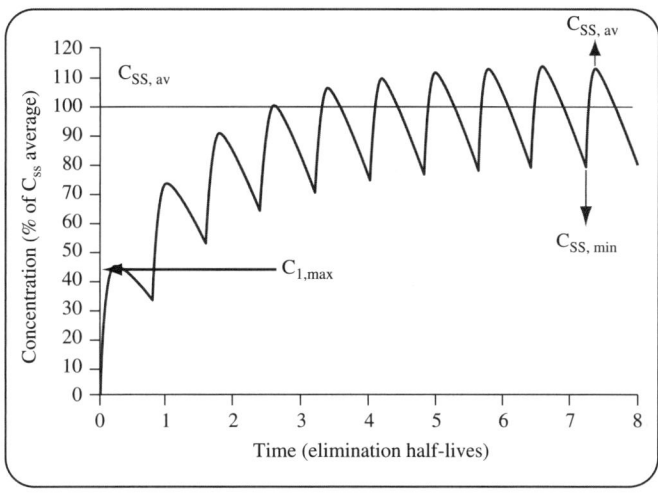

concentration will tend toward the steady-state concentration as t approaches infinity. Also, regardless of the drug, 50% of the plateau concentration is attained in one half-life of the drug; 75, 87.5, and 93.75% of the plateau concentration are reached in two, three, and four half-lives, respectively. For all practical purposes, steady state is achieved in three to five half-lives. Thus, the time required to reach steady state depends only on the drug's half-life. The shorter the half-life, the more rapidly the steady state is reached. The size of the dose and the route of administration have little effect. Figure 3.7b shows the concentration–time profile of a successively administered oral drug. The profile parallels that observed for the constant rate infusion. However, fluctuations occur within each dosing interval, as a dose is absorbed and eliminated, leading to a $C_{ss,max}$ and a $C_{ss,min}$. $C_{ss,av}$ is the average of $C_{ss,max}$ and $C_{ss,min}$. The C_{ss} after an IV infusion or $C_{ss,av}$ following multiple oral dose administrations are simply given by the ratio of dosing rate to clearance and given by

$$C_{ss,av} = \frac{\text{dosing rate}}{\text{clearance}} = \frac{R_0}{CL} \quad C_{ss,av} = \frac{\text{dose} \times F}{\tau \times CL} \quad (3.20)$$

$$C_{SS,av,u} = f_{up} \times \frac{\text{dose} \times F}{\tau \times CL} \quad (3.21)$$

where R_0 is the infusion rate, F is the bioavailability of a non-IV drug, and τ is the dosing interval, which is 24 h for a once-daily drug. $C_{ss,av,u}$ is the unbound steady-state drug concentration and f_{up} is the fraction unbound in plasma. According to equation 3.21, the steady-state concentration of the drug increases as clearance decreases, as in the case of nonlinear kinetics arising from enzyme saturation (Section 3.3.3). Consequently, small increases in dose result in large increases in total and unbound steady-state drug concentration. The magnitude of drug concentrations at steady state compared with that after the first dose is determined by the relationship between dosing interval and the half-life. The ratio of maximum drug concentration under steady-state conditions ($C_{ss,max}$) to the maximum drug concentration after the first dose ($C_{1,max}$) is called the accumulation ratio. A drug with a long half-life compared with its dosing interval is likely to accumulate, while a drug with a short half-life compared with its dosing interval will get eliminated between doses and is likely to diminish drug efficacy. Thus, dosing intervals should ideally equal the half-lives of drugs. However, for drugs following nonlinear clearance at therapeutic doses, clearance decreases with increasing doses, which means that the apparent half-life increases with increasing dose. Consequently, the time to reach steady state is longer than that expected from the half-life predicted for a single dose. Another parameter closely related to the half-life of a drug is its mean residence time (MRT), defined as the inverse of the elimination rate constant. It is estimated from the concentration–time profile as the ratio of the area under the first moment curve ($AUMC$) to the AUC:

3.6 DOSE ESTIMATIONS

$$\mathrm{MRT} = \frac{1}{k_{el}} = \frac{V_{ss}}{CL} = \frac{AUMC}{AUC} = \frac{\int_{t=0}^{\infty} Ct\, dt}{\int_{t=0}^{\infty} C\, dt} \qquad (3.22)$$

Knowing the MRT and CL, the steady-state volume of distribution (V_{ss}) of a drug can be estimated.

Since a drug normally requires at least three to five half-lives to reach steady state, effective plasma levels may be achieved more rapidly by the administration of a single large dose, called the loading dose, to bring the concentration in plasma quickly to the steady-state levels followed by maintenance doses. The loading dose required to achieve the plasma levels present at steady state can be determined from the fraction of drug eliminated during the dosing interval and the maintenance dose:

$$\text{Loading dose} = \frac{\text{Maintenance dose}}{1 - e^{-k_{el}t}} \qquad (3.23)$$

3.6 DOSE ESTIMATIONS

Dose estimations use equation 3.21, with $C_{ss,av,u}$ obtained from chronic dosing and equated to the effective concentration. An appropriate dosing interval for most drugs depends on the distance between the maximum and the minimum target drug concentration or therapeutic range (Fig. 3.8). Dosing intervals

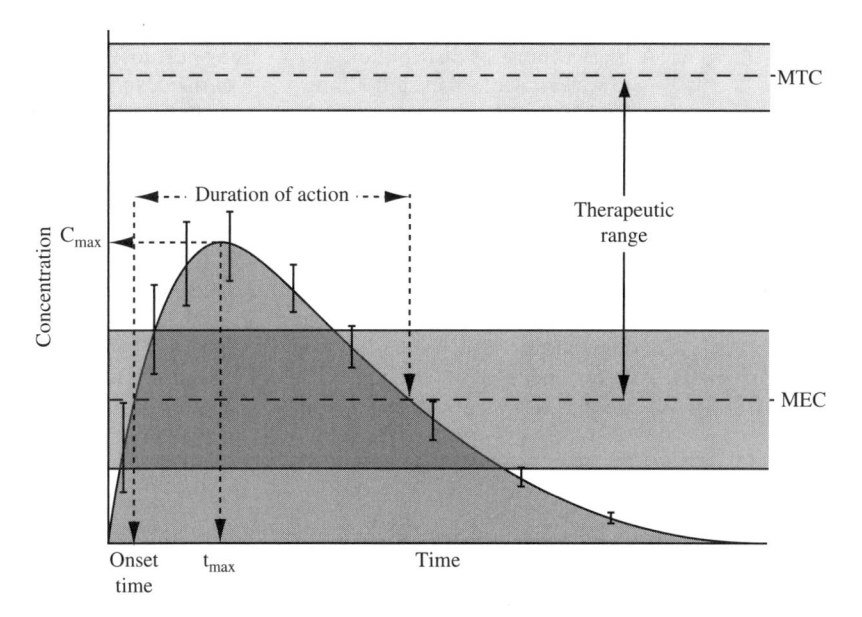

Figure 3.8. Therapeutic range with margins of uncertainty. MTC: Minimum toxic concentration; MEC: minimum effective concentration.

TABLE 3.3. Optimization of Pharmacokinetics—What?, How? and Why?

PK Optimization	How?	Need for Optimization
Gut bioavailability	Balance lipophilicity and solubility to achieve good absorption. Reduce potential for CYP3A metabolism and glucuronidation, the two major pathways for gut extraction.	To enhance bioavailability.
Clearance	Low clearance can be achieved by reducing lipophilicity, by avoiding functional groups that are known targets for metabolism and by reducing the activity of potential sites of metabolism through steric hindrance.	Reduce hepatic extraction.
	Avoid reliance on single elimination pathway especially the high-affinity, low-capacity CYPs (2C9 and 2C19) and CYP3A4 (common pathway for many drugs).	Reduce potential for being a victim of DDI.
	Avoid clearance through polymorphic enzymes (such as CYP2D6) or transporters (OATP1B1).	Minimize the interindividual variability in exposure.
Volume of distribution	The greater the lipophilicity and greater the fraction unbound in plasma, the greater the V_{ss}. Bases generally have a high V_{ss}, followed by neutrals and then acids.	Ensures long duration of the drug.
Half-life	A large volume of distribution and low clearance will ensure a long half-life.	Long postdose duration will ensure a simplified dosing regimen of once daily and promote patient compliance.
Biotransformation	Avoid carboxylic acids that are likely to form reactive acyl glucuronides. Avoid reactive metabolites.	To improve drug safety.
Transporters	Ensure sufficient permeability to reduce interplay of transporter and metabolism.	To reduce potential for DDI.
Pharmacokinetics	Aim for linear PK by keeping the dose as low as possible.	To minimize uncertainty in predictions of disposition.

equal to the half-lives are impractical for drugs with short half-lives. In most cases, either high doses of a relatively nontoxic drug are given to attain therapeutic concentrations for a sufficient time period or potentially harmful drugs are administered by careful IV infusion. Another approach is to use dosage

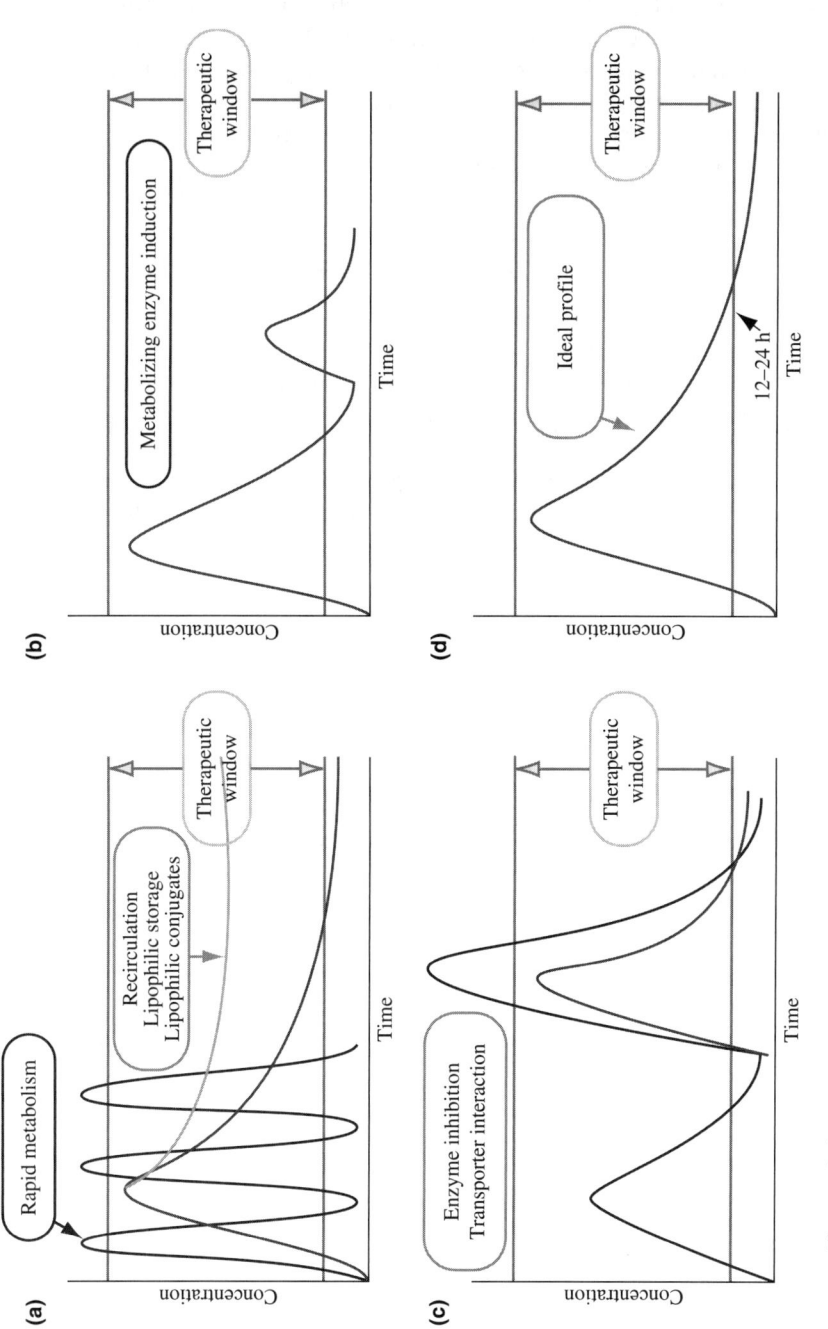

Figure 3.9. Problematic and ideal pharmacokinetic profiles. (a) Rapid metabolism or prolonged duration. Nonlinear PK and lack of repeatability arising from (b) induction of drug metabolizing enzyme and (c) drug-metabolizing enzyme/transporter inhibition. (d) An ideal pharmacokinetic profile.

39

formulations or devices that allow for a more gradual release of the drug into the systemic circulation. The dosing regimen should take into consideration the target characteristics such as half-life response. The disease area plays a role too. For example, painkillers need to be fast acting, and, therefore, half-life need not be high, but the dose should rapidly reach its target in sufficient concentrations. The age, severity of disease, and compliance of the target population should also be taken into consideration.

3.7 SUCCESSFUL PK OPTIMIZATION IN DRUG DISCOVERY

A successful lead optimization of PK aims at maximizing the bioavailability and half-life, reducing clearance and toxic metabolites and minimizing the risk for drug–drug and food–drug interactions. A comprehensive preclinical pharmacokinetic evaluation would ensure that compounds do not fail in the clinic.[33] Table 3.3 summarizes all of the PK optimizations and describes how and why they should be done. Figure 3.9 shows different PK profiles resulting from unoptimized PK.

KEYWORDS

Glomerular Fitration Rate (GFR): Rate at which water is filtered out of the plasma through the glomerular capillary walls into Bowman's capsules, usually measured by the rate of clearance of creatinine.

Enterohepatic Recirculation (EHR): Refers to the circulation of bile components secreted from the liver, into the small intestine, from where it can be reabsorbed again. EHR contributes to an increased half-life of a drug. Some phase II conjugates are also secreted into bile and converted back to the parent in the distal small intestine and reabsobed as parent.

Lead Optimization: Stage of the drug discovery phase when a candidate drug is identified from one or more lead series through iterative cycles of compound synthesis, testing, and design. The identified lead should satisfy a predefined criteria with respect to novelty, potency, selectivity, physicochemical characteristics, pharmacokinetics, and toxicity.

Polymorphism, Polymorphic Enzymes: A discontinuous genetic variation in proteins/enzymes that occurs in at least 1% of the population, leading to different types of individuals in a single species.

Steric Hindrance: Caused by the introduction of bulky chemical groups close to the site of metabolism in a new chemical entity in an attempt to protect the site from metabolism.

REFERENCES

1. Wright JD, Boudinot FD, Ujhelyi MR. Measurement and analysis of unbound drug concentrations. *Clin Pharmacokinet*. 1996;30(6):445–462.
2. Benet LZ, Hoener BA. Changes in plasma protein binding have little clinical relevance. *Clin Pharmacol Ther*. 2002;71(3):115–121.
3. Mager DE, Woo S, Jusko WJ. Scaling pharmacodynamics from in vitro and preclinical animal studies to humans. *Drug Metab Pharmacokinet*. 2009;24(1):16–24.
4. Uchimura T, Kato M, Saito T, Kinoshita H. Prediction of human blood-to-plasma drug concentration ratio. *Biopharm Drug Dispos*. 2010;31(5–6):286–297.
5. Hinderling PH. Red blood cells: A neglected compartment in pharmacokinetics and pharmacodynamics. *Pharmacol Rev*. 1997;49(3):279–295.
6. Jusko WJ, et al. Pharmacokinetics of tacrolimus in liver transplant patients. *Clin Pharmacol Ther*. 1995;57(3):281–290.
7. Paixao P, Gouveia LF, Morais JA. Prediction of drug distribution within blood. *Eur J Pharm Sci*. 2009;36(4–5):544–554.
8. Varma MV, et al. Physicochemical determinants of human renal clearance. *J Med Chem*. 2009;52(15):4844–4852.
9. Ho RH, Kim RB. Transporters and drug therapy: Implications for drug disposition and disease. *Clin Pharmacol Ther*. 2005;78(3):260–277.
10. Shen S, Zhang W. ABC transporters and drug efflux at the blood-brain barrier. *Rev Neurosci*. 2010;21(1):29–53.
11. Fredriksson R, Nordstrom KJ, Stephansson O, Hagglund MG, Schioth HB. The solute carrier (SLC) complement of the human genome: Phylogenetic classification reveals four major families. *FEBS Lett*. 2008;582(27):3811–3816.
12. Duan P, You G. Short-term regulation of organic anion transporters. *Pharmacol Ther*. 2010;125(1):55–61.
13. Hagenbuch B, Gui C. Xenobiotic transporters of the human organic anion transporting polypeptides (OATP) family. *Xenobiotica*. 2008;38(7–8):778–801.
14. Zhang EY, Knipp GT, Ekins S, Swaan PW. Structural biology and function of solute transporters: Implications for identifying and designing substrates. *Drug Metab Rev*. 2002;34(4):709–750.
15. Hediger MA, Romero MF, Peng JB, Rolfs A, Takanaga H, Bruford EA. The ABCs of solute carriers: Physiological, pathological and therapeutic implications of human membrane transport proteins. Introduction. *Pflugers Arch*. 2004;447(5):465–468.
16. Schinkel AH, Jonker JW. Mammalian drug efflux transporters of the ATP binding cassette (ABC) family: An overview. *Adv Drug Deliv Rev*. 2003;55(1):3–29.
17. van de Water FM, Masereeuw R, Russel FG. Function and regulation of multidrug resistance proteins (MRPs) in the renal elimination of organic anions. *Drug Metab Rev*. 2005;37(3):443–471.
18. Szakacs G, Varadi A, Ozvegy-Laczka C, Sarkadi B. The role of ABC transporters in drug absorption, distribution, metabolism, excretion and toxicity (ADME-tox). *Drug Discov Today*. 2008;13(9–10):379–393.

19. Kusuhara H, Sugiyama Y. In vitro—in vivo extrapolation of transporter-mediated clearance in the liver and kidney. *Drug Metab Pharmacokinet.* 2009;24(1):37—52.
20. Ahn SY, Nigam SK. Toward a systems level understanding of organic anion and other multispecific drug transporters: A remote sensing and signaling hypothesis. *Mol Pharmacol.* 2009;76(3):481—90.
21. El-Sheikh AA, Masereeuw R, Russel FG. Mechanisms of renal anionic drug transport. *Eur J Pharmacol.* 2008;585(2—3):245—255.
22. Wright SH, Dantzler WH. Molecular and cellular physiology of renal organic cation and anion transport. *Physiol Rev.* 2004;84(3):987—1049.
23. Nies AT, Schwab M, Keppler D. Interplay of conjugating enzymes with OATP uptake transporters and ABCC/MRP efflux pumps in the elimination of drugs. *Expert Opin Drug Metab Toxicol.* 2008;4(5):545—568.
24. Benet LZ. The drug transporter-metabolism alliance: Uncovering and defining the interplay. *Mol Pharm.* 2009;6(6):1631—1643.
25. Masereeuw R, Russel FG. Therapeutic implications of renal anionic drug transporters. *Pharmacol Ther.* 2010;126(2):200—216.
26. Dresser MJ, Leabman MK, Giacomini KM. Transporters involved in the elimination of drugs in the kidney: Organic anion transporters and organic cation transporters. *J Pharm Sci.* 2001;90(4):397—421.
27. Zhang L, Strong JM, Qiu W, Lesko LJ, Huang SM. Scientific perspectives on drug transporters and their role in drug interactions. *Mol Pharm.* 2006;3(1):62—69.
28. Lai Y, Sampson KE, Stevens JC. Evaluation of drug transporter interactions in drug discovery and development. *Comb Chem High Throughput Screen.* 2010;13(2):112—134.
29. Mizuno N, Niwa T, Yotsumoto Y, Sugiyama Y. Impact of drug transporter studies on drug discovery and development. *Pharmacol Rev.* 2003;55(3):425—461.
30. International Transporter Consortium, Giacomini KM, et al. Membrane transporters in drug development. *Nat Rev Drug Discov.* 2010;9(3):215—236.
31. European Medicines Agency. Available at: http://www.ema.europa.eu/docs/en_GB/document_library/Scientific_guideline/2010/05/WC500090112.pdf.
32. Ludden TM. Nonlinear pharmacokinetics: Clinical implications. *Clin Pharmacokinet.* 1991;20(6):429—446.
33. Singh SS. Preclinical pharmacokinetics: An approach towards safer and efficacious drugs. *Curr Drug Metab.* 2006;7(2):165—182.

4

PHYSIOLOGICAL MODEL FOR ABSORPTION

CONTENTS

4.1	Introduction	44
4.2	Drug Absorption and Gut Bioavailability	44
	4.2.1 Solubility and Dissolution Rate	44
	4.2.2 Permeability: Transcellular, Paracellular, and Carrier-Mediated Pathways	51
	4.2.3 Barriers to Membrane Transport—Luminal Degradation, Efflux, and Gut Metabolism	53
4.3	Factors Affecting Drug Absorption and Gut Bioavailability	56
	4.3.1 Physiological Factors Affecting Oral Drug Absorption and Species Differences in Physiology	56
	4.3.2 Compound-Dependent Factors	62
	4.3.3 Formulation-Dependent Factors	63
4.4	*In Silico* Predictions of Passive Permeability and Solubility	66
	4.4.1 *In Silico* Models for Permeability	66
	4.4.2 *In Silico* Models for Solubility	67
4.5	Measurement of Permeability, Solubility, Luminal Stability, Efflux, and Intestinal Metabolism	67

Physiologically-Based Pharmacokinetic (PBPK) Modeling and Simulations: Principles, Methods, and Applications in the Pharmaceutical Industry, First Edition. Sheila Annie Peters.
© 2012 John Wiley & Sons, Inc. Published 2012 by John Wiley & Sons, Inc.

	4.5.1	*In Vitro*, *in Situ*, and *in Vivo* Assays for Permeability......	67
	4.5.2	Measurement of Thermodynamic or Equilibrium Solubility .	72
	4.5.3	Luminal Stability................................	74
	4.5.4	Efflux..	74
	4.5.5	*In Vitro* Models for Estimating Extent of Gut Metabolism..	76
4.6	Absorption Modeling....................................		76
	Keywords...		83
	References..		84

4.1 INTRODUCTION

A drug administered in any non-IV route needs to get absorbed into systemic circulation before it can be pharmacologically active. The oral route is the most convenient mode of drug administration. This chapter will provide an overview of the physiological, compound-, and formulation-dependent factors impacting the solubility, dissolution rate, permeability, and **intestinal loss** of an orally dosed compound, which in turn contribute to its absorption and systemic bioavailability. *In silico* and *in vitro* models for solubility, dissolution rate, permeability, and intestinal loss that enable the prediction of **fraction absorbed** and **gut bioavailability** will be discussed. The differential equations of a physiologically-based mathematical model and the integration of compound- and formulation-dependent data generated from *in vitro* and *in silico* models into the mathematical model will then follow.

4.2 DRUG ABSORPTION AND GUT BIOAVAILABILITY

The fraction of an orally administered dose that gets absorbed (f_{abs}) in the gastrointestinal tract is determined by fundamental physical phenomena such as dissolution rate, solubility, and permeability of the drug (Fig. 4.1). Gut bioavailability, F_g, is defined as the fraction of an orally administered drug that escapes first-pass gut extraction. Gut extraction of a drug can be due to luminal degradation, efflux, or gut metabolism. If the fraction of absorbed that escapes gut extraction is f_{gut}, then F_g is given by the product of f_{abs} and f_{gut}. Depending on the extent of metabolism in the liver, an oral drug is also subjected to hepatic first-pass extraction before it is systemically available. Thus, the systemic bioavailability (F) is a product of gut bioavailability and the fraction of dose escaping hepatic extraction (f_{hep}):

$$F = f_{abs} \times f_{gut} \times f_{hep} = F_g \times f_{hep} \qquad (4.1)$$

4.2.1 Solubility and Dissolution Rate

Drug molecules ingested orally molecules have to get dissolved and go past an unstirred water layer before they can permeate the enterocytes (Fig. 4.2). This

4.2 DRUG ABSORPTION AND GUT BIOAVAILABILITY

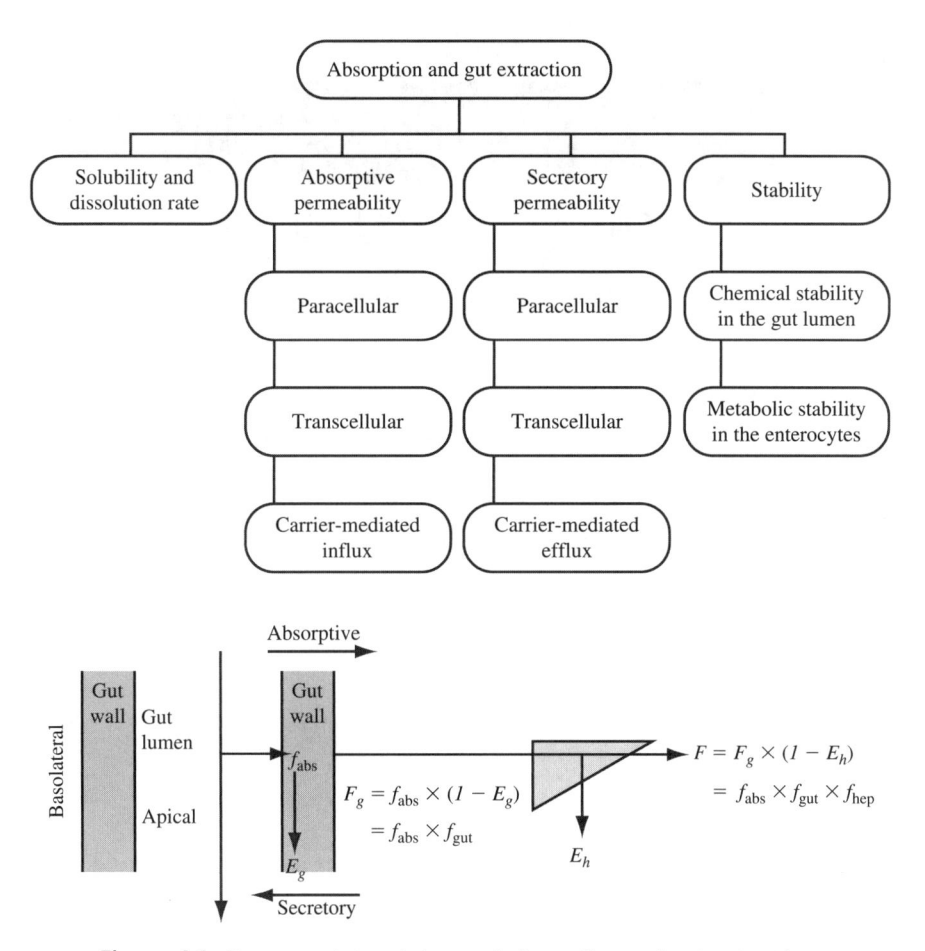

Figure 4.1. Processes determining oral absorption and gut extraction.

requirement implies that they should be reasonably hydrophilic. Solubility and dissolution rate represent the extent and rate of solubilization respectively. While solubility of a drug is determined by its lipophilicity, the dissolution rate depends on its solid-state packing and crystallinity, which are difficult to predict. Isotretinoin, cilostazol, danazol, dipyridamole, gefitinib, nitrendipine, phenytoin, spironolactone, and griseofulvin are examples of compounds with low dissolution rates. According to the Noyes–Brunner equation,[1,2] the dissolution rate (DR) of a drug is given by

$$\mathrm{DR} = \frac{dA}{dt} = \frac{D \times \mathrm{SA}}{d} \times (S - C) \qquad (4.2)$$

where SA is the surface area of the particle, d is the diffusion layer thickness or effective boundary layer thickness, S is the saturation solubility of the drug, and C is the concentration of the dissolved drug in the bulk; D is the diffusion

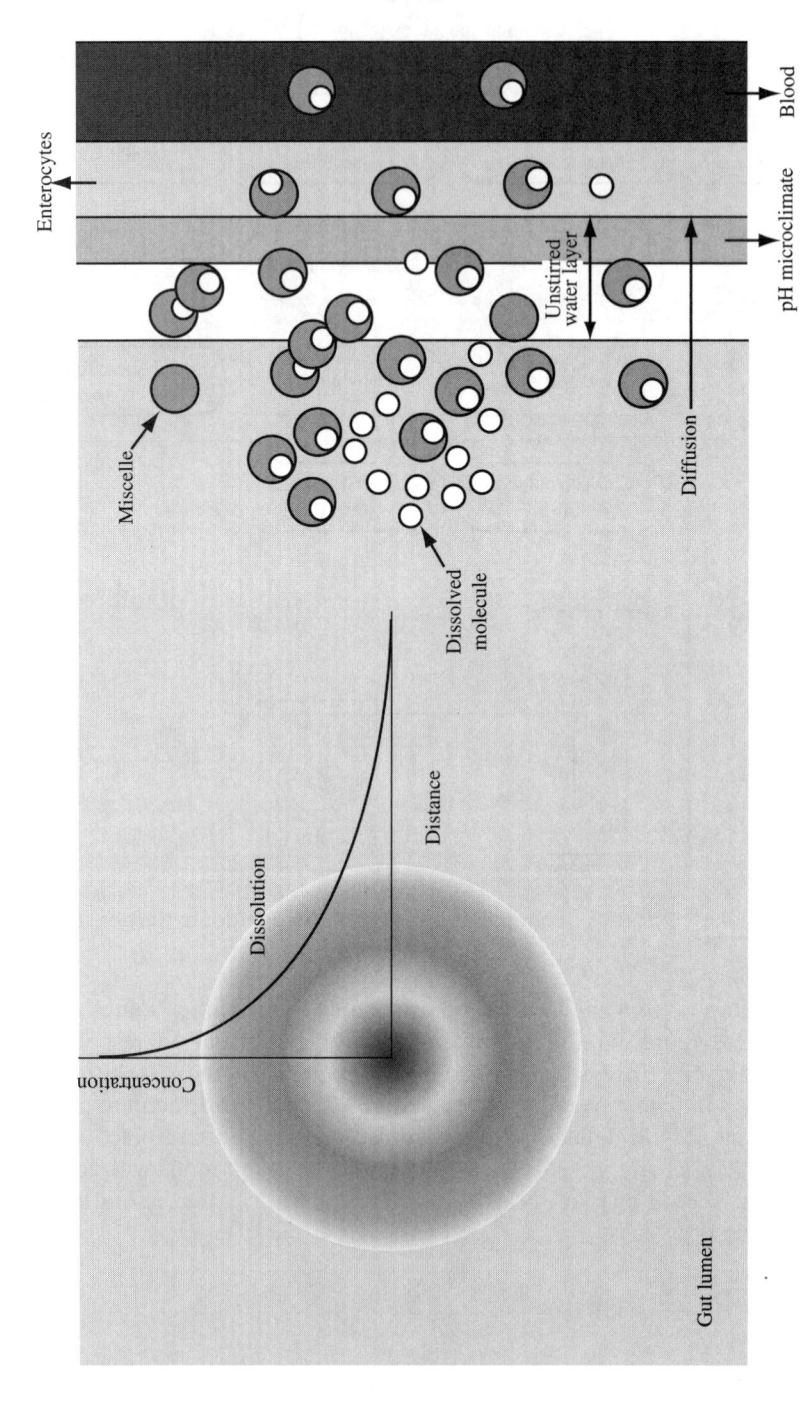

Figure 4.2. Fick's law and drug dissolution. Diffusion of a drug across the unstirred water layer aided by micelles.

coefficient, a proportionality constant relating the molar flux due to molecular diffusion to the concentration gradient according to Fick's first law. When the particles move independent of each other and possess the same mean kinetic energy as gas molecules at any given temperature, D can be obtained from the Stokes–Einstein equation (Eq. 4.3), according to which the frictional resistance of a hard sphere as it travels through a viscous liquid is inversely proportional to the diffusion rate of the sphere.

$$D = \frac{k \times T}{6\pi r \times \eta} \qquad (4.3)$$

where k is the Boltzmann constant, T is the temperature, r is the radius of the diffusing particle, and η is the viscosity of the medium. D ranges between 0.5 and 1×10^{-5} cm²/s. D can also be obtained using the Hayduk–Laudie equation, which was developed for measuring the diffusion coefficients of nonelectrolytes in dilute aqueous solutions. This equation is suitable for estimating diffusivity of organic compounds in water and is given by:

$$D(\text{cm}^2/\text{s}) = 13.26 \times 10^{-5}/\eta_W^{1.14} \times V^{0.589}$$

where, η_W is the viscosity of water in centipoise at 37 degrees, V is the molar volume of the substance. V can be given in terms of molecular weight and density. Dissolution rates are largely influenced by the chemical form, crystal form, wettability, particle size, and surface area. An increase in surface area is brought about by micronization. Noyes–Brunner equation assumes that the solubility remains constant with time, as dissolution progresses. However, solubility tends to decrease with time for many powders. Conditions in the gastrointestinal (GI) tract can also change with time. Thus, the assumption is not valid. An alternative to the Noyes–Brunner equation is the three-parameter Weibull distribution in which the fraction of drug released from the stomach is given by the first-order equation:

$$A_{\text{ur},t} = A_0 \times \exp\left[-\left(\frac{t-T}{t_D}\right)^b\right] \qquad (4.4)$$

where $A_{\text{ur},t}$ is the amount unreleased from the stomach at any time t, A_0 is the initial amount, T is the location parameter or time lag in case of a delayed emptying, t_D is a scaling factor, and b is a shape parameter generally less than 20. The parameters of the Weibull equation are obtained by fitting to experimental data. Hintz and Johnson[3] modified the Nernst–Brunner equation for a set number of monodispersed spherical particles to relate the dissolution rate of a particle to its size (r) and density (ρ) as follows:

$$\text{DR} = \frac{dA}{dt} = \frac{3 \times D \times A_0^{1/3} \times A_t^{2/3}}{\rho \times r \times d} \times (S - C) \qquad (4.5)$$

where A_0 and A_t are the amounts undissolved at time zero and at any time t, respectively. Polydispersed powders can be simulated as a number of individual monodisperse fractions.[3,4] Further improvements include changes in the effective boundary layer thickness over time,[3,4] assuming a cylindrical geometry instead of spherical[4] for the nonlinear concentration gradient across the diffusion layer of spherical particles and reducing particle radius with time as dissolution progresses as proposed by Wang and Flanagan[5,6] and shown below:

$$\text{DR} = -4\pi r(t)^2 \times D \times \left[\frac{1}{r(t)} + \frac{1}{d_{\text{eff}}}\right] \times [S - C(t)] \quad (4.6)$$

where d_{eff} is generally taken to be r when $r < 30$ μm and set to 30 μm when r exceeds 30 μm. $r(t)$ represents the time dependent r, which reduces with increasing dissolution. Solubility in the above models is generally considered a constant. However, as described by the Freudlich–Ostwald equation, an increase in compound saturation solubility is also expected at very small particle sizes. Specifically, the effect of particle size on solubility is given by[7]

$$S = S_\infty \times \exp\left(\frac{2\gamma M}{r\rho RT}\right) \quad (4.7)$$

where S is the saturation solubility of the nanosized particle, S_∞ is saturation solubility of an infinitely large crystal, γ is the interfacial tension of the medium, M is the molecular weight of the compound, r the particle radius, ρ the density, R is a gas constant, and T is the temperature. For a typical drug development candidate with a molecular weight of 500 and assuming $\rho = 1$ and a γ value of 0.015–0.020 N/m, this equation would predict an approximately 10–15% increase in solubility at a particle size of 100 nm compared to the solubility of large particles.

The **thermodynamic** or **equilibrium solubility** of a compound at any given temperature and pressure is the solubility of its most stable crystalline form under equilibrium conditions and is therefore the least value. The **kinetic solubility** on the other hand is defined as the solubility of a compound at any given time when the compound is in amorphous state or metastable **polymorphic forms** (Fig. 4.3). Only the thermodynamic solubility from the most stable crystalline form is directly correlated to the dissolution rate (Noyes–Brunner equation). Being related to the compound structure, thermodynamic solubility allows comparison between compounds and therefore better able to guide chemistry in the lead generation stage. In drug development, it is the preferred form for a drug, as there is less uncertainty associated with its solubility and therefore its bioavailability, especially if its bioavailability is solubility limited. At the screening stage of drug discovery, the thermodynamic or equilibrium solubility of a compound is what is measured. The solubility of a compound in its nonionized form is its **intrinsic solubility**, S_0. Acids and bases have a higher solubility in their ionized forms. The greater the ionized fraction at any given

4.2 DRUG ABSORPTION AND GUT BIOAVAILABILITY

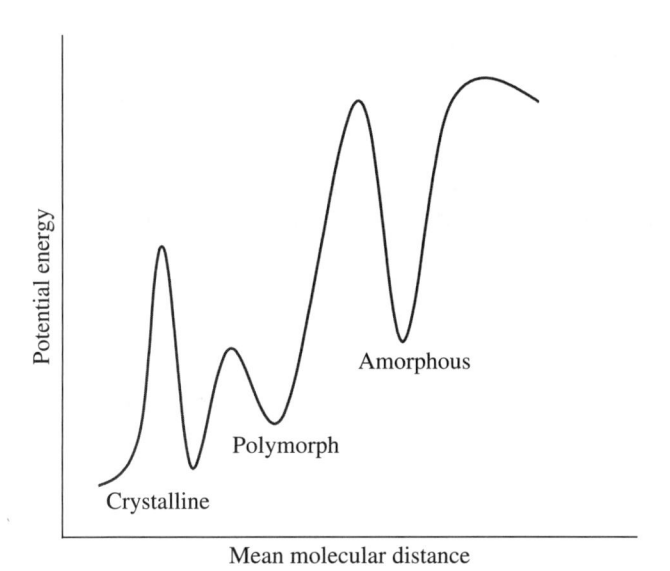

Figure 4.3. Potential energies corresponding to different solid states. The greater the potential energy, the greater is the solubility.

pH, the greater its solubility. The ionization equilibrium of an acid, HA, at a given pH is represented as

$$HA \xrightleftharpoons{K_a} H^+ + A^-$$

$$K_a = \frac{[H^+] \times [A^-]}{[HA] - [A^-]} \tag{4.8}$$

where K_a is the acid dissociation constant; $[H^+]$, $[A^-]$, and $[HA]$ are the equilibrium concentrations of the H^+ A^- ions, and the unionized acid, respectively. Taking negative logarithms on both sides and substituting $-\log[H^+]$ and $-\log K_a$ as pH and pK_a respectively, we arrive at the Henderson–Hasselbalch equation:

$$pK_a = pH - \log\left(\frac{[A^-]}{[HA] - [A^-]}\right)$$

or

$$\frac{[A^-]}{[HA] - [A^-]} = 10^{pH - pK_a}$$

Adding 1 to both sides,

$$1 + \frac{[A^-]}{[HA] - [A^-]} = 1 + 10^{pH - pK_a}$$

After simplification and rearrangement

$$[HA] = ([HA] - [A^-]) \times (1 + 10^{pH-pK_a}) \quad (4.9)$$

Recognizing that [HA] is the total concentration of the acid or the total solubility S and [HA]−[A$^-$] is the concentration of the un-ionized acid defined earlier as the intrinsic solubility (S_0) of the acid, we get

$$S = S_0 \times (1 + 10^{pH-pK_a}) \quad (4.10)$$

It also follows that the fraction of un-ionized acid, ([HA]−[A$^-$])/[HA] can be obtained as $1/(1 + 10^{(pH-pK_a)})$. Similarly, fraction of base and ampholyte un-ionized can also be obtained as $1/(1 + 10^{(pK_a-pH)})$ and $1/(1 + 10^{(pH-pK_a)} + 10^{(pK_a-pH)})$, respectively. The pH-dependent solubilities of a base and ampholyte can be obtained as

$$S = S_0 \times (1 + 10^{pK_a-pH}) \quad (4.11)$$

$$S = S_0 \times (1 + 10^{pK_a-pH} + 10^{pH-pK_a}) \quad (4.12)$$

Figure 4.4 illustrates the pH dependence of the fraction ionized and solubilities of an acid, base, and an ampholyte. Although Figure 4.4c suggests that the

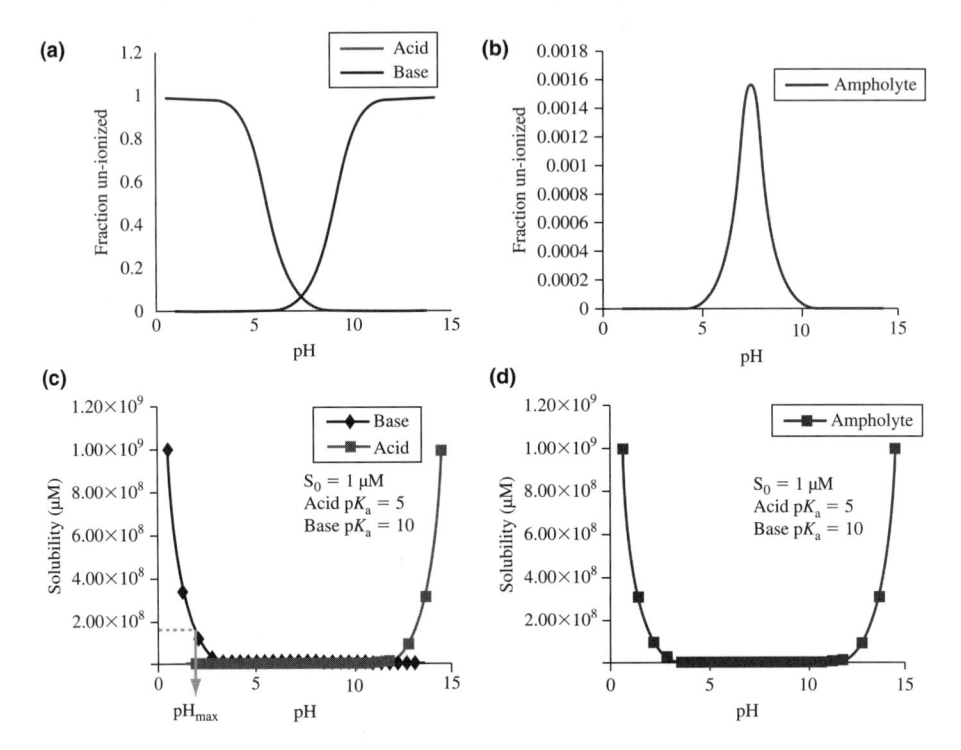

Figure 4.4. pH-dependent solubility of an acid, base, and an ampholyte. pH$_{max}$ due to salt formation.

4.2 DRUG ABSORPTION AND GUT BIOAVAILABILITY

solubility of a base increases steeply at low pH, in practice, the formation of ion pairs/ salt below a certain pH (pH$_{max}$) restricts the solubility (shown by the dotted lines) to the limiting solubility value. At this point, the solubility exceeds the solubility product (K_{sp}) of the ions. pH$_{max}$ depends on the K_{sp} of the salt as follows:

$$\text{pH}_{max} = pK_a + \log\frac{S_0}{\sqrt{K_{sp}}} \qquad (4.13)$$

K_{sp} and, therefore, pH$_{max}$ depend on the counterion.

4.2.2 Permeability: Transcellular, Paracellular, and Carrier-Mediated Pathways

A compound can permeate the gut wall either transcellularly or paracellularly (Fig. 4.5). A majority of drugs rely on the transcellular pathway for absorption.

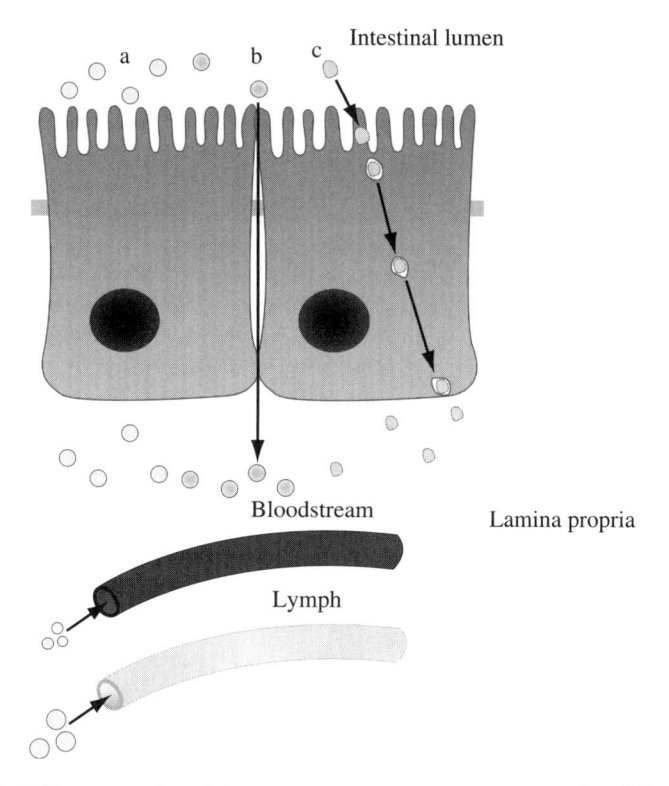

Figure 4.5. Different modes of drug transport across enterocytes. Lipophilic molecules have a transcellular drug absorption (a) Small hydrophilic drugs rely on the paracellular route (b), while large hydrophilic molecules rely on transcytosis and receptor-mediated endocytosis (c) for absorption. Molecules possessing certain special groups like peptide linkages are transported across the membrane by carrier transporters. Large lipophilic molecules are absorbed into lymph vessels rather than into blood vessels.

However, for hydrophilic and ionizable compounds, other mechanisms are needed to enhance transport across the enterocytes. Paracellular absorption is reliant on small intestinal aqueous channels of diameter 30–60 nm in humans[8] and is therefore restricted to small and hydrophilic molecules such as chlorothiazide, cimetidine, and furosemide. Large molecules (greater than 5000 Da) are transported by endocytosis across the enterocytes. Large and/or hydrophilic molecules with certain specific functional groups are assisted by carrier proteins across the enterocytes. Carrier proteins are transmembrane proteins that bind to physiologically important molecules (amino acids, oligopeptides, bile, glucose, monocarboxylic acids, etc.) on the extracellular surface and transport them to the intracellular side by changing their shape, either by a passive process, in the direction of electrochemical gradient (facilitated diffusion), or by an active process, against the electrochemical gradient. Drugs that are structurally similar to the substrates of the carrier proteins can be transported using this mechanism. Examples of drugs transported by carrier proteins are listed in Table 4.1. An escalation of dose can saturate the carrier transporters, leading to altered bioavailability. The carrier-mediated influx of a drug, P_{CM}, is given by

$$P_{CM} = \frac{J_{max}}{K_m + C_W} \qquad (4.14)$$

where J_{max} is the maximum transport rate, K_m is the molar concentration of the substrate at half-maximum substrate concentration, and C_W is the concentration of the drug at the gut wall.

Molecules need to be sufficiently lipophilic in order to get past the enterocytic membrane. Thus, an optimal drug should have a balanced hydrophilic and lipophilic nature to get absorbed. Very often multiple mechanisms act

TABLE 4.1. Influx or Uptake Transporter Families

Carrier Protein	Drugs Transported
Amino acid transporter	Gabapentin, λ-methyl-dopa, D-cycloserine, L-dopa, baclofen
Oligopeptide transporter (PepT1)	Peptidomimetics like cefadroxil, captopril, lisinopril, ceftibuten
Na^+/phosphate co-transporter	Fosfomycin, foscarnet, phosphonoacetic acid
Vitamin transporters	Salicylic acid, penicillins, methotrexate
Monocarboxylic acid transporter (MCT1)	Valproic acid, pravastatin, lovastatin, salicylic acid, fluoroquinolines, nonsteroidal anti-inflammatory drugs (NSAIDs)
Oraganic cation transporters (OCT)	Epinephrine, guanidine, choline, dopamine, cimetidine, verapamil
Nucleoside transporters	Antiviral (idoxuridine and didanosine) and anticancer nucleoside analogues
Lipid and bile acid transporters (LBAT)	Fatty acids

simultaneously.[9] The overall permeability (P) of a compound is related to the rate constant of absorption (k_a) as follows:

$$k_a = P \times \frac{2}{r}$$

$$\frac{\text{Intestinal surface area}}{\text{Intestinal volume}} = \frac{2\pi r l}{\pi r^2 l} = \frac{2}{r} \quad (4.15)$$

where r and l are the radius and length of the intestinal section, respectively. The small intestine (SI) is much more permeable, has less tight (or more porous) intercellular junctions and has higher expression of carrier proteins, compared to the colon. The SI is, therefore, roughly 10 times more permeable compared to the colon.

4.2.3 Barriers to Membrane Transport—Luminal Degradation, Efflux, and Gut Metabolism

A drug can undergo luminal degradation, or gut metabolism in the erythrocytes mediated by cytochrome P450 (CYP) or uridine 5′-diphospho-glucuronosyl-transferase (UGT) enzymes or get effluxed back into the lumen, all of which result in an intestinal loss of the drug before it can reach the liver. Efflux and intestinal metabolism are highly regulated, saturable biological processes that can lead to low and variable bioavailability of substrate drugs.

The intestinal mucosa is the most important extrahepatic site of drug biotransformation, as enterocytes express a wide variety of drug-metabolizing enzymes (DMEs)—cytochrome P450s (CYPs),[10] uridine 5′-diphospho-glucuronosyltransferases (UGTs), sulfotransferases (SULTs), glutathione S-transferases (GSTs), esterases, amidases, and epoxide hydrolases. Among the CYPs, CYP3A is most abundant in the gut, accounting for 80% of all CYPs. Unlike CYPs, UGT expression levels in the gut wall can even exceed that in the liver. In the human intestine, for example, the isoforms UGT1A1 and UGT2B7 are abundant. UGT1A8 and UGT1A10 are said to be expressed only in the intestine. In addition to the CYPs and UGTs, esterases also play an important role in gut metabolism. Table 4.2 lists the abundances of some of the drug metabolizing enzymes (DME) in the human gut. Apart from reductive and hydrolytic reactions mediated in the gut lumen by microflora, compounds can get substantially metabolized by oxidative, conjugative, and hydrolytic enzymes in the enterocytes. However, because of its broad substrate specificity and its high abundance in the intestine, CYP3A is the subfamily that is most frequently implicated in gut metabolism.

Orally administered drugs that are metabolized in the gut are generally associated with a low and variable systemic bioavailability. Studies in rat precision-cut intestinal and liver slices have shown that the rates of intestinal glucuronidation and sulfation can be one to three times greater than in the liver. The role of the gut in CYP-mediated first-pass metabolism is comparable to that in the liver. This is despite the fact that intestinal CYP expression and

TABLE 4.2. Abundance of Drug-Metabolizing Enzymes (DME) in the Human Gut

DME	Isoform	Location/Abundance	Source
CYPs	3A4	82% of CYPs in small intestine	Paine et al.[11]
	2C9	14% of all CYPs in small intestine	
	Others (2C19, 2J2 and 2D6)	4% of all CYPs in small intestine	
UGTs (gene expression 3 times > that of CYP)	1A1 and 1A4	Stomach and colon	Fisher et al.[12]
	1A8 and 1A10	Present only in the gut (small intestine and/or colon)	Yoshigae et al.[13]
	Others (1A3, 1A6)	Stomach and colon	
	2B7	Throughout the GI	
	2B15	Esophagus	
Esterases	hCE1	Less abundant in gut compared to liver	
	hCE2	More abundant in gut compared to liver	Glatt et al.[14]
SULTs	1B1	36% of total intestinal SULTs	
	1A3	31% of total intestinal SULTs	
	Others (1A1, 1E1, 2A1)	6–19% of total intestinal SULTs	

cofactor levels in the small intestine (SI) are only a small fraction compared to that in the liver. Additionally, as CYPs are concentrated in the villus tips, the high drug concentrations that are likely in the gut could easily saturate the CYPs. Yet, several factors favor a greater extraction in SI compared to liver:

1. In the liver, only the unbound drug in plasma is available for metabolism. There is no such restriction in the gut for an oral drug.
2. While presystemic gut extraction has the potential to be high, systemic gut extraction is generally considered negligible, as functionally mature enterocytes are only in the villus tips. Therefore, unlike in the liver, intestinal metabolism is not limited by blood perfusion rate.
3. Recycling and diluting effects of P-gp ensures suppression of CYP3A enzyme saturation and, consequently, a higher intestinal metabolism than would have been possible without P-gp. The likelihood of such synergistic effects is heightened by the overlapping substrate specificities of CYP3A and P-gp.
4. Long transit times in SI and lower blood flow to the intestine compared to the liver provides an increased opportunity for drug presentation to DME in enterocytes compared to hepatocytes.

The extent of metabolism and efflux depends on the intracellular drug concentrations in the enterocytes and the competing rates of permeability, metabolism, and efflux (Fig. 4.6). The variability in human gut extraction stems

4.2 DRUG ABSORPTION AND GUT BIOAVAILABILITY

Metabolism rate increases

P-gp likely to saturate

Efflux: High-permeability compounds have lower efflux. Metabolism: Permeability limited, for low-permeability compounds. With increasing permeability, access to drug metabolizing increases. However, competition from permeability also increases.

Permeability

Solubility limited — Therapeutic concentration — No solubility limitation or high dose

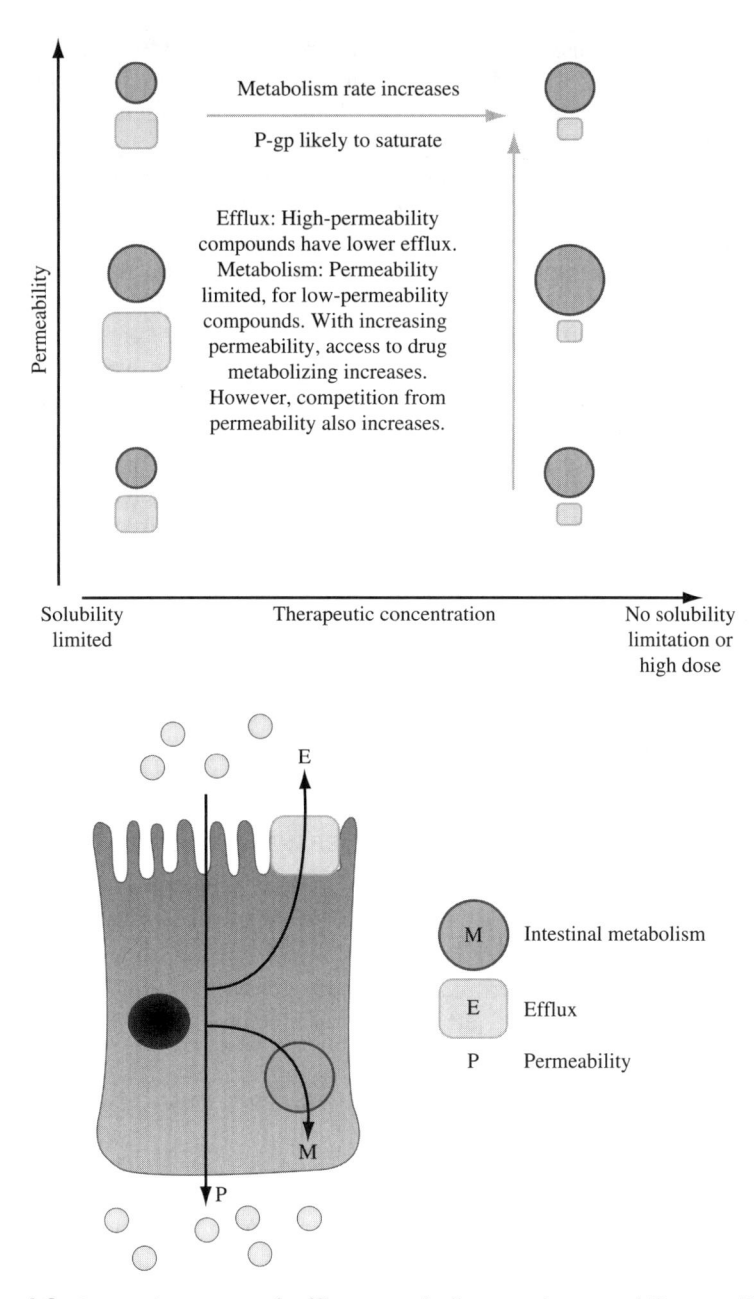

M Intestinal metabolism
E Efflux
P Permeability

Figure 4.6. Competing rates of efflux, metabolism, and permeability at different therapeutic concentrations.

from the polymorphism of some of the intestinal DMEs (e.g., CYP3A5, UGT1A1, and UGT2B7) and the large interindividual differences in expression levels of many DMEs along the gut. In addition to these factors, the rate and extent of intestinal metabolism depends on the exposure to enzyme, the permeability of the drug, and the region of gut where absorption occurs, all of which are determined by a multitude of variable physiological parameters such as gut emptying, transit down the GI, and solubility of drug to name a few. An assessment of the enzymes involved in gut metabolism and their relevance in humans is therefore important to ensure that uncertainties in human dose predictions are minimized. The heterogenous distribution of enzymes along the GI tract can lead to a regional variation in gut extraction and possibly to a narrow window of absorption. For highly permeable compounds, this narrow therapeutic window combined with the higher doses required to achieve therapeutic levels, translates to peaks in plasma concentrations in the PK profile of the compound. For a compound metabolized by enzymes such as UGT2B7 and UGT2B15, which are more abundant in the colon compared to small intestine,[15] the feasibility of extended release formulation is also limited. So, it is important that the drug has a large therapeutic window.

Once a drug has passed the gut wall, it can pass into the lymphatic system or into the blood stream (Fig. 4.5). For highly lipophilic compounds, the lymph route can be an additional route to systemic circulation. Lymphatic transport of a drug protects it from the hepatic first-pass extraction and, therefore, offers an attractive method for improving its bioavailability.

4.3 FACTORS AFFECTING DRUG ABSORPTION AND GUT BIOAVAILABILITY

Several physiological, compound-, and formulation-dependent factors can influence the properties determining fraction absorbed (solubility, dissolution rate, and permeability),[16] and gut extraction (intestinal metabolism and efflux).

4.3.1 Physiological Factors Affecting Oral Drug Absorption and Species Differences in Physiology

A number of physiological factors affect drug absorption across enterocytes. These are gastric emptying, gut transit, secretions in fed and fasted states, transporters (carrier proteins and efflux proteins), drug-metabolizing enzymes, pH, and porosity of tight junctions. An orally administered drug has to be emptied into the proximal duodenum (gastric emptying) before absorption can commence in the SI. Neural and hormonal signals control gastric emptying while nutrient content, **osmolarity**, and pH of chyme influences the rate of gastric emptying. Tablets that disintegrate to particle size less than 12 mm in the stomach empty along with the digestible nutrients, while nondisintegrating large capsules empty along with nondigestible solids in phase III of the migrating motor complex (MMC). MMC is a cyclically occurring (repeats every 75–90 min) pattern of electrical and mechanical activity that is initiated in the stomach and duodenum almost simultaneously. From there, it propagates the length of the small intestine. MMC is the predominant pattern of activity during fasting. The purpose of MMC may be to sweep the SI clean and to keep colonic bacteria from migrating into ileum. Cyclic motor activity also occurs in the lower esophageal sphincter, the gallbladder, and the sphincter of Oddi with a duration that is related to the MMC in the small intestine. In humans, following the ingestion of a meal, the MMC is disrupted and replaced by irregular contractions. Segmental contractions cause the mixing and turning of chyme. Peristaltic contractions move the chyme down the GI tract, typically over small distances of about 10 cm. Fed-state motility prepares the food for absorption. Its duration depends on the caloric content and physical properties of the food. Thus, gastric emptying of a drug is highly dependent on the amount of food in the stomach—much faster in the fasted compared to the fed state. Solid meals generally empty slowly compared to liquid meals. The rate of absorption of drugs with high solubility and permeability are limited only by the rate of gastric emptying.

Once the drug empties from the stomach, it mixes with bile and pancreatic juice in the duodenum and continues to disintegrate, dissolve, and to get absorbed as it transits down the small intestine. Basic secretions into the duodenum increase the pH of the intestinal fluids, which in turn can affect the solubility of an acidic or basic drug. These secretions are critical for digestion. The optimal pH for enzymes varies between the stomach and the duodenum in keeping with the changes in pH. Bile is a complex fluid containing water, electrolytes, and a battery of organic molecules including bile acids, cholesterol, phospholipids, and bilirubin, which flows through the biliary tract into the small intestine. The flow of bile is lowest during fasting, and a majority of that is diverted into the gallbladder for concentration. When a meal is ingested, the partially digested fats and proteins stimulate secretion of cholecystokinin and secretin, the 2 enteric hormones that have important effects on pancreatic exocrine secretion and are important for secretion and flow of bile. Cholecystokinin, secreted in response to fat, stimulates contractions of the gallbladder and common bile duct, resulting in the delivery of bile into the gut. Secretin, secreted in response to acid in the duodenum, simulates biliary duct cells to secrete bicarbonate and water, which expands the volume of bile and increases its flow out into the intestine. Bile acid is one of the components of bile. It is amphipathic, having both hydrophilic and hydrophobic domains, which is important in its function as lipid emulsifier by the formation of micelles. Bile micelles play an important role in oral absorption of poorly soluble compounds. Primary bile acids made in liver are converted to secondary bile acids by enteric bacteria. Bile salts are bile acids conjugated to glycine or taurine. In humans, taurocholic acid and glycocholic acid (derivatives of cholic acid) represent approximately 80% of all bile salts. The fasted- and

TABLE 4.3. SI Secretions Impacting Drug Absorption

Secretion	Function
Bicarbonate	Increase the intestinal pH, which ranges from 6 to 7.4 in humans
Bile acids	Emulsification of fat by forming micelles
	Elimination of cholesterol, bile pigments originating from the breakdown of red blood cells, and also some drugs

fed-state bile salt concentrations are 2–5 and 10–20 mM, respectively.[17] The presence of these bile salts in the intestinal fluids enables more efficient dissolution of lipophilic drugs in the intestinal media compared to water. The equations for dissolution rate and pH-dependent solubility discussed in Section 4.2.1 need to be modified to consider effective diffusion coefficient and to accommodate the bile–miscelle partitioning.[18] The diffusion coefficient of bile micelles is 8- to 18-fold less compared to monomer molecules. Micelle partitioning significantly increases permeability due to a reduced diffusion coefficient across the unstirred water layer. SI secretions that are important for drug absorption and their functions are summarized in Table 4.3.

For high-permeability drugs, with a reasonably good dissolution rate, absorption is complete even in the small intestine. Since the SI is the main region of absorption, for drugs whose absorption is limited by solubility, the longer the transit time, the higher the fraction absorbed. While gastric emptying and colon transit times are generally associated with a high degree of interindividual variability, the small intestinal transit time is relatively constant.

In humans, anaerobic bacterial microflora are abundant in distal SI and in the large intestines. Aerobic bacteria can thrive in the stomach and proximal SI, but to a much lesser degree compared to the anaerobic bacteria probably because of the low pH in these regions. Bacterial microflora acts on xenobiotics through hydrolysis, dehydroxylation, deamidation, decarboxylation, and reduction of azide groups. Drugs that have a prolonged absorption phase, either because of poor dissolution rate or permeability are likely to be affected most. Gut microflora produce β-glucuronidase, which hydrolyze the glucuronidated drugs originating from bile or from the gut, converting them to parent molecules, which are again available for absorption. This reconversion of glucuronide conjugates to parent molecules by the gut microflora is much more effective in rats compared to humans, as microflora is more spread out in the rat, starting even in the stomach.

The small intestinal, mucosal epithelial cells express a variety of transporters (carrier proteins and efflux proteins) as well as phase I and II DMEs. P-gp, a 170-kDa transmembrane protein and CYP3A are often co-located on the apical side of SI villi. They have overlapping substrate specificity and share inhibitors and inducers. Examples include verapamil and ketoconazole. Cooperative action of P-gp and CYP3A is effective in keeping out drugs from reaching systemic circulation. P-gp efflux ensures the dilution of drug within the

4.3 FACTORS AFFECTING DRUG ABSORPTION AND GUT BIOAVAILABILITY

enterocytes, thus preventing the saturation of CYP3A. It also guarantees repeated exposure of the drug to DME, maximizing the intestinal first-pass extraction. While the expression levels of P-gp increase from proximal to distal SI, the expression levels of CYP3A decrease in that direction.[19] P-gp is also expressed in the colon but to a much less degree. Apart from P-gp, intestine also expresses other efflux transporters such as MRP2 and BCRP and uptake transporters OATP, OCT1, MCT1, ASBT, and PEPT1[20] (see Chapter 3). MRP2 has the highest expression in the intestines. The gut also contains many other DMEs, the most notable among them being the UGTs.

Small hydrophilic drugs, such as antibiotics and some β-blockers, can pass through structures called tight junctions between the epithelial cells lining the GI tract and be absorbed. Tight junctions comprise proteins such as claudins and occludins. Each of these protein molecules interact with a corresponding molecule in the adjacent cell by a loop-shaped bond consisting of peptides. To let more substances pass through, researchers have tried to interfere with the bonds between these proteins and increase the "porosity" of the intestinal wall. Research has been directed toward manipulating the porosity via the claudins, which are easier to manipulate. However, such a manipulation is associated with undesirable and irreversible effects that increase the risk of cell damage. Tight junctions linking epithelial cells vary considerably in permeability along the GI tract, with the effective pore size decreasing going from proximal to distal SI and to colon.[8] Even within a given segment, cells in the crypts transport very differently than cells on the tips of villi. The contribution of the paracellular pathway even for small, hydrophilic molecules is not very high because the tight junctions represent only 0.1% of the surface area. It can, however, translate into reasonably good fraction absorbed for some compounds (50% fraction absorbed for atenolol).

Food can affect absorption as it increases blood flow rate to the GI tract, increases bile and pancreatic secretions, increases the viscosity of the contents, increases the pH in the SI, and decreases gastric emptying. A drug can also bind to food components, thereby reducing its availability for absorption. Drug–food interactions can also cause a reduced availability of carrier proteins, efflux proteins, or DME to the drug. Table 4.4 summarizes the effects of food on drug absorption.

Many GI diseases lead to malabsorption of nutrients, caused by changes in gastric and intestinal motility, absorptive surface area, and luminal content. Some of these are inherited diseases such as celiac disease (gluten intolerance), Crohn's disease (inflammation), and Hartcup disease (impaired amino acid absorption). These may also impact drug absorption.

The influence of ethnicity, age, and gender on the physiological features such as gastric motility, gastric emptying, transit time, regional pHs in the SI, and expression levels of proteins (efflux and uptake transporters and DMEs) should be well understood in order to adjust dosing to different populations.

Differences in gut physiology, anatomy, and biochemistry between preclinical species (Table 4.5) lead to differences in gut absorption and extraction.

TABLE 4.4. Effects of Food on Drug Absorption

Food	Effect on Physiological Function	Consequence	Other Considerations
Meal-induced increase in viscosity in duodenum	Reduced diffusivity and altered fluid flow dynamics	Decreased gastrointestinal transit time	Species with short transit time and BCS class II compounds are the most affected
Food-induced changes in osmolarity, viscosity, calories, stomach volume, and temperature	Reduced gastric emptying	Delayed absorption	Affects the BCS class I compounds to a large extent, as the absorption rate of these compounds are further limited by gastric emptying
Food-induced increase in blood flow rate to GI and liver	Increases permeability rate and thereby reduces competition from DME and efflux proteins	Increases the gut bioavailability	Propranolol F increases 30% when taken with food
Drug adsorption or binding to food components such as ions, proteins, pectins, and fiber	Drug not available for absorption	Reduced bioavailability	Mg^{2+} and Ca^{2+} in dairy products and SH group in proteins can bind to drugs
Increase in bile and pancreatic secretions by high-fat meals	Higher solubility of lipophilic drugs	Increased bioavailabilty of solubility-limited lipophilic drugs	Enhanced dissolution rate favors absorption of BCS class II
Dietary deficiency in iron	Rapid decrease in CYP enzymes	Reduced metabolism	CYP levels quickly restored with supplemented iron
Grapefruit juice	Inhibits P-gp and CYP3A	Increased intestinal/hepatic metabolism	Causes food–drug interaction
Watercress	Inhibits CYP2E1		
Red and black pepper	Inhibits CYPs 1A, 2B, and 2E1		

TABLE 4.4 (Continued)

Food	Effect on Physiological Function	Consequence	Other Considerations
Charcoal-broiled and smoked food	Potent inducers of CYP1A1 and 1A2	Decreased bioavailability of drugs metabolized by these enzymes and/or elimination	
Cruciferous vegetables	Induces several enzymes including phase II enzymes such as glutathione-s-transferase and quininone oxidoreductase		
Sodium chloride	High levels increases CYP 3A activity	Decreases gut bioavailability of CYP3A substrate drugs	

The higher permeability and shorter transit time in dogs compared to humans means that the t_{max} in dogs for many drugs is shorter compared to that in humans.[21] The rat is a better model compared to the dog for drug absorption and gut bioavailability in humans.[22] The suitability of the monkey as a model for absorption, volume of distribution, and clearance[23–25] has been evaluated. The monkey has been shown to have much lower bioavailability compared to humans, especially for CYP3A4 substrates. Since the hepatic extraction in the monkey is similar to that of humans, it is assumed that differences in bioavailability are due to the product of $f_{abs} \times f_{gut}$. As the absorption characteristics in monkeys have been shown to be similar to humans, the intestinal loss in monkeys should be higher than in humans. This can be due to the higher expression levels of efflux proteins (MRP2, BCRP, and P-gp) and DMEs (especially CYP3A) in monkeys. In animals and in humans, CYPs can be found in virtually all organs, notably the liver, intestine, skin, nasal epithelia, lung, and kidney but also in testis and brain. However, the liver (300 pmol of total CYPs/mg microsomal protein) and the intestinal epithelia (~ 20 pmol of total CYPs/mg microsomal protein) are the predominant sites for P450-mediated drug elimination, while the other tissues contribute to a much smaller extent to drug elimination. Species differences should be taken into account when using animals as models for humun intestinal absorption and gut extraction. The physiological parameters for common preclinical species and humans are provided in Appendices A and B, respectively.

TABLE 4.5. Physiological and Anatomical Features of Preclinical Species Compared to Humans

	Rat	Dog	Monkey
Gastric emptying	Faster compared to human	Similar to human in fasted but more delayed emptying of granules in fed state	Similar to human
Intestinal transit times	Similar to human in SI	Short (111±17 min) and variable compared to human (3–5 h)	Similar to human; 2.2–3.2 h in fasted and 2.2–4.2 h in fed states
Intestinal pH	Higher gastric pH	Higher intestinal pH compared to human in fasted and fed states. pH in dog can be manipulated	Gastric pH—similar to human. Higher intestinal pH in SI
Unstirred water layer	100 μm	40 μm—similar to human	
Villi length and shape	Leaflike and similar to human	Long and slender compared to human	Ridges and leaflike
Intestinal secretions	More bile secreted into duodenum 48–200 mL/kg/day compared to human (2–22 mL/kg/day)	Higher bile salt concentrations compared to human. Bile flow: 14.4 mL/kg/day	Bile flow rate, secretion, and composition similar to human
Tight junction porosity	Similar to human	Higher frequency and channel pore size compared to human	
Enterocytic DME	CYPs 2C and 2B more important in rat compared to 3A. In human CYP3A is the most important	CYP3A and carboxyl esterases present in SI. All phase II DMEs present except NAT	Intestinal metabolism as important as hepatic. High CYP3A levels in SI. CYPs, carboxylesterases, and phase II DMEs like in human
Transporters		MDR1 very low in dog GI	

4.3.2 Compound-Dependent Factors

Lipophilicity of a compound is a measure of its hydrophobicity and determines both the permeability and solubility of compounds. The octanol–water partition coefficient (P) is a good surrogate for membrane partitioning, as a membrane is not a completely hydrophobic environment. For acidic and basic molecules, the partitioning into octanol will be lesser in its ionic form compared to its neutral form. The partitioning of anionic form of an acid or base into a second phase is termed as distribution, and it is determined by the extent of ionization, which depends on the pH of the medium and the pK_a of the compound. Thus, for an acid, its distribution (D) into octanol from water is related to its partition coefficient by

$$D = P \times \frac{1}{1 + 10^{pH - pK_a}} \qquad (4.16)$$

For a neutral compound, $D = P$. Partitioning into membranes determine both absorption and distribution of compounds. Although the log D of a compound is a good surrogate for partitioning into membranes, it should be remembered that the cell membrane is certainly more permissive to ionized molecules than suggested by log D. Thus, although bases are expected to be more ionized in the lower pH of SI and therefore less absorbed, this is not strictly so.

The molecular weight of a compound is a measure of its molecular volume, an important determinant of paracellular permeability. Hydrogen-bonded donors (HBD), lipophilicity, and **polar surface area (PSA)** are the drug properties that correlate well with permeability. HBD is the number of hydrogens attached to oxygen or nitrogen and it inversely correlates with membrane permeability. PSA is defined as the sum of surface contributions from all polar atoms such as oxygen and nitrogen, including those attached to hydrogens. Transcellular, passive permeability is highly correlated to PSA and therefore, to the fraction absorbed of transcellular-permeability-driven absorption. Compounds with PSA < 60 have good permeability and those with PSA > 140 have poor permeability. Methods to determine PSA are available in the literature.[26,27]

The presence of certain functional groups in compounds can make them substrates to uptake proteins or DMEs in the enterocytes, thus increasing their absorption or gut extraction.

4.3.3 Formulation-Dependent Factors

Drug formulations are needed to address one or more of the following problems:

- Short half-life drugs that need twice (or more) daily dosing or quickly metabolized drugs with peak concentrations that are close to safety limits or drugs that are unstable or form insoluble complexes in the SI lumen or enterocytes. An extended release (ER) formulation can help circumvent these problems.

- Extent of absorption is limited by dissolution rate. Micronization affords reduced particle size and increased surface area and thereby an improved dissolution rate. Use of wetting agents such as polyethylene glycol (PEG), polyvinyl pyrrolidine, or complexing agents such as **cyclodextrins (CD)** can reduce the surface tension of the dissolving media, thus increasing the surface area of contact between drug particles and media and therefore the dissolution rate (see Noyes–Brunner equation). Often, micronized drug particles tend to aggregate even more. It is, therefore, important to stabilize them with surfactants. Cyclodextrins are cyclic oligosaccharides of glucose consisting of six, seven, or eight glucose units (α-, β-, and γ-cyclodextrin, respectively). A lipophilic drug of appropriate size can enter the apolar cavity in the middle of a cyclodextrin molecule, forming a 1:1 inclusion complex and prevent the agglomeration of particles by keeping them apart. Cyclodextrins and other modifications of CDs such as hydroxypropyl-β-CD (HP-β-CD) can alter the PK of the drug, if they do not dissociate from the drug and get metabolized as soon as they are in systemic circulation. By reducing binding of the drug to the plasma proteins and favoring enhanced tissue distribution, they can increase metabolism and the volume of distribution. They are not suitable for high-dose drugs as the molar ratio requirement could make them toxic. Nanosuspensions, which are submicron colloidal dispersions of drug particles stabilized by surfactants, combine the advantages of micronization and aggregate prevention.
- The fraction absorbed of unstable amorphous or polymorphic forms can be very variable, if limited by dissolution, as the variable conditions in the lumen may favor different forms, and solubility depends on the solid state. The crystalline form has high lattice energy and low free energy compared to amorphous or polymorphic forms. It is, therefore, very stable but has lower solubility and dissolution rate compared to amorphous. Crystalline nanosuspensions aim to take advantage of the stability of the crystalline form but with improved dissolution characteristics. Nanosuspensions have low amounts of additives and do not induce toxic side effects. They are, therefore, suitable for *in vivo* toxicological studies.

Dosage forms include solution, capsule, tablet, or suspension. A drug is profiled to establish its **biopharmaceutics classification system (BCS)** class and to identify the issues that need to be addressed through formulation. The type of formulation required for a drug is dependent on the BCS class (Fig. 4.7) and the issues—PK-driven twice daily dosage or peak concentrations, solubility or dissolution-limited absorption, a need to have reduced uncertainty in disposition, and the like. Once the issues are well understood, the choice of formulation then depends on the size, lipophilicity, and acid/base characteristics of the drug and whether the dosage requirements are high (see Table 4.6). Apart from these considerations, the need for rapid turn around in early discovery,

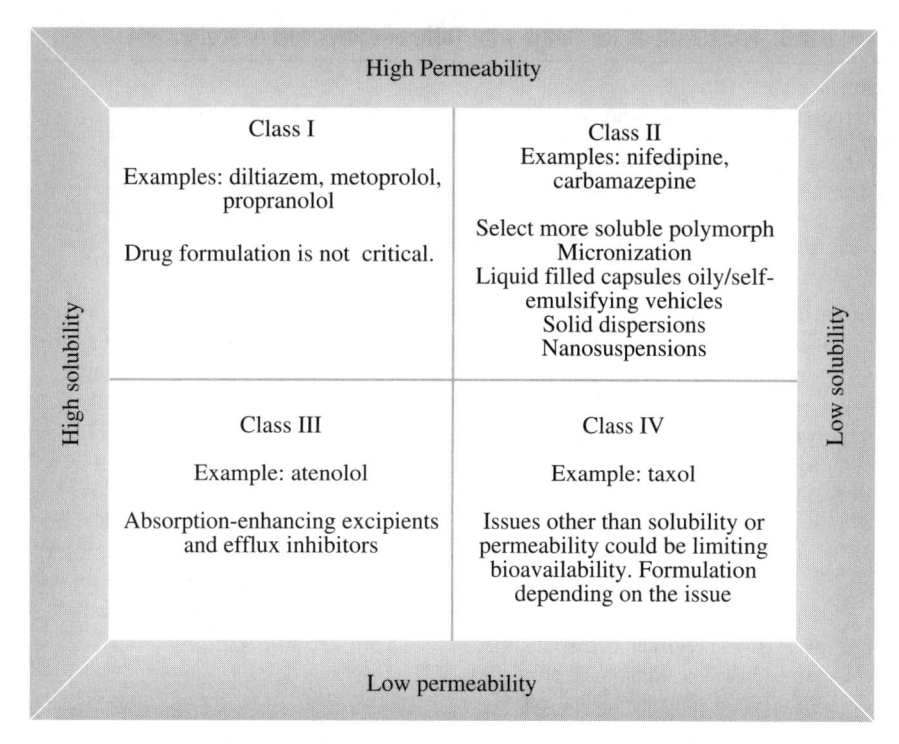

4.3 FACTORS AFFECTING DRUG ABSORPTION AND GUT BIOAVAILABILITY

	High Permeability	
High solubility	**Class I** Examples: diltiazem, metoprolol, propranolol Drug formulation is not critical.	**Class II** Examples: nifedipine, carbamazepine Select more soluble polymorph Micronization Liquid filled capsules oily/self-emulsifying vehicles Solid dispersions Nanosuspensions
	Class III Example: atenolol Absorption-enhancing excipients and efflux inhibitors	**Class IV** Example: taxol Issues other than solubility or permeability could be limiting bioavailability. Formulation depending on the issue
	Low permeability	

Figure 4.7. Formulation choices for the different classes of drugs.

cost, the route of administration of the drug, and the disease area could also influence the choice of formulation. Lipid vesicles and particularly liposomes have been extensively used to target tumors. Liposomes are tiny vesicles, made out of the same material as a cell membrane (lipid bilayer) that can be filled with drugs. They deliver their contents to sites of action, and the lipid bilayer can fuse with other cell membrane bilayers, thus delivering the liposome contents. Endothelial cells, bound together by tight junctions, are leakier for tumor vessels compared to healthy vasculature. Liposomes of diameter less than 400 nm take advantage of this ability (known as the enhanced permeability and retention effect) to deliver the drug to the targeted tumor. Anticancer drugs such as doxorubicin (Doxil), Camptothecin, and daunorubicin (DaunoXome) are currently being marketed in liposome delivery systems. Liposomes can be carriers of both hydrophilic and lipophilic drugs. Hydrophilic drugs are trapped between the bilayers. However, the high levels of lipid in the formulation can pose an analytical challenge. Choice of formulation can be very different for different preclinical species, as some of the excipients used in formulations can be toxic to animals.

TABLE 4.6. Formulations for Drugs with Different Physicochemical Profiles

Drug Characteristic	Formulation	Potential Drawbacks
Acidic or basic drug	Solubility enhancement through salt formation or pH adjustment with HCl/NaOH (preferred pH range 4–9)	Precipitation
Low or high log P, low dose	Use of solubility-enhancing water-miscible co-solvents such as ethanol, propylene glycol, PEG400, or DMA	Precipitation and hemolysis impact PK profile. Altered PK makes it unsuitable for early screening
High crystallinity, high log P, low dose, suitable size	Molecular complexation with CD or HP-β-CD (20%); or surfactants like Cremophor EL, Tween 80, or solutol HS 15	Precipitation and hemolysis impact on PK profile. Entrapment into CD or miscellar cavity alters free fraction available for RBC and tissue partitioning. High molar ratios needed could cause toxic side effects for high-dose drugs
High log P, low dose and low crystallinity	Miscellar dispersions, liposomes, emulsions, other lipid systems	Impact on PK profile
High log P, high dose and high crystallinity	Crystalline nanosuspension	Slow dissolution

4.4 IN SILICO PREDICTIONS OF PASSIVE PERMEABILITY AND SOLUBILITY

4.4.1 In Silico Models for Permeability

4.4.1.1 Passive Transcellular Permeability. There are many *in silico* models for oral drug absorption that is not limited by solubility.[26, 28–30] Winiwarter et al.[31] have proposed a model based on 13 compounds to predict human jejunal **effective permeability** (P_{eff}) by the combined use of simple structural (hydrogen bond donors, HBD) and physicochemical properties such as polar surface area (PSA) and calculated log of octanol–water partition coefficient ($C \log P$):

$$\log P_{eff} = -3.067 + 0.162 \times C \log P - 0.01 \times \text{PSA} - 0.235 \times \text{HBD} \quad (4.17)$$

4.4.1.2 Paracellular Permeability. A paracellular absorption rate constant ($k_{a,p}$) is added to the k_a for small, hydrophilic compounds. Paracellular absorption is dependent on the physiological flux of water from the mucosal to the serosal side of the lumen (J_{ms}) as well as on the compound-dependent fraction ϕ, which ranges from 0 to 1 depending on the lipophilicity and the size of the molecule, represented by the molecular weight (M). Small hydrophilic compounds have a ϕ value close to 1, while large lipophilic compounds have a ϕ value of 0.[32] For all other compounds, it ranges between 0 and 1, according to the following model:

$$\log P > 0.7: \phi = 0$$
$$\log P < 0.7:$$
$$M > 200 \text{ but} < 360: \phi = 0.1$$
$$M < 200: \phi = -0.0045 \cdot M + 1$$

The $k_{a,p}$ is then given by

$$k_{a,p} = \phi \times J_{ms}/V_{GI} \qquad (4.18)$$

where V_{GI} is the volume of the gastrointestinal compartment under consideration.

4.4.2 In Silico Models for Solubility

The aqueous solubility of liquids and solids has been correlated with an amended solvation equation[33] that incorporates a crystal-packing energy term in Sigma alpha(2)(H) × Sigma beta(2)(H), where the latter are the hydrogen bond acidity and basicity of the solutes, respectively. The cavity-forming energy term related to size parameters and the solvation energy terms, dipolar and hydrogen bonding, are the other components of the equation. Yalkowsky's model[34] for solubility uses the melting point (MP) of the compound as a surrogate for solid state and uses the octanol–water partition coefficient (P) as a measure of hydrophobicity, which is inversely correlated to solubility:

$$\log S_W = -1.0 \times \log P - 0.01 \times MP + 1.05 \qquad (4.19)$$

Higher melting point reflects a stable crystalline state and less ability to get into aqueous phase. The melting point of a compound is, however, a poor surrogate and does not adequately represent the complexity of the solid state.

4.5 MEASUREMENT OF PERMEABILITY, SOLUBILITY, LUMINAL STABILITY, EFFLUX, AND INTESTINAL METABOLISM

4.5.1 In Vitro, in Situ, and in Vivo Assays for Permeability

Artificial membranes, "intestine-like" cell lines (Caco-2, MDCK, TC-7, HT29-MTX, 2/4/A1), everted sac, Ussing chamber, and *in situ* rat intestinal perfusion techniques are some of the models available for human permeability.[35] Human effective permeability data based on direct, *in vivo* determinations in the human GI tract with a single-pass perfusion are available for very few compounds, and they serve as the basis for correlation with other models for permeability. Any permeability model should be able to mimic the sink conditions *in vivo*.

Different *in vitro* models provide different types of information.[36,37] Thus, a combination of models is often more predictive than any one. The parallel artificial membrane assay (PAMPA), first introduced by Kansy in 1998,[38] uses a hydrophobic filter material coated with a mixture of lecithin/phospholipids dissolved in an inert organic solvent such as dodecane creating an artificial lipid membrane barrier that mimics the intestinal epithelium. An excipient that binds to the permeated drug on the receiver side and a pH gradient from donor to acceptor ensures that the *in vivo* sink conditions are reproduced. PAMPA assay (Fig. 4.8) is a good model for passive permeability. It is more reproducible and a lot less resource intensive than cell culture methods and has a high-throughput capability. However, PAMPA permeability is not reliable for compounds that are actively absorbed or effluxed via drug transporters. It is not a good model for paracellular permeability. Due to its low cost, rapid operation, the assay can serve as a rapid permeability screen during early drug discovery.[38,39] Caco-2 cells derived from human colon adenocarcinoma lineage exhibit the required apical to basolateral (A→B) polarity after 14–21 days culture (Fig. 4.9). They have been widely employed in the pharmaceutical and food industries for permeability screening. Caco-2 permeability correlates

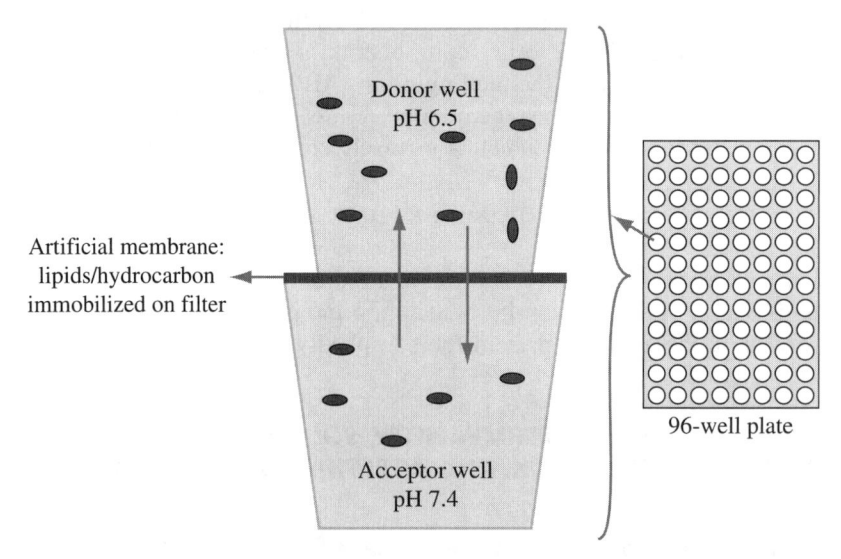

Figure 4.8. Parallel artificial membrane assay (PAMPA) experimental setup. Sample is allowed to permeate for 4–18 h. The plate is separated and each well is analyzed with LC-MS. Apparent permeability (P_{app}) is calculated from relative concentrations.

4.5 MEASUREMENT OF PERMEABILITY, SOLUBILITY, LUMINAL STABILITY

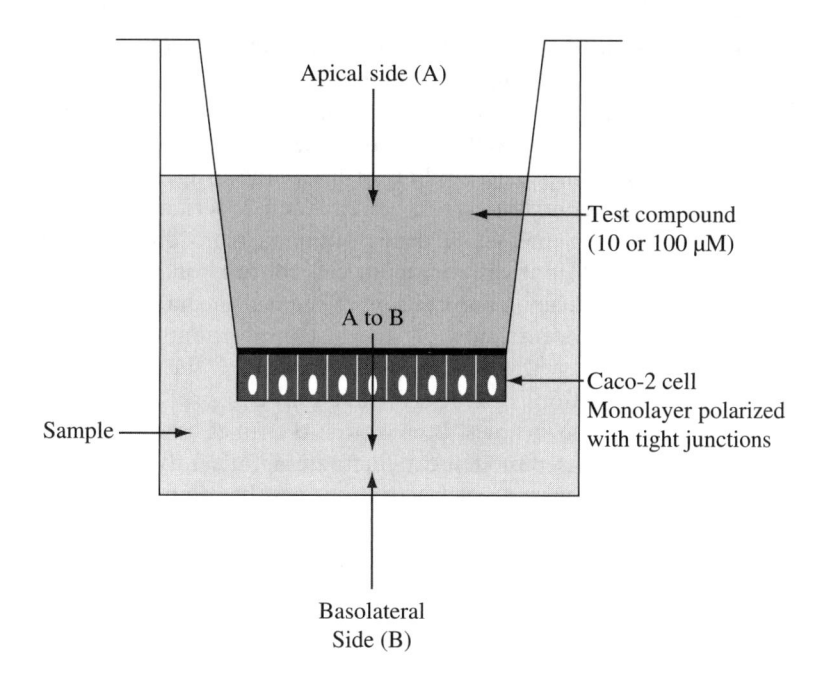

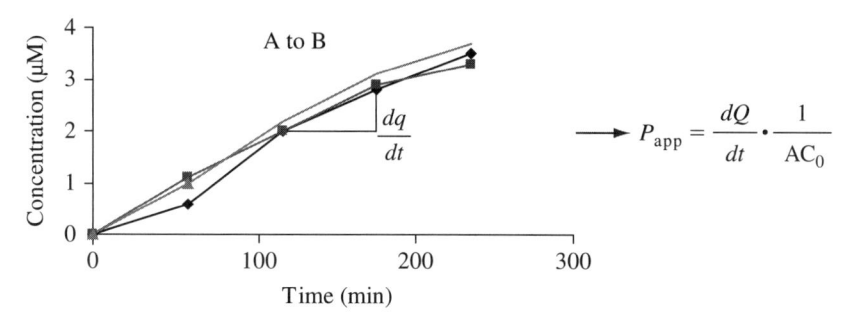

Figure 4.9. Cultured cell-line Caco-2 in the determination of permeability.

well with fraction absorbed in humans.[40] Although pharmaceutically important drug transporters such as PEPT1, OCT, and OAT and DME such as CYP3A are functionally expressed in Caco-2 cells, they are quantitatively underexpressed when compared to *in vivo* situation. For example, angiotensin-converting enzyme (ACE) inhibitors and the β-lactam antibiotics cephalexin and amoxicillin, which are known substrates of transporters are poorly permeable across the Caco-2 cell monolayer despite the fact that they are completely absorbed *in vivo*. Caco-2 cells have a 10-fold thicker unstirred water layer and significantly tighter cell junctions compared with human intestine, and thus Caco-2 cells normally underpredict paracellular permeability. Stirring can reduce the thickness of the unstirred water layer. Low-molecular-weight hydrophilic compounds such as metformin, ranitidine, atenolol, furosemide, and hydrochlorothiazide showed poor permeability in Caco-2 cells despite adequate absorption in humans. Precipitation of poorly soluble drugs in the donor compartment (cells do not tolerate the typical organic co-solvents such as dimethyl sulfoxide (DMSO) and propylene glycol that can improve compound solubility), nonspecific binding of lipophilic drugs, and P-gp efflux of P-gp substrates all contribute to a lower Caco-2 permeability. Caco-2 cell permeabilities of the same set of drugs obtained from different laboratories vary significantly.[41] Minor differences in cell culture conditions (e.g., seeding density, feeding frequency, composition of the cell media), experimental protocol (e.g., initial concentration of drugs, composition of the permeability buffer, pH, monolayer washing steps), and age of the cells (e.g., passage number, culture duration, tightness of junction) can produce dramatic differences in the permeability values. In addition, the function of drug transporters expressed in the cell-based models can fluctuate significantly with a difference in culture conditions. However, such variations can be minimized by adopting a good protocol. Other cell lines include Madin-Darby canine kidney (MDCK),[42] 2/4/A1, and TC7. TC7 is a Caco-2 subclone with higher expression levels of functional markers such as CYP3A4.

Ussing chamber[43] (Fig. 4.10) studies provide a physiological measure of the transport of drugs, ions, and nutrients across a small section of epithelial tissue.[44] To study the permeability of a drug, a section of intestinal mucosa is clamped between two chambers containing buffer and maintained at 37°C. Mannitol is used as a marker for surface damage and water flux, while glucose is used as a marker for viability and cell activity with respect to glucose transporters. Oxygen containing 5% CO_2 is used to circulate the fluid. Since the intestinal mucosa is metabolically active, an adequate energy supply from glucose or glutamine should be maintained.[45] Under steady-state

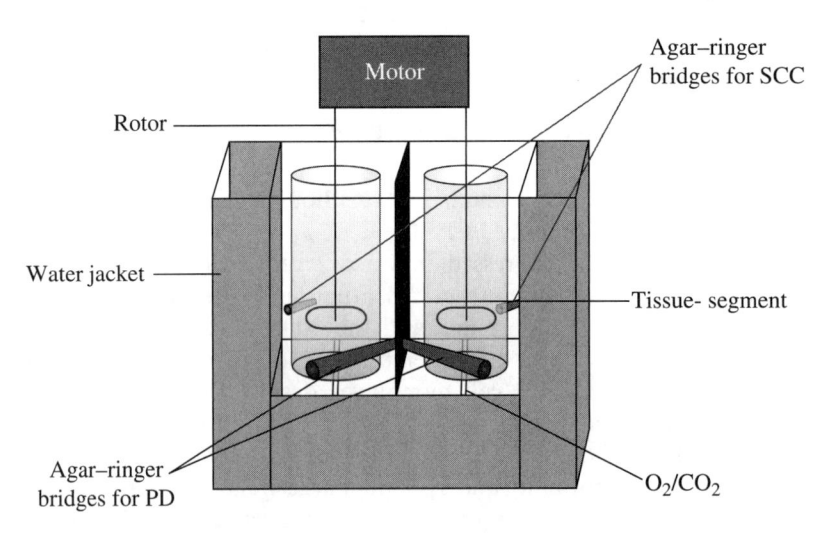

Figure 4.10. Ussing chambers in the determination of permeability.

4.5 MEASUREMENT OF PERMEABILITY, SOLUBILITY, LUMINAL STABILITY

conditions, the unidirectional mucosal to serosal absorptive flux normalized by area and the initial concentration of the test compound (C_0) in the donor compartment gives the apparent permeability (P_{app}) in 10^{-6} cm/s:

$$P_{app} = \frac{1}{SA \times C_0} \times \frac{dQ}{dt} \tag{4.20}$$

where SA is the exposed surface area of the cell intestinal segment, and dQ represents the amount of test compound accumulated in the receiver fluid during the time interval dt. The advantages of the Ussing experiment over the Caco-2 monolayer include the presence of a mucous layer, availability of different cell types, and the presence of DMEs and transporters. Passive diffusion is characterized by an equal transport of test compound from mucosal to serosal and serosal to mucosal sides. With active uptake or efflux, the fluxes are unequal. The measurement of potential difference and the current flow across the tissue due to flow of inorganic ions across the epithelium allows the determination of transepithelial electrical resistance (TEER) for the continuous monitoring of tissue viability and integrity.[46] Provided a sensitive drug–metabolite assay is available, the method can be used to study drug metabolism.[47] The mounted tissue can also be from any other site of the intestine other than jejunum. The dilution and sensitivity limitations of the measurement can be overcome by the use of isotopes. P_{app} from *in vitro* experiments should be converted to *in vivo* permeability (P) using a relevant correlation, before they are employed in the calculation of absorption rate constant in equation 4.15.

In the rat *in situ* model (Fig. 4.11), a 10-cm jejunal segment was isolated in fasted rats and cannulated at both ends with plastic tubing. The segment was rinsed with saline, and approximately 10 cm of inlet tubing placed inside the abdominal cavity to achieve an inlet perfusion solution at a temperature of 37°C. The experiment was initiated by filling the segment with a 4-mL bolus of the perfusion solution containing phosphate buffer, membrane leakage marker, and the test compound. A perfusion rate of 0.2 mL/min was maintained by means of an infusion pump. The outlet perfusate samples were collected at

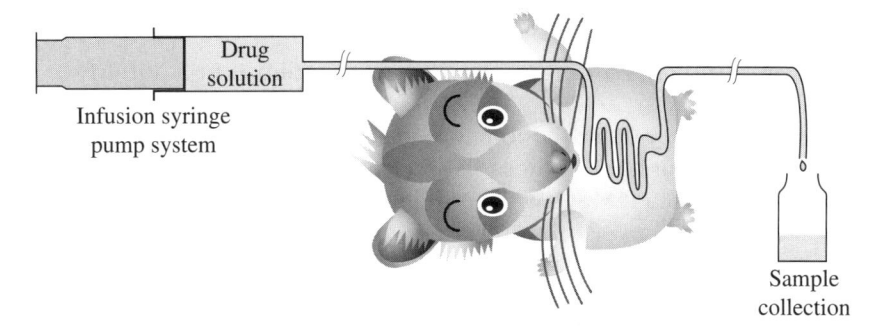

Figure 4.11. Perfused rat intestinal loop.

different time intervals, its drug content analyzed, and corrected for water flux based on carbon radiolabeled PEG4000 (membrane leakage marker). Binding to the tubing and stability of the compound in the buffer and jejunal fluid at the temperature of study are measured. Knowing the rate of disappearance of the drug from the perfusion solution, and considering the intestinal segment to be a well-mixed compartment, the effective permeability P_{eff} of the compound is calculated using the following equation:

$$P_{eff} = \ln\left(\frac{C_{out}}{C_{in}}\right) \times \frac{-Q_{in}}{2\pi rL} \qquad (4.21)$$

where C_{in} and C_{out} are the concentrations of the inlet perfusion solution and the outlet perfusate, Q_{in} is the perfusion rate. r, L and $2\pi rL$ are the radius, length, and the surface area of the isolated segment. Intestinal effective permeability estimated from *in situ* perfused rat intestine has been shown to correlate well with the extent of *in vivo* absorption in humans of highly soluble, stable, passively, and carrier-mediated absorbed drugs.[48]

A similar regional intestinal perfusion technique, Loc-I-Gut[49] (see Fig. 4.12) in humans has been used to study the effective permeability of drugs in humans, investigate transport mechanism, food/excipient interaction with drugs, presystemic metabolism, and *in vivo* drug dissolution. A solution of test compound is infused along with a nonabsorbable marker such as PEG4000 into a jejunal section. The difference in the inlet and outlet concentrations of the compound in the solution is assumed to be absorbed. The P_{eff} of the test compound is calculated as follows:

$$P_{eff} = \frac{C_{in} - C_{out}}{C_{out}} \times \frac{Q_{in}}{2\pi rL} \qquad (4.22)$$

The *in vivo* drug intestinal permeability in humans ($P_{eff, human}$) is roughly 3.6 times the rat permeability ($P_{eff, rat}$) values according to Fagerholm et al.[50]:

$$P_{eff, human} = 3.6 \times P_{eff, rat} + 0.03 \times 10^{-4} \qquad (4.23)$$

4.5.2 Measurement of Thermodynamic or Equilibrium Solubility

In the drug discovery stage, the equilibrium solubility of a compound is usually determined by diluting 100 times, a stock solution of a compound that is predissolved in 100% DMSO, with a suitable aqueous buffer (pH 7.4), allowing it to stand for 8 or 24 h and analyzing the supernatant solution in HPLC–ultraviolet (UV). Disadvantages of this method include loss of solid-state information and errors due to the presence of DMSO. A dry solid sample of the compound or solid obtained from evaporating the DMSO from the stock solution can be used instead for determining the solubility. The determination

4.5 MEASUREMENT OF PERMEABILITY, SOLUBILITY, LUMINAL STABILITY

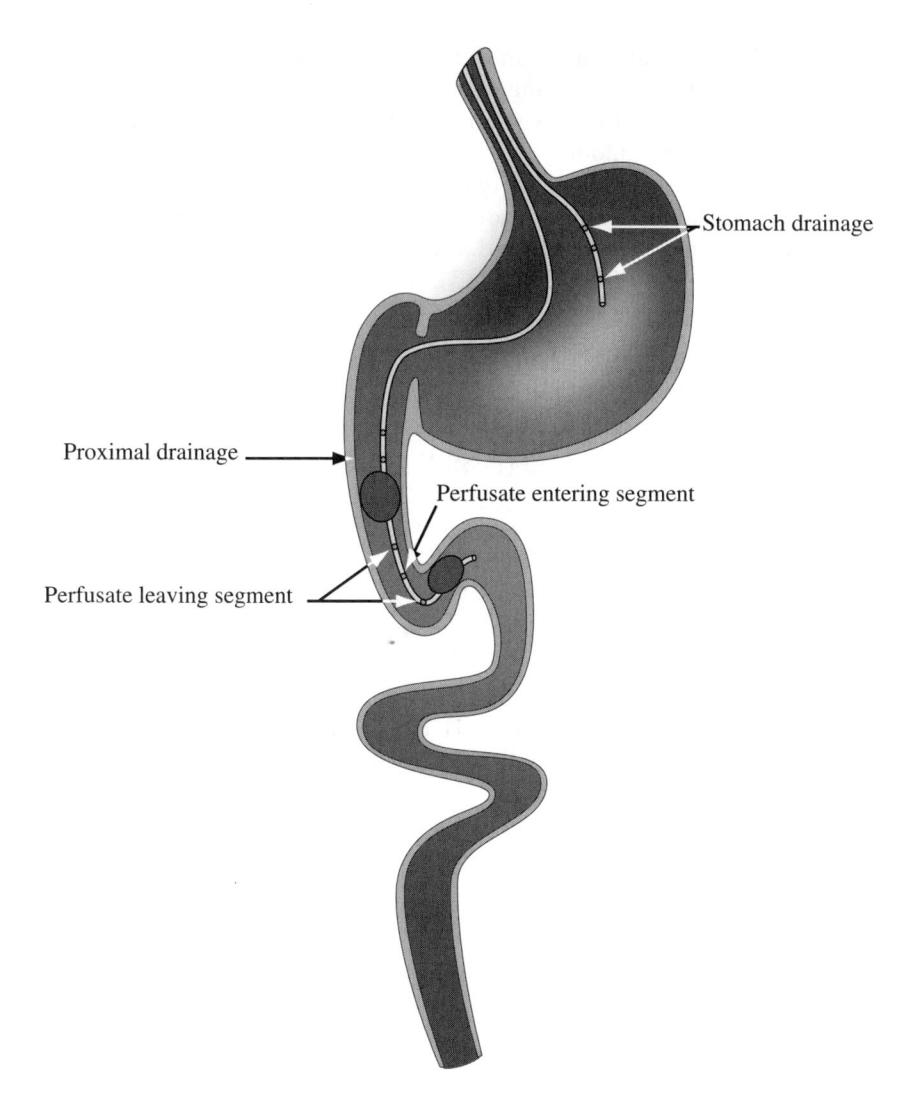

Figure 4.12. Determination of effective permeability with Loc-I-Gut.

of thermodynamic solubility[51] by measuring the concentration of the supernatant solution is likely to be overestimated, if sufficient time for equilibration is not allowed.

In discovery settings, high-throughput screening and rank ordering a large number of compounds is done on the basis of kinetic solubility, which precludes the need for equilibration time. A compound in DMSO stock solution is added stepwise to a buffer until a change in the intensity of scattered light caused by the precipitation is detected by a nephlometer or UV spectrophotometer. The neglect of equilibration means that the measured solubility could be much higher than that for the most stable crystalline form. For compounds with ionizable groups, the Chasing equilibrium solubility (CheqSol)[52] provides a measure of intrinsic thermodynamic solubility, using the potentiometric acid–base titration technique.

Apart from buffers, other media that better reflect the composition of the GI fluids are used in solubility and dissolution experiments.[53,54] Biorelevant media mimic mucous and gastric juices and are composed of lipids, surfactants, and buffers to maintain physiological pH. They are listed below:

Simulated gastric fluid (SGF)
Simulated intestinal fluid (SIF)
Fasted-state small intestinal fluid (FaSSIF)
Fed-state small intestinal fluid (FeSSIF)
Human intestinal fluid (HIF)[54]
Human gastric fluid (HGF)[17]

4.5.3 Luminal Stability

If a compound is degraded to less than 5% when it is in contact with HIF for up to 3 h, it is considered to be stable in HIF. If a compound is degraded to less than 5% when it is in contact with SGF for up to 1 h, it is considered to be stable in the gastric environment.

4.5.4 Efflux

The bidirectional Caco-2 permeability assay, where the basolateral to apical (secretory direction, B to A) permeability is compared with the apical to basolateral (absorptive direction, A to B) permeability, is the gold standard in identifying efflux substrates (Fig. 4.13). Compounds with an efflux ratio (ratio of B to A/A to B) greater than 2 to 3 are typically considered as efflux substrates. Since efflux transporters such as MRP2 and BCRP are also functionally expressed in the Caco-2 cells, apart from P-gp,[20,55] a simple bidirectional difference may not ascertain that the compounds being tested are indeed P-gp substrates. As a confirmatory study, a follow-up bidirectional experiment is routinely repeated in the presence of known inhibitors of P-gp, MRP2, and BCRP. Drugs such as famotidine and ranitidine are substrates for secretory transporter proteins but are not detected as P-gp substrates due to their low passive permeability. A better *in vitro* tool to examine the drug–transporter interactions are the engineered cell lines that selectively express the transporter of interest and facilitate interaction studies with a specific transporter in isolation. Follow-up studies can be conducted in the presence of a selective inhibitor to further confirm the involvement of the transporter. MDCK and

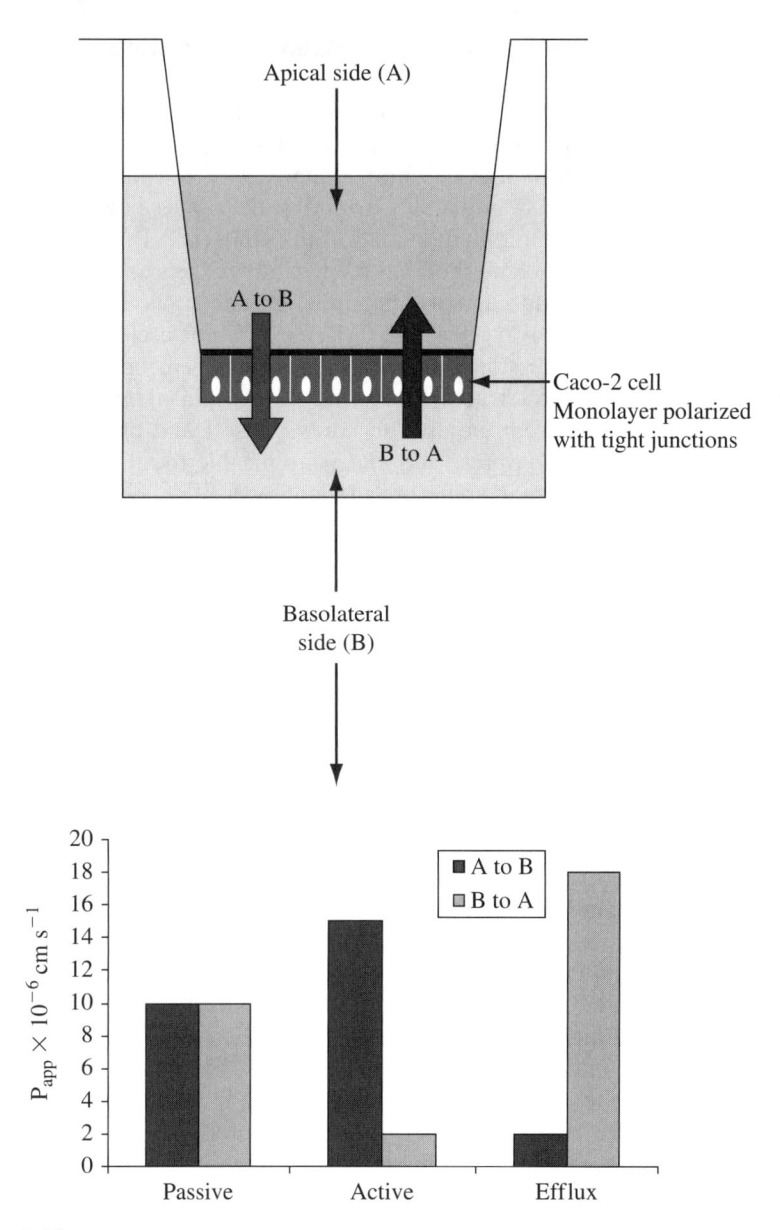

4.5 MEASUREMENT OF PERMEABILITY, SOLUBILITY, LUMINAL STABILITY

Figure 4.13. Bidirectional Caco-2 assay for determining active uptake or efflux potential of a drug.

LLC-PK1 cell lines stably transfected with a specific efflux transporter (e.g., MDR1, mdr1, MRP2, BCRP) may be used to tease out transporter interactions at early discovery stage. The functional expression of transporters in these cell lines appears to be more stable compared to Caco-2 cells.

4.5.5 In Vitro Models for Estimating Extent of Gut Metabolism

Methods to study intestinal drug metabolism *in vitro* include intact tissues (Ussing chamber and precision cut slices), primary cells, subcellular fractions (S9 fractions, microsomes, and cell homogenate), and cell lines (Caco-2 TC7 and LS-180).[35,47] Methods that employ nonintestinal tissues have recently been proposed and include recombinant cytochrome P-450s (rCYP) and human liver microsomes (HLM).[56] Each method has its own advantages and disadvantages. Low cost, high speed, and capacity demands exclude tissue and primary cell methods in the drug discovery process. Cell lines are not useful because of the large differences in expression levels of metabolizing enzymes compared to *in vivo* conditions.[35,47] S9 fractions consist of cytoplasm and endoplasmic reticulum fragments and can be used to study phase I and phase II intestinal metabolism. S9 can be cryopreserved and is applicable to all species.[47,57] The limitations with S9 are that cofactors are not present in physiological concentrations[47] and the abundance of S9 protein in the intestine is unknown, which requires empirical scaling factors for *in vitro in vivo* extrapolation (IVIVE). The rCYP and HLM methods have the advantage of known abundances, which gives the opportunity for direct scaling, but are only applicable to substrates of CYP3A or CYP substrates, respectively. Using the model of competing rates (Fig. 4.14), the *in vitro* CL_{int} obtained from these systems are combined with their permeability CL (CL_{perm}) to arrive at estimates of fraction of dose escaping gut extraction (f_{gut}). Other models employed for the prediction of f_{gut} from *in vitro* data have been reported in the literature.[58,59] The Q_{gut} model reduces to the model of competing rates, when the blood flow to enterocytes is very much higher than the clearance from permeability. A group of 25 CYP3A4 substrates were used to establish that *in vitro* clearance can be extrapolated between hepatic and intestinal microsomes, if enzyme abundance data are available and contribution of the different CYP isoforms to drug metabolism is known. The prediction accuracy of the Q_{gut} model was much better for low extraction drugs ($f_{gut} < 0.5$) compared to the high extraction drugs. Circumventing the need to calculate CL_{perm}, a simplified model, valid under the conditions of high permeability ($>1 \times 10^{-6}$ cm/s) has been validated. This model relies on an empirical scaling factor derived from fitting CL_{int} from *in vitro* data to f_{gut} of 9 model compounds. By virtue of its assumptions, this model overpredicts the f_{gut} of low-permeability compounds, whose long residence time in the enterocytes results in high extraction.

4.6 ABSORPTION MODELING

A reliable prediction of f_{abs} and f_{gut} is critical for candidate selection in the discovery phase and for pharmaceutical development. Models of varying complexity have been built to predict absorption. The simplest of these is the Lipinski rule of 5,[60] according to which an orally active drug satisfies at least 3 of the following criteria:

4.6 ABSORPTION MODELING

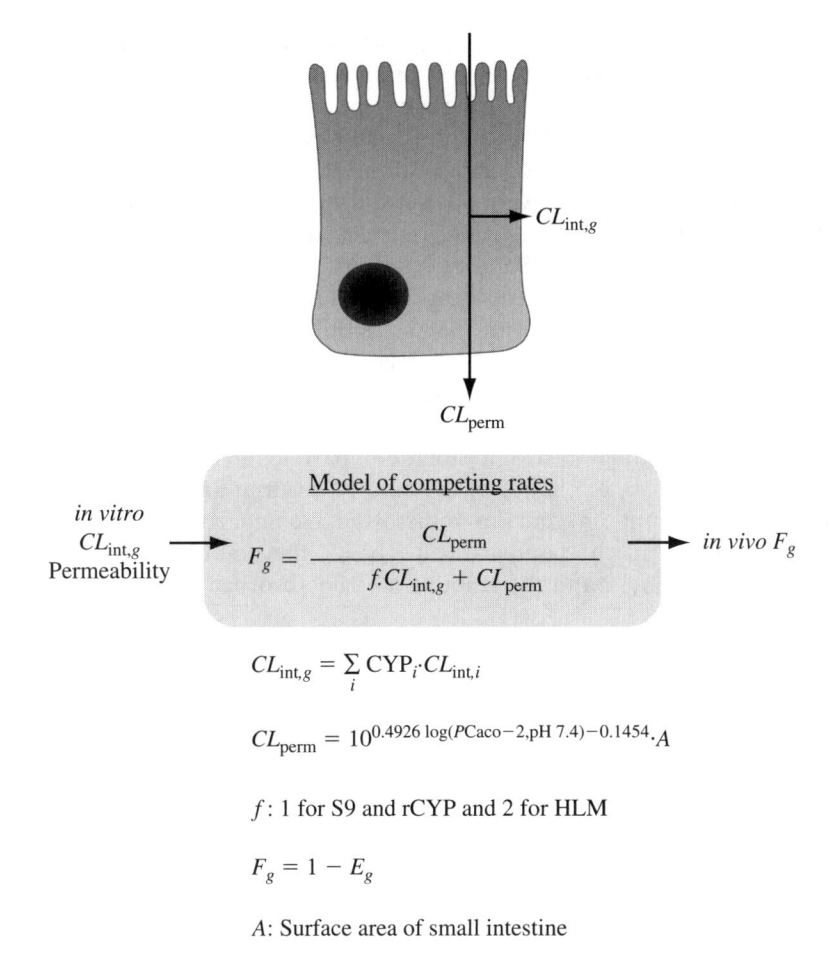

Figure 4.14. Model of competing rates for the determination of gut extraction.

$$F_g = \frac{CL_{perm}}{f \cdot CL_{int,g} + CL_{perm}}$$

$$CL_{int,g} = \sum_i CYP_i \cdot CL_{int,i}$$

$$CL_{perm} = 10^{0.4926 \log(PCaco-2, pH\ 7.4) - 0.1454} \cdot A$$

f: 1 for S9 and rCYP and 2 for HLM

$$F_g = 1 - E_g$$

A: Surface area of small intestine

- Not more than 5 hydrogen bond donors (nitrogen or oxygen atoms with one or more hydrogen atoms)
- Not more than 10 hydrogen bond acceptors (nitrogen or oxygen atoms)
- A molecular weight under 500 Da
- An octanol–water partition coefficient log P of less than 5

All the numbers involved are multiples of 5 and therefore the name. Quantitative structure bioavailability relationships (QSBR) further refined predictions of the drug absorption.[61,62] Alternatives to the QSBR approach are the prediction of absorption potential or the estimation of a **maximum absorbable dose**. However, these simple models do not provide the detailed understanding of the drug absorption process that is needed in the advanced phase of development. To address issues of incomplete absorption, nonlinear exposure, and food effects, physiologically-based pharmacokinetic models are better suited. The compartmental absorption and transit model (CAT)[63,64] was the first physiologically-based drug absorption model that tried to bring together many compound-, formulation-, and physiology-dependent properties to predict absorption. It compartmentalizes the small intestine into 7 equal transit compartments. However, since it did not include dissolution of solid particles, the simulations were valid only for drugs whose absorption were not dissolution limited. An advanced CAT (ACAT) model[65] that considered a pH-dependent solubility, drug absorption from the stomach and colon, as well as physiological characteristics of the absorption surface area and drug transporters formed the basis of the commercial software GastroPlus (Simulations Plus, USA). A schematic representation of a physiology-based absorption model is shown in Figure 4.15.

For each of the nine gastrointestinal compartments, differential equations representing the rates of change of the amount of drug are shown below with respect to the amount that remains undissolved in compartment C, $A_{UND,C}$; the amount dissolved in the compartment, A_{DISC}; the amount degraded in the compartment, $A_{DEG,C}$, and the amount of drug absorbed from that compartment, $A_{ABS,C}$.

1. *Stomach Compartment*

$$\frac{dA_{UND,ST}}{dt} = -\text{GER} \times A_{UND,ST} - \text{DR} \quad (4.24)$$

The solubility S in DR is the pH-dependent solubility in the compartment (Section 4.2.1). Under sink conditions, the C in DR in Eq. 4.5 is zero and A_t is $A_{UND,C}$, the amount of undissolved drug in the stomach, where C refers to the compartment, in this case stomach. GER is the gastric emptying rate. The first term is the rate at which the undissolved drug is lost from the stomach as it transits to the SI, while the second term is the rate of dissolution.

$$\frac{dA_{DIS,ST}}{dt} = -\text{GER} \times A_{DIS,ST} - k_{il,ST} A_{DIS,ST} - k_{a,ST} A_{DIS,ST} \quad (4.25)$$

The first term refers to the rate at which the the dissolved compound is lost from the compartment due to transit to the SI compartment. The second term refers to the to luminal degradation and the third to the absorption rate from the stomach.

$$\frac{dA_{DEG,ST}}{dt} = -\text{GER} \times A_{DEG,ST} + k_{il,ST} A_{DIS,ST} \quad (4.26)$$

4.6 ABSORPTION MODELING

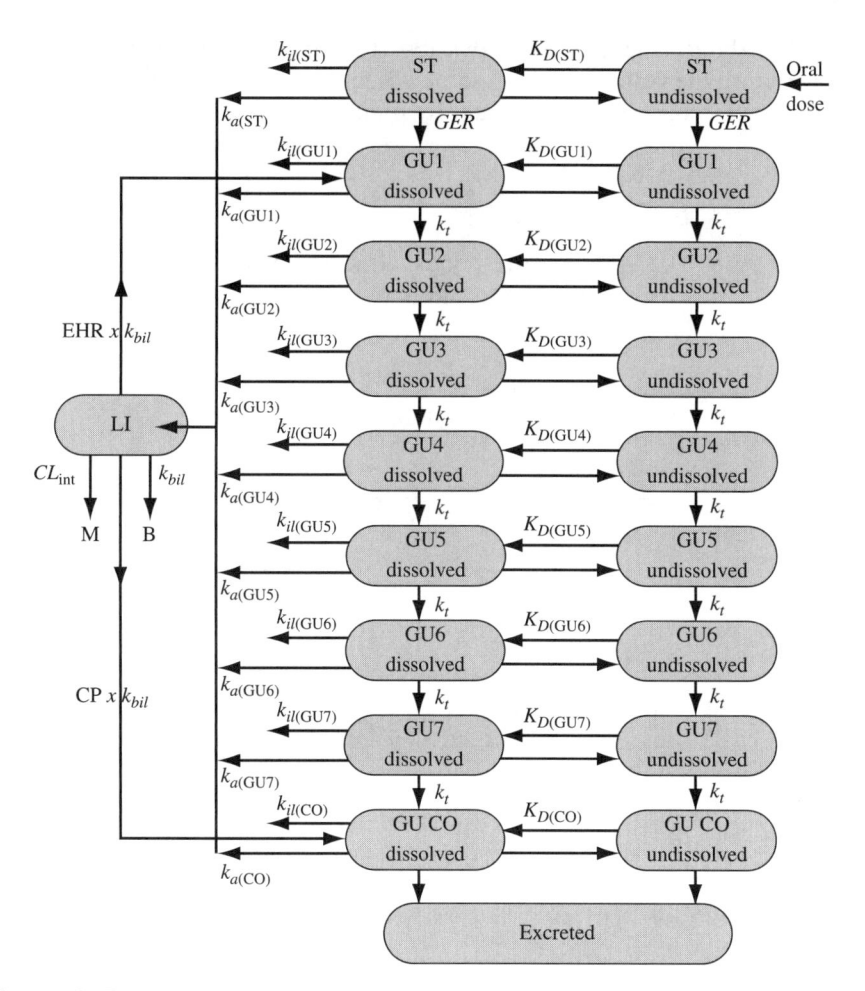

Figure 4.15. Physiology-based oral drug absorption model. Colonic (GUCO) and stomach (ST) compartments to consider colonic and gastric absorptions. K_D is the dissolution rate constant. k_t is transition rate constant. GER is the gastric emptying rate constant. k_a and k_{il} represent rates of absorption and intestinal loss for the different GI compartments. k_{bil} is the rate of emptying of a parent compound or its metabolite into bile. EHR is the enterohepatic recirculation constant that switches between 1 and 0 to include or exclude emptying of parent compound into the duodenal compartment (GU1). M and B are the amounts of compound eliminated by metabolism or into bile. A nonzero fraction to consider metabolite conversion to parent (CP) in the distal SI may also be used if necessary.

Metabolic degradation of the absorbed drug in the enterocytes would require the volume of the enterocytes in the compartment, in order to get the concentration of the drug. It is not considered here.

$$\frac{dA_{\text{ABS, ST}}}{dt} = k_{a,\text{ST}} A_{\text{DIS, ST}} = \text{gastric absorption} \qquad (4.27)$$

At time = 0, $A_{\text{UND,ST}}$ corresponds to the oral dose; k_a and k_{il} are the first-order rate constants of absorption and intestinal loss.

2. *GU1 Small Intestinal Compartment*

$$\frac{dA_{\text{UND, GU1}}}{dt} = \text{GER} \times A_{\text{UND, ST}} - k_t A_{\text{UND, GU1}} - \text{DR} \qquad (4.28)$$

$$\frac{dA_{\text{DIS, GU1}}}{dt} = \text{GER} \times A_{\text{DIS, ST}} - k_t A_{\text{DIS, GU1}} + \text{DR} - k_{a,\text{GU1}} A_{\text{DIS, GU1}}$$

$$+ \frac{\text{EHR} \times V_{\max,\text{bile}}}{K_{m,\text{bile}} + f_{u,\text{LI}} \times C_{\text{LI}}^{\text{IC}}} \times f_{u,\text{LI}} \times C_{\text{LI}}^{\text{IC}} - k_{il,\text{GU1}} A_{\text{DIS, GU1}}$$

$$(4.29)$$

The last term in this equation is the rate of transport of lipophilic acidic or basic drugs into bile or directly into duodenal compartment, when EHR = 1.

$$\frac{dA_{\text{DEG, GU1}}}{dt} = \text{GER} \times A_{\text{DEG, ST}} - k_t A_{\text{DEG, GU1}} + k_{il,\text{GU1}} A_{\text{DIS, GU1}} \qquad (4.30)$$

$$\frac{dA_{\text{ABS, GU1}}}{dt} = k_{a,\text{GU1}} A_{\text{DIS, GU1}} \qquad (4.31)$$

where k_t is the first-order transit rate constant from one compartment to the next, EHR is the enterohepatic recirculation of lipophilic acids or bases. To consider the recirculation of the parent, EHR takes a value of 1 in the first small intestinal compartment (GU1). C is the drug concentration in the intracellular (IC) compartment of liver (LI). $V_{\max}$ and K_m for biliary transport can be replaced by k_{bil} under linear conditions. k_{bil} is the first-order rate constant of transfer of bile content into the duodenal compartment:

$$\text{Rate of bile emptying} = \frac{\text{EHR} \times k_{\text{bil}} \times A_{\text{LI}}}{f_{\text{up}} \times K_{\text{pu, LI}}}$$

where f_{up} is the fraction unbound in plasma, K_{pu} is the unbound tissue partition coefficient. K_p, the tissue partition coefficient is the product $f_{\text{up}} \times K_{\text{pu}}$.

4.6 ABSORPTION MODELING

3. *Other Small Intestinal Compartments* (GU2–GU7)

$$\frac{dA_{\text{UND,GU}i}}{dt} = k_t A_{\text{UND,GU}i-1} - k_t A_{\text{UND,GU}i} - \text{DR} \tag{4.32}$$

$$\frac{dA_{\text{DIS,GU}i}}{dt} = k_t A_{\text{DIS,GU}i-1} - k_t A_{\text{DIS,GU}i} + \text{DR} \times k_{\text{il,GU}i} \times A_{\text{DIS,GU}i}$$
$$- k_{a,\text{GU}i} A_{\text{DIS,GU}i} \tag{4.33}$$

$$\frac{dA_{\text{DEG,GU}i}}{dt} = k_t A_{\text{DEG,GU}i-1} - k_t A_{\text{DEG,GU}i} + k_{\text{il,GU}i} A_{\text{DIS,GU}i} \tag{4.34}$$

$$\frac{dA_{\text{ABS,GU}i}}{dt} = k_{a,\text{GU}i} A_{\text{DIS,GU}i} \tag{4.35}$$

where $i = 2$–7.

4. *Colon Compartment* (GUCO)

$$\frac{dA_{\text{UND,CO}}}{dt} = k_t A_{\text{UND,GU7}} - k_t A_{\text{UND,CO}} - \text{DR} \tag{4.36}$$

$$\frac{dA_{\text{DIS,CO}}}{dt} = k_t A_{\text{DIS,G7}} - k_t A_{\text{DIS,CO}} + \text{DR}$$
$$- k_{a,\text{CO}} A_{\text{DIS,CO}} - k_{\text{il,GU}i} \times A_{\text{DIS,GU}i}$$
$$+ \frac{\text{CP} \times V_{\text{max,bile},M}}{K_{m,\text{bile}} + f_{uM,\text{LI}} \times M_{\text{LI}}^{\text{IC}}} \times f_{uM,\text{LI}} \times M_{\text{LI}}^{\text{IC}} \tag{4.37}$$

$$\frac{dA_{\text{DEG,CO}}}{dt} = k_t A_{\text{DEG,G7}} - k_t A_{\text{DEG,CO}} + k_{\text{il,CO}} A_{\text{DIS,CO}} \tag{4.38}$$

where $\text{CP} = 1$ for a drug whose metabolite is reconverted to the parent in the distal SI or colon. M is the metabolite concentration in the intracellular (IC) compartment of the liver (LI). V_{max} and K_m for biliary emptying rates can be replaced by $k_{\text{bile},M}$ under linear conditions:

$$\frac{dA_{\text{ABS,CO}}}{dt} = k_{a,\text{CO}} A_{\text{DIS,CO}} \tag{4.39}$$

5. *Total Intestinal Absorption*

$$\frac{dA_{\text{IA}}}{dt} = \sum (k_{a,\text{GU}i} A_{\text{DIS,GU}i}) \tag{4.40}$$

The compound-dependent parameters (DR, k_a, etc.) in the above equations can be calculated as indicated in Section 4.2, with the measured data like solubility, permeability, and so on. The physiological parameters (GER, k_t, etc.) needed in the above equations are provided in the Appendix. f_{up} is measured *in vitro* or *ex vivo*. The intestinal loss considered in the above equations is only luminal degradation which depends on the amount of dissolved drug in the lumen. To consider loss due intestinal metabolism, an additional term that depends on the amount of absorbed drug should be introduced in the intestinal compartments.

$$\frac{dA_{M,i}}{dt} = \sum_{\text{isoform}} \left(\frac{k_{\text{ilm,GU}i} \times E_{\text{GU}i,\text{isoform}} \times (A_{\text{ABS,GU}i}/V_{\text{ent,GU}i})}{K_m + (A_{\text{ABS,GU}i}/V_{\text{ent,GU}i})} \right) \quad (4.41)$$

$A_{M,i}$ is the amount of drug metabolized in the ith compartment and $k_{\text{ilm,GU}i}$ is the first order rate constant for intestinal loss by metabolism in ith compartment of the gut. $E_{\text{GU}i,\text{isoform}}$ is the abundance of the enzyme isoform that metabolizes the drug in the ith compartment of the gut. $V_{\text{ent,GU}i}$ is the volume of the enterocytes in that compartment. $A_{\text{Abs,GU}i}$ is the amount absorbed in the ith compartment of the gut. K_m is the Michaelis–Menten constant.

The rate of change of metabolite concentration (M) in the portal vein (PV) is given by equation 4.42.

6. *Metabolites in Portal Vein*

$$\frac{dM_{PV}}{dt} = \frac{1}{V_{PV}} \left(\sum_{\text{GU}i} \left(\sum_{\text{isoform}} \left(\frac{k_{\text{ilm,GU}i} \times E_{\text{GU}i,\text{isoform}} \times (A_{\text{ABS,GU}i}/V_{\text{ent,GU}i})}{K_m + (A_{\text{ABS,GU}i}/V_{\text{ent,GU}i})} \right) \right) - Q_{PV} \times M_{PV} \right) \quad (4.42)$$

If the metabolites formed are conjugates of the drug such as sulphates or glucuronides, part of the metabolite formed is absorbed and the rest transported into the lumen by efflux transporters. Once in the lumen, the metabolite can get converted to parent in distal SI which is thereafter available for reabsorption as parent. Depending on the permeability of the compound, the significance of this reabsorption could vary. The greater the absorption in the colon, the greater will be the extent of reabsorption. Compounds with low permeability will transit further down in the gut and get excreted. Equation 4.42 can be suitably modified to take into account the reconversion of metabolite to parent. An additional term is subtracted from equation 4.41, which should be added to the parent in the distal SI compartments (equation 4.33).

In early discovery, physiological models allow an integrated view of all measured *in vitro* compound and formulation parameters as well as physicochemical properties in a physiological context. In the lead optimization stage, sensitivity analysis and hypothesis testing can provide valuable information to aid the design of new compounds. Predictions of f_{abs} are useful in the screening stage of lead optimization. Mechanistic understanding and hypothesis-driven simulations are of special interest in formulation development in the preclinical phase and for first-in-human studies. Once initial pharmacokinetic information is available, a refined model can also support formulation development for clinical phases II and III or even after the compound is marketed.[66] The integration of physiology-based drug absorption models with somatic physiologically-based pharmacokinetic (PBPK) model widens the scope of applications, as will be shown in later chapters.

KEYWORDS

Biopharmaceutics Classification System (BCS): Classification system, based on solubility and permeation, that places drugs in four classes. Class I molecules have high permeation/high solubility, BCS class II molecules have high permeation/low solubility, BCS class III molecules have low permeation/high solubility, and BCS class IV molecules have low permeation/low solubility.

Cyclodextrins: Cyclic oligosaccharides of glucose consisting of 6–8 glucose or amylose units with a hydrophobic interior to accommodate a lipophillic molecule and a hydrophilic exterior to facilitate solubilization in water.

Effective Permeability: *In vivo* permeability measured in perfusion experiments.

Fraction Absorbed (f_{abs}): Fraction of an orally administered dose that enters the enterocytes, after any degradation in the lumen of the gut.

Gut Bioavailability ($F_g = f_{abs} \times f_{gut}$): The fraction of an orally administered dose that reaches the hepatic portal vein.

Intestinal Loss: The loss of an orally administered dose by any mechanism. A compound can get degraded in the gut lumen, metabolically converted in the enterocytes, or get effuxed back into the lumen by the action of efflux transporters, all of which can lead to its intestinal loss.

Intrinsic Solubility: The solubility of a compound in its nonionized form.

Kinetic Solubility: The solubility of a compound in its amorphous state or in any of its metastable polymorphic forms.

Lead Generation: Phase in drug discovery in which the objective is to identify one or more chemical series with potential drug activity, reduced off-target toxicity, and a physicochemical and a metabolic profile that is compatible with acceptable *in vivo* bioavailability.

Osmolarity: The distribution of water through the generation of osmotic pressure among the different fluid compartments, particularly between the extracellular and intracellular fluids. The osmotic pressure generated by a solution is proportional to the number of particles per unit volume of solvent, not to the type, valence, or weight of the particles.

Polar Surface Area (PSA): Defined as the sum of surface area contributions from all polar atoms such as oxygen and nitrogen, including those attached to hydrogens. Transcellular, passive permeability is highly correlated to PSA.

Polymorphic Forms: Distinct crystal-packing arrangements of a compound associated with different formation energies and melting points.

Maximum Absorbable Dose (MAD): MAD of a drug is the maximum amount of drug that can be absorbed by increasing the dose of a drug, as the amount absorbed gets limited by the solubility and transit time in the small intestine. It is given by the product of absorption rate constant, solubility, volume of small intestinal fluid, and small intestinal transit time.

Thermodynamic or Equilibrium Solubility: At any given temperature and pressure, it is the solubility of its most stable crystalline form under equilibrium conditions. It is lower than the kinetic solubility of any of its metastable or amorphous forms.

REFERENCES

1. Noyes A, Whitney WR. The rate of solution of solid substances in their own solutions. *J Am Chem Soc.* 1897;19:930–934.
2. Brunner E. Reaktionsgeschwindigkeit in heterogenen systemen. *Zeit Phys Chem.* 1904;47:56–102.
3. Hintz RJ, Johnson KC. The effect of particle size on dissolutiona rate and oral absorption. *Int J pharm.* 1989;51:9.
4. Lu AT, Frisella ME, Johnson KC. Dissolution modeling: Factors affecting the dissolution rates of polydisperse powders. *Pharm Res.* 1993;10(9):1308–1314.
5. Wang J, Flanagan DR. General solution for diffusion-controlled dissolution of spherical particles. 1. Theory. *J Pharm Sci.* 1999;88(7):731–738.
6. Wang J, Flanagan DR. General solution for diffusion-controlled dissolution of spherical particles. 2. Evaluation of experimental data. *J Pharm Sci.* 2002;91(2):534–542.
7. Muller RH, and Peters K. Nanosuspensions for the formulation of poorly soluble drugs I. Preparation by a size-reduction technique. *Int J Pharm.* 1998;160:229.
8. Linnankoski J, et al. Paracellular porosity and pore size of the human intestinal epithelium in tissue and cell culture models. *J Pharm Sci.* 2010;99(4):2166–2175.
9. Sugano K, et al. Coexistence of passive and carrier-mediated processes in drug transport. *Nat Rev Drug Discov.* 2010;9(8):597–614.

REFERENCES

10. Thelen K, Dressman JB. Cytochrome P450-mediated metabolism in the human gut wall. *J Pharm Pharmacol.* 2009;61(5):541–558.
11. Paine MF, Hart HL, Ludington SS, Haining RL, Rettie AE, Zeldin DC. The human intestinal cytochrome P450 "pie." *Drug Metab Dispos.* 2006;34(5):880–886.
12. Fisher MB, Paine MF, Strelevitz TJ, Wrighton SA. The role of hepatic and extrahepatic UDP-glucuronosyltransferases in human drug metabolism. *Drug Metab Rev.* 2001;33(3–4):273–297.
13. Yoshigae Y, Imai T, Horita A, Matsukane H, Otagiri M. Species differences in stereoselective hydrolase activity in intestinal mucosa. *Pharm Res.* 1998;15(4):626–631.
14. Glatt H, et al. Human cytosolic sulphotransferases: Genetics, characteristics, toxicological aspects. *Mutat Res.* 2001;482(1–2):27–40.
15. Ritter JK. Intestinal UGTs as potential modifiers of pharmacokinetics and biological responses to drugs and xenobiotics. *Expert Opin Drug Metab Toxicol.* 2007;3(1): 93–107.
16. Martinez MN, Amidon GL. A mechanistic approach to understanding the factors affecting drug absorption: A review of fundamentals. *J Clin Pharmacol.* 2002;42 (6):620–643.
17. Lindahl A, Ungell AL, Knutson L, Lennernas H. Characterization of fluids from the stomach and proximal jejunum in men and women. *Pharm Res.* 1997;14 (4):497–502.
18. Sugano K. Computational oral absorption simulation for low-solubility compounds. *Chem Biodiv.* 2009;6:2014.
19. Bruyere A, et al. Effect of variations in the amounts of P-glycoprotein (ABCB1), BCRP (ABCG2) and CYP3A4 along the human small intestine on PBPK models for predicting intestinal first-pass. *Mol Pharm.* 2010;7(5):1596–1607.
20. Hilgendorf C, Ahlin G, Seithel A, Artursson P, Ungell AL, Karlsson J. Expression of thirty-six drug transporter genes in human intestine, liver, kidney, and organotypic cell lines. *Drug Metab Dispos.* 2007;35(8):1333–1340.
21. Chiou WL, Jeong HY, Chung SM, Wu TC. Evaluation of using dog as an animal model to study the fraction of oral dose absorbed of 43 drugs in humans. *Pharm Res.* 2000;17(2):135–140.
22. Chiou WL, Barve A. Linear correlation of the fraction of oral dose absorbed of 64 drugs between humans and rats. *Pharm Res.* 1998;15(11):1792–1795.
23. Ikegami K, Tagawa K, Narisawa S, Osawa T. Suitability of the cynomolgus monkey as an animal model for drug absorption studies of oral dosage forms from the viewpoint of gastrointestinal physiology. *Biol Pharm Bull.* 2003;26(10):1442–1447.
24. Ward KW, Smith BR. A comprehensive quantitative and qualitative evaluation of extrapolation of intravenous pharmacokinetic parameters from rat, dog, and monkey to humans. II. Volume of distribution and mean residence time. *Drug Metab Dispos.* 2004;32(6):612–619.
25. Ward KW, Smith BR. A comprehensive quantitative and qualitative evaluation of extrapolation of intravenous pharmacokinetic parameters from rat, dog, and monkey to humans. I. Clearance. *Drug Metab Dispos.* 2004;32(6):603–611.
26. Palm K, et al. Evaluation of dynamic polar molecular surface area as predictor of drug absorption: Comparison with other computational and experimental predictors. *J Med Chem.* 1998;41(27):5382–5392.

27. Ertl P, Rohde B, Selzer P. Fast calculation of molecular polar surface area as a sum of fragment-based contributions and its application to the prediction of drug transport properties. *J Med Chem.* 2000;43(20):3714–3717.
28. Palm K, Luthman K, Ungell AL, Strandlund G, Artursson P. Correlation of drug absorption with molecular surface properties. *J Pharm Sci.* 1996;85(1):32–39.
29. Stenberg P, Norinder U, Luthman K, Artursson P. Experimental and computational screening models for the prediction of intestinal drug absorption. *J Med Chem.* 2001;44(12):1927–1937.
30. Oprea TI, Gottfries J. Toward minimalistic modeling of oral drug absorption. *J Mol Graph Model.* 1999;17(5–6):261–274, 329.
31. Winiwarter S, Bonham NM, Ax F, Hallberg A, Lennernas H, Karlen A. Correlation of human jejunal permeability (in vivo) of drugs with experimentally and theoretically derived parameters. A multivariate data analysis approach. *J Med Chem.* 1998;41(25):4939–4949.
32. Leahy DE, Lynch J, Finney RE, Taylor DC. Estimation of sieving coefficients of convective absorption of drugs in perfused rat jejunum. *J Pharmacokinet Biopharm.* 1994;22(5):411–429.
33. Abraham MH, Le J. The correlation and prediction of the solubility of compounds in water using an amended solvation energy relationship. *J Pharm Sci.* 1999;88(9):868–880.
34. Yalkowsky SH, Valvani SC. Solubility and partitioning I: Solubility of nonelectrolytes in water. *J Pharm Sci.* 1980;69(8):912–922.
35. van de Kerkhof EG, Ungell AL, Sjoberg AK, et al. Innovative methods to study human intestinal drug metabolism in vitro: Precision-cut slices compared with Ussing chamber preparations. *Drug Metab Dispos.* 2006;34(11):1893–1902.
36. Nigsch F, Klaffke W, Miret S. In vitro models for processes involved in intestinal absorption. *Expert Opin Drug Metab Toxicol.* 2007;3(4):545–556.
37. Balimane PV, Han YH, Chong S. Current industrial practices of assessing permeability and P-glycoprotein interaction. *AAPS J.* 2006;8(1):E1–13.
38. Kansy M, Senner F, Gubernator K. Physicochemical high throughput screening: Parallel artificial membrane permeation assay in the description of passive absorption processes. *J Med Chem.* 1998;41(7):1007–1010.
39. Kerns EH. High throughput physicochemical profiling for drug discovery. *J Pharm Sci.* 2001;90(11):1838–1858.
40. Artursson P, Karlsson J. Correlation between oral drug absorption in humans and apparent drug permeability coefficients in human intestinal epithelial (caco-2) cells. *Biochem Biophys Res Commun.* 1991;175(3):880–885.
41. Briske-Anderson MJ, Finley JW, Newman SM. The influence of culture time and passage number on the morphological and physiological development of caco-2 cells. *Proc Soc Exp Biol Med.* 1997;214(3):248–257.
42. Irvine JD, et al. MDCK (madin-darby canine kidney) cells: A tool for membrane permeability screening. *J Pharm Sci.* 1999;88(1):28–33.
43. Using HH, Zrahn K. Active transport of sodium as the source of electric current in the short-circuited isolated frog skin. *Acta Physiol Scand.* 1951;23(2–3):110–127.

REFERENCES

44. Lennernas H, Nylander S, Ungell AL. Jejunal permeability: A comparison between the Ussing chamber technique and the single-pass perfusion in humans. *Pharm Res.* 1997;14(5):667–671.
45. Yang H, et al. Glutamine effects on permeability and ATP content of jejunal mucosa in starved rats. *Clin Nutr.* 1999;18(5):301–306.
46. Polentarutti BI, Peterson AL, Sjoberg AK, Anderberg EK, Utter LM, Ungell AL. Evaluation of viability of excised rat intestinal segments in the Ussing chamber: Investigation of morphology, electrical parameters, and permeability characteristics. *Pharm Res.* 1999;16(3):446–454.
47. van de Kerkhof EG, de Graaf IA, Groothuis GM. In vitro methods to study intestinal drug metabolism. *Curr Drug Metab.* 2007;8(7):658–675.
48. Amidon GL, Sinko PJ, Fleisher D. Estimating human oral fraction dose absorbed: A correlation using rat intestinal membrane permeability for passive and carrier-mediated compounds. *Pharm Res.* 1988;5(10):651–654.
49. Lennernas H, Ahrenstedt O, Hallgren R, Knutson L, Ryde M, Paalzow LK. Regional jejunal perfusion, a new in vivo approach to study oral drug absorption in man. *Pharm Res.* 1992;9(10):1243–1251.
50. Fagerholm U, Johansson M, Lennernas H. Comparison between permeability coefficients in rat and human jejunum. *Pharm Res.* 1996;13(9):1336–1342.
51. Palmer DS, et al. Predicting intrinsic aqueous solubility by a thermodynamic cycle. *Mol Pharm.* 2008;5(2):266–279.
52. Stuart M, Box K. Chasing equilibrium: Measuring the intrinsic solubility of weak acids and bases. *Anal Chem.* 2005;77(4):983–990.
53. Jantratid E, Janssen N, Reppas C, Dressman JB. Dissolution media simulating conditions in the proximal human gastrointestinal tract: An update. *Pharm Res.* 2008;25(7):1663–1676.
54. Dressman JB, Amidon GL, Reppas C, Shah VP. Dissolution testing as a prognostic tool for oral drug absorption: Immediate release dosage forms. *Pharm Res.* 1998;15(1):11–22.
55. Taipalensuu J, Tavelin S, Lazorova L, Svensson AC, Artursson P. Exploring the quantitative relationship between the level of MDR1 transcript, protein and function using digoxin as a marker of MDR1-dependent drug efflux activity. *Eur J Pharm Sci.* 2004;21(1):69–75.
56. Yang J, Jamei M, Yeo KR, Tucker GT, Rostami-Hodjegan A. Prediction of intestinal first-pass drug metabolism. *Curr Drug Metab.* 2007;8(7):676–684.
57. Sohlenius-Sternbeck AK, Orzechowski A. Characterization of the rates of testosterone metabolism to various products and of glutathione transferase and sulfotransferase activities in rat intestine and comparison to the corresponding hepatic and renal drug-metabolizing enzymes. *Chem Biol Interact.* 2004;148(1–2):49–56.
58. Gertz M, Harrison A, Houston JB, Galetin A. Prediction of human intestinal first-pass metabolism of 25 CYP3A substrates from in vitro clearance and permeability data. *Drug Metab Dispos.* 2010;38(7):1147–1158.
59. Kadono K, Akabane T, Tabata K, Gato K, Terashita S, Teramura T. Quantitative prediction of intestinal metabolism in humans from a simplified intestinal availability model and empirical scaling factor. *Drug Metab Dispos.* 2010;38(7):1230–1237.

60. Lipinski CA, Lombardo F, Dominy BW, Feeney PJ. Experimental and computational approaches to estimate solubility and permeability in drug discovery and development settings. *Adv Drug Deliv Rev.* 2001;46(1–3):3–26.
61. Andrews CW, Bennett L, Yu LX. Predicting human oral bioavailability of a compound: Development of a novel quantitative structure-bioavailability relationship. *Pharm Res.* 2000;17(6):639–644.
62. Hou T, Wang J, Li Y. ADME evaluation in drug discovery. 8. The prediction of human intestinal absorption by a support vector machine. *J Chem Inf Model.* 2007;47(6):2408–2415.
63. Yuh L, et al. Population pharmacokinetic/pharmacodynamic methodology and applications: A bibliography. *Biometrics.* 1994;50(2):566–575.
64. Yu LX, Amidon GL. Saturable small intestinal drug absorption in humans: Modeling and interpretation of cefatrizine data. *Eur J Pharm Biopharm.* 1998;45(2):199–203.
65. Agoram B, Woltosz WS, Bolger MB. Predicting the impact of physiological and biochemical processes on oral drug bioavailability. *Adv Drug Deliv Rev.* 2001;50 Suppl 1:S41–67.
66. Kuentz M. Drug absorption modeling as a tool to define the strategy in clinical formulation development. *AAPS J.* 2008;10(3):473–479.

5

PHYSIOLOGICAL MODEL FOR DISTRIBUTION

CONTENTS

5.1	Introduction	90
5.2	Factors Affecting Tissue Distribution of Xenobiotics	91
	5.2.1 Physiological Factors and Species Differences in Physiology.	91
	5.2.2 Compound-Dependent Factors.	98
5.3	*In Silico* Models of Tissue Partition Coefficients.	98
5.4	Measurement of Parameters Representing Rate and Extent of Tissue Distribution.	105
	5.4.1 Assessment of Rate and Extent of Brain Penetration	105
5.5	Physiological Model for Drug Distribution	110
5.6	Drug Concentrations at Site of Action	111
	Keywords	114
	References	115

Physiologically-Based Pharmacokinetic (PBPK) Modeling and Simulations: Principles, Methods, and Applications in the Pharmaceutical Industry, First Edition. Sheila Annie Peters.
© 2012 John Wiley & Sons, Inc. Published 2012 by John Wiley & Sons, Inc.

5.1 INTRODUCTION

Drug distribution refers to the reversible partitioning of a drug into the various tissues of the body from systemic circulation, driven by blood flow rates or the ability of a drug to cross membrane barrier, leading to defined proportions in the different tissues at steady state, depending on the compound characteristics and composition of the various tissues. Drugs are often confined to total body water, extracellular volume, or plasma volume, depending on their physicochemical nature (Fig. 5.1). Hydrophillic drugs tend to stay in plasma or in the interstitial fluids surrounding the tissue cells, while lipophilic bases have the greatest tendency to cross the membrane barrier and distribute into cells constituting tissues, especially into the more fatty tissues. However, although the brain is a fatty tissue, distribution into this organ is restricted by the **blood–brain barrier (BBB)**. Tissues act as a reservoir for a drug, slowly releasing it into the bloodstream as more and more of it is eliminated from the blood. Lipophilic drugs that tend to accumulate in the tissues can remain in the body long after the discontinuation of drug treatment and consequently may have prolonged pharmacological action. The apparent, hypothetical volumes of distribution for these drugs are large. Table 5.1 lists the physiological volumes available for drugs with different physicochemical properties. This chapter will focus on the factors affecting drug distribution into various tissues and briefly discuss the *in silico* and *in vitro* methods available to measure them. The

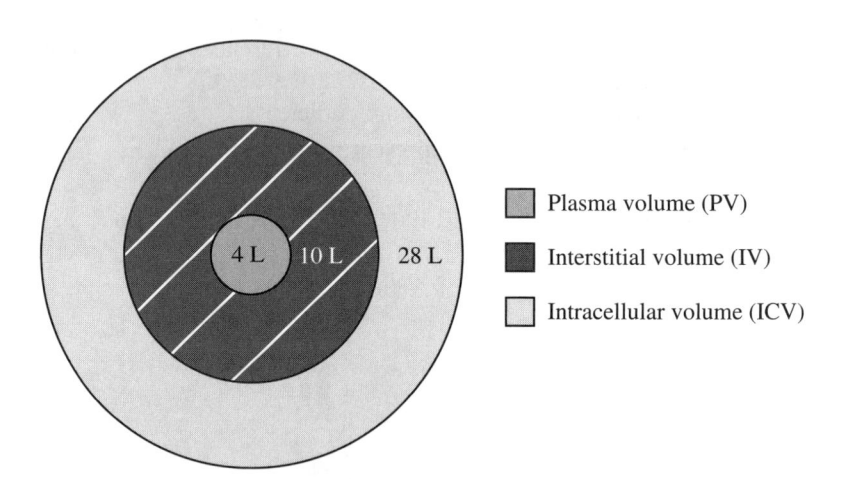

Figure 5.1. Distribution volumes available to a drug. Large molecules cannot cross the endothelial barrier; they are confined to plasma volume. Acids and small, hydrophilic molecules are confined to extracellular volume. Lipophilic drugs can easily cross the membrane barrier and distribute themselves into the total body water. Some drugs binding to intracellular components can have a distribution volume greater than the total body water.

TABLE 5.1. Drugs Confined to Different Physiological Volumes

Physiological Volume	Distribution Volume (L)	Types of Drug Molecules	Examples
Plasma volume (PV)	≤ 4	Large molecules	Heparin (4 L)
Extracellular volume (ECV) = PV + interstitial volume	4–14	Small, hydrophilic molecules and acids	Warfarin (8 L)
Total body water = ECV + intracellular volume	14–42	Lipophilic compounds that diffuse through cell membranes	Ethanol (34–41 L) Alfentanil (35–77 L)
Greater than total body water	> 42	Lipophilic bases and drugs binding strongly to tissues	Fentanyl (280 L) Propofol (560 L)

integration of *in silico* and measured *in vitro* properties into physiological models will follow. The importance of estimating tissue concentrations at the site of pharmacological action and the application of PBPK models to obtain time profiles of tissue concentrations will be outlined.

5.2 FACTORS AFFECTING TISSUE DISTRIBUTION OF XENOBIOTICS

As with other pharmacokinetic properties, the rate and extent of tissue distribution of xenobiotics is influenced by both compound-specific and physiology-dependent factors (Fig. 5.2).

5.2.1 Physiological Factors and Species Differences in Physiology

The rate of drug distribution into tissues is determined by blood flow rates, transporter-driven uptake rate, and/or membrane permeability depending on the nature of the drug. The extent of tissue distribution is determined by **partition coefficients** as well as binding to plasma and tissue proteins. Thus, physiology and compound characteristics together determine the rate and extent of tissue distribution of a drug.

5.2.1.1 Blood Flow Rate. Drug distribution occurs rapidly into highly perfused tissues and with **discontinuous capillaries** such as liver, spleen, and intestine. Tissues with low perfusion rates and with continuous capillaries such as muscle and skin are represented by larger and slowly exchanging extravascular compartments.

5.2.1.2 Membrane Permeability. The walls of capillaries are very thin, consisting of only a single layer of endothelial cells, making them highly

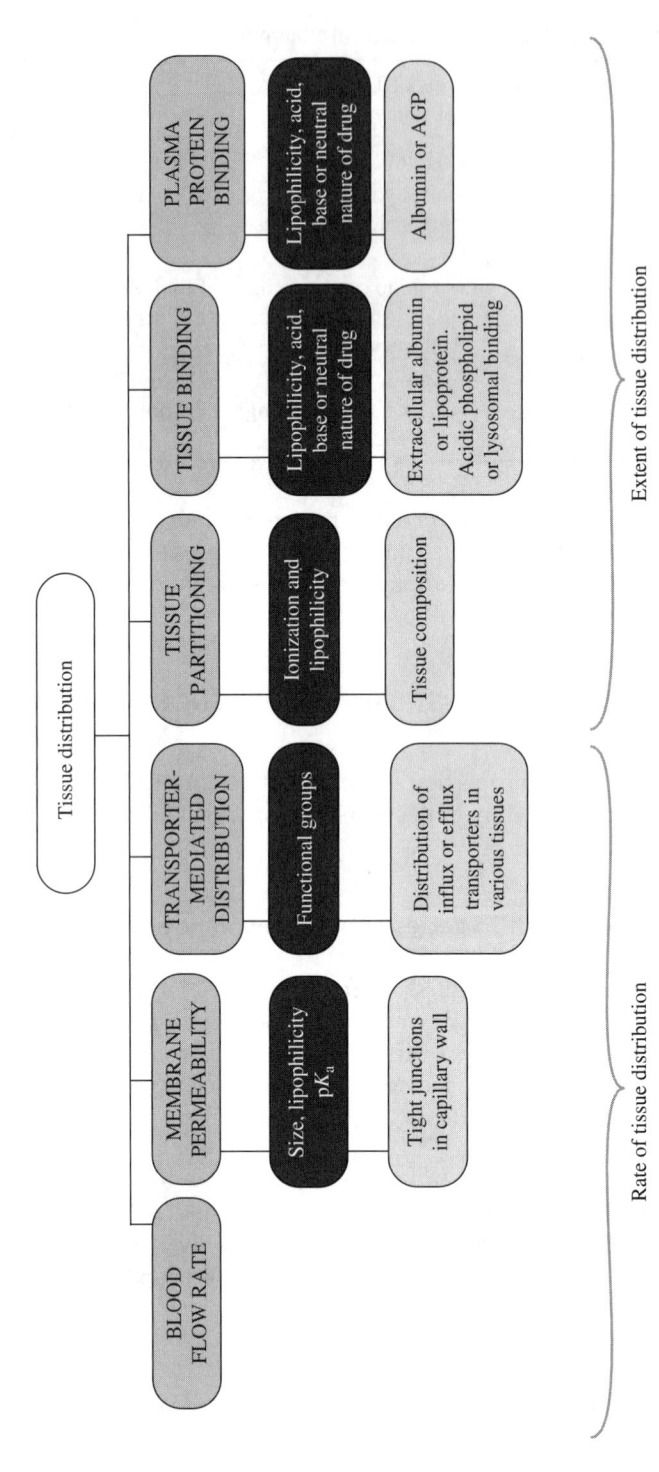

Figure 5.2. Compound- and physiology-dependent factors affecting rate and extent of tissue distribution. Except blood flow rate, which is a physiological parameter, all other determinants of tissue distribution are dependent on both compound characteristics (first level) as well as the physiological properties (second level) of the species.

permeable especially to small, lipophilic drugs. Renal capillaries and hepatic sinusoids allow extensive transfer of drug into interstitial space. Distribution into intracellular space is dependent on cell membrane permeability. Cell membranes are made up of phospholipid bilayers that readily allow lipophilic compounds to permeate through them. Due to the negative polar heads of the lipid bilayers, bases have a preferential entry into cell membranes compared to acids and neutrals. For large molecules, the membrane permeability limits the rate of distribution into tissues.

5.2.1.3 Transporters. Well-perfused organs do not necessarily have a high rate of tissue distribution. For example, the brain is a well-perfused organ, but the rate of brain penetration is rather low for most drugs, except for small, lipophilic drugs. The tightly packed endothelial cells of the narrow continuous capillaries of the brain, along with the multitude of influx and efflux transporters (Fig. 3.6), constitute the BBB (Fig. 5.3), making it exceedingly selective in what gets through it. P-gp and other efflux transporters are located on the apical surface of the endothelial cells of the brain capillaries toward the vascular lumen and contribute to the poor BBB penetration of a number of drugs. Although many drugs can still cross the BBB, only small, lipophilic molecules and hydrophilic molecules such as alcohol, caffeine, and nicotine are able to cross the barrier to a larger extent. Neuropathic agents and drugs meant for neurological disorders such as stroke, Alzheimer's disease, and Parkinson's disease need to cross the BBB in order to be pharmacologically active. Understanding mechanisms and factors that regulate the permeability of the BBB could contribute significantly to the development of new therapeutic approaches for improving the bioavailability of drugs to the brain. A drug that manages to cross the BBB is available in the interstitial fluid from where it can enter the brain cells to get distributed throughout the brain or get metabolized by brain enzymes or eliminated via ventricular cerebrospinal fluid (CSF). Attempts to predict brain penetration *in silico* have shown that many central nervous system (CNS)-active compounds are bases, tightly clustered around an apparent optimum log D of 2, with a molecular weight below 450 and a total polar surface area of at least 90 Å.[1-3] Various polar compounds, ions, sugars, amino acids, and even certain peptides such as insulin cross the BBB by using passive and secondary active transporters. Examples of transporters include, Glucose transporter 1 (GLUT-1), a glucose carrier non-dependant on insulin, which can also transfer some glycopeptides, LNA system (large neutral amino acids), which allows the influx of certain amino acids and drugs having a similar structure, such as L-dopa, melphalan, gabapentin, and baclofen, and peptide carriers. Certain compounds, such as bradykinin, distend intercellular tight junctions and facilitate the transfer of various compounds through the BBB. Several Na^+-dependent carriers of amino acids exist on the abluminal membrane of the BBB to transfer amino acids from the extracellular fluid of the brain to the endothelial cells. These carriers remove nonessential, nitrogen-rich, or acidic (excitatory) amino acids, all of which may be detrimental to brain function.

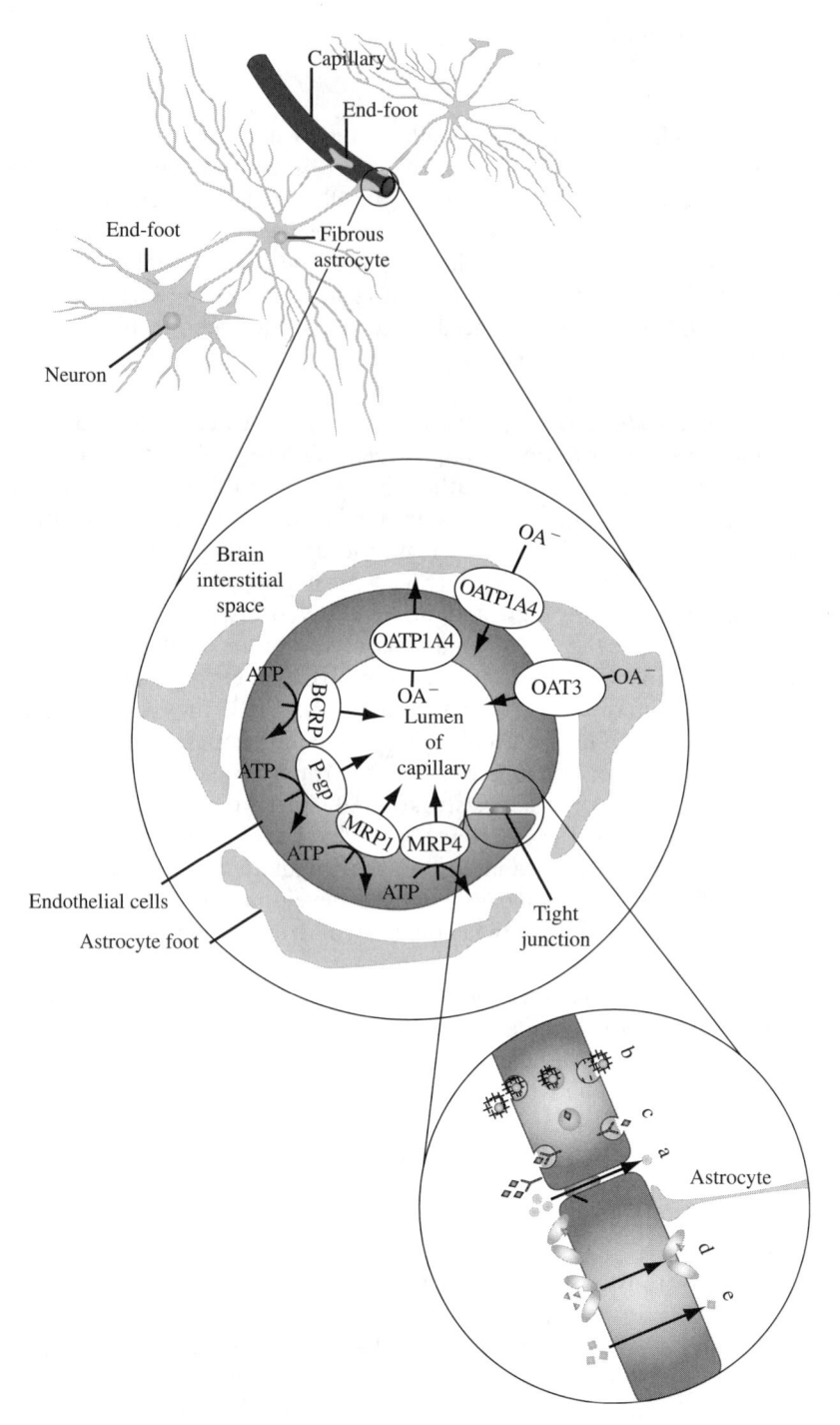

Figure 5.3. Important transporters constituting the blood–brain barrier and different types of drug transport across the capillary wall in the brain: (a) paracellular permeability, (b) adsorptive transcytosis, (c) receptor-mediated transcytosis, (d) transporter-mediated uptake, and (e) transcellular passive permeability.

Besides the BBB, the brain also has a brain–CSF barrier that is located at the level of choroid plexus, originating at the tight epithelium lining the ventricle rather than the endothelium as in the cerebral capillaries of the BBB. The cerebrospinal fluid secreted by the choroid plexus has a volume of approximately 150–200 mL with a rate of renewal of approximately 0.5 mL/min. Normally, it contains only about 0.5% proteins as that in the plasma with approximately similar composition. The concentration of a drug that diffuses passively in the CSF is approximately equal to the concentration of its free form in plasma. A third barrier to consider is the arachnoid barrier. However, since the relative surface area of the BBB is much higher than that of other barriers in the brain and the density of the brain parenchymal capillaries is very high, the BBB is considered to play a much greater role. Excellent reviews on the BBB are available in the literature.[4-7] Inclusion of BBB model within PBPK facilitates the prediction of CNS exposure of drugs that target CNS or those that need to be kept out for safety reasons.

The influx or efflux activity of transporters can be an important factor driving the drug distribution not just in the brain. The liver and kidney are two other organs where transporters play a major role in determining the rate and extent of distribution and, therefore, the rates of elimination. Whereas in the brain the unbound concentrations of a drug in plasma exceeds that in the brain because of efflux transporters, in the liver and in the kidney the unbound concentrations in plasma is less than the intracellular concentrations for drugs that are substrates for uptake transporters. For example, the unbound hepatic concentrations of pravastatin, a substrate of the uptake transporter OATP1B1 is much higher compared to its unbound systemic levels. This accumulation in the target organ serves to enhance its pharmacological activity of inhibiting 3-hydroxy-3-methyl-glutaryl-CoA (HMG-CoA) reductase, an enzyme that catalyzes the rate-limiting step in the de novo synthesis of cholesterol in the liver. The hepatic levels of biguanides such as metformin are increased by the uptake transporter OCT1 but can lead to lactic acidosis, a concentration-dependent toxic effect of metformin.[8] In general, the rates of tissue distribution of many ionizable compounds (acids and bases) are determined by their uptake rates in organic anion or cation transporters such as OATP1B1, OATP1B3, OATP2B1, OATs, MCT1, OCT1, OCT2, and OCTNs (see Fig. 3.6).

Tissue compositions affect the tissue partitioning and thereby the extent of distribution of drugs. Lipophilic drugs distribute into fat-rich organs such as adipose, liver, brain, and kidney, while hydrophilic drugs distribute into water-rich organs such as muscle.

Binding to plasma proteins and tissue components also determines the extent of distribution into tissues. Drugs bind reversibly to plasma proteins (discussed in Chapter 3). Only the unbound fraction in blood distributes into tissues from systemic circulation. Thus, acids that are extensively and strongly bound to albumin generally have a low fraction unbound in plasma and consequently a low **apparent volume of distribution** (V_{app}). Binding to tissue components such as membrane phospholipids, deoxyribonucleic acid (DNA), and proteins can be reversible or irreversible. Irreversible binding leads to drug accumulation. For example, chloroquine binds to DNA and tends to concentrate in white blood cells (WBCs) and liver cells 1000 times more than in plasma, leading to accumulation. Lipophilic drugs tend to distribute quickly into highly perfused organs such as brain and liver but gradually redistribute into adipose tissue. As the plasma levels go down due to their elimination, more get released in to the blood from the fat storage. Examples include thiopental, polychlorobiphenyls, and barbiturates. Specific binding of drugs to certain tissues either because of high affinity to these tissues or because of a special nature of the tissues to retain certain drugs can increase drug distribution by several-fold in those tissues. Examples include absorption of tetracycline into bone and teeth and iodine in the thyroid gland. Many lipophilic bases bind to the membrane or cellular components of the erythrocytes (red blood cells, RBCs). Partitioning into RBCs can be measured as **blood-plasma ratio**. For most drugs, however, the binding to plasma is greater than to tissues. Basic compounds also tend to accumulate in lysosomes (Fig. 5.4). Lysosomes are subcellular organelles that are capable of trapping lipophilic basic compounds such as chlorpromazine, biperiden, and imipramine[9] due to the low pH associated with them. Lysosomal uptake is even more profound for dibasic compounds. For example, azithromycin, a dibasic compound achieves high *in vivo* tissue–plasma ratios and, therefore, has a high V_{app} compared to its monobasic structural analog,

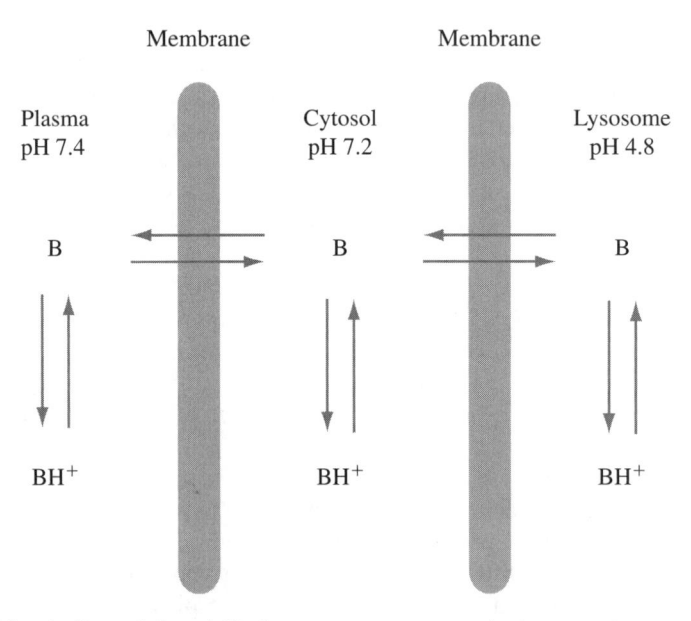

Figure 5.4. Binding of lipophilic bases to components in lysosomal compartment is favored by the low pH.

5.2 FACTORS AFFECTING TISSUE DISTRIBUTION OF XENOBIOTICS

erythromycin. The apparent volume of lysosome-rich tissues such as lung and liver may be several times its physical volume.

The volume of distribution, V, is a measure of the extent of tissue distribution and is related to the protein binding and physiological volumes of plasma (V_p) and tissues (V_T) as follows:

$$V = V_p + \sum_T \left(V_T \times \frac{f_{up}}{f_{uT}} \right) \quad (5.1)$$

Where f_{up} is the fraction of drug unbound in plasma and f_{uT} is the fraction unbound in the various tissues. Tissue and plasma volumes normalized by weight are similar across species, but could still be different as f_{up} could vary across species. Thus, the unbound volume of distribution per kilogram body weight can be expected to be the same across species.

For a drug bound to plasma proteins, the unbound drug concentration in plasma (C_{up}) is related to its total concentration in plasma ($C_{total,p}$) by

$$C_{up} = C_{total,p} \times f_{up} \quad (5.2)$$

Similarly, the unbound drug concentration in a tissue T (C_{uT}) is related to its total concentration in that tissue ($C_{total,T}$) by

$$C_{uT} = C_{total,T} \times f_{uT} \quad (5.3)$$

where C_{uT} is the pharmacologically relevant concentration that correlates to pharmacological activity of the drug in a target tissue and should therefore, be linked to pharmacodynamic data. Dividing the total amount of drug in a tissue, A_T, by C_{uT} gives the unbound volume of distribution within that tissue, $V_{u,T}$, which may be equal to only the interstitial volume of that tissue or a sum of interstitial and its intracellular volumes. When distribution of drug into tissues has reached its equilibrium, the free (unbound) drug concentrations in plasma and tissues are the same and, therefore,

$$C_{total,T} \times f_{uT} = C_{total,p} \times f_{up} \quad (5.4)$$

$$\frac{C_{total,T}}{C_{total,p}} = K_{Tp} = \frac{f_{up}}{f_{uT}} \quad (5.5)$$

The ratio of total concentrations of a drug in a tissue (T) and plasma is defined as the tissue partition coefficient K_{Tp}. Tissue–plasma partition coefficients are a measure of the steady-state distribution of the total drug concentration in tissue. Equation 5.5 does not address the dependence of K_{Tp} on tissue compositions. Substituting for f_{up}/f_{uT} in equation 5.1, in terms of K_{Tp},

$$V = V_p + \sum_T (V_T \times K_{\text{Tp}}) \qquad (5.6)$$

where K_{Tp} equals $f_{\text{up}}/f_{\text{uT}}$ or $K_{\text{Tp}} \times f_{\text{uT}}/f_{\text{up}} = 1$, only in the absence of efflux and influx transporter involvement (Fig. 5.5), when tissue distribution is driven by passive diffusion alone. The unbound tissue partition coefficient $K_{\text{Tp}} \times f_{\text{uT}}/f_{\text{up}}$ (or $K_{\text{Tp,uu}}$) represents the extent of distribution equilibrium of a compound between unbound fractions in tissue and in plasma. It is 1, <1, or >1 for passive diffusion, efflux, and uptake, respectively.

Drug distribution could vary between different populations. For instance, obese people may store large amounts of fat-soluble drugs, whereas very thin people may store relatively little. Yet, the weight normalized volume of distribution is not very much higher in the obese.[10,11] Older people, even when thin, may store large amounts of fat-soluble drugs because the proportion of body fat increases with aging. The BBB becomes less effective with aging, allowing more compounds into the brain.

5.2.2 Compound-Dependent Factors

The influence of size, log P, pK_a, acid–base–neutral character, and the chemistry of molecules along side the species physiology in determining the membrane permeability, transporter-driven influx and efflux, tissue partitioning, as well as plasma and tissue binding has already been discussed in the preceding section. The unbound tissue partition coefficients are a measure of the extent of tissue distribution into tissue, and they depend on both compound characteristics and the tissue compositions.

The rate of tissue distribution (determined by blood flow rate for small lipophilic molecules, by membrane permeability for large, hydrophilic molecules, and by the rate of transporter uptake for ionizable compounds), the extent of tissue distribution (determined by influx and efflux transporters and plasma and tissue binding of the compound), and the unbound distribution volume within the tissue are three distinct, yet interdependent, properties. The parameters that represent these properties, namely tissue partition coefficients, blood–plasma and brain–plasma ratios, and fractions unbound in plasma and tissues can be obtained from *in silico*, *in vitro*, or *in vivo* models as outlined in the following section. These parameters can then be integrated into a physiological model for distribution.

5.3 *IN SILICO* MODELS OF TISSUE PARTITION COEFFICIENTS

In silico models to estimate tissue partition coefficients abound in the literature. Two of these stand out for their wide applicability. The first one, proposed by Poulin and Thiel in 2002,[12] considered the plasma and the tissues of a species to

5.3 IN SILICO MODELS OF TISSUE PARTITION COEFFICIENTS

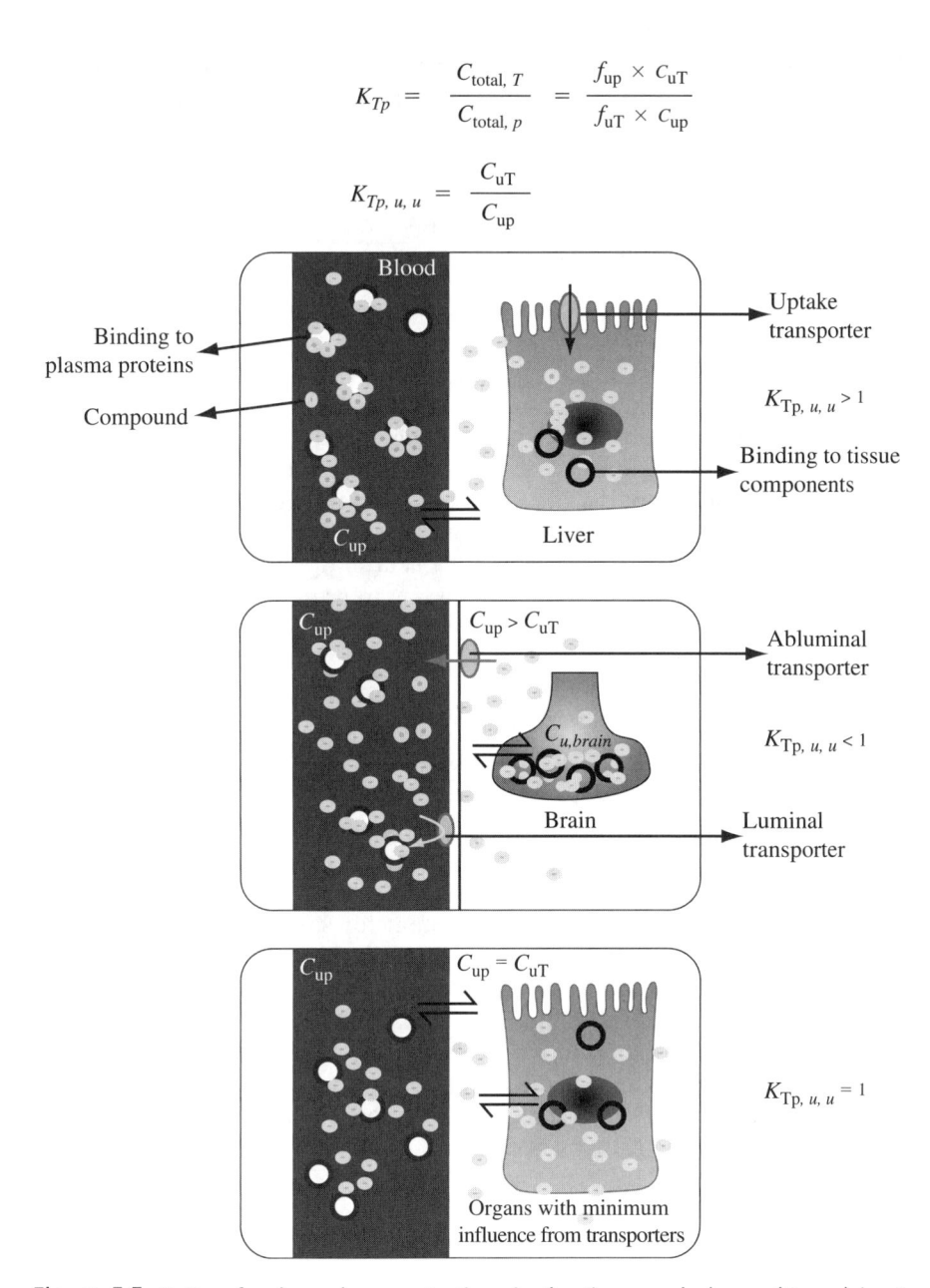

$$K_{Tp} = \frac{C_{total,\,T}}{C_{total,\,p}} = \frac{f_{up} \times C_{uT}}{f_{uT} \times C_{up}}$$

$$K_{Tp,\,u,\,u} = \frac{C_{uT}}{C_{up}}$$

Figure 5.5. Ratio of unbound concentrations in the tissue and plasma ($K_{Tp,u,u}$) is >1, when uptake transporters act to accumulate the drug within the cells. This is valid for the liver. The extracellular drug concentrations are in equilibrium with the plasma concentrations. The middle panel shows the lower tissue unbound concentrations in the brain compared to the plasma due to the abluminal and luminal transporters that constitute the blood–brain barrier. The last panel shows the equilibrium between the extracellular, intracellular, and plasma unbound concentrations, making the $K_{Tp,u,u} = 1$.

be composed of water, lipid, and phospholipids (Fig. 3.2). A compound gets distributed between plasma and the tissues depending on its lipophilicity. The tissue–plasma partition coefficient, K_{Tp}, of a drug in a tissue T is given as follows:

$$K_{T(\text{nonadipose}),p} = \frac{P_{\text{ow}} \times (f_{\text{nlT}} + 0.3 f_{\text{plT}}) + (f_{\text{wT}} + 0.7 f_{\text{plT}})}{P_{\text{ow}} \times (f_{\text{nlP}} + 0.3 f_{\text{plp}}) + (f_{\text{wP}} + 0.7 f_{\text{plp}})} \times \frac{f_{\text{up}}}{f_{\text{uT}}} \quad (5.7)$$

$$K_{T(\text{adipose}),p} = \frac{D_{\text{vow}} \times (f_{\text{nlT}} + 0.3 f_{\text{plT}}) + (f_{\text{wT}} + 0.7 f_{\text{plT}})}{D_{\text{vow}} \times (f_{\text{nlp}} + 0.3 f_{\text{plp}}) + (f_{\text{wp}} + 0.7 f_{\text{plp}})} \times \frac{f_{\text{up}}}{1} \quad (5.8)$$

Where P_{ow} is the octanol–water partitioning of a drug and D_{vow} is the partitioning of the neutral form of an ionizable drug between vegetable oil and water; f_{nl}, f_{pl}, and f_w are the fractional volume content of neutral lipid, phospholipids, and water in plasma or in a tissue. Neutral phospholipids are assumed to behave like a mixture of 30% neutral lipids and 70% water. f_{uT} is assumed to be 1 for adipose tissue, as binding to proteins in that compartment can be expected to be minimal. For all other tissues, f_{uT} is derived from f_{up} as follows:

$$f_{\text{uT}} = \frac{1}{1 + 0.5 \times \left(\dfrac{1 - f_{\text{up}}}{f_{\text{up}}}\right)} \quad (5.9)$$

Equations 5.7 and 5.8 consider lipophilicity-driven passive transmembrane diffusion as well as binding to plasma and tissue components, assuming uniform distribution of a drug within a tissue. However, they do not consider paracellular permeability, which is important for small, hydrophilic drugs or influx/efflux transporter-influenced tissue distribution. Additionally, it cannot handle membrane-permeability-limited distribution, important for large molecules. The equation does not distinguish between acids, bases, and neutrals, which is very important in determining the extent of distribution as explained in earlier sections. To overcome this drawback, Rodgers et al. proposed two mechanistic models (Fig. 5.6), one for moderate to strong bases and type 1 zwitterions (at least one basic $pK_a > 7$)[13] and another for very weak bases, neutrals, acids, and type 2 zwitterions (no $pK_a > 7$).[13–15] Both of these address the preferential binding of a drug to the tissue components, averting the need for f_{uT}. Acidic drugs bind preferentially to albumin and lipophilic neutrals to lipoproteins. Both albumin and lipoproteins are present in appreciable amounts in the extracellular fluids in a tissue. On the other hand, a strong or moderate base binds preferentially to α_1-acid glycoprotein, or AGP, which is largely restricted to plasma. The concentration of a strong or moderate base in a tissue is then simply the sum of the amounts of the compound unbound in extracellular and intracellular water (ew and iw respectively), and the amounts bound to intracellular neutral lipids (nl), neutral phospholipid (npl), and acidic

Figure 5.6. Binding of (a) moderate to strong bases to intracellular neutral lipids (nl), neutral phospholipids (npl) and acidic phospholipids (apl) and (b) acids to nl, npl, and albumin. Weak bases also bind to albumin. Lipophilic neutrals on the other hand bind to extracellular lipoproteins (ECP).

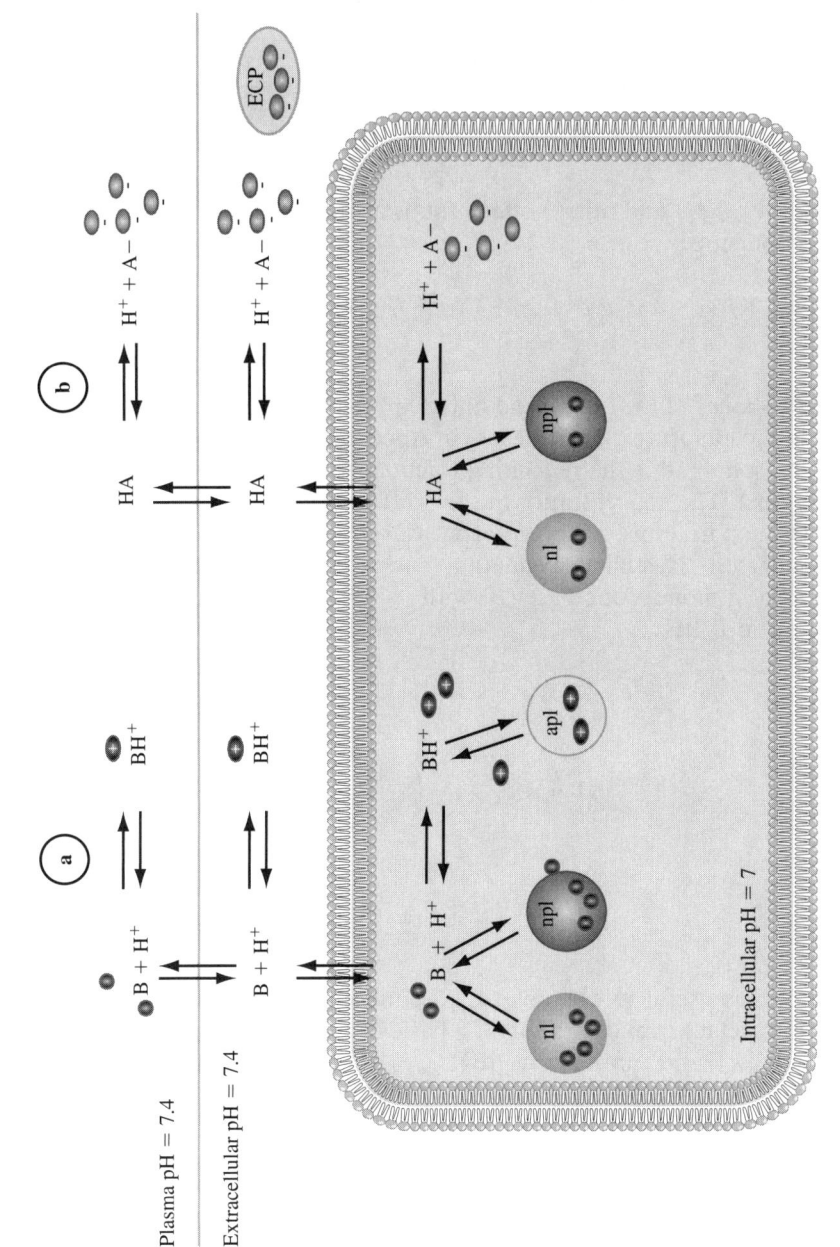

101

phospholipids (apl) (see Fig. 5.6) divided by the volume of the tissue (T). The amount of drug in a cellular component is the product of drug concentration in that component and the volume of that component. Therefore,

$$C_T = \frac{C_{u,\text{iwT}} \times V_{\text{iwT}} + C_{u,\text{ewT}} \times V_{\text{ewT}} + C_{\text{nlT}} \times V_{\text{nlT}} + C_{\text{nplT}} \times V_{\text{nplT}} + C_{\text{aplT}} \times V_{\text{aplT}}}{V_T}$$

(5.10)

Recognizing V_{iwT}/V_T and other volume ratios as the fractional volumes of the cellular components, we get

$$C_T = C_{u,\text{iwT}} \times f_{\text{iwT}} + C_{u,\text{ewT}} \times f_{\text{ewT}} + C_{\text{nlT}} \times f_{\text{nlT}} + C_{\text{nplT}} \times f_{\text{nplT}} + C_{\text{aplT}} \times f_{\text{rem}}$$

(5.11)

Fractional volumes of iw, ew, nl, and npl are physiological parameters that are available in the literature (see references in Appendices). The fractional volume of acidic phospholipid is the remaining fraction, f_{rem}, given by 1 − (sum of all other fractions). The concentrations in the different cellular components need to be substituted in terms of known or measurable parameters.

The Henderson–Hasselbalch equation (equation 5.13) can be applied to the equilibrium of a monoprotic base (B) with its ions (BH$^+$) to obtain the concentration ratio B/BH$^+$:

$$B + H^+ \leftrightarrow BH^+$$

(5.12)

$$\text{pH} = \text{p}K_a + \log\left(\frac{B}{BH^+}\right)$$

or

(5.13)

$$\frac{B}{BH^+} = 10^{\text{pH}-\text{p}K_a}$$

Equation 5.13 is useful in obtaining $C_{u,\text{iwT}}$ and $C_{u,\text{ewT}}$ in terms of pK_a. The unbound intracellular concentration of a basic drug in a tissue is the sum of its ionized ($[B]_{u,\text{iwT}}$) and un-ionized ($[BH^+]_{u,\text{iwT}}$) concentrations that remain unbound in the intracellular water of that tissue:

$$C_{u,\text{iwT}} = [B]_{u,\text{iwT}} + [BH^+]_{u,\text{iwT}}$$

(5.14)

Dividing throughout by [BH$^+$], applying the Henderson–Hasselbalch equation to substitute for B/[BH$^+$], and rearranging the resulting equation,

$$C_{u,\text{iwT}} = [BH^+]_{u,\text{iwT}} \times (1 + 10^{\text{pHiwT}-\text{p}K_a}) = B_{u,\text{iwT}} \times (1 + 10^{\text{p}K_a-\text{pHiwT}})$$ (5.15)

5.3 IN SILICO MODELS OF TISSUE PARTITION COEFFICIENTS

Similarly, the unbound concentration of a basic drug which is the sum of its ionized and unionized concentrations in plasma is given by

$$C_{up} = B_{up} \times (1 + 10^{pK_a - pHp}) \tag{5.16}$$

Dividing equations 5.15 by 5.16 and recognizing that the un-ionized, unbound concentrations of a base in plasma ($B_{u,p}$) and in the various tissues ($B_{u,iwT}$) are in equilibrium and are therefore equal, we get

$$C_{u,iwT} = C_{up} \times \left(\frac{1 + 10^{pK_a - pHiwT}}{1 + 10^{pK_a - pHp}} \right) \tag{5.17}$$

Thus, $C_{u,iwT}$ has now been obtained in terms of the pK_a of the compound and the pH of the medium. One can arrive at a similar equation for $C_{u,ewT}$. However, since the pH of the extracelluar water and plasma are the same (7.4), $C_{u,ewT}$ equals C_{up}.

The concentration of base bound to acidic phospholipids, C_{apIT}, is obtained from its equilibrium constant, K_{apIT}:

$$BH^+ + apl \leftrightarrow BH^+ : apl$$

$$K_{apIT} = \frac{C_{apIT}}{[BH^+]_{u,iwT} \times [apl]_{uT}} \tag{5.18}$$

where $[apl]_{uT}$ is the concentration of unbound acidic phospholipids in the tissue that is still available for binding and is given by $[apl]_T/f_{rem}$. $[apl]_T$ is the concentration of acidic phospholipids in the tissue. Substituting for $[BH^+]_{u,iwT}$ from equation 5.13 and recognizing that the equilibrium concentrations $[B]_{u,iwT}$ and $[B]_{up}$ are the same and substituting for $[B]_{up}$ using equation 5.16:

$$\begin{aligned} C_{apIT} &= \frac{K_{apIT} \times [apl]_{u,T} \times C_{up} \times 10^{pK_a - pHiwT}}{1 + 10^{pK_a - pHp}} \\ &= \frac{K_{apIT} \times ([apl]_T / f_{rem}) \times C_{up} \times 10^{pK_a - pHiwT}}{1 + 10^{pK_a - pHp}} \end{aligned} \tag{5.19}$$

The partitioning of the unbound, un-ionized base between neutral lipids and intracellular water can be equaled to partitioning between octanol and water. Using similar arguments as those used for equation 5.19,

$$C_{nIT} = P_{ow} \times [B]_{u,iwT} = P_{ow} \times [B]_{up} = \frac{P_{ow} \times C_{up}}{1 + 10^{pK_a - pHp}} \tag{5.20}$$

The concentrations of an un-ionized base in neutral phospholipids (which can be roughly regarded as similar to a combination of 30% neutral lipids and 70% water) is

$$C_{npIT} = 0.3 \times C_{nIT} + 0.7 \times [B]_{u,\,iwT}$$
$$= \frac{C_{up}}{1 + 10^{pK_a - pHp}} \times (0.3 P_{ow} + 0.7) \quad (5.21)$$

Now, substituting the concentrations of the drug in iw, ew, apl, and npl in various components back into equation 5.11,

$$C_T = C_{up} \left(\begin{array}{c} \left(\dfrac{1 + 10^{pK_a - pHiwT}}{1 + 10^{pK_a - pHp}} \right) \times f_{iwT} + f_{ewT} + \left(\dfrac{P_{ow}}{1 + 10^{pK_a - pHp}} \times f_{nIT} \right) \\ + \left(\dfrac{C_{up}}{1 + 10^{pK_a - pHp}} \times (0.3 P_{ow} + 0.7) \times f_{npIT} \right) \\ + \left(\dfrac{K_{apIT} \times ([apl]_T / f_{rem}) \times C_{up} \times 10^{pK_a - pHiwT}}{1 + 10^{pK_a - pHp}} \right) \end{array} \right)$$

(5.22)

From equation 5.22, $K_{Tp,u}$ can be obtained from C_T/C_{up}. K_{apIT} can be obtained by applying equation 5.22 to blood cells for which $f_{ewT} = 0$, as blood cells do not possess an extracellular space. $K_{blood\ cells\ p,u}$ is the blood plasma ratio $C_{blood\ cells}/C_{up}$. The $K_{Tp,u}$ for acids, given by equation 5.23,[13–15] can be derived with similar reasoning, defining X and Y as follows:

For very weak monoprotic bases:

$$X = 1 + 10^{pK_a - pHiw} \qquad Y = 1 + 10^{pK_a - pHp}$$

For monoprotic acids:

$$X = 1 + 10^{pHiw - pK_a} \qquad Y = 1 + 10^{pHp - pK_a}$$

For neutrals, since there is no ionization:

$$x = y = 1$$

$$K_{Tpu} = \frac{X f_{iwT}}{Y} + f_{ewT} + \left(\frac{P_{ow} \times f_{nIT} + (0.3 P_{ow} + 0.7) \times f_{npIT}}{Y} \right)$$
$$+ \frac{1}{f_{up}} - 1 - \left(\frac{P_{ow} \times f_{nlpT} + (0.3 P_{ow} + 0.7) \times f_{nplp}}{Y} \right) \times \frac{[protein]_T}{[protein]_p}$$

(5.23)

where [protein] refers to albumin or lipoprotein concentration in tissue or plasma. *In silico* models for tissue partitioning can only handle partitioning driven by the physicochemical nature of the drug or the protein/lipid/water composition of the tissues. They cannot address transporter-driven influx/

efflux of drugs and specialized binding to intracellular components, like the binding of chloroquine to DNA in the liver. Thus, partitioning into the brain is difficult to model as well as some drug- or organ-specific interactions.

A unified method to predict tissue partition coefficients for drugs and for environmental models has also been proposed.[16]

5.4 MEASUREMENT OF PARAMETERS REPRESENTING RATE AND EXTENT OF TISSUE DISTRIBUTION

In vitro, ex vivo, and *in vivo* methods to assess membrane permeability, transporter involvement in drug distribution, tissue partitioning, and binding to plasma and tissues are summarized in Table 5.2. The interstitial concentrations in the brain are among the most studied tissue concentrations, as transporter involvement makes it difficult to predict. The unbound brain interstitial fluid concentrations ($C_{u,brain}$) of a CNS-active drug are the pharmacologically relevant concentrations. For non-CNS drugs, it is important to ensure that $C_{u,brain}$ is kept to a minimum to avoid side effects.

5.4.1 Assessment of Rate and Extent of Brain Penetration

In silico models for BBB have been described in the literature.[17] Assays that help evaluate the permeability of a compound of interest across the BBB have been reviewed[18,19] in the literature.

Due to the tightness of the MDCK (Madine–Darby canine kidney) cell monolayers, *in vitro* permeability measured in these cell lines, serve as surrogate for brain permeability even though they are not endothelial cells originating from the brain.

In situ brain perfusion method measures the rate of brain penetration across brain endothelium in whole animals. The method utilizes catheterization of the common carotid artery in the anesthetized rat, together with ligation of the external carotid artery. The brain is then perfused via the internal carotid using an oxygenated physiological saline buffer containing the test substance. Once perfusion is complete, the brain is removed for analysis and the volume of distribution, V_d, within the brain is determined using the tissue and perfusate concentrations. The rate of uptake (k_{in}) is then simply V_d/perfusion time.

The brain–plasma ratio in preclinical species can be determined by administering the test compound and sampling the brain and plasma at different time intervals. $K_{brain,p}$ is then the ratio of areas under the brain and plasma concentration time profiles. This requires at least one animal per sampling time. Alternatively, the test compound is infused intravenously until steady state can be assumed and the brain and plasma sampled at one time point. $K_{brain,p}$ is calculated as the ratio of concentrations in brain and plasma at that time point. At each time point, the brain is homogenized and the total brain concentrations determined by liquid chromatography/mass spectrometry

TABLE 5.2. Methods to Determine the Unbound Volume, Rate, and Extent of Drug Distribution into Brain and Other Tissues

Measure	In Vitro Method	In Situ	In Vivo Method
Endothelial membrane permeability (rate of brain penetration)	Permeability in Madine Darby canine kidney (MDCK) cells.	In situ brain perfusion	
Transporter involvement (extent of drug distribution)	Efflux ratio in MDCK transfected cells		
Amount in brain (A_{brain}) and brain–plasma ratio ($K_{brain,p}$) (extent of total drug partitioning into brain)			Compound dosed in vivo to rodent and concentrations in homogenized tissues measured by LC/MS
Fraction unbound in brain ($f_{u,brain}$) (extent of drug distribution)	Equilibrium dialysis with brain homogenate; brain slices. $f_{u,brain}$ from equilibrium brain homogenate can be used along with V_{brain} to get $V_{u,brain}$ $V_{u,brain} = V_{brain}/f_{u,brain}$		
$V_{u,brain}$ and $C_{u,brain}$ in interstitial space as a function of time	Drug uptake in brain slice. $V_{u,brain}$ is the amount in slice divided by the buffer concentration at equilibrium. $C_{u,brain}$ = amount in brain (A_{brain}) from tissue homogenization / $V_{u,brain}$		$C_{u,brain}$ from brain microdialysis. $V_{u,brain}$ = amount in brain (A_{brain}) from tissue homogenization / $C_{u,brain}$

(LC/MS). The homogenization of the brain destroys the compartmental structures of the brain, and therefore the brain concentrations obtained by this method may not be pharmacologically relevant. The key compartment for a CNS drug is the brain interstitial fluid (Fig. 5.7a), and unbound concentrations in the interstitial fluid can be assessed by tissue microdialysis (Fig. 5.7b), following the administration of a test substance to a preclinical species. Microdialysis uses the principle of dialysis. A hollow-fiber semipermeable dialysis membrane or a microdialysis probe, permeable to water and small solutes, is introduced into the interstitial space. The membrane is perfused with a liquid (perfusate), which equilibrates with the fluid outside the membrane by diffusion. Following equilibration of the tissue with the perfusion fluid, the dialysate can be analyzed for concentrations of the test drug. Tissue

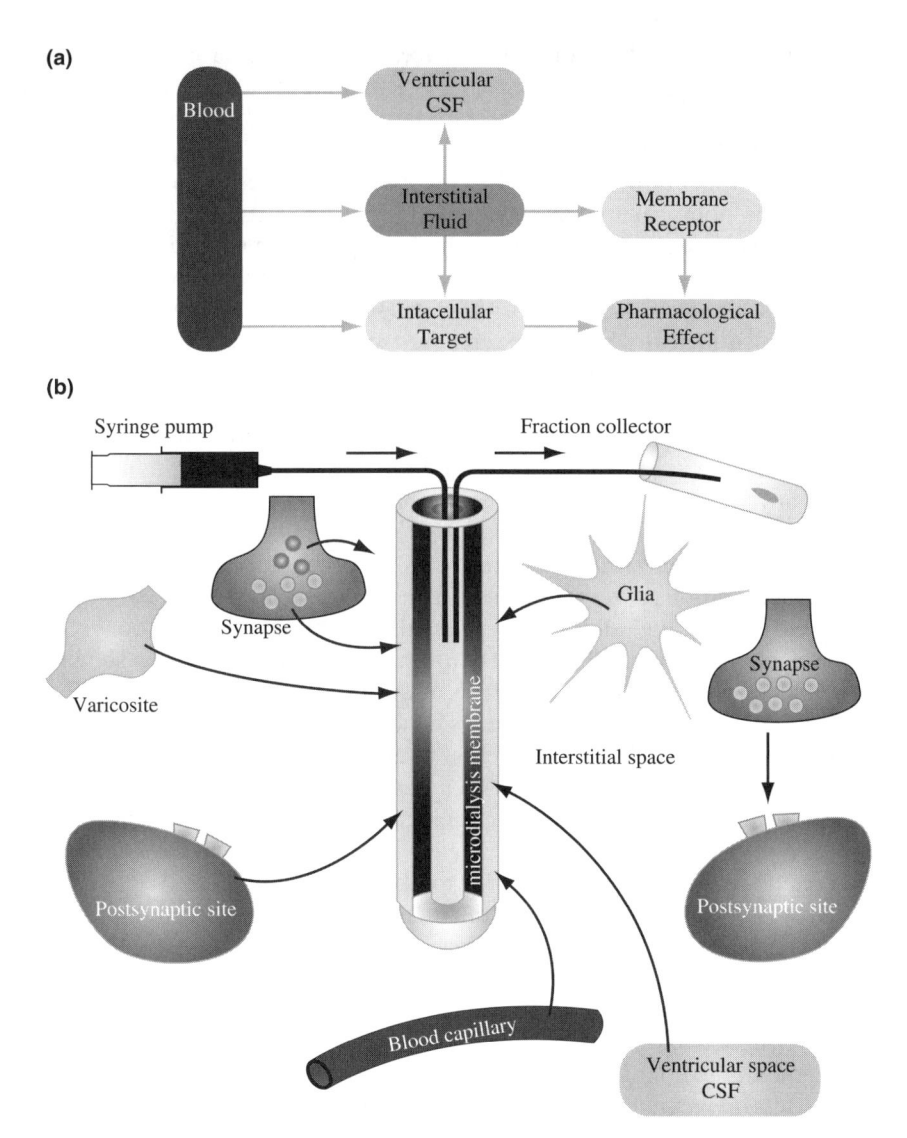

5.4 MEASUREMENT OF PARAMETERS REPRESENTING RATE

Figure 5.7. (a) Interstitial fluid is the key compartment for the pharmacological action of a compound targeting CNS. (b) Microdialysis.

microdialysis permits measurement of the unbound concentration of a compound in brain interstitial fluid over time, allowing C_{max}, $t_{1/2}$, and area under the concentration–time curve (AUC) to be calculated. Microdialysis also permits a level of anatomical precision that is not available with any other method, since the probe can be placed into the brain region of interest; this is important because systemically administered compounds do not distribute evenly across the various compartments of the brain, namely cerebral blood vessels, interstitial fluid, intracellular space, and the like. However, the use of microdialysis in drug discovery programs is limited by the time requirements and by specific technical difficulties with lipophilic drugs. Methods other than microdialysis that have been used for estimating the unbound volume of distribution in the brain ($V_{u,\text{brain}}$) include the brain slice uptake technique[20] and the brain homogenate binding method.[21,22] The fraction unbound in brain, $f_{u,\text{brain}}$, although useful in the estimation of $V_{u,\text{brain}}$,[23,24] has the limitation of the homogenate method, where no distinction is made between intra- and extracellular distribution. A comparison of these *in vitro* methods with microdialysis[25] identified the brain slice uptake technique in combination with the total brain amounts (see Fig. 5.8), as being the closest to microdialysis. To improve the throughput of the slice method for better utility in drug discovery, optimizing the dimensions of the incubation vessel, mode of stirring (to reduce the equilibration time), and the amount of brain tissue per unit of buffer volume have been suggested.[26] The authors have also recommended the use of cassette experiments for investigating $V_{u,\text{brain}}$.

Cerebrospinal fluid drug concentrations are potentially more closely related to the concentrations of unbound drug in brain interstitial fluid (ISF) ($C_{u,\text{brain}}$) as the ependymal lining of the ventricles allows diffusional and convectional exchange of CSF with the brain interstitium.[19] However, it has been shown that the CSF drug concentrations are not necessarily equal to those in brain ISF.[26–28]

The brain is a very special organ, as concentrations in the plasma and brain interstitial space can be very different due to the BBB. Other organs with brain–tissue barriers are the retina,[29] mammary glands, and testis. Techniques for measurement of $V_{u,\text{T}}$ and $C_{u,\text{T}}$ are specific to organs with brain–tissue barriers. However, methods that describe parameters such as K_{Tp} and $f_{u\text{T}}$ can be applied to other organs as well. A measurement of total drug concentration at steady state in the tissue is all that is needed to estimate K_{Tp} for organs lacking brain–tissue barriers. Unless transporters play a key role in determining intracellular concentrations in a tissue, the interstitial and intracellular concentrations are likely to be the same as unbound drug concentrations in plasma at steady state and K_{Tp} is given by equation 5.5. Therefore, a measurement of $f_{u\text{T}}$ and f_{up} should also provide K_{Tp}. Eliminating organs such as the liver and kidney have uptake and efflux drug transporters on parenchymal cells, facilitating the drug uptake and toxin removal, but have fenestrated capillary. In these organs, the intracellular unbound drug concentrations can be different from that in interstitial space. The unbound concentrations in the interstitial space and plasma are, however, the same. Thus, K_{Tp}, which is still the ratio of total concentrations in tissue and plasma is no longer equal to the ratio of unbound fractions in plasma and tissue. There are many sophisticated, expensive techniques to measure total tissue concentrations[30] at the target tissue. Most of these are employed in drug development rather than in drug discovery. Some common techniques are outlined below.

The **whole-body autoradiography (WBA)**[31,32] is an imaging technique used to determine the tissue distribution of ^{14}C or ^{3}H radiolabeled compounds in preclinical species. In earlier experiments, quantification was difficult to achieve due to the limited linear response of the films used to capture autoradiographic

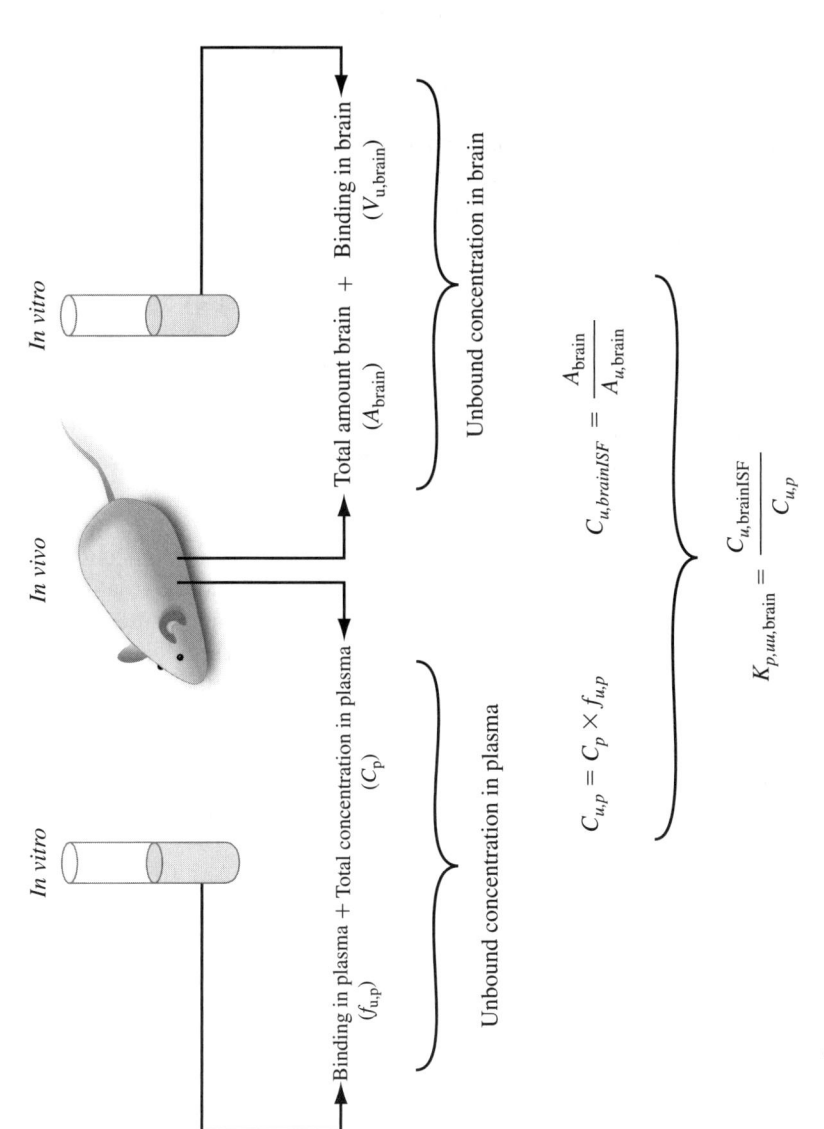

Figure 5.8. Brain slice uptake technique in combination with the total brain amounts (A_{brain}) from tissue homogenization will give the unbound brain concentrations in the interstitial fluid (ISF). Combining this with the unbound drug plasma concentration will yield $K_{Tp,u,u}$. (Figure kindly donated by Marcus Friden, Discovery DMPK, AstraZeneca R&D, Mölndal, Sweden.)

images. More recently, WBA is used in combination with phosphor imaging technologies. This quantitative WBA can provide reliable, quantitative tissue distribution data. Animals are dosed intravenously or orally with a single dose of test compound such that each animal receives 200–400 μCi of ^{3}H or 10–60 μCi of ^{14}C. The rate and extent of tissue distribution are then obtained from the time-dependent data on radioactivity. A major drawback of this technique is the lack of specificity, as it does not distinguish between the radioactivity from a drug and its metabolites.

Matrix–assisted laser desorption/ionization–imaging mass spectrometry (MALDI–IMS)[33–35] is a label-free, high-resolution technique that allows location-specific measurement of endogenous or exogenous molecules, their metabolites, and their targets.

The methods described so far in this section are all in preclinical species. Although tissue partitioning is likely to be similar across species due to similar tissue compositions, species differences in drug distribution could arise from differences in membrane permeability, drug influx/efflux transporters, and tissue binding. The study of biodistribution of drugs in humans then becomes important. **Positron emission tomography (PET)** is a noninvasive imaging technique that can be used to study biodistribution following the microdosing of humans.

Positron emission tomography microdosing[36] is a noninvasive tomographic imaging method that relies on the administration (usually as an IV injection) of a low dose of a drug labeled with a positron-emitting radionuclide to a human volunteer. Positron emitted from the labeled drug as it distributes to the different tissues is eventually annihilated by an electron capture. The resultant γ photons are externally detected and visualized as tomographic images of the drug's distribution *in vivo*. The half-life of most positron-emitting nuclides are short (e.g., ^{11}C has a half-life of 20 min), making exposure of humans to radioactivity minimal. This expensive technique is currently used in pharmaceutical companies to carry out clinical studies in humans in order to ensure safety and efficacy of CNS drugs.

Drug properties impacting tissue distribution and derived from *in silico* models or from experiments can be integrated into physiological models to obtain realistic measure of tissue distribution of the drug *in vivo*.

5.5 PHYSIOLOGICAL MODEL FOR DRUG DISTRIBUTION

For a non-eliminating organ other than the lung, the rate of change of drug concentration (C_T) in the tissue, T, is (dC_T/dt) and is described by the following differential equation:

$$\frac{dC_T}{dt} = \frac{1}{V_T}\left(Q_{ART} \times C_{ART} - Q_{VEN} \times \frac{C_T \times f_{uT} \times R}{f_{up} \times K_{Tp,uu}}\right) \quad (5.24)$$

where V_T is the volume of the tissue, Q_{ART} is the blood flow rate into the tissue, and Q_{VEN} is the blood flow rate out of the tissue. These physiological parameters are available in the literature for common preclinical species and for humans (see references in Appendices). The term $K_{Tp,uu} \times f_{up}/f_{uT}$ is K_{Tp}. Dividing the K_{Tp} by R, the blood–plasma ratio, gives the tissue–blood partition coefficient in that tissue. The first term in the above equation represents the rate of increase in the drug concentrations arriving from the arterial blood, and the second term represents the rate of decrease in the drug concentrations due to the draining venous blood, after distribution of the drug into the tissue, dictated by K_{Tp}. Methods to determine f_{uT}, R, and $K_{Tp,uu}$ have been described in previous sections. Alternatively, the use of the Rodgers and Rowland equation circumvents the need to use f_{uT}:

$$\frac{dC_T}{dt} = \frac{1}{V_T}\left(Q_{ART} \times C_{ART} - Q_{VEN} \times \frac{C_T \times R}{f_{up} \times K_{Tp,u}}\right) \quad (5.25)$$

Equations 5.24 and 5.25 apply only to drugs whose distribution into tissues is limited by the blood flow rate to the tissue, that is, those that are perfusion rate limited, which implies that the drug diffuses into the interstitial space and into intracellular space without any difficulty—the endothelial membrane offers practically no resistance to the diffusion of the drug across it. While this is true for many lipophilic small-molecule drugs, the distribution of large hydrophilic drugs like proteins and other macromolecules into tissues is limited by their permeability across the endothelial barrier. They are said to have membrane-permeability-limited distribution kinetics, which can best be described by further sub-compartmentalization of a tissue compartment (Fig. 5.9). Sub-compartmentalization is also required for tissues such as the brain, which cannot be considered well stirred on account of the several barriers to uniform distribution, such as the presence of transporters and tight junctions in the vasculature. Sub-compartmentalization of tissues will be considered for biologics in Chapter 10.

5.6 DRUG CONCENTRATIONS AT SITE OF ACTION

The concentration of a drug at the target protein in a target tissue (site of action) is the one that generally drives pharmacological response. If the targeted protein is localized on the cell surface, such as a membrane receptor, ion channel, or a transporter, the unbound interstitial concentration at the target organ is the most relevant. For cytoplasmic receptors and enzymes, as well as for proteins at specific cell organelles, it is the unbound intracellular drug concentration that is relevant. In tissues such as the brain, kidney, and liver, where interstitial and intracellular drug concentrations are driven by transporters, plasma drug concentration is a poor surrogate measure of tissue

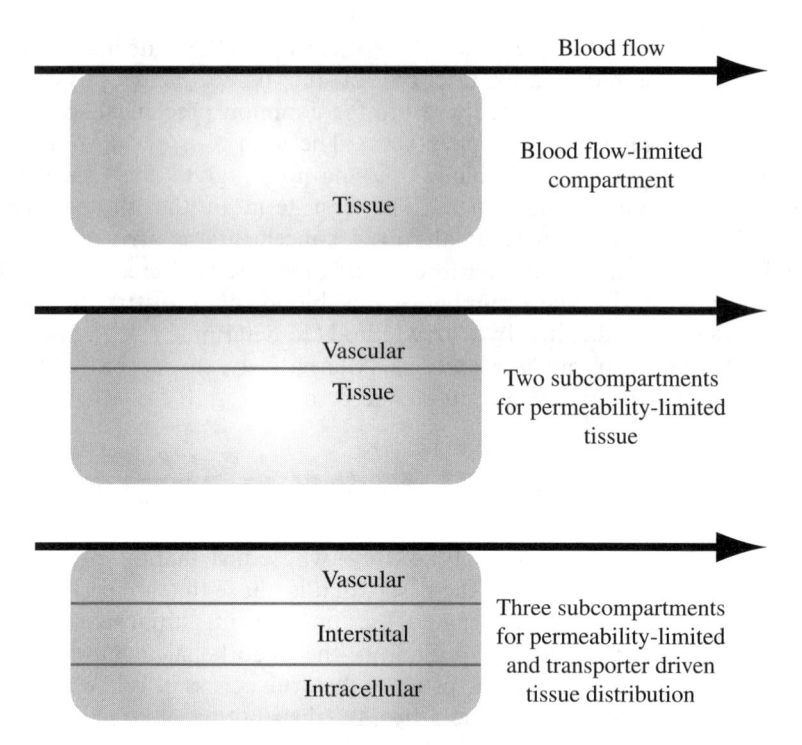

Figure 5.9. Compartmentalization of a tissue.

concentration, and linking it to PD data may not always correlate to the observed pharmacological response. In this case, unbound partition coefficients derived from animal experiments can be applied for the PBPK simulation of human PK and tissue concentrations (Fig. 5.10). Even in tissues such as the liver and kidney, where transporters can hugely influence interstitial and intracellular concentrations, functional similarity of transporters across species can be assumed. However one hurdle remains—obtaining a reliable estimate of f_{uT} for humans. $f_{uT,human}$ can be obtained either from $f_{uT,preclinical}$ or from $f_{up,human}$. Depending on the tissue of interest, $f_{uT,human}$ from $f_{uT,preclinical}$ or from $f_{up,human}$ may be more appropriate. The brain, for example, has very different lipid and protein composition compared to plasma, as plasma is known to contain twice as much protein as brain and the brain has 20-fold more lipid compared to plasma. In this case, the use of $f_{uT,preclinical}$ may be more appropriate.[37] However, as pointed out in Section 5.4, the method of brain–plasma ratio in combination with f_{uT} does not consider the different compartmentalization in the brain and, therefore, the brain slice uptake method should be combined with the brain–plasma ratio instead. The time profiles of tissue concentrations obtained from human PBPK simulations of NCEs can be combined with the pharmacodynamic (PD) data to provide a better PK/PD relationship than what might result from the use of plasma drug concentrations.

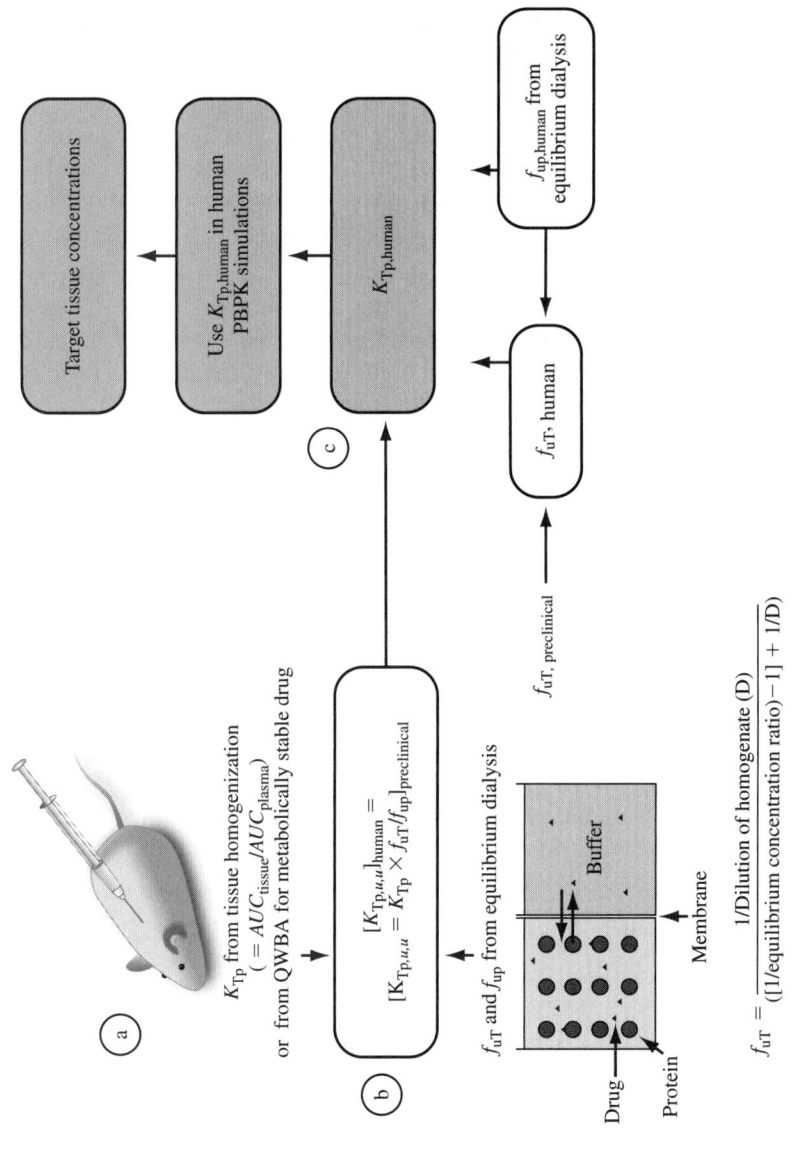

Figure 5.10. Scheme illustrating the determination of target tissue concentrations: (a) K_{Tp} is experimentally determined for a preclinical species. (b) $K_{Tp,u,u}$ is assumed to be the same for different species. (c) $K_{Tp,human}$ is then obtained from $K_{Tp,u,u,preclinical}$ using human-specific f_{ut} and f_{up}.

$$f_{uT} = \frac{1/\text{Dilution of homogenate (D)}}{([1/(\text{equilibrium concentration ratio})-1] + 1/D)}$$

KEYWORDS

Apparent Volume of Distribution: Volume of fluid required to contain all of a drug in the body at the same concentration as observed in plasma.

Blood–Brain Barrier: Protective network of blood vessels and cells that filters blood flowing to the brain, thereby limiting the exchange of drugs between the brain and the peripheral blood.

Blood–Plasma Ratio: Ratio of concentrations of drug in blood and plasma. It is a measure of the extent of binding of a drug to erythrocytes and is equal to 1 if the binding to plasma and erythrocyte proteins are the same. Lipophilic bases tend to distribute into erythrocytes to a large extent.

Discontinuous Capillaries: Also called sinusoidal capillaries, they are located in the liver, spleen, bone marrow, lymph nodes, and adrenal gland. They are a special type of fenestrated capillaries characterized by larger openings (30–40 μm in diameter) in the endothelium, allowing red and white blood cells (7.5–25 μm diameter) to pass through them.

Distribution Coefficient (D): Ratio of the concentrations of all ionized and unionized forms of the compound in each of the two phases. Since the extent of ionized form will depend on the pK_a of the compound and the pH of the medium containing the compound, measurements of D are done such that the pH of the aqueous phase is buffered to a specific value, usually the physiological pH of 7.4. For nonionizable compounds, $P = D$ at any pH.

Matrix–Assisted Laser Desorption/Ionization–Imaging Mass Spectrometry (MALDI-IMS): Powerful tool for investigating the distribution of proteins and small molecules within biological systems through the *in situ* analysis of tissue sections. MALDI-IMS can determine the distribution of hundreds of unknown compounds in a single measurement and enables the acquisition of cellular expression profiles while maintaining the cellular and molecular integrity. Many advances in recent years have made the technique more sensitive, robust, and useful.

Partition Coefficient (P): Partition coefficient of a compound is the ratio of concentrations of unionized compound between the two liquid phases. It is generally measured at a neutral pH when the unionized form is expected to be predominant.

Positron Emission Tomography (PET) Microdosing: Noninvasive tomographic imaging method that relies on the administration of a low dose (microdose) of a drug labeled with a positron-emitting radionuclide (^{11}C) to a human volunteer.

Whole-Body Autoradiography (WBA): Traditionally uses X-ray films for the detection of radioactive-labeled compounds in a preclinical species. Quantification of radioactivity has been hampered by the limited linear range of X-ray films and the laborious densitometry measurements of the exposed films. Radioluminography (RLG) enables a fast and reliable quantification of the radioactivity distribution in whole-body sections.

REFERENCES

1. van de Waterbeemd H, Camenisch G, Folkers G, Chretien JR, Raevsky OA. Estimation of blood-brain barrier crossing of drugs using molecular size and shape, and H-bonding descriptors. *J Drug Target.* 1998;6(2):151–165.
2. Friden M, et al. Structure-brain exposure relationships in rat and human using a novel data set of unbound drug concentrations in brain interstitial and cerebrospinal fluids. *J Med Chem.* 2009;52(20):6233–6243.
3. Ajay, Bemis GW, Murcko MA. Designing libraries with CNS activity. *J Med Chem.* 1999;42(24):4942–4951.
4. Hammarlund-Udenaes M, Bredberg U, Friden M. Methodologies to assess brain drug delivery in lead optimization. *Curr Top Med Chem.* 2009;9(2):148–162.
5. Hammarlund-Udenaes M, Friden M, Syvanen S, Gupta A. On the rate and extent of drug delivery to the brain. *Pharm Res.* 2008;25(8):1737–1750.
6. Reichel A. Addressing central nervous system (CNS) penetration in drug discovery: Basics and implications of the evolving new concept. *Chem Biodivers.* 2009;6(11): 2030–2049.
7. Eyal S, Hsiao P, Unadkat JD. Drug interactions at the blood-brain barrier: Fact or fantasy? *Pharmacol Ther.* 2009;123(1):80–104.
8. Wang DS, Kusuhara H, Kato Y, Jonker JW, Schinkel AH, Sugiyama Y. Involvement of organic cation transporter 1 in the lactic acidosis caused by metformin. *Mol Pharmacol.* 2003;63(4):844–848.
9. Yokogawa K, Ishizaki J, Ohkuma S, Miyamoto K. Influence of lipophilicity and lysosomal accumulation on tissue distribution kinetics of basic drugs: A physiologically-based pharmacokinetic model. *Methods Find Exp Clin Pharmacol.* 2002;24(2): 81–93.
10. Hanley MJ, Abernethy DR, Greenblatt DJ. Effect of obesity on the pharmacokinetics of drugs in humans. *Clin Pharmacokinet.* 2010;49(2):71–87.
11. Cheymol G. Effects of obesity on pharmacokinetics implications for drug therapy. *Clin Pharmacokinet.* 2000;39(3):215–231.
12. Poulin P, Theil FP. Prediction of pharmacokinetics prior to in vivo studies. 1. Mechanism-based prediction of volume of distribution. *J Pharm Sci.* 2002;91(1): 129–156.
13. Rodgers T, Leahy D, Rowland M. Physiologically-based pharmacokinetic modeling 1: Predicting the tissue distribution of moderate-to-strong bases. *J Pharm Sci.* 2005;94(6):1259–1276.
14. Rodgers T, Rowland M. Physiologically-based pharmacokinetic modelling 2: Predicting the tissue distribution of acids, very weak bases, neutrals and zwitterions. *J Pharm Sci.* 2006;95(6):1238–1257.
15. Rodgers T, Leahy D, Rowland M. Tissue distribution of basic drugs: Accounting for enantiomeric, compound and regional differences amongst beta-blocking drugs in rat. *J Pharm Sci.* 2005;94(6):1237–1248.
16. Peyret T, Poulin P, Krishnan K. A unified algorithm for predicting partition coefficients for PBPK modeling of drugs and environmental chemicals. *Toxicol Appl Pharmacol.* 2010.

17. Goodwin JT, Clark DE. In silico predictions of blood-brain barrier penetration: Considerations to "keep in mind." *J Pharmacol Exp Ther*. 2005;315(2):477–483.
18. Abbott NJ, Dolman DE, Patabendige AK. Assays to predict drug permeation across the blood-brain barrier, and distribution to brain. *Curr Drug Metab*. 2008;9(9):901–910.
19. Abbott NJ. Prediction of blood-brain barrier permeation in drug discovery from in vivo, in vitro and in silico models. *Drug Discovery Today: Technologies*. 2004;1:407.
20. Kakee A, Terasaki T, Sugiyama Y. aBrain efflux index as a novel method of analyzing efflux transport at the blood-brain barrier. *J Pharmacol Exp Ther*. 1996;277(3):1550–1559.
21. Kalvass JC, Maurer TS. Influence of nonspecific brain and plasma binding on CNS exposure: Implications for rational drug discovery. *Biopharm Drug Dispos*. 2002;23(8):327–338.
22. Mano Y, Higuchi S, Kamimura H. Investigation of the high partition of YM992, a novel antidepressant, in rat brain—in vitro and in vivo evidence for the high binding in brain and the high permeability at the BBB. *Biopharm Drug Dispos*. 2002;23(9):351–360.
23. Becker S, Liu X. Evaluation of the utility of brain slice methods to study brain penetration. *Drug Metab Dispos*. 2006;34(5):855–861.
24. Liu X, et al. Evaluation of cerebrospinal fluid concentration and plasma free concentration as a surrogate measurement for brain free concentration. *Drug Metab Dispos*. 2006;34(9):1443–1447.
25. Friden M, Gupta A, Antonsson M, Bredberg U, Hammarlund-Udenaes M. In vitro methods for estimating unbound drug concentrations in the brain interstitial and intracellular fluids. *Drug Metab Dispos*. 2007;35(9):1711–1719.
26. Friden M, Ducrozet F, Middleton B, Antonsson M, Bredberg U, Hammarlund-Udenaes M. Development of a high-throughput brain slice method for studying drug distribution in the central nervous system. *Drug Metab Dispos*. 2009;37(6):1226–1233.
27. de Lange EC, Danhof M. Considerations in the use of cerebrospinal fluid pharmacokinetics to predict brain target concentrations in the clinical setting: Implications of the barriers between blood and brain. *Clin Pharmacokinet*. 2002;41(10):691–703.
28. Shen DD, Artru AA, Adkison KK. Principles and applicability of CSF sampling for the assessment of CNS drug delivery and pharmacodynamics. *Adv Drug Deliv Rev*. 2004;56(12):1825–1857.
29. Tomi M, Hosoya K. The role of blood-ocular barrier transporters in retinal drug disposition: An overview. *Expert Opin Drug Metab Toxicol*. 2010;6(9):1111–1124.
30. Pelkonen O, Kapitulnik J, Gundert-Remy U, Boobis AR, Stockis A. Local kinetics and dynamics of xenobiotics. *Crit Rev Toxicol*. 2008;38(8):697–720.
31. Solon EG, Balani SK, Lee FW. Whole-body autoradiography in drug discovery. *Curr Drug Metab*. 2002;3(5):451–462.
32. Jacob S, Ahmed AE. Effect of route of administration on the disposition of acrylonitrile: Quantitative whole-body autoradiographic study in rats. *Pharmacol Res*. 2003;48(5):479–488.

REFERENCES

33. Khatib-Shahidi S, Andersson M, Herman JL, Gillespie TA, Caprioli RM. Direct molecular analysis of whole-body animal tissue sections by imaging MALDI mass spectrometry. *Anal Chem.* 2006;78(18):6448–6456.
34. Stoeckli M, Staab D, Schweitzer A. Compound and metabolite distribution measured by MALDI mass spectrometric imaging in whole-body tissue sections. *Int. J. Mass Spec.* 2007;260:195.
35. Nilsson A, et al. Fine mapping the spatial distribution and concentration of unlabeled drugs within tissue micro-compartments using imaging mass spectrometry. *PLoS One.* 2010;5(7):e11411.
36. Bergstrom M, Grahnen A, Langstrom B. Positron emission tomography microdosing: A new concept with application in tracer and early clinical drug development. *Eur J Clin Pharmacol.* 2003;59(5–6):357–366.
37. Summerfield SG, et al. Toward an improved prediction of human in vivo brain penetration. *Xenobiotica.* 2008;38(12):1518–1535.

6

PHYSIOLOGICAL MODELS FOR DRUG METABOLISM AND EXCRETION

CONTENTS

6.1 Introduction 119
6.2 Factors Affecting Drug Metabolism and Excretion of Xenobiotics . 120
6.3 Models for Hepatobiliary Elimination and Renal Excretion 124
    6.3.1 *In Silico* Models 124
    6.3.2 *In Vitro* Models for Hepatic Metabolism 125
    6.3.3 *In Vitro* Models for Transporters 127
6.4 Physiological Models 136
    6.4.1 Hepatobiliary Elimination of Parent Drug and Metabolites . 136
    6.4.2 Renal Excretion............................. 141
    References................................... 144

6.1 INTRODUCTION

The rate and extent of metabolism and excretion of an administered drug is central to its pharmacokinetics. Drug metabolism refers to the enzymatic

---

Physiologically-Based Pharmacokinetic (PBPK) Modeling and Simulations: Principles, Methods, and Applications in the Pharmaceutical Industry, First Edition. Sheila Annie Peters.
© 2012 John Wiley & Sons, Inc. Published 2012 by John Wiley & Sons, Inc.

biotransformation of a drug, the likely consequences of which are accelerated renal excretion, drug activation or inactivation, and increased or decreased safety. Excretion of a drug refers to the removal of a drug and its metabolites from the body and can occur either via the biliary or renal route, depending on the physicochemical and structural properties of the drug. Several *in vitro* and animal *in vivo* models in the drug discovery phase are focused on metabolism and excretion. These models aim to provide an understanding of metabolic stability, drug-metabolizing enzymes (DMEs) involved in the metabolism, metabolites formed, role of transporters in metabolism, and drug–drug interaction potential of a new chemical entity (NCE). *In vivo* animal models allow estimation of the fraction of an NCE excreted in bile and urine. As with other PK properties, drug metabolism and excretion are influenced by both compound properties and the physiology of the species. In addition, enzymology plays a vital role in determining metabolism and this can vary significantly across different species. Physiologically-based pharmacokinetic models enable the integration of all available measures of metabolism and excretion in a physiological context, in order to provide realistic estimates of human exposure in the target organ. This chapter will discuss the factors affecting drug metabolism and excretion, provide an overview of the methods available to determine metabolism and excretion parameters, and enable an understanding of the integration of measured parameters into physiology-based models. The differential equations necessary to tackle linear and nonlinear drug elimination will be outlined.

6.2 FACTORS AFFECTING DRUG METABOLISM AND EXCRETION OF XENOBIOTICS

The rate and extent of drug metabolism and excretion are influenced by all of the physiology- and compound-dependent factors that determine the rate and extent of tissue distribution. These were outlined in Chapter 5. The greater the distribution of a drug into eliminating organs such as the liver and kidney, the greater is the probability of its metabolism or excretion. The liver and kidney are highly perfused organs. This guarantees a high rate of delivery of xenobiotics to these organs for small, lipophilic compounds with good permeability. However, for large or hydrophilic compounds, the membrane permeability rather than blood flow rate to organ could limit the rate of delivery. In the liver, good distribution of the drug into hepatocytes ensures better access to DMEs. Further, access to DMEs is also limited by plasma protein binding, as only the unbound drug is available for distribution into cells and, therefore for metabolism or biliary excretion. Similarly, only the unbound drug is available for renal filtration.

In the liver, once the drug becomes available within the hepatocytes, the extent of drug metabolism is determined by enzymology and transporters. The hepatocytes contain a variety of DMEs (Fig. 6.1) that can metabolize

Phase I Metabolism

- **Oxidation:** Cytochrome P450 (**CYPs** 1A2, 2A6*, 2C8*, 2C9*, 2C19*, 2D6*, 2E1, 3A4, 3A5*) Aldehyse dehydrogenase (**ALDH**) Monoamine oxidase flavinmonoamine oxidase (**FMO**)
- **Reduction:** Azo and nitro reductase
- **Hydrolysis:** Esterases (**hCE1** and **hCE2**), amidases, epoxide hydrolase

Phase II Metabolism

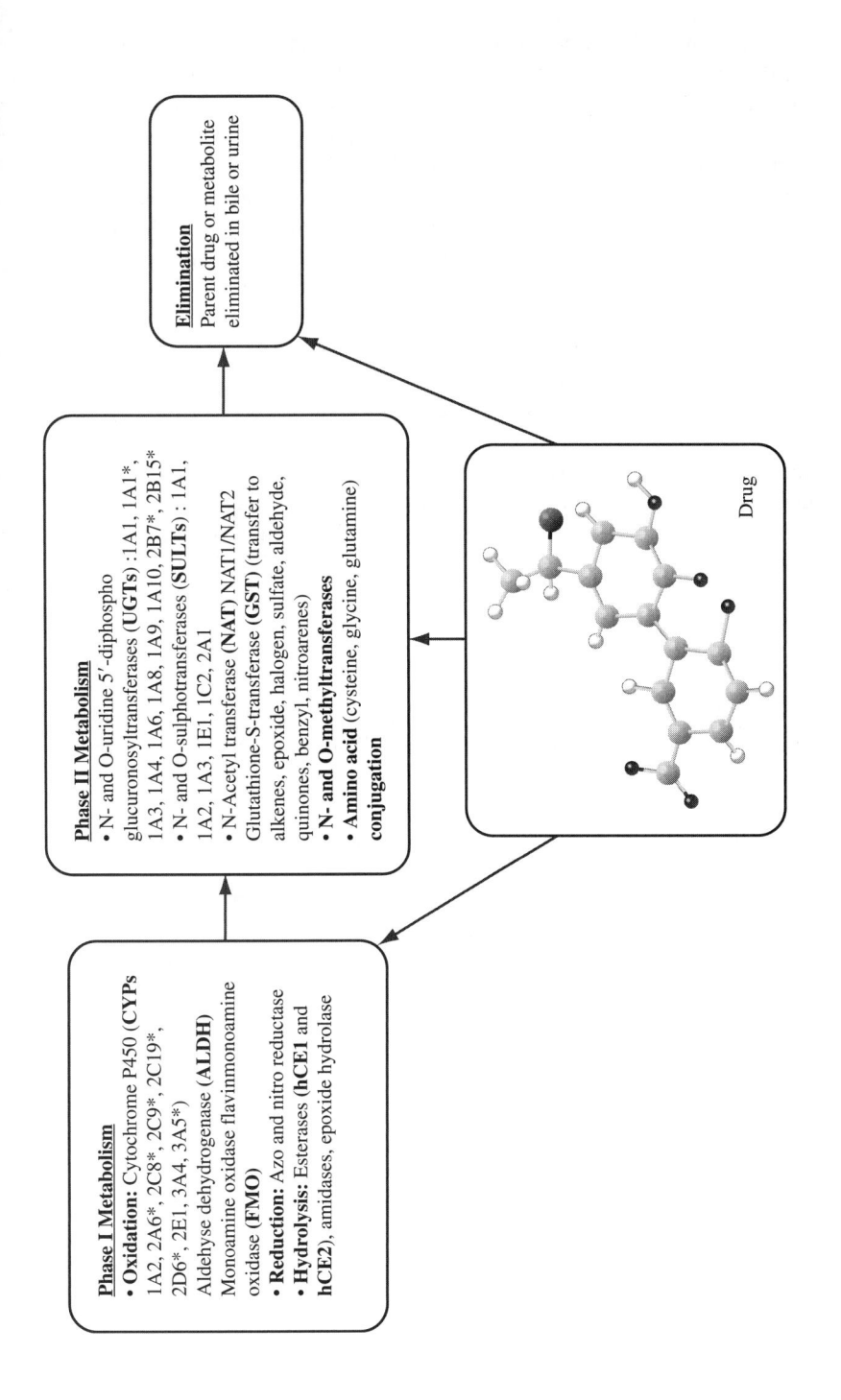

- N- and O-uridine 5′-diphospho glucuronosyltransferases (**UGTs**) :1A1, 1A1*, 1A3, 1A4, 1A6, 1A8, 1A9, 1A10, 2B7*, 2B15*
- N- and O-sulphotransferases (**SULTs**) : 1A1, 1A2, 1A3, 1E1, 1C2, 2A1
- N-Acetyl transferase (**NAT**) NAT1/NAT2 Glutathione-S-transferase (**GST**) (transfer to alkenes, epoxide, halogen, sulfate, aldehyde, quinones, benzyl, nitroarenes)
- **N- and O-methyltransferases**
- **Amino acid** (cysteine, glycine, glutamine) **conjugation**

Elimination
Parent drug or metabolite eliminated in bile or urine

Drug

Figure 6.1. Fate of a xenobiotic. Molecules can directly undergo phase I metabolism, phase II conjugation, or elimination. The enzymes indicated by an asterisk are polymorphic enzymes.

xenobiotics through a phase I and/or phase II metabolism. The capacity of an enzyme to catalyze a reaction (see Chapter 3) is described by its intrinsic clearance ($CL_{int,met}$), defined as the rate of an enzyme-catalyzed reaction divided by the substrate concentration at the site of the enzyme. In addition to enzymes, transporters can play an important role in determining the rate and extent of metabolism of ionizable compounds, whose membrane permeability may be poor. Uptake/efflux transporters can regulate the intracellular concentrations of their substrates. The OATPs, OAT2, and OCT1 are the principal uptake transporters in the liver for organic anions and cations. The ATP-binding cassette (ABC) transporters are a family of membrane proteins that mediate efflux in various organs, including the intestine, liver, kidney, and blood–brain barrier. These transporters have ATPase activity and mediate ATP-dependent transport of various drugs, fats, sterols, and endogenous metabolites. Forty-eight ABC transporters have so far been identified, and approximately 11 are known to transport compounds across cell membranes. Equation 6.1 summarizes the effects of uptake, diffusion, metabolism, and efflux on the total intrinsic clearance of the substrate from liver ($CL_{int,LI}$):

$$CL_{int,Li} = \frac{CL_{int,uptake} + CL_{int,diffusion}}{CL_{int,met} + CL_{int,diffusion} + CL_{int,efflux} + CL_{int,bile}} \times (CL_{int,met} + CL_{int,bile}) \quad (6.1)$$

Where $CL_{int,uptake}$, $CL_{int,diffusion}$, $CL_{int,met}$, $CL_{int,efflux}$, and $CL_{int,bile}$ represent the intrinsic clearances corresponding to uptake, diffusion, metabolism, sinusoidal efflux, and biliary excretion (Fig. 6.2). Although equation 6.1 has not been derived here, it can be understood by recognizing that the numerator and denominator of the ratio preceding the sum ($CL_{int,met} + CL_{int,bile}$), represent the rates of appearance and disappearance from the intracellular space in hepatocytes, respectively. This ratio, when multiplied by the total rate of elimination from the intracellular space in hepatocytes ($CL_{int,met} + CL_{int,bile}$), gives the net rate of elimination from the liver. According to equation 6.1, if the sum of the rates of the metabolic and biliary clearances is high, then the organ clearance will be limited by the sum $CL_{int,uptake} + CL_{int,diffusion}$. Uptake transporters have the potential to alter the intracellular disposition of their substrates only if influx (transported-mediated uptake and diffusion) is the rate-limiting step and $CL_{int,diffusion}$ is low. The functional groups in a molecule heavily influences the substrate selectivity of various DMEs and transporters and thereby the rates of metabolism as well as renal and biliary excretion. DMEs and transporters could show large affinity and capacity variations across different species that need to be considered during extrapolation to humans.

Transporters also play an important role in the rate and extent of renal excretion. Equation 6.1 can be suitably modified to apply to the kidney. Uptake transporters (OATs, OCTs, OATPs, and PEPTs) and efflux transporters (P-gp, MRP2, and MRP4) are expressed in the proximal tubule of the nephron to regulate the uptake and secretion of endogenous and exogenous organic

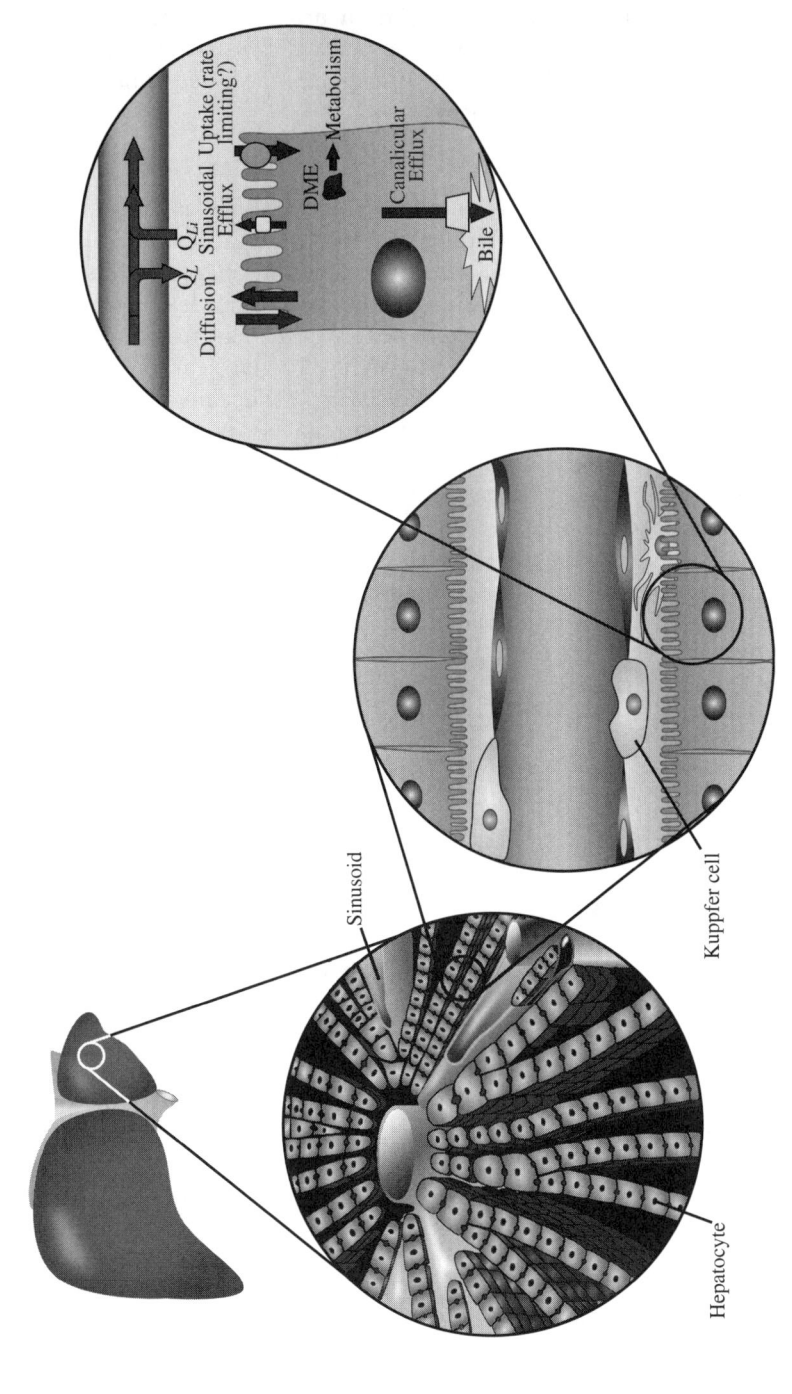

Figure 6.2. Cellular processes in hepatocytes that determine the disposition of a xenobiotic. DME: drug metabolizing enzyme.

anions, cations, peptides, nucleosides and a broad range of drugs and their metabolites. An expression for renal clearance (CL_r) that considers the combined effect of glomerular filtration, transporter-mediated tubular secretion, and passive reabsorption is given by

$$CL_r = (f_{ub} \times \text{GFR} + CL_{sec}) \times (1 - F_r) \qquad (6.2)$$

where f_{ub} is the fraction of drug unbound in blood, GFR is the glomerular filtration rate, F_r is the fraction of drug reabsorbed, and CL_{sec} is the clearance rate of renal secretion. This equation reflects the fact that only the unbound drug is filtered through the glomerulus and that renal clearance is the fraction of the filtered and secreted drug that is not reabsorbed. The reabsorption process through the length of the nephron results in lower renal clearance than expected based on glomerular filtration ($CL_r < \text{GFR} \times f_{ub}$), while involvement of uptake and efflux transporters will result in a higher renal clearance than $\text{GFR} \times f_{ub}$.

A variety of demographic and genetic factors, pathological conditions, coadministration of other drugs, alcohol consumption, and nutritional status can modulate the levels of active DMEs and drug transporters as well as other physiological factors impacting the rates of metabolism and excretion. Consequently, the disposition of a compound in systemic circulation is highly variable across different populations. Chronic renal failure can affect disposition of drugs that are predominantly cleared renally. These variabilities result in adverse reactions or loss of efficacy, especially for drugs with a narrow therapeutic window and will be discussed in detail in Chapter 9.

6.3 MODELS FOR HEPATOBILIARY ELIMINATION AND RENAL EXCRETION

6.3.1 *In Silico* Models

In silico models for drug metabolism include predicting metabolic stability or identifying the substrate/inhibitor specificity for different DMEs, sites of metabolism in a molecule, and predicting the likelihood for an NCE to be an inhibitor or inducer of drug metabolism.

Quantitative structure–activity relationship (QSAR) models and molecular modeling approaches for predicting metabolic stability offer a possible alternative to *in vitro* metabolism studies. *In silico* approaches to predicting metabolism can be divided into QSAR and three-dimensional (3D) QSAR[1] studies, protein- and pharmacophore[2]-based modeling, and predictive databases that can be used as rapid filters for the screening of virtual libraries, for example, to test for CYP3A4 liability.[3] The availability of crystal structures of the mammalian cytochrome P450s (CYPs) has prompted structure-based modeling studies. Early predictions of the vulnerability to metabolism of certain positions in the molecule might help in the drug design stage to eliminate metabolic liabilities.[4–9] These models are based on a pharmacophore representation obtained from interaction fields for the protein structure and a pharmacophoric fingerprint for the potential substrate.

Ligand and/or transporter protein structure-based models involving 3D QSARs, pharmacophore modeling, homology modeling, and molecular dynamics have been successfully employed to understand drug transporters.[10] Models to distinguish between competitive, noncompetitive, and uncompetitive, inhibition of transporters have met with limited success.[11] *In silico* models for renal excretion predict f_e and CL_R using physicochemical properties.[12-14]

6.3.2 *In Vitro* Models for Hepatic Metabolism

The ever-increasing number of NCEs synthesized by pharmaceutical companies has prompted the development of research methods for rapid screening. A series of *in vitro* assays employing intact cell systems[15] (primary hepatocytes, precision-cut liver slices), subcellular fractions (S9 fractions, microsomes), and transfected cell lines (MDCK and HEK cells) have been developed to serve as reliable indicators of disposition. Cytochrome P450 enzymes (CYPs), the heme-containing enzymes that constitute the major enzymatic system for metabolism of xenobiotics are primarily located in the smooth endoplasmic reticulum. The liver microsomes (Fig. 6.3) are a rich source of CYPs and are useful in the study of CYP-based xenobiotic metabolism and drug interactions. The choice of system for the measurement of intrinsic clearance of metabolism ($CL_{int, met}$) in the drug discovery phase is influenced by the nature of the compounds.

Recombinant CYPs. The complementary DNA (cDNA) for the common CYPs has been cloned. Recombinant enzymes have been expressed in a variety of cells. Recombinant human CYPs[16] are incubated individually with a drug under the same conditions as the liver microsomal studies to assess the ability of a particular CYP to metabolize the drug. Recombinant enzymes provide an effective way to screen metabolic activity per picomole of protein and to scale up to *in vivo* using enzyme abundances. It is also the preferred system for CYP reaction phenotyping (CRP) to identify major CYPs involved in the metabolism of an NCE.

Hepatic S9. Hepatic S9 pools represent the postmitochondrial supernatant fraction from homogenized liver. Known to be a rich source of DMEs including CYPs, flavin-monooxygenases (FMOs), carboxylesterases, epoxide hydrolase, uridine 5′-diphospho-glucuronosyltransferases (UGTs), sulfotransferases (SULTs), methyltransferases, acetyltransferases, glutathione S-transferases (GSTs), and other DMEs, they are useful in the study of xenobiotic metabolism. Microsomes are derived from S9 by an ultracentrifugation process that separates the cytosolic fraction from the microsomal fraction. CYP activity in the S9 fraction is approximately 20–25% of those in the corresponding microsomal fraction due to the dilution.

Pooled cytosol. Pooled cytosol from human liver consists of soluble human conjugative or phase II enzymes from a pool of mixed human liver donors.

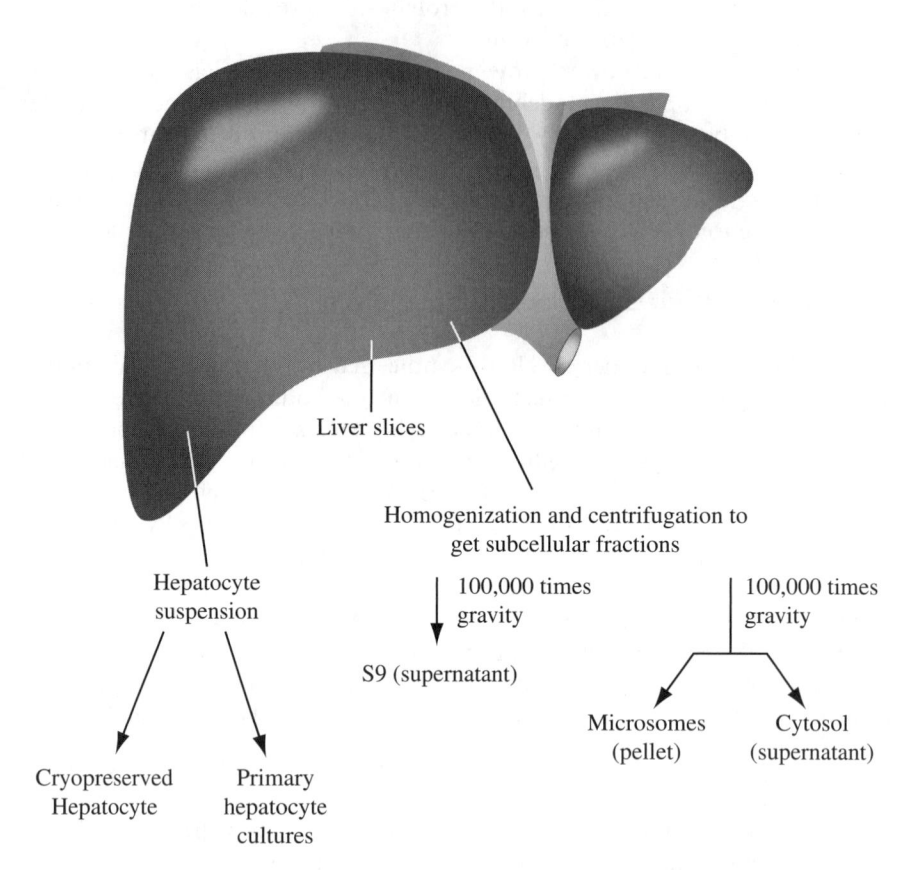

Figure 6.3. *In vitro* systems for metabolic stability.

Cytosolic preparations can be used for screening potential substrates, inhibitors, or drugs and drug candidates to determine phase II DME interactions.

Primary hepatocytes. Primary hepatocytes are effective tools for the evaluation of metabolism, drug–drug interactions, hepatotoxicity, and transporter activity as they contain the full complement of phase I and phase II DMEs, do not rely on artificially high concentrations of cofactors, and express cell membrane transporters.[17,18] Primary cells are cells taken directly from living tissue such as biopsy material and grown *in vitro*. Unlike artificially immortalized cell lines, primary hepatocytes cannot be cultured indefinitely. As they have undergone very few population doublings, they are considered to be more functionally representative of the tissue from which they are derived in comparison to continuous tumor or immortalized cell lines. Since the availability of freshly isolated human hepatocytes could limit experiments in a drug discovery environment, the use of cryopreserved hepatocytes is

6.3 MODELS FOR HEPATOBILIARY ELIMINATION AND RENAL EXCRETION

recommended.[19,20] Cryopreserved hepatocytes retain their full activity for more than one year in liquid N_2 and are thus a flexible resource of hepatocytes for *in vitro* assays. They have been shown to predict human hepatic metabolism just as well as those freshly isolated.[21] Their suitability for the assessment of metabolism, hepatotoxicity, drug–drug interaction,[22] and induction studies have also been demonstrated.[23] Suspended hepatocytes are suitable for most short-term metabolism studies. However, they lose viability, drug-metabolizing capacity, and other liver-specific physiological functions within a few hours of isolation,[24] making them unsuitable for long-term studies, as required for induction studies. Several types of culture systems have been developed to extend the life span of isolated hepatocytes and potentially allow a more accurate assessment of low-clearance drugs. Monolayer cultures on collagen, sandwich cultures, and immobilized cultures[25–27] can all maintain liver-specific function to differing extents. However, none of these systems can completely restore liver-specific function to the levels seen *in vivo* but simply slow the rate of decline. Although these culture systems are widely used for induction studies, estimation of intrinsic clearance from monocultures have been shown to be uptake rate limited.[28] Primary cells from different species may be used to identify potential differences between humans and preclinical test species. The use of hepatocytes in drug discovery has been reviewed.[29]

Liver slices.[30–35] Liver slices from a variety of species (including humans) are prepared using mechanical slicers that produce reproducible slices of a uniform thickness (0.2 μm but ideally 1 cell thick), which allows optimum exchange of nutrients, waste, and gases. Slices are incubated in dynamic systems to maintain viability in culture for 1–10 days. The viability of slices can be assessed by ion content (K^+, Na^+ ATPase status), intermediary metabolism, energy status (ATP), respiration, biosynthetic ability, or biotransformation activity. In addition, liver tissue slices allow the opportunity for extensive evaluation by microscope (light and electron) as well as by newer technologies such as confocal microscopy. Assessment of the toxic potential of a chemical can be performed after a short-term or constant exposure by evaluating the viability parameters. Liver slices have been used extensively for rank ordering the toxicity of chemicals as well as for examining the mechanisms of liver injury. Liver slices in culture can also be used for an examination of the induction of enzymes such as cytochrome P450 and the expression of stress proteins or peroxisomal enzymes. Finally, whole or cryopreserved liver slices offer a system for evaluating liver function and the regeneration of liver tissue after toxic insult. Thus, liver slices are a valid *in vitro* system that serves as a bridge between *in vivo* and cell culture systems.

6.3.3 *In Vitro* Models for Transporters

In human hepatocytes, the sodium-independent uptake of amphiphilic organic anions is mediated by organic anion transporters (OAT2 and OAT7)[36,37] and

3 of the 11 isoforms of the organic anion transporter polypeptides (OATPs 1B1, 1B3, and 2B1).[37] The substrate spectrum of OATPs includes bile salts, conjugates of steroid hormones, statins, thyroid hormones, and many other amphiphilic organic anion drugs. OCT1 mediates transport of organic cations. Sodium-dependent taurocholate co-transporting polypeptide (NTCP) is another uptake transporter that transports conjugated and unconjugated bile acids,[38,39] as well as other acids such as thyroxine[40] and estrone-3-sulfate[39] but has a limited role in the transport of drugs. The export of organic ions into bile canaliculi is predominantly mediated by P-glycoprotein (P-gp), multidrug and toxin extrusion protein (MATE), bile salt export pump (BSEP), breast cancer resistance protein (BCRP), and by multidrug resistance associated protein (MRP2). Although freshly isolated hepatocytes are the widely accepted gold standard for providing reliable data on drug uptake across the sinusoidal membrane,[41] its suitability for the assessment of canalicular efflux transport is poor, as efflux proteins have been shown to be rapidly internalized after collagenase digestion and/or mechanical disruption during isolation.[42] In the rat hepatocytes, the internalized proteins have been shown to remain confined to cell–couplet junctions or reside on "legacy" canalicular networks of cells, which represent regions with concentrated tight junction protein expression where cell-to-cell contact and a functional canalicular network[43] was once present. Several in vitro systems can be employed to measure $CL_{int,uptake}$ (basolateral uptake) and $CL_{int,bile}$ (canalicular, apical secretion into bile) for transporter-dependent drugs. Similarly, renal transporters can be studied with a variety of in vitro systems including human kidney slices, primary cultures of human kidney cells, human immortalized cells, transient or stably transfected cells, vesicles.

Bidirectional transport through polarized cell monolayers grown on a permeable filter support, such as Caco-2 and HT29 cells, provide a means of identifying substrates and inhibitors of efflux proteins, BCRP, MRP2, and P-gp. Caco-2 cells, however, are not a pure model for identification of single transporter interaction but are valuable tools for understanding the mechanistic function. Plated hepatocytes cannot distinguish between basolateral and biliary efflux, due to loss of polarity by internalization.

6.3.3.1 Transfected Cell Lines. Single transporter interactions are better studied with transfected mammalian HEK293 cells[41,44,45] or *Xenopus laevis* oocyte expression systems injected with complementary ribonucleic acid (RNA),[46–48] expressing a single exogenous recombinant transport protein. The concerted action of many basolateral uptake and canalicular efflux transporter pairs such as OATP1B1 or 1B3 and MRP2 are better studied with double transfected polarized Madin–Darby canine kidney cells (MDCK) of defined uptake transporters and export pumps[49–53] or even quadruple transfected cell lines.[54]

6.3.3.2 ATPase Assay for ABC Efflux Transporters. ABC transporter membranes may be utilized in an ATPase assay to determine possible

6.3 MODELS FOR HEPATOBILIARY ELIMINATION AND RENAL EXCRETION

interactions between an efflux transporter and test compound. As the ABC superfamily of transporters are reliant on ATP hydrolysis for energy to mediate drug efflux from cells, a measurement of ATP conversion through colorimetric detection of phosphate ion liberation[55–57] or the spectrophotometric monitoring of NADH turnover in a coupled enzyme cycling assay[57,58] gives an indirect measure of transporter function. The assay has successfully been used for determining the P-gp affinity[59] in different species. However, some drugs can stimulate phosphate ion release at low concentrations, but inhibit the effect at higher concentrations[60,61] or result in a false negative due to significant nonspecific binding.[61]

6.3.3.3 Membrane Vesiclular Transporter Assays for ABC Efflux Transporters.
Membranes and vesicles prepared by technically advanced methodologies, using plasma membrane purified from an insect cell system (Sf9 cells transfected with baculovirus), over-expressing ABC transporters are effective for the *in vitro* screening of transporter substrates and inhibitors in a 96-well plate format. ABC transporter vesicles, which have an inside-out structure, are prepared from ABC transporter membranes for use in vesicular transport assays. While ABC transporters typically mediate the efflux of substrates from cells, transporters expressed on these inside-out vesicles import substrates into the vesicles. It is therefore possible to quantitatively evaluate transport activity for a drug candidate by determining the amount of compound incorporated into the vesicles. Vesicular transport assays using ABC transporter vesicles are simple, effective tools for *in vitro* screening of substrates and inhibitors of transporters involved in hepatic and renal excretion. Sensitive fluorescent techniques have been employed to investigate the intracellular disposition of drugs in tubular cells in real time.[62–67] Initial ATP-dependent uptake rate by canalicular membrane vesicles (CMVs) provides the canalicular efflux rate of a compound.

6.3.3.4 Sandwich Hepatocytes.
Primary mammalian hepatocytes largely retain their liver-specific functions when freshly derived. However, long-term cultures of functional hepatocytes are difficult to establish. To increase the longevity and maintain differentiated functions of hepatocytes in primary cultures, cells are cultured in a sandwich configuration between double layers of extracellular matrix (ECM) rather than on single layers of collagen.[68–70] A serum-free culture medium,[71] supplemented with the physiological glucocorticoid hormone and dexamethasone, is recommended for the long-term preservation of hepatocyte-specific functions such as polygonal hepatocyte morphology and structural integrity features (cytoplasmic membranes and bile canaliculi-like features). Hepatocytes cultured in a sandwich configuration reorganize to form an architecture similar to that found in the liver and are able to form functional bile canalicular networks and gap junctions with re-established hepatic polarity and transporter activity. Markers of differentiation

include morphology, expression of plasma proteins, hepatic nuclear factors, as well as phase I and II metabolic enzymes. Functionally, these culture conditions also preserve hepatic stress response pathways, such as the stress-activated protein kinase (SAPK) and mitogen-activated protein kinase (MAPK) pathways, as well as prototypical xenobiotic induction responses. Retention of differentiated functions makes it a useful tool for evaluating hepatotoxicity caused by interference with hepatic transporters. LeCluyse et al.[72] first demonstrated the repolarization of rat hepatocytes over time. Applications of the ECM-based sandwich culture are limited by the mass transfer barriers induced by the top gelled ECM layer, complex molecular composition of ECM with batch-to-batch variation, and uncontrollable coating of the ECM double layers. These can be overcome by using bioactive synthetic materials instead of ECM.[73] Functional bile canalicular networks in sandwich culture allow determination of *in vitro* biliary clearance.[74] A quantitative estimation is still elusive.

6.3.3.5 Hepatic Uptake Assays. Although the molecular cloning of the major hepatic OATP isoforms and their expression in mammalian cell lines generated a wealth of knowledge concerning the substrate specificity of OATPs,[37,75–77] quantification to *in vivo* is more problematic than it was for the CYPs, due to lack of scaling factors. Plated or suspended hepatocytes have, therefore, become the system of choice for obtaining quantitative information regarding hepatic drug uptake.[41,78,79] Investigation of hepatic uptake involves measuring the rate of appearance of radiolabeled substrate into cells, determined after a centrifugation step through oil.[41,79,80] This method provides robust, mechanistic data on individual compounds, but it is clearly not amenable for use within an early discovery setting, as radiolabeled compounds are not routinely available. A nonradiolabeled method to assess the impact of hepatic uptake on unbound drug intrinsic clearance *in vivo* has been developed in which the uptake (loss or disappearance) of parent compound from the incubation medium into suspended hepatocytes ("media loss")[81,82] and its appearance in cells[83] are determined at different time points within the first few minutes of incubation using tandem mass spectrometry (LC/MS/MS) quantification with standard curve (Fig. 6.4). Cell and media concentration–time data were simultaneously fitted to a model incorporating active uptake, passive permeation, binding, and metabolism but not efflux into canaliculi, as efflux cannot be considered in suspended hepatocytes.

At steady state, the rate of loss of compound from the suspended hepatocyte incubation media equals the rate of disappearance from the cells[83,84]:

$$CL_{med} \, f_{u,med} \, C_{med} = CL_{int,met} \, f_{u,cell} \, C_{cell} \quad (6.3)$$

where CL_{med} is the clearance from the media, obtained from the measured half-life and volume of the media, V_{med}. $CL_{int,met}$ is the intrinsic clearance of metabolism measured in a standard hepatocyte assay. And, $f_{u,med}$ and $f_{u,cell}$ are the fractions of compound unbound in media and cell, respectively. C_{med}

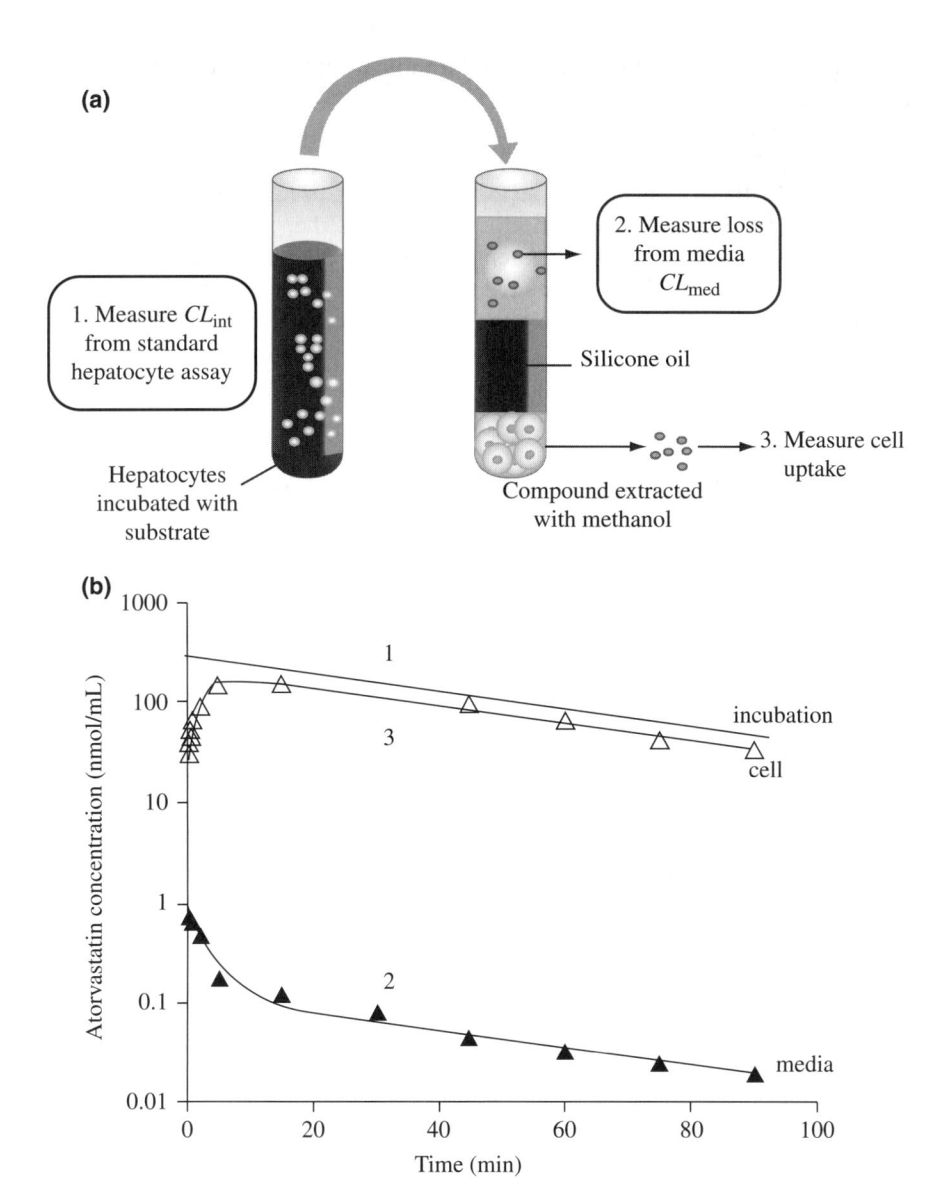

6.3 MODELS FOR HEPATOBILIARY ELIMINATION AND RENAL EXCRETION

Figure 6.4. (a) Suspended hepatocytes hepatic uptake assay relies on the quantification of compound concentrations from 1. incubation, 2. media and 3. cells. (b) Concentration–time profiles from incubation, media and cells.

and C_{cell} are the concentrations of compound in media and cell, respectively. Rearranging the above equation,

$$\frac{CL_{med}}{CL_{int,met}} = \frac{f_{u,cell}C_{cell}}{f_{u,med}C_{med}} = \frac{C_{u,cell}}{C_{u,med}} = \psi \quad (6.4)$$

The ratio of $C_{u,\text{cell}}$ and $C_{u,\text{med}}$, the unbound concentrations of compound in cell and media, is denoted as ψ. It represents the extent of drug accumulation or depletion in the cell compared to the medium, depending on whether the cell uptake or efflux is the more important driver of unbound concentration differences. In the absence of transporter-mediated uptake or efflux, ψ tends to 1 at steady state, except when permeability limits the rate of metabolism, in which case a steady state is not reachable. In the *in vivo* situation, ψ can be related to the various processes, by recognizing that at steady state, the rate of compound entering the liver equals the rate at which it leaves the liver.

$$\frac{C_{u,\text{cell}}}{C_{u,\text{med}}} = \Psi = \frac{CL_{\text{int, uptake}} + CL_{\text{int, diffusion}}}{CL_{\text{int, met}} + CL_{\text{int, diffusion}} + CL_{\text{int, efflux}} + CL_{\text{int, bile}}} \quad (6.5)$$

If both sinusoidal and biliary efflux can be neglected, then the last two terms in the denominator become zero.

Also, $f_{u,\text{cell}}$ can be obtained from a measurement of fraction of compound unbound in suspended hepatocyte incubation ($f_{u,\text{inc}}$), when uptake is incapacitated (see Fig. 6.4). To understand the relationship between $f_{u,\text{cell}}$ and $f_{u,\text{inc}}$, we need to first define k_{mem} as the proportionality constant relating the amount of drug bound to the cell membrane (A_{mem}) to the unbound concentration in the medium ($C_{u,\text{med}}$). Thus,

$$A_{\text{mem}} = k_{\text{mem}} \times C_{u,\text{med}} \quad (6.6)$$

and $f_{u,\text{inc}}$ is the ratio of unbound to total concentration of a compound that should be employed for correction of observed clearance from hepatocyte incubation (CL_{inc}) in the absence of uptake:

$$CL_{\text{med}} = \frac{CL_{\text{inc}}}{f_{u,\text{inc}}} = CL_{\text{int, met}} \quad (6.7)$$

Furthermore, $f_{u,\text{inc}}$ is measured by dialysis using uptake-incapacitated hepatocytes for uptake substrates. Therefore, the unbound concentration in the cells equals that in the medium or in the incubation. Defining A_{inc} as the total amount of compound in the incubation and V_{inc} as the volume of the incubation and recognizing that A_{inc} is the sum of the amounts of compound in the medium, membrane, and cells,

$$f_{u,\text{inc}} = \frac{V_{\text{inc}} \times C_{u,\text{med}}}{A_{\text{inc}}} = \frac{V_{\text{inc}} C_{u,\text{med}}}{C_{u,\text{med}} V_{\text{med}} + C_{u,\text{med}} k_{\text{mem}} + C_{\text{cell}} V_{\text{cell}}} \quad (6.8)$$

Dividing both the numerator and denominator of the right-hand side of the above equation by C_{cell} and substituting $C_{u,\text{med}}/C_{\text{cell}}$ as $f_{u,\text{cell}}$, we get

6.3 MODELS FOR HEPATOBILIARY ELIMINATION AND RENAL EXCRETION

$$f_{u,\text{inc}} = \frac{V_{\text{inc}} f_{u,\text{cell}}}{f_{u,\text{cell}} V_{\text{med}} + f_{u,\text{cell}} k_{\text{mem}} + V_{\text{cell}}} \quad (6.9)$$

Making $f_{u,\text{cell}}$ the subject, we get

$$f_{u,\text{cell}} = \frac{V_{\text{cell}}}{V_{\text{inc}}/f_{u,\text{inc}} - V_{\text{med}} - k_{\text{mem}}} \quad (6.10)$$

where V_{cell}, the volume of the cells in the incubation medium, is taken as $4\ \mu\text{L}/10^6$ cells.[80,84]

When uptake activity is retained in hepatocytes, the fraction of compound in the incubation media at steady state ($f_{\text{med, SS}}$) is the key parameter to be employed for appropriate correction of observed clearance of an uptake substrate in hepatocyte incubation (CL_{inc}) obtained from a standard hepatocyte assay. This is because the media concentration is expected to be lower compared to that in the total incubation, due to higher uptake-driven intracellular concentrations and binding to membrane (see Fig. 6.4). Thus,

$$CL_{\text{med}} = \frac{CL_{\text{inc}}}{f_{\text{med, SS}}} \quad (6.11)$$

From Figure 6.4, it is clear that the terminal elimination half-lives measured in the media and incubation are same. Therefore,

$$\frac{CL_{\text{med}}}{V_{\text{SS, med}}} = \frac{CL_{\text{inc}}}{V_{\text{inc}}} \quad (6.12)$$

where $V_{\text{ss,med}}$ is the steady-state volume of distribution as viewed from the media. The above two equations can be combined to give

$$f_{\text{med, SS}} = \frac{CL_{\text{inc}}}{CL_{\text{med}}} = \frac{V_{\text{inc}}}{V_{\text{SS, med}}} \quad (6.13)$$

where $V_{\text{SS,med}}$ is given by

$$V_{\text{SS, med}} = V_{\text{med}} + k_{\text{mem}} + \frac{V_{\text{cell}} \Psi}{f_{u,\text{cell}}} \quad (6.14)$$

which can be divided throughout by V_{med} and inversed to give

$$\frac{V_{\text{med}}}{V_{\text{SS, med}}} = \frac{1}{1 + \dfrac{k_{\text{mem}}}{V_{\text{med}}} + \dfrac{V_{\text{cell}} \Psi}{f_{u,\text{cell}} V_{\text{med}}}} \quad (6.15)$$

When V_{med} can be approximated to V_{inc}, it follows that

$$f_{med,ss} = \frac{1}{1 + \dfrac{k_{mem}}{V_{med}} + \dfrac{V_{cell}\Psi}{f_{u,cell}V_{med}}} \qquad (6.16)$$

The compound concentrations measured in the cells and incubation media as a function of time represent changes arising from the composite terms of intrinsic metabolism ($CL_{int,met}$), uptake ($CL_{int,uptake}$), binding to cell membranes (k_{mem}), and passive diffusion ($CL_{int,diffusion}$) in and out of cells, in the absence of efflux. These parameters can be obtained by fitting the observed cell and media concentration–time data from hepatocyte incubations to solutions from the following differential equations:

$$\begin{aligned}\frac{dC_{cell}}{dt} &= f_{med,ss}C_{media}CL_{int,diffusion} + f_{med,ss}C_{media}CL_{int,uptake} \\ &\quad - f_{u,cell}C_{cell}CL_{int,met} - f_{u,cell}C_{cell}CL_{int,diffusion} \\ \frac{dC_{media}}{dt} &= -f_{med,ss}C_{media}CL_{int,diffusion} - f_{med,ss}C_{media}CL_{int,uptake} \\ &\quad + f_{u,cell}C_{cell}CL_{int,diffusion}\end{aligned} \qquad (6.17)$$

The parameters $CL_{int,met}$, $CL_{int,diffusion}$, and $CL_{int,uptake}$ can be suitably scaled and employed in whole body physiologically-based pharmacokinetic (WB-PBPK) models.

Poirier et al.[85,86] described a mechanistic two-compartmental model to obtain the saturable, active, transporter-mediated uptake and the nonsaturable, passive diffusion parameters ($K_{m,uptake}$, $V_{max,uptake}$, and $CL_{int,diffusion}$) from transporter studies in hepatocytes and overexpressed cells.

6.3.3.6 Liver Slices. Uptake clearance measured in kidney[87] and liver slices are generally underpredicted due to the multiple layers even in thin slices, resulting in diffusion-limited uptake. Poorly perfused interiors leads to ischemic deterioration and necrosis with possible changes in flow path, cellular volume, accessible sinusoidal surface area, homeostatic metabolism, and microarchitecture, all of which impact the enzyme and transporter functions.

Unlike uptake and canalicular efflux discussed above, *in vitro* measurement of sinusoidal efflux clearance $CL_{int,efflux}$ has not been developed and is generally assumed to be the same as $CL_{int,diffusion}$. A summary of all *in vitro* systems used to study hepatobiliary and renal excretions is provided in Table 6.1.

The clearances of metabolism, canalicular efflux, and saturable (transporter-mediated uptake) and unsaturable (diffusion) influx are incorporated into equation 6.1, to get an estimate of total organ intrinsic clearance. Alternatively,

TABLE 6.1. *In Vitro* Models for Assessment of Hepatobiliary and Renal Excretion of Drugs

In vitro Model	Applications	Strengths	Limitations
Recombinant CYPs	Metabolic stability; CYP reaction phenotyping	Amenable to high-throughput screening	Only CYPs. Absence of competing enzymes
Microsomes	Metabolic stability; metabolite identification	Easy to use, cheap. Amenable to high-throughput screening	Oxidative phase I enzymes and UGTs only. Induction not modeled. No intact cell membranes. Needs cofactors
S9	Metabolic stability; metabolite identification	Phase I and phase II enzymes	Lower enzyme activity compared to microsomes. Induction not modeled. No intact cell membranes. Needs cofactors
Freshly isolated or cryopreserved hepatocytes in suspension	Metabolic stability; inhibition or substrate potential; metabolite identification; CYP reaction phenotyping	Phase I and phase II enzymes. Contain cell membrane and full complement of DMEs and cofactors. Intact cells. Induction modeled well in fresh but poorly in cryopreserved	Expensive. Low throughput. Unstable enzyme activity. Viability to be checked. Transporters may be rapidly down-regulated after isolation and by cryopreservation. Loss of polarity due to internalization
Hepatocyte monocultures	Induction and transporter activity	Intact cells, resembling the environment to which drugs are exposed in the liver	Support matrices may introduce artifacts like additional collagen diffusion barrier or loss of enzyme activity. Lacks a 3D extracellular matrix environment and hepatocytes have to quickly adopt their shape to establish cell–cell contacts
Hepatocyte sandwich culture	Induction; biliary efflux; hepatotoxicity	Morphological stability. Retains functional differentiation for a longer time. Preserve hepatic stress–response pathways	Support matrices may introduce artifacts like additional collagen diffusion barrier or loss of enzyme activity. Not quantitative
Precision-cut liver slices	Metabolic stability; CYP induction; hepatotoxicity	Physiologically relevant with complete set of DMEs	Diffusion barriers. Functional deterioration of enzymes and transporters
Single/double transfected cell lines	Uptake and efflux transporters	Focus on one transporter in single and on vectorial transport in multiple transfections	Not quantitative
Transporter membrane vesicles	Renal and biliary efflux transporter substrates and inhibitors	Simple and effective screening tool for mechanistic understanding	Applicable only for ABC efflux proteins. Not sensitive enough to be quantitative

they can be scaled to *in vivo* parameters by multiplying with physiological scaling factors as well as with *in vitro* to *in vivo* scaling factors derived from comparing *in vitro* and *in vivo* data in the rat (or other preclinical species). Thus, a parameter X in humans *in vivo* can be estimated from its *in vitro* measure using a scaling factor in a preclinical species such as the rat, in the following manner:

$$X(\text{human}, \textit{in vivo}) = \frac{X(\text{rat}, \textit{in vivo})}{X(\text{rat}, \textit{in vitro})} \times X(\text{human}, \textit{in vitro}) \qquad (6.18)$$

These human *in vivo* parameters can then be employed in PBPK models to predict disposition of NCEs in humans in the whole-body physiological context.[88]

Isolated kidney cells cannot be used to estimate renal clearance. Instead thin slices of kidney can be employed. There are no reliable *in vitro* methods to evaluate tubular reabsorption in the kidney, as tubular transit and water reabsorption cannot be mimicked. However, using the lipophilicity of the drug and physiological parameters, a physiology-based model can still predict reabsorption.

6.4 PHYSIOLOGICAL MODELS

Physiological models employing results from *in vitro* assays provide the best estimates of hepatobiliary and renal excretion and therefore human exposure in a target organ. Such models account for the complex interplay of transporters and enzymes under physiological conditions as well as dynamically changing drug concentrations *in vivo*.

6.4.1 Hepatobiliary Elimination of Parent Drug and Metabolites

Recognizing the role of transporters is critical to understanding the hepatobiliary elimination of a parent drug and its metabolites in the liver. Transporter-mediated cellular uptake and efflux as well as metabolism are saturable processes requiring Michaelis–Menton type of equations, while diffusion is a nonsaturable, linear process. In order to consider transporter-mediated cellular uptake, the liver is compartmentalized into the well-perfused interstitial (I) space and the intracellular space (IC) (see Fig. 6.5), where the following differential equations apply for parent drug and metabolites.[89]

6.4.1.1 Parent Drug.

Interstitial (I) *Liver* (LI) *Compartment.* The rate of change of drug concentration (C) in this compartment is given by

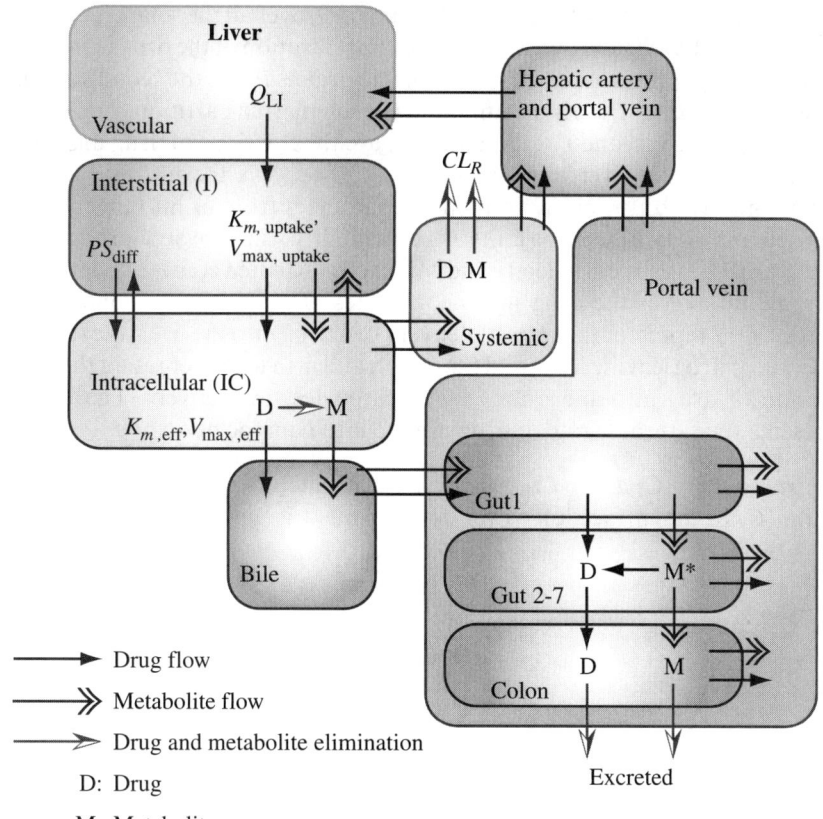

6.4 PHYSIOLOGICAL MODELS

Figure 6.5. Schematic illustration of a physiological model of hepatobiliary elimination.

$$\frac{dC_{LI}^I}{dt} = \frac{1}{V_{LI}^I}\left(\begin{array}{c} Q_{HA} \times C_{ART} + \sum \dfrac{Q_i \times C_i \times R}{f_{up} \times K_{i,p,u}} - \dfrac{Q_{LI} \times C_{LI} \times R}{f_{up} \times K_{LI:p,u}} \\ - CL_{\text{int, diffusion}} \times f_{ub} \times C_{LI}^I + CL_{\text{int, diffusion}} \times f_{u,LI} \times C_{LI}^{IC} \\ - \dfrac{V_{\max,\text{uptake}}}{K_{m,\text{uptake}} + f_{ub} \times C_{LI}^I} \times f_{ub} \times C_{LI}^I \end{array}\right)$$

(6.19)

where V_{LI} is the volume of the interstitial compartment of the liver, Q_{HA} is the blood flow to the liver from the hepatic artery, Q_i is the blood flow to the liver from the splanchic organ i, representing gut, pancreas, spleen, or stomach, and

Q_{LI} is the blood flow out of the liver. The sum of Q_i over all the splanchic organs is the portal vein blood flow Q_{pv}; C_{ART} is the concentration of the drug in the artery; C_i is the concentration in any of the splanchic organs i; R is the blood-to-plasma concentration ratio; $K_{i,p,u}$ is the unbound tissue–plasma partition coefficients of the splanchic organs, and f_{up}, f_{ub}, and $f_{u,LI}$ are the fractions of drug unbound in plasma, blood, and liver, respectively. $CL_{int, diffusion}$ is the intrinsic clearance pertaining to the diffusion of drug into hepatocytes from or into the interstitial compartment. It has the same units as permeability–surface area product. V_{max} and K_m are the maximum velocity and Michaelis–Menten constant related to the saturable uptake of the drug into hepatocytes. The first and second terms of equation 6.19 represent the rates of delivery of a drug into the liver, the third term is the rate of drug leaving liver, and the fourth and fifth terms represent the rates of diffusion of drug into and out of the hepatocytes, respectively. The last term represents the saturable rate of drug uptake into hepatocytes.

Intracellular (IC) *Liver Compartment.* The rate of change of drug concentration (C) in the intracellular liver compartment is given by

$$\frac{dC_{LI}^{IC}}{dt} = \frac{1}{V_{LI}^{IC}} \left(\begin{array}{c} CL_{int, diffusion} \times f_{ub} \times C_{LI}^{I} - CL_{int, diffusion} \times f_{u,LI} \times C_{LI}^{IC} \\ + \dfrac{V_{max, uptake}}{K_{m, uptake} + f_{ub} \times C_{LI}^{I}} \times f_{ub} \times C_{LI}^{I} - \dfrac{V_{max, met}}{K_{m, met} + f_{u,LI} \times C_{LI}^{IC}} \\ \times f_{u,LI} \times C_{LI}^{IC} - \dfrac{V_{max, bile}}{K_{m, bile} + f_{u,LI} \times C_{LI}^{IC}} \times f_{u,LI} \times C_{LI}^{IC} \end{array} \right)$$

(6.20)

The first and second terms represent the rates of diffusion of drug into and out of the hepatocytes, respectively; the third, fourth, and fifth terms represent the saturable rates of drug uptake into hepatocytes, metabolism, and efflux into bile. V_{max} and K_m can be replaced by CL_{int} under linear conditions.

6.4.1.2 Metabolite.

Intracellular Liver Compartment. The rate of change of metabolite concentration (M) in the intracellular liver compartment is given by

$$\frac{dM_{LI}^{IC}}{dt} = \frac{1}{V_{LI}^{IC}} \left(\begin{array}{c} Q_{pv} \times M_{pv} + Q_{HA} \times M_{SYS} + \dfrac{V_{max, met}}{K_{m, met} + f_{ub} \times C_{LI}^{IC}} \times f_{ub} \times C_{LI}^{IC} \\ - \dfrac{V_{max, met, M}}{K_{m, met, M} + f_{uM,LI} \times M_{LI}^{IC}} \times f_{uM,LI} \times M_{LI}^{IC} \\ - \dfrac{V_{max, bile, M}}{K_{m, bile} + f_{uM,LI} \times M_{LI}^{IC}} \times f_{uM,LI} \times M_{LI}^{IC} - Q_{LI} \times M_{LI}^{IC} \end{array} \right)$$

(6.21)

6.4 PHYSIOLOGICAL MODELS

where Q_{Li} is the sum of Q_{pv} and Q_{HA}. The first two terms represent the rate of entry of metabolite into the hepatic portal vein and subsequently the systemic circulation. The third term is the rate of formation of the metabolite, the fourth is the rate of further metabolism of the metabolite, the fifth represents the rate of excretion of the metabolite into bile, and the last term is the rate of metabolite leaving the liver.

Systemic (SYS) Circulation.

$$\frac{dM_{SYS}}{dt} = \frac{1}{V_{M,SYS}} \left(Q_{LI} \times M_{LI}^{IC} - \frac{V_{max,renal,M}}{K_{m,renal,M} + f_{uM,b} \times M_{SYS}} \times f_{uM,b} \times M_{SYS} - Q_{HA} \times M_{SYS} \right) \quad (6.22)$$

where $V_{M,SYS}$ is the volume of distribution of the metabolite. The second term refers to the renal excretion of the metabolite, the last term is the rate of metabolite leaving systemic to enter liver.

For metabolites generated in the gut, the rate of change of metabolite concentration (M) in the hepatic portal vein (PV) is given by

$$\frac{dM_{PV}}{dt} = \frac{1}{V_{PV}} \left(\sum_{GUi} \left(\sum_{isoform} \left(\frac{k_{ilm,GUi} \times E_{GUi,isoform} \times (A_{ABS,GUi}/V_{ent,GUi})}{K_m + (A_{ABS,GUi}/V_{ent,GUi})} \right) \right) - Q_{PV} \times M_{PV} \right) \quad (6.23)$$

$k_{ilm,GUi}$ is the first order rate constant for intestinal loss by metabolism in ith compartment of the gut. $E_{GUi,isoform}$ is the abundance of the enzyme isoform that metabolizes the drug in the ith compartment of the gut. $V_{ent,GUi}$ is the volume of the enterocytes in that compartment. $A_{Abs,GUi}$ is the amount absorbed in the ith compartment of the gut. K_m is the Michaelis–Menten constant.

The first summation term refers to the rate of incoming metabolite from the gut. The second term refers to the rate of metabolite leaving the portal vein into the liver. The parameters required for the model are derived from *in vitro* assays or preclinical *in vivo* experiments as shown in Figure 6.6. A physiological model that combines drug and metabolite kinetics and considers the interplay of enzymes and transporters has been reported in the literature.[90] Inclusion of primary and secondary metabolite kinetics in a PBPK model allows for treatment of metabolite reconversion to parent drug that could be important for some phase II metabolites such as glucuronides. This reconversion and parent reabsorption leads to enterohepatic recirculation. In the elderly and in the renally impaired population, enterohepatic recirculation can result in altered disposition of the parent drug as the removal of the glucuronide is less efficient due to a reduced glomerular filtration rate. The sequential metabolism of a drug to form primary

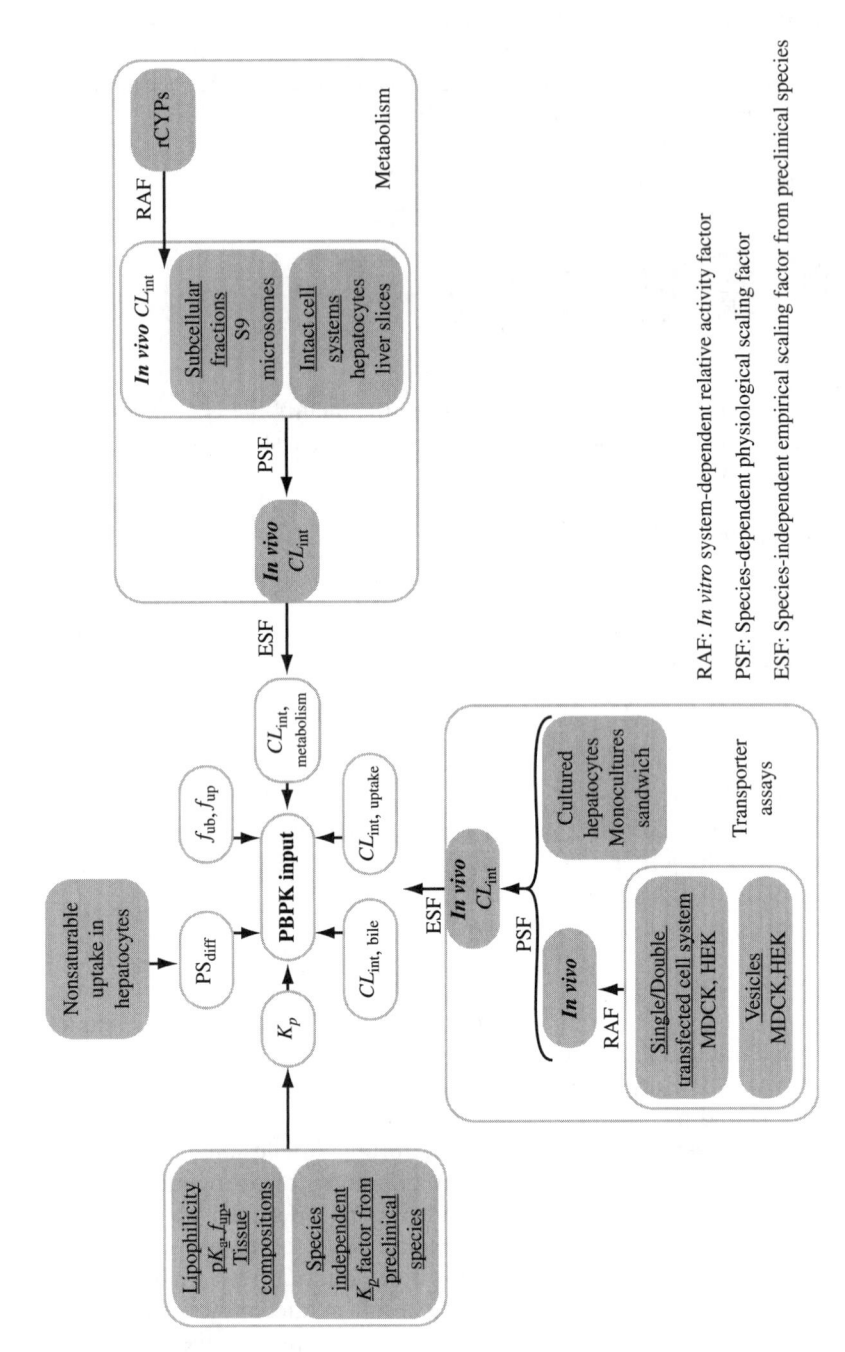

Figure 6.6. Input to PBPK model from parameters generated in *in vitro* assays.

6.4 PHYSIOLOGICAL MODELS

and secondary metabolites in the first-pass organs has been considered using a segregated flow model of the intestine.[91,92] Treatment of metabolite kinetics is also essential for prodrugs and drugs with active metabolites. Metabolites can be important as inhibitors of enzymes and receptors of pharmacological relevance, thereby playing a role in drug–drug interactions. This will be considered in Chapter 9.

6.4.2 Renal Excretion

Allometric models for renal excretion are based on species differences in the nephron count. Mechanistic elements that determine renal excretion are rate of blood flow to the kidney, plasma protein binding, glomerular filtration rate, active secretion in proximal tubules, and passive reabsorption. Reabsorption by active processes is important only for endogenous nutrients. A schematic illustration of a physiologically-based kidney model is shown in Figure 6.7. The corresponding differential equations are shown below.

6.4.2.1 Kidney.

Vascular (V).

$$\frac{dC_{KI}^V}{dt} = \frac{1}{V_{KI}^V} \left(Q_{KI} \times C_{ART} - GFR \times C_{KI}^V - (Q_{KI} - GFR) \times C_{KI}^V \right) \quad (6.24)$$

where C_{KI}^V is the concentration of a drug in the vascular compartment of the kidney, V_{KI}^V is the volume of the vascular compartment of the kidney, Q_{KI} is the arterial blood flow to the kidney, and GFR is the glomerular filtration rate.

Interstitial (I).

$$\frac{dC_{KI}^I}{dt} = \frac{1}{V_{KI}^I} \left(\begin{array}{c} (Q_{KI} - GFR) \times C_{KI}^V - \dfrac{V_{max,\,uptake}}{K_{m,\,uptake} + f_{ub} \times C_{KI}^I} \\ \times f_{ub} \times C_{KI}^I - Q_{KI} \times C_{KI}^I \end{array} \right) \\ + \frac{2P}{r^{PTL}} \times \left(\frac{A_{KI}^{PTL}}{V_{KI}^{PTL}} - C_{KI}^I \right) + \sum_i \frac{2P}{r^{TL_i}} \times \left(\frac{A_{KI}^{TL_i}}{V_{KI}^{TL_i}} - C_{KI}^I \right) \quad (6.25)$$

where C_{KI}^I and C_{KI}^{IC} are the concentrations of a drug in the interstitial and intracellular compartments of the kidney, f_{ub} and $f_{u,\,LI}$ are the fractions of drug unbound in plasma, blood and tubular cells, respectively; V_{max} and K_m are the maximum velocity and Michaelis–Menten constant related to the saturable uptake of the drug into renal tubular cells; P is any measure of membrane permeability, r^{PTL} and r^{TL_i}, are the radii of the proximal tubule

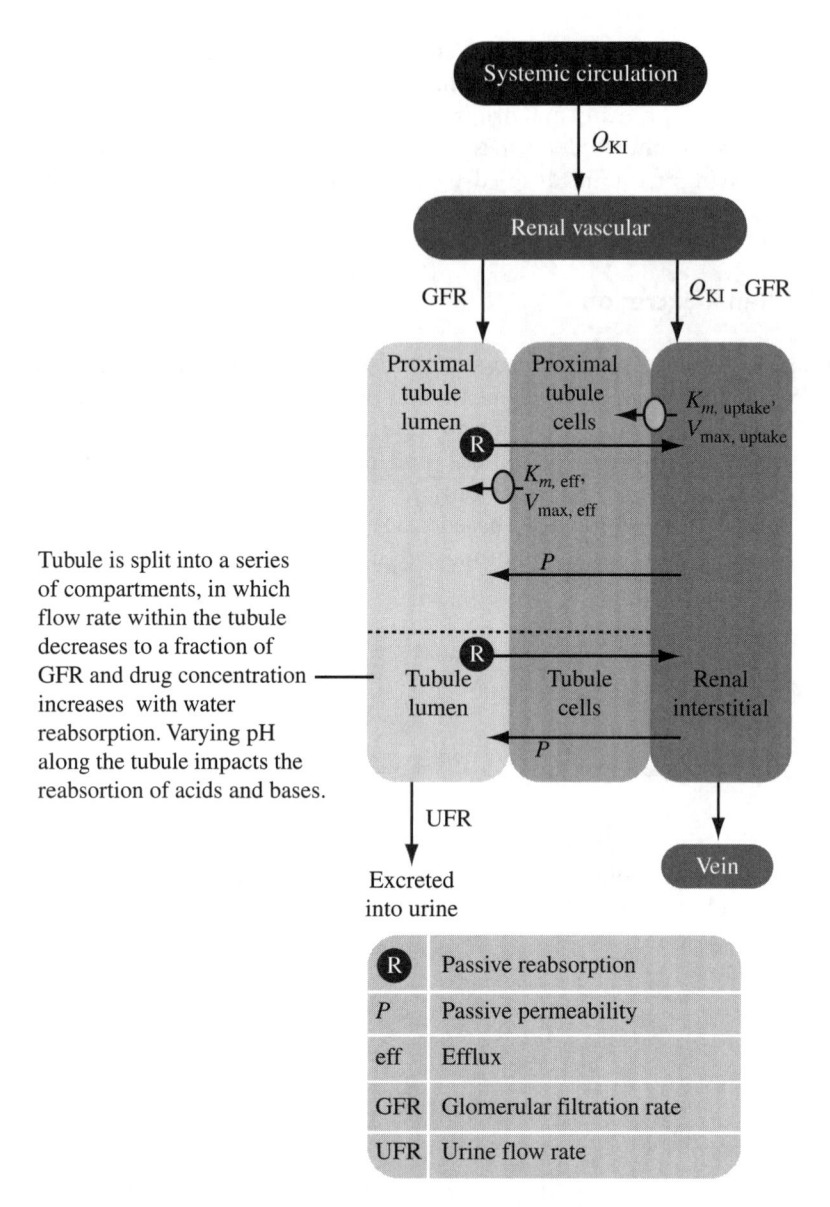

Figure 6.7. Schematic representation of a physiological model of the kidney.

lumen (PTL) and latter tubule compartments (TL$_i$); A is the amount of drug. The first term of equation 6.25 represents the rate of delivery of a drug into renal interstitium. The second term is the rate of saturable uptake into tubule cells. The third term is the rate of drug leaving the kidney. Reabsorption is modeled by diffusion-driven permeability. The last two terms represent the

6.4 PHYSIOLOGICAL MODELS

rates of passive diffusion from the capillary to the tubule or reabsorption of the drug from the tubule into capillary, depending on whether the terms are negative or positive, from PTL and TL$_i$.

Changes in Volume of Water along the Tubule Lumen.

$$\frac{dV_{KI}^{PTL}}{dt} = (GFR - FR_{KI}^{PTL} - RW_{KI}^{PTL}) \quad (6.26)$$

where FR_{KI}^{PTL} is flow rate down the tubule, RW_{KI}^{PTL} is the rate of reabsorption of water in the PTL compartment of the kidney:

$$\frac{dV_{KI}^{TL_1}}{dt} = (FR_{KI}^{PTL} - FR_{KI}^{TL_1} - RW_{KI}^{TL_1}) \quad (6.27)$$

where $FR_{KI}^{TL_1}$ is flow rate down the tubule, and $RW_{KI}^{TL_1}$ is the rate of reabsorption of water in the first compartment following proximal tubule:

$$\frac{dV_{KI}^{TL_i}}{dt} = \left(FR_{KI}^{TL(i-1)} - FR_{KI}^{TL_i} - RW_{KI}^{TL_i}\right) \quad (6.28)$$

where $FR_{KI}^{TL_i}$ is flow rate down the tubule, and $RW_{KI}^{TL_i}$ is the rate of reabsorption of water in the ith compartment following the proximal tubule:

In the last compartment (collecting duct), the rate of change of volume of water is given by

$$\frac{dV_{KI}^I}{dt} = GFR - UFR = \sum_i RW_{KI}^{TL_i} \quad (6.29)$$

The rate of transit out of the collecting duct is simply the urine flow rate UFR

Tubule Lumen.

$$\frac{dC_{KI}^{PTL}}{dt} = \frac{1}{V_{KI}^{PTL}}\left(GFR \times C_{KI}^V + \frac{V_{max,eff}}{K_{m,eff} + f_{u,KI} \times C_{KI}^{IC}} \times f_{u,KI} \times C_{KI}^{IC}\right)$$
$$- \frac{2P}{r^{PTL}} \times \left(\frac{A_{KI}^{PTL}}{V_{KI}^{PTL}} - C_{KI}^I\right) - k_t^{PTL} \times C_{KI}^{PTL} \quad (6.30)$$

$$\frac{dC_{KI}^{TL_1}}{dt} = FR_{KI}^{PTL} \times C_{KI}^{PTL} - \frac{2P}{r^{TL_1}} \times \left(\frac{A_{KI}^{TL_1}}{V_{KI}^{TL_1}} - C_{KI}^I\right) - FR_{KI}^{TL_1} \times C_{KI}^{TL_1} \quad (6.31)$$

$$\frac{dC_{KI}^{TL_i}}{dt} = FR_{KI}^{TL(i-1)} \times C_{KI}^{TL(i-1)} - \frac{2P}{r^{TL_i}} \times \left(\frac{A_{KI}^{TL_i}}{V_{KI}^{TL_i}} - C_{KI}^I\right) - FR_{KI}^{TL_i} \times C_{KI}^{TL_i} \quad (6.32)$$

Intracellular.

$$\frac{dC_{KI}^{IC}}{dt} = \frac{V_{max, uptake}}{K_{m, uptake} + f_{ub} \times C_{KI}^{I}} \times f_{ub} \times C_{KI}^{I} - \frac{V_{max, eff}}{K_{m, eff} + f_{u, KI} \times C_{KI}^{IC}} \times f_{u, KI} \times C_{KI}^{IC}$$
$$- CL_{int, KI} \times f_{u, KI} \times C_{KI}^{IC} \qquad (6.33)$$

The kidney also contains a variety of phase I DMEs such as CYPs and FMOs and phase II DMEs such as SULTs, UGTs and GSTs. The enzyme levels are much lower compared to those in the liver. $CL_{int,KI}$ is the measured intrinsic clearance of metabolism in the kidney, if this is measured and available. Otherwise, it is set to zero. Kinetic parameters related to active basolateral uptake and efflux (eff) into renal tubule are obtained from assays described in Section 6.3.

De Lannoy et al.[93] developed a physiologically-based kidney model incorporating clearance terms for diffusion and transport of drug and metabolite across the basolateral and luminal membranes of the renal cells and intrinsic clearance for renal drug metabolism. Additional inputs include physiological variables such as rate of kidney blood flow (Q_{KI}), glomerular filtration rate (generally, a fifth of the renal plasma flow in human), and urine flow rate. This model was used to describe the metabolism of enalapril. The model explains the observed discrepancies between generated and preformed enalaprilat (metabolite) elimination in the constant-flow single-pass and recirculating isolated perfused rat kidney animal models after simultaneous delivery of ^{14}C-enalapril and ^{3}H-enalaprilat. The model output accounts for the differing points of origin of the metabolite within the kidney. Other physiological renal models have described the clearance of drugs predominantly cleared by the kidney.[94–106]

Several compound- and system-specific factors influence drug elimination. Some of these can be captured by *in silico* and *in vitro* models. Physiological models for hepatobiliary and renal drug elimination employ physicochemical properties and data generated from *in vitro* experiments in a physiological context, thus providing a mechanistic framework for predictions and extrapolations.

REFERENCES

1. Ekins S, et al. Three- and four-dimensional-quantitative structure activity relationship (3D/4D-QSAR) analyses of CYP2C9 inhibitors. *Drug Metab Dispos.* 2000; 28(8):994–1002.
2. Ekins S, de Groot MJ, Jones JP. Pharmacophore and three-dimensional quantitative structure activity relationship methods for modeling cytochrome p450 active sites. *Drug Metab Dispos.* 2001;29(7):936–944.
3. Roche O, et al. Development of a virtual screening method for identification of "frequent hitters" in compound libraries. *J Med Chem.* 2002;45(1):137–142.

REFERENCES

4. Oh WS, Kim DN, Jung J, Cho KH, No KT. New combined model for the prediction of regioselectivity in cytochrome P450/3A4 mediated metabolism. *J Chem Inf Model.* 2008;48(3):591–601.
5. Vaz RJ, Zamora I, Li Y, Reiling S, Shen J, Cruciani G. The challenges of in silico contributions to drug metabolism in lead optimization. *Expert Opin Drug Metab Toxicol.* 2010;6(7):851–861.
6. Caron G, Ermondi G, Testa B. Predicting the oxidative metabolism of statins: An application of the MetaSite algorithm. *Pharm Res.* 2007;24(3):480–501.
7. Cruciani G, et al. MetaSite: Understanding metabolism in human cytochromes from the perspective of the chemist. *J Med Chem.* 2005;48(22):6970–6979.
8. Sheridan RP, Korzekwa KR, Torres RA, Walker MJ. Empirical regioselectivity models for human cytochromes P450 3A4, 2D6, and 2C9. *J Med Chem.* 2007;50(14):3173–3184.
9. Zhou D, Afzelius L, Grimm SW, Andersson TB, Zauhar RJ, Zamora I. Comparison of methods for the prediction of the metabolic sites for CYP3A4-mediated metabolic reactions. *Drug Metab Dispos.* 2006;34(6):976–983.
10. Winiwarter S, Hilgendorf C. Modeling of drug-transporter interactions using structural information. *Curr Opin Drug Discov Devel.* 2008;11(1):95–103.
11. Kolhatkar V, Polli JE. Reliability of inhibition models to correctly identify type of inhibition. *Pharm Res.* 2010;27(11):2433–2445.
12. Doddareddy MR, Cho YS, Koh HY, Kim DH, Pae AN. In silico renal clearance model using classical volsurf approach. *J Chem Inf Model.* 2006;46(3):1312–1320.
13. Paine SW, et al. A rapid computational filter for predicting the rate of human renal clearance. *J Mol Graph Model.* 2010;29(4):529–537.
14. Manga N DJ, Rowe PH, Cromin MTD. A hierarchial QSAR model for urinary excretion of drugs in humans as a predictive tool for biotransformation. *QSAR Comb Sci.* 2003;22:263–273.
15. Vermeir M, Annaert P, Mamidi RN, Roymans D, Meuldermans W, Mannens G. Cell-based models to study hepatic drug metabolism and enzyme induction in humans. *Expert Opin Drug Metab Toxicol.* 2005;1(1):75–90.
16. Jia L, Noker PE, Coward L, Gorman GS, Protopopova M, Tomaszewski JE. Interspecies pharmacokinetics and in vitro metabolism of SQ109. *Br J Pharmacol.* 2006;147(5):476–485.
17. Li AP. Human hepatocytes as an effective alternative experimental system for the evaluation of human drug properties: General concepts and assay procedures. *ALTEX.* 2008;25(1):33–42.
18. Li AP. Human hepatocytes: Isolation, cryopreservation and applications in drug development. *Chem Biol Interact.* 2007;168(1):16–29.
19. Li AP, et al. Cryopreserved human hepatocytes: Characterization of drug-metabolizing enzyme activities and applications in higher throughput screening assays for hepatotoxicity, metabolic stability, and drug-drug interaction potential. *Chem Biol Interact.* 1999;121(1):17–35.
20. Li AP. Overview: Hepatocytes and cryopreservation—a personal historical perspective. *Chem Biol Interact.* 1999;121(1):1–5.
21. McGinnity DF, Soars MG, Urbanowicz RA, Riley RJ. Evaluation of fresh and cryopreserved hepatocytes as *in vitro* drug metabolism tools for the prediction of metabolic clearance. *Drug Metab Dispos.* 2004;32(11):1247–1253.

22. Li AP. Evaluation of drug metabolism, drug-drug interactions, and *in vitro* hepatotoxicity with cryopreserved human hepatocytes. *Methods Mol Biol.* 2010;640:281–294.
23. Fahmi OA, Boldt S, Kish M, Obach RS, Tremaine LM. Prediction of drug-drug interactions from *in vitro* induction data: Application of the relative induction score approach using cryopreserved human hepatocytes. *Drug Metab Dispos.* 2008;36(9): 1971–1974.
24. Skett P. Problems in using isolated and cultured hepatocytes for xenobiotic metabolism/metabolism-based toxicity testing-solutions? *Toxicol In Vitro.* 1994;8(3):491–504.
25. Dunn JC, Yarmush ML, Koebe HG, Tompkins RG. Hepatocyte function and extracellular matrix geometry: Long-term culture in a sandwich configuration. *FASEB J.* 1989;3(2):174–177.
26. Koebe HG, Pahernik S, Eyer P, Schildberg FW. Collagen gel immobilization: A useful cell culture technique for long-term metabolic studies on human hepatocytes. *Xenobiotica.* 1994;24(2):95–107.
27. Lave T, et al. The use of human hepatocytes to select compounds based on their expected hepatic extraction ratios in humans. *Pharm Res.* 1997;14(2):152–155.
28. Griffin SJ, Houston JB. Prediction of *in vitro* intrinsic clearance from hepatocytes: Comparison of suspensions and monolayer cultures. *Drug Metab Dispos.* 2005;33(1):115–120.
29. Soars MG, McGinnity DF, Grime K, Riley RJ. The pivotal role of hepatocytes in drug discovery. *Chem Biol Interact.* 2007;168(1):2–15.
30. Parrish AR, Gandolfi AJ, Brendel K. Precision-cut tissue slices: Applications in pharmacology and toxicology. *Life Sci.* 1995;57(21):1887–1901.
31. Lerche-Langrand C, Toutain HJ. Precision-cut liver slices: Characteristics and use for in vitro pharmaco-toxicology. *Toxicology.* 2000;153(1–3):221–253.
32. Ekins S. Past, present, and future applications of precision-cut liver slices for *in vitro* xenobiotic metabolism. *Drug Metab Rev.* 1996;28(4):591–623.
33. Gandolfi AJ, Wijeweera J, Brendel K. Use of precision-cut liver slices as an *in vitro* tool for evaluating liver function. *Toxicol Pathol.* 1996;24(1):58–61.
34. Edwards RJ, et al. Induction of cytochrome P450 enzymes in cultured precision-cut human liver slices. *Drug Metab Dispos.* 2003;31(3):282–288.
35. Lake BG, Charzat C, Tredger JM, Renwick AB, Beamand JA, Price RJ. Induction of cytochrome P450 isoenzymes in cultured precision-cut rat and human liver slices. *Xenobiotica.* 1996;26(3):297–306.
36. Miyazaki H, Sekine T, Endou H. The multispecific organic anion transporter family: Properties and pharmacological significance. *Trends Pharmacol Sci.* 2004; 25(12):654–662.
37. Hagenbuch B, Meier PJ. Organic anion transporting polypeptides of the OATP/ SLC21 family: Phylogenetic classification as OATP/ SLCO superfamily, new nomenclature and molecular/functional properties. *Pflugers Arch.* 2004;447(5):653–665.
38. Hagenbuch B, Stieger B, Foguet M, Lubbert H, Meier PJ. Functional expression cloning and characterization of the hepatocyte Na+/bile acid cotransport system. *Proc Natl Acad Sci USA.* 1991;88(23):10629–10633.

39. Meier PJ, Eckhardt U, Schroeder A, Hagenbuch B, Stieger B. Substrate specificity of sinusoidal bile acid and organic anion uptake systems in rat and human liver. *Hepatology*. 1997;26(6):1667–1677.
40. Friesema EC, et al. Identification of thyroid hormone transporters. *Biochem Biophys Res Commun*. 1999;254(2):497–501.
41. Hirano M, Maeda K, Shitara Y, Sugiyama Y. Contribution of OATP2 (OATP1B1) and OATP8 (OATP1B3) to the hepatic uptake of pitavastatin in humans. *J Pharmacol Exp Ther*. 2004;311(1):139–146.
42. Groothuis GM, Hulstaert CE, Kalicharan D, Hardonk MJ. Plasma membrane specialization and intracellular polarity of freshly isolated rat hepatocytes. *Eur J Cell Biol*. 1981;26(1):43–51.
43. Bow DA, Perry JL, Miller DS, Pritchard JB, Brouwer KL. Localization of P-gp (Abcb1) and Mrp2 (Abcc2) in freshly isolated rat hepatocytes. *Drug Metab Dispos*. 2008;36(1):198–202.
44. Cui Y, Konig J, Leier I, Buchholz U, Keppler D. Hepatic uptake of bilirubin and its conjugates by the human organic anion transporter SLC21A6. *J Biol Chem*. 2001;276(13):9626–9630.
45. Shimizu M, et al. Contribution of OATP (organic anion-transporting polypeptide) family transporters to the hepatic uptake of fexofenadine in humans. *Drug Metab Dispos*. 2005;33(10):1477–1481.
46. Hagenbuch B, Scharschmidt BF, Meier PJ. Effect of antisense oligonucleotides on the expression of hepatocellular bile acid and organic anion uptake systems in xenopus laevis oocytes. *Biochem J*. 1996;316 (Pt 3)(Pt 3):901–904.
47. Kullak-Ublick GA, et al. Organic anion-transporting polypeptide B (OATP-B) and its functional comparison with three other OATPs of human liver. *Gastroenterology*. 2001;120(2):525–533.
48. Ismair MG, et al. Hepatic uptake of cholecystokinin octapeptide by organic anion-transporting polypeptides OATP4 and OATP8 of rat and human liver. *Gastroenterology*. 2001;121(5):1185–1190.
49. Ishiguro N, et al. Establishment of a set of double transfectants coexpressing organic anion transporting polypeptide 1B3 and hepatic efflux transporters for the characterization of the hepatobiliary transport of telmisartan acylglucuronide. *Drug Metab Dispos*. 2008;36(4):796–805.
50. Letschert K, Komatsu M, Hummel-Eisenbeiss J, Keppler D. Vectorial transport of the peptide CCK-8 by double-transfected MDCKII cells stably expressing the organic anion transporter OATP1B3 (OATP8) and the export pump ABCC2. *J Pharmacol Exp Ther*. 2005;313(2):549–556.
51. Sasaki M, Suzuki H, Ito K, Abe T, Sugiyama Y. Transcellular transport of organic anions across a double-transfected madin-darby canine kidney II cell monolayer expressing both human organic anion-transporting polypeptide (OATP2/SLC21A6) and multidrug resistanceassociated protein 2 (MRP2/ABCC2). *J Biol Chem*. 2002;277(8):6497–6503.
52. Spears KJ, et al. Directional trans-epithelial transport of organic anions in porcine LLC-PK1 cells that co-express human OATP1B1 (OATP-C) and MRP2. *Biochem Pharmacol*. 2005;69(3):415–423.

53. Cui Y, Konig J, Keppler D. Vectorial transport by double-transfected cells expressing the human uptake transporter SLC21A8 and the apical export pump ABCC2. *Mol Pharmacol.* 2001;60(5):934–943.
54. Kopplow K, Letschert K, Konig J, Walter B, Keppler D. Human hepatobiliary transport of organic anions analyzed by quadruple-transfected cells. *Mol Pharmacol.* 2005;68(4):1031–1038.
55. Sarkadi B, Price EM, Boucher RC, Germann UA, Scarborough GA. Expression of the human multidrug resistance cDNA in insect cells generates a high activity drug-stimulated membrane ATPase. *J Biol Chem.* 1992;267(7):4854–4858.
56. Drueckes P, Schinzel R, Palm D. Photometric microtiter assay of inorganic phosphate in the presence of acid-labile organic phosphates. *Anal Biochem.* 1995;230(1):173–177.
57. Sauna ZE, Smith MM, Muller M, Kerr KM, Ambudkar SV. The mechanism of action of multidrug-resistance-linked P-glycoprotein. *J Bioenerg Biomembr.* 2001; 33(6):481–491.
58. Garrigues A, Nugier J, Orlowski S, Ezan E. A high-throughput screening microplate test for the interaction of drugs with P-glycoprotein. *Anal Biochem.* 2002;305 (1):106–114.
59. Xia CQ, et al. Comparison of species differences of P-glycoproteins in beagle dog, rhesus monkey, and human using atpase activity assays. *Mol Pharm.* 2006;3(1): 78–86.
60. Markowitz JS, Devane CL, Liston HL, Boulton DW, Risch SC. The effects of probenecid on the disposition of risperidone and olanzapine in healthy volunteers. *Clin Pharmacol Ther.* 2002;71(1):30–38.
61. Polli JW, et al. Rational use of *in vitro* P-glycoprotein assays in drug discovery. *J Pharmacol Exp Ther.* 2001;299(2):620–628.
62. Masereeuw R, Russel FG, Miller DS. Multiple pathways of organic anion secretion in renal proximal tubule revealed by confocal microscopy. *Am J Physiol.* 1996;271(6 Pt 2):F1173–1182.
63. Masereeuw R, Saleming WC, Miller DS, Russel FG. Interaction of fluorescein with the dicarboxylate carrier in rat kidney cortex mitochondria. *J Pharmacol Exp Ther.* 1996;279(3):1559–1565.
64. Masereeuw R, van den Bergh EJ, Bindels RJ, Russel FG. Characterization of fluorescein transport in isolated proximal tubular cells of the rat: Evidence for mitochondrial accumulation. *J Pharmacol Exp Ther.* 1994;269(3):1261–1267.
65. Miller DS, Letcher S, Barnes DM. Fluorescence imaging study of organic anion transport from renal proximal tubule cell to lumen. *Am J Physiol.* 1996;271(3 Pt 2): F508–520.
66. Miller DS, Barnes DM, Pritchard JB. Confocal microscopic analysis of fluorescein compartmentation within crab urinary bladder cells. *Am J Physiol.* 1994;267(1 Pt 2): R16–25.
67. Miller DS, Pritchard JB. Nocodazole inhibition of organic anion secretion in teleost renal proximal tubules. *Am J Physiol.* 1994;267(3 Pt 2):R695–704.
68. LeCluyse EL, Fix JA, Audus KL, Hochman JH. Regeneration and maintenance of bile canalicular networks in collagen-sandwiched hepatocytes. *Toxicol In Vitro.* 2000;14(2):117–132.

# REFERENCES

69. Boess F, et al. Gene expression in two hepatic cell lines, cultured primary hepatocytes, and liver slices compared to the *in vivo* liver gene expression in rats: Possible implications for toxicogenomics use of in vitro systems. *Toxicol Sci.* 2003;73(2):386–402.
70. Kienhuis AS, et al. A sandwich-cultured rat hepatocyte system with increased metabolic competence evaluated by gene expression profiling. *Toxicol In Vitro.* 2007;21(5):892–901.
71. Tuschl G, Mueller SO. Effects of cell culture conditions on primary rat hepatocytes-cell morphology and differential gene expression. *Toxicology.* 2006;218(2–3):205–215.
72. LeCluyse EL, Audus KL, Hochman JH. Formation of extensive canalicular networks by rat hepatocytes cultured in collagen-sandwich configuration. *Am J Physiol.* 1994;266(6 Pt 1):C1764–1774.
73. Du Y, et al. Synthetic sandwich culture of 3D hepatocyte monolayer. *Biomaterials.* 2008;29(3):290–301.
74. Ghibellini G, Leslie EM, Brouwer KL. Methods to evaluate biliary excretion of drugs in humans: An updated review. *Mol Pharm.* 2006;3(3):198–211.
75. Hagenbuch B, Meier PJ. The superfamily of organic anion transporting polypeptides. *Biochim Biophys Acta.* 2003;1609(1):1–18.
76. Mizuno N, Niwa T, Yotsumoto Y, Sugiyama Y. Impact of drug transporter studies on drug discovery and development. *Pharmacol Rev.* 2003;55(3):425–461.
77. Shitara Y, Horie T, Sugiyama Y. Transporters as a determinant of drug clearance and tissue distribution. *Eur J Pharm Sci.* 2006;27(5):425–446.
78. Olinga P, et al. Characterization of the uptake of rocuronium and digoxin in human hepatocytes: Carrier specificity and comparison with *in vivo* data. *J Pharmacol Exp Ther.* 1998;285(2):506–510.
79. Shitara Y, et al. Function of uptake transporters for taurocholate and estradiol 17beta-D-glucuronide in cryopreserved human hepatocytes. *Drug Metab Pharmacokinet.* 2003;18(1):33–41.
80. Petzinger E, Fuckel D. Evidence for a saturable, energy-dependent and carrier-mediated uptake of oral antidiabetics into rat hepatocytes. *Eur J Pharmacol.* 1992;213(3):381–391.
81. Soars MG, Webborn PJ, Riley RJ. Impact of hepatic uptake transporters on pharmacokinetics and drug-drug interactions: Use of assays and models for decision making in the pharmaceutical industry. *Mol Pharm.* 2009;6(6):1662–1677.
82. Soars MG, Grime K, Sproston JL, Webborn PJ, Riley RJ. Use of hepatocytes to assess the contribution of hepatic uptake to clearance *in vivo*. *Drug Metab Dispos.* 2007;35(6):859–865.
83. Paine SW, Parker AJ, Gardiner P, Webborn PJ, Riley RJ. Prediction of the pharmacokinetics of atorvastatin, cerivastatin, and indomethacin using kinetic models applied to isolated rat hepatocytes. *Drug Metab Dispos.* 2008;36(7):1365–1374.
84. Reinoso RF, Telfer BA, Brennan BS, Rowland M. Uptake of teicoplanin by isolated rat hepatocytes: Comparison with *in vivo* hepatic distribution. *Drug Metab Dispos.* 2001;29(4 Pt 1):453–459.
85. Poirier A, Cascais AC, Funk C, Lave T. Prediction of pharmacokinetic profile of valsartan in human based on *in vitro* uptake transport data. *J Pharmacokinet Pharmacodyn.* 2009;36(6):585–611.

86. Poirier A, et al. Design, data analysis, and simulation of *in vitro* drug transport kinetic experiments using a mechanistic *in vitro* model. *Drug Metab Dispos.* 2008; 36(12):2434–2444.
87. Hasegawa M, Kusuhara H, Endou H, Sugiyama Y. Contribution of organic anion transporters to the renal uptake of anionic compounds and nucleoside derivatives in rat. *J Pharmacol Exp Ther.* 2003;305(3):1087–1097.
88. Watanabe T, Kusuhara H, Maeda K, Shitara Y, Sugiyama Y. Physiologically-based pharmacokinetic modeling to predict transporter-mediated clearance and distribution of pravastatin in humans. *J Pharmacol Exp Ther.* 2009;328(2):652–662.
89. Pang KS, Maeng HJ, Fan J. Interplay of transporters and enzymes in drug and metabolite processing. *Mol Pharm.* 2009;6(6):1734–1755.
90. Sun H, Pang KS. Physiological modeling to understand the impact of enzymes and transporters on drug and metabolite data and bioavailability estimates. *Pharm Res.* 2010;27(7):1237–1254.
91. Fan J, Chen S, Chow EC, Pang KS. PBPK modeling of intestinal and liver enzymes and transporters in drug absorption and sequential metabolism. *Curr Drug Metab.* 2010;11(9):743–761.
92. Pang KS, Durk MR. Physiologically-based pharmacokinetic modeling for absorption, transport, metabolism and excretion. *J Pharmacokinet Pharmacodyn.* 2010; 37(6):591–615.
93. de Lannoy IA, Hirayama H, Pang KS. A physiological model for renal drug metabolism: Enalapril esterolysis to enalaprilat in the isolated perfused rat kidney. *J Pharmacokinet Biopharm.* 1990;18(6):561–587.
94. Blakey GE, Nestorov IA, Arundel PA, Aarons LJ, Rowland M. Quantitative structure–pharmacokinetics relationships: I. Development of a whole-body physiologically-based model to characterize changes in pharmacokinetics across a homologous series of barbiturates in the rat. *J Pharmacokinet Biopharm.* 1997; 25(3):277–312.
95. Boom SP, Meyer I, Wouterse AC, Russel FG. A physiologically-based kidney model for the renal clearance of ranitidine and the interaction with cimetidine and probenecid in the dog. *Biopharm Drug Dispos.* 1998;19(3):199–208.
96. Harashima H, Sawada Y, Sugiyama Y, Iga T, Hanano M. Prediction of serum concentration time course of quinidine in human using a physiologically-based pharmacokinetic model developed from the rat. *J Pharmacobiodyn.* 1986;9(2): 132–138.
97. Janku I, Zvara K. Quantitative analysis of drug handling by the kidney using a physiological model of renal drug clearance. *Eur J Clin Pharmacol.* 1993;44(6): 521–524.
98. Kawahara M, Nanbo T, Tsuji A. Physiologically-based pharmacokinetic prediction of p-phenylbenzoic acid disposition in the pregnant rat. *Biopharm Drug Dispos.* 1998;19(7):445–453.
99. Komiya I. Urine flow dependence of renal clearance and interrelation of renal reabsorption and physicochemical properties of drugs. *Drug Metab Dispos.* 1986; 14(2):239–245.

REFERENCES

100. Levitt DG, Schoemaker RC. Human physiologically-based pharmacokinetic model for ACE inhibitors: Ramipril and ramiprilat. *BMC Clin Pharmacol*. 2006;6:1.
101. Nestorov IA, Aarons LJ, Rowland M. Physiologically-based pharmacokinetic modeling of a homologous series of barbiturates in the rat: A sensitivity analysis. *J Pharmacokinet Biopharm*. 1997;25(4):413–447.
102. Plowchalk DR, Andersen ME, deBethizy JD. A physiologically-based pharmacokinetic model for nicotine disposition in the Sprague-Dawley rat. *Toxicol Appl Pharmacol*. 1992;116(2):177–188.
103. Russel FG, Wouterse AC, Van Ginneken CA. Physiologically-based pharmacokinetic model for the renal clearance of iodopyracet and the interaction with probenecid in the dog. *Biopharm Drug Dispos*. 1989;10(2):137–152.
104. Russel FG, Wouterse AC, van Ginneken CA. Physiologically-based pharmacokinetic model for the renal clearance of salicyluric acid and the interaction with phenolsulfonphthalein in the dog. *Drug Metab Dispos*. 1987;15(5):695–701.
105. Russel FG, Wouterse AC, van Ginneken CA. Physiologically-based pharmacokinetic model for the renal clearance of phenolsulfonphthalein and the interaction with probenecid and salicyluric acid in the dog. *J Pharmacokinet Biopharm*. 1987;15(4):349–368.
106. Tsuji A, Nishide K, Minami H, Nakashima E, Terasaki T, Yamana T. Physiologically-based pharmacokinetic model for cefazolin in rabbits and its preliminary extrapolation to man. *Drug Metab Dispos*. 1985;13(6):729–739.

7

GENERIC WHOLE-BODY PHYSIOLOGICALLY-BASED PHARMACOKINETIC MODELING

CONTENTS

7.1 Introduction 153
7.2 Structure of a Generic Whole Body PBPK Model 154
7.3 Model Assumptions 157
7.4 Commercial PBPK Software 158
 References 159

7.1 INTRODUCTION

Chapters 4–6 provided descriptions of physiological models for absorption, distribution, and hepatobiliary elimination. This chapter integrates these organ models into a single whole-body physiologically-based pharmacokinetic (PBPK) model. A detailed description of the physiological elements affecting the absorption, distribution, metabolism, and excretion (ADME) will aid mechanistic understanding through hypothesis testing and provide improved predictions of PK. The value of the model may be limited by the availability of reliable

---

Physiologically-Based Pharmacokinetic (PBPK) Modeling and Simulations: Principles, Methods, and Applications in the Pharmaceutical Industry, First Edition. Sheila Annie Peters.
© 2012 John Wiley & Sons, Inc. Published 2012 by John Wiley & Sons, Inc.

physiological and compound-dependent data generated through experiments. The complexity of a PBPK model should, therefore, depend on the intended purpose of the model and availability of the necessary input data. A generic PBPK model that incorporates several PK mechanisms is described in this chapter. In the absence of all the required input parameters, reduction in model complexity is achieved through the use of appropriate default values, which serve to eliminate the terms for which input parameters were not provided.

7.2 STRUCTURE OF A GENERIC WHOLE BODY PBPK MODEL

Generic whole-body PBPK modeling and simulation methods aim to simulate the concentration–time profiles of compounds administered orally, intravenously, or through any other route of interest to any species, by integrating within a physiological framework, measured or calculated physicochemical properties such as log P, pK_a, solubility, and *in vitro* PK data such as Caco-2 permeability, protein binding, and intrinsic clearance. As compounds progress through drug discovery and more *in vivo* data from animals become available, these can be integrated into the models after appropriate scaling to make quantitative predictions for humans. A generic whole-body PBPK model can be built by combining the physiological models of the GI tract, brain, liver, kidney, and other distribution tissues discussed in the preceding chapters with a somatic 14-compartment PBPK model. The schematic structure of a generic whole-body PBPK model is illustrated in Figure 7.1. Somatic PBPK models usually comprise 14 compartments in which tissues such as adipose, heart, lung, brain, muscle, bone, skin, stomach, spleen, pancreas, gut, liver, kidney, and thymus are represented as compartments. These compartments are linked together by the arterial and venous compartments. Organs that are critical for blood circulation (heart and lung), eliminating organs (liver, kidney, and intestine), large and/or lipophilic organs (muscle, skin, adipose) that are important for drug distribution, and the pharmacodynamic effect organ (lung for respiratory drugs, heart for cardiovascular drugs) are generally included as compartments in a PBPK model. Each compartment has an associated blood flow rate, volume, and a tissue partition coefficient. For cancer drugs, the tumor tissue is introduced as a separate compartment whose associated characteristics vary with time.[1] The structure of blood vessels in tumors is markedly different from those in normal tissues. The quick development of tumor cells is possible only if they stimulate the formation of new blood vessels for nutritional and oxygen supply. Thus, tumor blood vessels have an abnormal architecture, as their endothelial cells are poorly aligned, with large pore sizes, wide fenestrae, and lacking a smooth muscle layer and effective lymphatic drainage. The enhanced permeability and retention (EPR) effect of tumor tissue justifies the need to treat it as a separate compartment. Simulation of drug concentrations in a tumor compartment in the long term would need to consider the time-dependent changes in the physiology

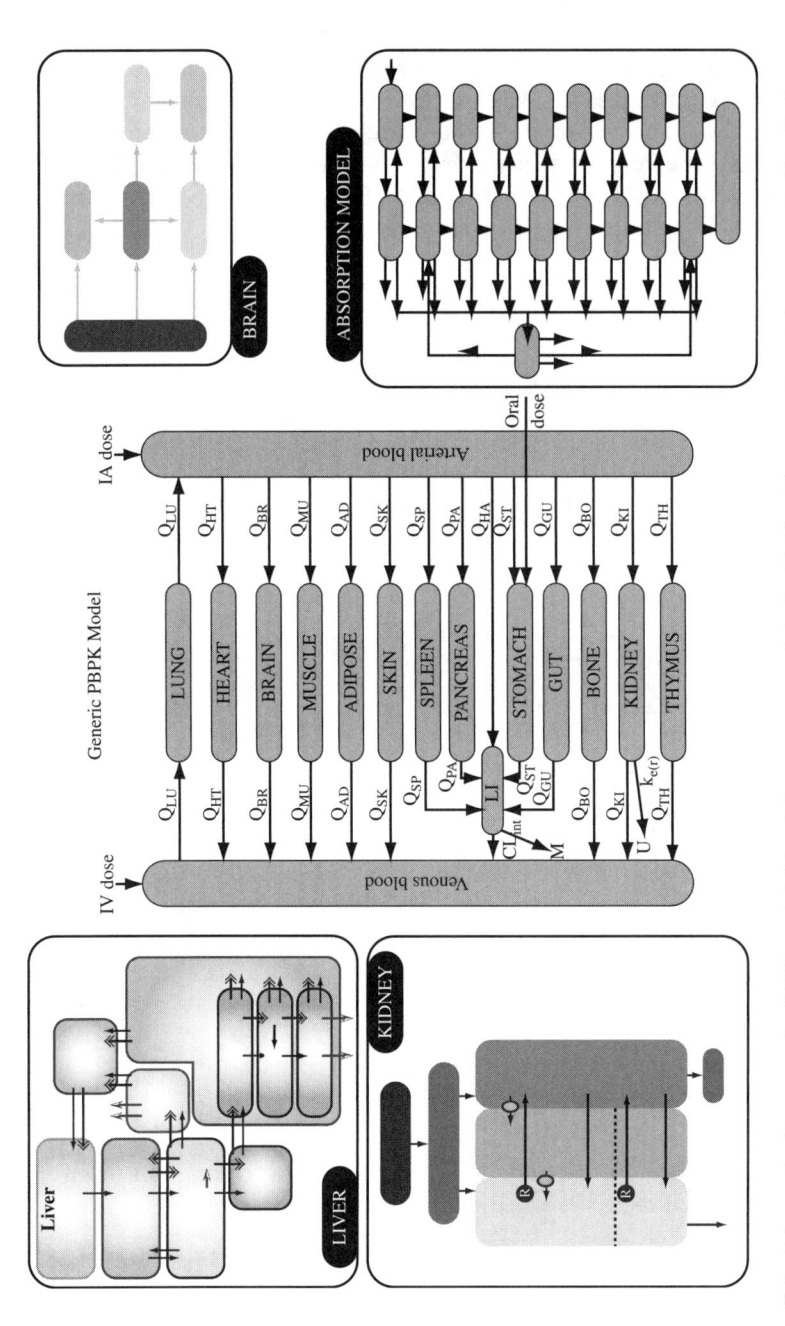

Figure 7.1. Schematic diagram of a PBPK model. Blood flow rates associated with the 14 compartments is represented by Q. Q_{HA} represents the blood flow rate from the hepatic artery. Intravenous (IV) dose enters the venous compartment and intraarterial (IA) dose enters the arterial compartment, while the oral dose enters the stomach compartment. Intrinsic clearance rate (CL_{int}) governs the rate of conversion to metabolite (M). k_{renal} is the renal elimination rate constant that determines the amount eliminated in the urine. The somatic model is linked to an absorption model through the liver compartment. Detailed models of liver, kidney, and brain can be incorporated into the somatic model as shown.

of the compartment as it grows. Similarly, for drugs targeting pregnant women, the growing foetus should be considered as a time-varying compartment.

Physiologically-based PK modeling attempts to mathematically model all the physical and biophysical processes that determine the fate of a drug in the body. The dynamics of a drug within each tissue, as it permeates the plasma membrane, gets eliminated in the tissue or carried forward are represented by differential equations. The rates of change of plasma drug concentration (C) dC/dt in the lungs (LU), arterial blood (ART), venous blood (VEN), stomach (ST), and gut (GU) are described by the following differential equations:

Somatic Compartments

1. *Lungs* (LU)

$$\frac{dC_{LU}}{dt} = \frac{Q_{LU}}{V_{LU}}\left(C_{VEN} - \frac{C_{LU}R}{f_{up}K_{LU:p,u}}\right)$$

where R is the blood–plasma concentration ratio. For inhaled drugs, an additional term representing the rate of absorption from the lungs should be introduced.

2. *Arterial Blood* (ART)

$$\frac{dC_{ART}}{dt} = \frac{1}{V_{ART}}\left[Q_{LU}\left(\frac{C_{LU}R}{f_{up}K_{LU:,p,u}} - C_{ART}\right) + \text{arterial infusion rate}\right]$$

Arterial infusion rate will be zero unless the drug is administered intra-arterially.

3. *Venous Blood* (VEN)

$$\frac{dC_{VEN}}{dt} = \frac{1}{V_{VEN}}\left(\sum\frac{Q_T C_T R}{f_{up}K_{T,p,u}} - Q_{LU}C_{VEN} + \text{venous infusion rate}\right)$$

where T represents all organs/tissues, other than gut, pancreas, spleen, stomach, and lung. Venous infusion rate will be zero unless the drug is administered intravenously.

4. *Stomach* (ST)

$$\frac{dC_{ST}}{dt} = \frac{1}{V_{ST}}\left[Q_{ST}\left(C_{ART} - \frac{C_{ST}R}{f_{up}K_{ST,p,u}}\right) + \text{gastric absorption}\right]$$

Gastric absorption represents the dynamic changes in the amount of an orally administered drug absorbed from stomach.

5. *Gut* (GU)

$$\frac{dC_{GU}}{dt} = \frac{1}{V_{GU}}\left[Q_{GU}\left(C_{ART} - \frac{C_{GU}R}{f_{up}K_{GU,p,u}}\right) + \text{total intestinal absorption}\right]$$

where V_T is the volume of the tissue, and Q is the blood flow rate. Intestinal absorption represents the dynamic changes in the amount of an orally administered drug absorbed from the small and large intestines. Apart from oral drugs, gastric and intestinal absorptions can be non-zero for inhaled drugs. This is because a considerable portion of the inhaled drug is swallowed, depending on the inhalation device used. These differential equations for somatic compartments are combined with the equations for GI compartments (Chapter 4), equations representing distribution into noneliminating organs (Chapter 5), and equations representing elimination in kidney and metabolism of parent drug and metabolites in the liver (Chapter 6). They are simultaneously solved with mathematical software such as MATLAB (The MathWorks, Natick, MA). It is important to ensure that certain basic constraints are respected. These constraints are listed below.

1. The sum of blood flow rates to all tissues making up the compartments should add up to the cardiac output.
2. The weights of individual tissues making up the compartments should be less than or equal to the body weight.
3. The mass balance of the compound of interest should be maintained. Thus, the total amount of a compound and its metabolites in the body at any time and the amount of compound and its metabolites eliminated by that time should add up to the initial dose administered.

7.3 MODEL ASSUMPTIONS

Most PBPK models assume that each tissue compartment is well-stirred and perfusion rate limited. The assumption that a tissue/organ is a single, well-stirred compartment with uniform concentration of the drug within it can fail for drugs whose intracellular concentrations can be different from their extracellular concentrations due to involvement of transporters in their uptake. Additionally, subcellular compartments such as mitochondria and lysosomes can assume importance for basic drugs, as the pH in these compartments is less compared to that in plasma. Many PBPK models (like the equations described in Section 7.2) also assume that drug distribution into different tissue compartments is driven by perfusion-limited rather than permeability-limited kinetics. The assumption of perfusion flow-limited kinetics is justified for small, lipophilic drugs that are neither too big or too hydrophilic to distribute into organs. Flow-limited distribution is not valid for drugs whose influx into cells is limited by permeability barriers. Thus, large, hydrophilic molecules are mostly confined to plasma. Drugs that are transporter dependent for overcoming the membrane barrier are not delivered to their intracellular targets at the organ perfusion rate. For biologics, permeability-limited distribution needs to be considered (see Chapter 10). In Chapter 6, transporter-mediated kinetics is discussed for liver and kidney, the two organs other than brain and intestine where transporters play an important role in drug disposition.

7.4 COMMERCIAL PBPK SOFTWARE

The lack of specialized skills needed for the development of PBPK modeling has been one of the main barriers to its wide applicability. However, in recent years, several commercial PBPK software packages, differing widely in their scope and application, have become available.[2,3] Notable among these are listed below.

- *Gastroplus* (http://www.simulations-plus.com), a physiologically-based absorption model, has an extended module called PBPKPlus that includes a perfusion- or permeability-limited distribution model and linear/saturable metabolism with passive/saturable active transport (uptake and efflux).
- *PKQuest* (http://www.pkquest.com) is a free, interactive PBPK software routine in which the user can employ a preprogrammed optimized human or rat PK data or an arbitrary data set for education and training purposes. Drugs are classified as "extracellular" or "highly fat soluble" and, therefore, do not require information about tissue/blood partition coefficients. PKQuest takes minimal user input parameters and is an excellent tool for teaching pharmacokinetics.
- *Cloe PK* (http://www.cyprotex.com/cloepredict/) is a server-based product from Cyprotex that predicts whole-body pharmacokinetics from simple *in vitro* ADME and physicochemical properties for decision making in drug discovery.
- *Simcyp* (http://www.simcyp.com) simulates and predicts the population variability of whole-body kinetics and drug–drug interactions by building virtual populations and incorporating extensive demographic, physiological, pathological, genetic, and ethnic variability.
- *PK-Sim* (http://www.systems-biology.com), from Bayer Technology Services GmbH, is an integrated whole-body PBPK software. It is built into two different modules—PK-Sim Preclinical and PK-Sim Clinical—that can be integrated with MoBi. PK-Sim Preclinical is meant for the analysis of mammalian pharmacokinetics following single or multiple intravenous or oral administrations. PK-Sim Clinical allows for the consideration of subject populations in clinical trials. The program supports metabolism as well as transporter-mediated uptake and efflux in any organ. MoBi is built for mechanistic modeling of biological processes and drug actions. Its integration with PK-Sim, therefore, permits simultaneous consideration of whole-body physiology-based PK models with PD models for drug action.

- *acslXtreme* (http://www.acslx.com) software enables modeling and simulation of continuous dynamic systems and processes. It is widely used for PBPK and PK/PD modeling.
- *Berkeley Madonna* (http://www.berkeleymadonna.com/) is a general-purpose differential equation solver serving as a high-level tool for scientific computing. It is generally useful as an educational aid.

Physiological Parameters for PBPK Modeling (P^3M) provides a unique source of data for human physiological parameters.[4]

Unlike the data-driven empirical models, PBPK models integrate compound-dependent and species-dependent parameters to predict the PK profile of compounds, thereby enabling an understanding of the underlying PK mechanisms in the absence of a good fit between the predicted and observed PK profiles. Through sensitivity analysis, PBPK models provide an understanding of which compound-dependent parameter needs to be altered in order to achieve the desired change in a pharmacokinetic property. The most important application of PBPK modeling is for extrapolations—from preclinical to human, from one route of administration to another, and from one population to another. A further level of integration with mechanistic pharmacodynamic models (PBPK/PD) extends the advantages of PBPK modeling, namely, gaining mechanistic insights and extrapolations, to pharmacological response. By considering polymorphism of enzymes and transporters as well as differences in physiological parameters between and within different subpopulations (pediatrics, elderly, obese, diabetic, or people with other disease conditions, and pregnant women), PBPK models can simulate the resulting variability in pharmacokinetic parameters in these populations. These applications will be dealt in detail in Section II of the book.

REFERENCES

1. Modok S, Hyde P, Mellor HR, Roose T, Callaghan R. Diffusivity and distribution of vinblastine in three-dimensional tumour tissue: Experimental and mathematical modeling. *Eur J Cancer*. 2006;42(14):2404–2413.
2. Schmitt W, Willmann S. Physiology-based pharmacokinetic modeling: Ready to be used. *Drug Disc Today: Tech*. 2004;1(4):449.
3. Rowland M, Peck C, Tucker G. Physiologically-based pharmacokinetics in drug development and regulatory science. *Annu Rev Pharmacol Toxicol*. 2011;51:45–73.
4. Price PS, et al. Modeling interindividual variation in physiological factors used in PBPK models of humans. *Crit Rev Toxicol*. 2003;33(5):469–503.

8

VARIABILITY, UNCERTAINTY, AND SENSITIVITY ANALYSIS

CONTENTS

8.1 Introduction 161
8.2 Need for Uncertainty Analysis 162
8.3 Sources of Physiological, Anatomical, Enzymatic, and Transporter Variability 163
8.4 Modeling Uncertainty and Population Variability With Monte Carlo Simulations 169
8.5 Sensitivity Analysis 172
8.6 Conclusions 174
    Keywords 174
    References 175

8.1 INTRODUCTION

Physiologically-based pharmacokinetic modeling involves a large number of compound- and physiology-dependent input parameters. Compound-dependent

Physiologically-Based Pharmacokinetic (PBPK) Modeling and Simulations: Principles, Methods, and Applications in the Pharmaceutical Industry, First Edition. Sheila Annie Peters.
© 2012 John Wiley & Sons, Inc. Published 2012 by John Wiley & Sons, Inc.

variables generated in assays have uncertainties associated with the measurements, while those predicted are likely to be associated with prediction errors. By incorporating the uncertainties in every compound parameter used as input to a PBPK model, one can predict the overall uncertainty in the predicted concentration–time profile and consequently in the PK parameters calculated using the profile, such as volume of distribution, clearance, and the like. An uncertainty assessment aids robust predictions. Similarly, all physiological parameters are generally variable across any given population and by incorporating all known **variability** in a population for each of the physiological and anatomical parameters as well as in enzyme and transporter levels, the resulting overall variability in the concentration–time profile that is expected in a population can be estimated. Several approaches to uncertainty and variability analysis are available[1]; notable among these are the Monte–Carlo method,[2–4] fuzzy simulation,[5,6] and Bayesian Markov chain Monte Carlo (MCMC)[7,8] simulation. Modeling of interindividual variability in the pharmacokinetics of drugs[9,10] and environmental chemicals[11,12] have been described in the literature. The extent to which an error/variability in an input parameter will affect the concentration–time profile as well as the PK parameters derived from it will depend on how sensitive the profile is to that input parameter. A **sensitivity analysis** provides a quantitative evaluation of the sensitivity of PK parameters to each of the input parameters and can, therefore, be employed at the design stage of drug discovery to identify parameters that have the most influence in bringing about a desired change in the PK parameters. In this chapter, the principles underlying Monte Carlo simulations and sensitivity analysis will be outlined. Applications of sensitivity and variability analysis will be discussed in Section II of the book.

8.2 NEED FOR UNCERTAINTY ANALYSIS

Lack of good estimates of predicted or measured compound-dependent input parameters and model inadequacies result in poor predictions from a PBPK model. Inadequacies in model structure need to be corrected to the extent possible and appropriate model validations should be done prior to using the PBPK model for any predictions. However, uncertainties in model input can differ from case to case. The greater the sensitivity of an outcome to an input parameter, the greater would be the impact of error in this input parameter on the outcome. A sensitivity analysis, would, therefore, provide insight into which of the parameters need to be accurately determined in order to substantially improve the predictive ability of the model. If further improvements in the measurement of the sensitive input parameters are not possible, at least an understanding of the quantitative impact of the error in the input needs to be in place. Such an understanding is the aim of an uncertainty analysis. If confidence in the predictions of a PBPK model is high, then any deviations of the observed from the predicted can be interpreted mechanistically. On the other hand, if the impact of error itself is likely to cause a large deviation from the observed, then mechanistic interpretations of PBPK predictions are baseless.

8.3 SOURCES OF PHYSIOLOGICAL, ANATOMICAL, ENZYMATIC, AND TRANSPORTER VARIABILITY

A population may be subdivided based on ethnicity. The mean values of different biological parameters (physiological, anatomical, enzymatic, and transporters) can be expected to be different between different ethnic populations.[13] For example, south Asian and African populations have lower CYP3A4 activities compared to Caucasians.[14,15] Asians have lower CYP2C19 and NAT2, while Africans have lower CYP2D6 levels compared to the Caucasians. The **genetic polymorphism** of enzymes (CYP2C9, CYP2C19, CYP2D6, UGT1A1, UGT2B7, and NAT2)[16–21] and transporters (BCRP, BSEP, MRP2, OATP1B1, OATP1B3, OCT1, and OCT2) should also be considered. MRP and OATP have 12 and 15 isoforms, respectively,[22,23] but all of them may not play a role in drug transport. Within an ethnic population, many biological parameters vary with age[24–29] and gender, justifying a shift in the average values of these parameters for pediatrics,[30–32] adults, and geriatrics (elderly populations) of either gender. Immaturity of CYP3A4, CYP1A2, glucuronidating, glycine conjugating enzymes in the neonates and the slower activity of renal excretion, CYP3A4, CYP2C19, and CYP2D6 in the elderly are examples of age-related differences. Models of continuous variation of different parameters with age can also be employed to set the mean values. Within an ethnic population of a certain age group and gender, further subdivisions based on their disease states should be considered to allow for a shift in the mean values for an appropriate subset of biological parameters. Liver cirrhosis patients have changes in blood flow due to the development of portacaval shunts. A subfunctional liver also leads to lower circulating levels of plasma proteins. Depending on the nature of a drug and the severity of the disease, varying degrees of impaired systemic clearance and first-pass metabolism can be expected.[33] Ethnicity, gender, age, and body mass index (BMI) will determine the average weights, while ethnicity, gender, and age will determine the average heights of a population. Weight and height together determine the body surface area (BSA).[34–38] Body surface area determines the cardiac output, proportion of blood flow rates to various organs, and organ volumes. An algorithm for building a virtual population has been proposed.[39] The obese subpopulation,[40] with a large BSA, is also expected to have an increased cardiac output, elevated levels of CYP2E1, and acid glycoprotein (AGP) with important consequences to the pharmacokinetic parameters[41,42] (Table 8.1).

Different subpopulations of defined ethnicity, gender, age, and disease state are each characterized by its own set of mean values for different biological parameters. Every biological parameter also has an associated variance within

TABLE 8.1. Impact of Changes in Biological Parameters on Pharmacokinetic Properties

PK Property	Change in Biological Parameter	Causes	Impact on PK
Absorption	Decrease in small intestinal surface area	Disease or age	Reduced absorption
	Reduced gastric emptying rate	Fed state; type of food	Slower rate of absorption
	Increase in gastric pH	Disease; age; some drugs Fed state	Reduced solubility of basic drugs
Distribution	Increased body fat relative total body water	Obesity	Increased volume of distribution of lipophilic drug
	Reduced albumin Increased AGP	Liver disease Obesity	Reduced protein binding of acidic drugs and increased protein binding of basic drugs Appropriate changes to both drug distribution and metabolism
Metabolism	Reduced CYP activity	Polymorphism Disease or age	Reduced metabolism
Elimination	Reduced GFR and tubular functions	Age	Altered elimination of drugs that are predominantly cleared by the kidney For compounds that are glucuronidated, the parent drug recirculates (enterohepatic recirculation) for longer due to reduced elimination of the glucuronide

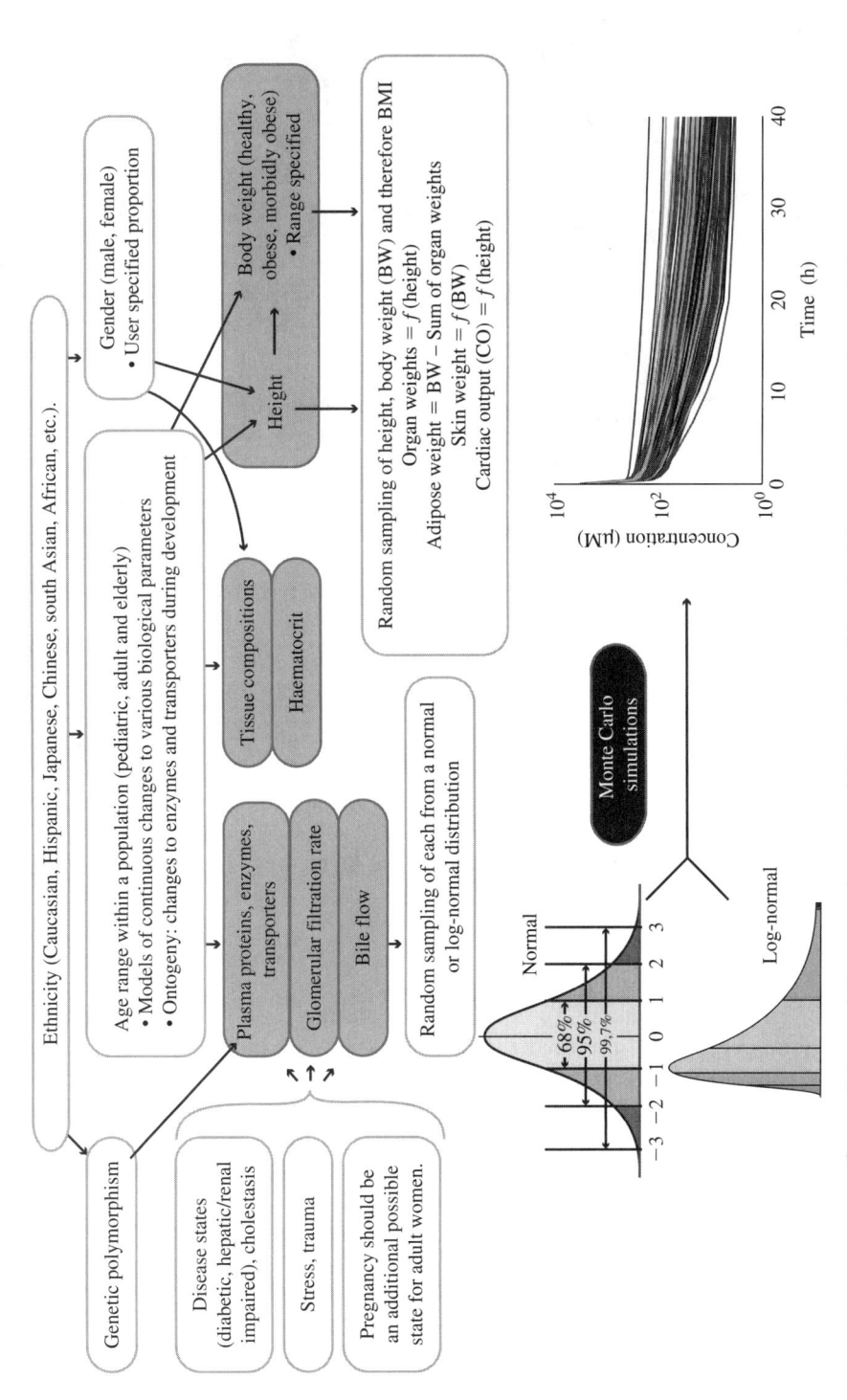

Figure 8.1. Modeling population variability: Every biological input parameter is sampled from a normal or log-normal distribution associated with a certain mean and variance. The Monte Carlo simulations result in several rather than a single pharmacokinetic profile.

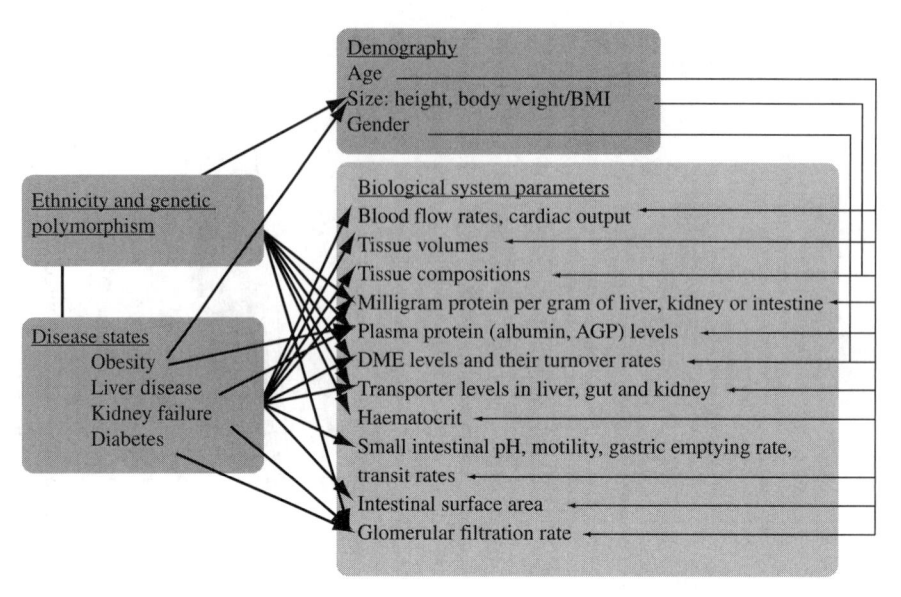

Figure 8.2. Sources of variability in the physiological parameters that impact pharmacokinetics. AGP: alpha glycoprotein; DME: drug metabolizing enzymes

a given subpopulation (see Fig. 8.1). This variance could simply be set to a plausible range of about 30%, representative of the reality or obtained from literature, if reported. The impact of ethnicity, gender, age, pregnancy, and health condition on the biological parameters is summarized in Figure 8.2. Sources of population data are obtained from the literature,[43–48] NHANES III study,[49] and the annals of the ICRP.[50] Table 8.2 summarizes the different sources of uncertainties and population variability.

For orally administered drugs, additional sources of variability[51] come from gastric emptying (influenced by fluid intake, dosage form, fed/fasted condition, and composition of formulation),[52–55] intestinal motility, and residence time.[56] In addition, gastric and intestinal pH can vary between individuals[57,58] and occasion,[59] between fed and fasted states,[60,61] and change with pathophysiological conditions[62] and disease[63,64]. When dose is staggered with respect to food intake, intestinal transit times were shown to be considerably shorter than for fed and fasted states.[65] The intestinal epithelial cells also express a variety of phase I and phase II drug-metabolizing enzymes, mainly CYPs[66–70] and UGTs,[71,72] whose abundance and distribution in the small intestine can vary between individuals due to age, genetic variability, diet, or disease.[73] **Variability** in intestinal transit time, pH, and enzyme/transporter differences along the gut[69,74,75] are likely to be reflected much more in the gut bioavailability of slow absorption drugs, (with poor solubility or permeability, or formulated as extended release) than of fast absorption drugs.

TABLE 8.2. Sources of Uncertainties and Population Variability

	Uncertainty in Measurement or Prediction	Variability in Physiological, Anatomical, Enzymatic, or Transporter
Absorption	Solubility and dissolution pH dependence of solubility and dissolution pK_a Particle size Particle disintegration Composition of formulation Permeability	Inter individual and inter occasion pH variation with fed/fasted conditions, food and its composition, pH variation with disease Gastric emptying rate variation with fed/fasted conditions, age, and disease Transit time variation with dose-staggering, fed/fasted conditions, age, and disease Variation of activity and abundance of gut transporters and DMEs[c] with age, genetics, diet, disease (celiac)
Distribution	Fraction unbound in plasma Lipophilicity pK_a	Variation of blood flow to various organs with body weight, age, ethnicity, and disease (hypertension, dislipidemia) Variation of albumin levels with age, gender, disease (cirrhosis) Tissue composition (body fat) variation with age, gender, and ethnicity Hematocrit
Metabolism	Intrinsic clearance MPPGL[a] or HPGL[b] Scaling factors Fraction unbound in plasma Food and other environmental factors	Hepatic blood flow variation with body weight, age, ethnicity, and disease (hypertension, dislipidaemia, cirrhosis) Impact of genetic polymorphism on the activity of hepatic enzymes and uptake transporters (OATP1B1)[76] Variation of enzyme abundance with age (CYPs 2C9, 2C19)[77] and ethnicity (CYPs 3A4, 3A5)[78,79] Hematocrit Variation of albumin levels with age, gender, disease (cirrhosis) Liver weight

(*Continued*)

TABLE 8.2. (*Continued*)

	Uncertainty in Measurement or Prediction	Variability in Physiological, Anatomical, Enzymatic, or Transporter
Renal elimination	Fraction unbound in plasma pK_a	Renal blood flow variation with body weight, age, ethnicity, and disease (hypertension, dislipidaemia) Genetic polymorphism of renal transporters Glomerular filtration rate variation with disease (hypertension and diabetes) and age Variation of albumin levels with age, gender, disease (cirrhosis) Urine flow rate variation with fluid in-take pH of urine
Biliary elimination	Efflux kinetics	Genetic polymorphism of hepatic efflux transporters Bile flow variation with disease (choestasis)

[a]MPPGL: milligram protein per gram liver.
[b]HPGL: million hepatocytes per gram liver.
[c]DME: Drug metabolizing enzymes.

8.4 MODELING UNCERTAINTY AND POPULATION VARIABILITY WITH MONTE CARLO SIMULATIONS

Variability in the pharmacokinetics of a drug in a population can translate to differences in its efficacy and safety in that population. Population PK modeling[80–82] aims to characterize the observed pharmacokinetic variability in a population in terms of patient demographic, pathophysiological, or genetic factors so that clinically significant differences can be identified and addressed through appropriate dosage corrections in any population. Population variability can be modeled by an **a posteriori** (top-down) approach in which empirical compartmental PK models aim to explain observed variability in terms of plausible covariates.[83,84] Alternatively, variability can be modeled by an **a priori** (bottom-up) approach, in which determinants of variability in each of the PK-determining parameters are used to estimate an overall variability in the concentration–time profile, which is then compared with the observed variability. A priori modeling allows for an early assessment of the impact of variability from different sources (genetics, ethnicity, age, gender, body weight, pregnancy, and disease states) on the pharmacokinetics and thereby on the efficacy or safety of the drug. Price et al,[85] describe the modeling of interindividual variability using the physiological parameters for PBPK modeling (P^3M) tool. The objective is the creation of a virtual population that is representative of a real population that is going to be included in a clinical study, so that the clinical study can become a confirmatory rather than an exploratory study. An a priori modeling can either be through a deterministic description of the determinants of variability[30] or through stochastic Monte Carlo simulations. Each physiological parameter can be characterized by a probability distribution with a population mean and variance. For parameters such as clearances and rate constants, the logarithms of individual values are normally distributed and therefore, a **log-normal distribution** may be appropriate. For others, a **normal** (Gaussian) **distribution** may be assumed. Uniform distribution and **weibull distribution** are other common distributions used. A weibull distribution can be used when an observed distribution for a biological parameter does not fit the standard normal or log-normal functions. It is given by

$$f(x) = \frac{\beta}{\theta} \times \left(\frac{x-\delta}{\theta}\right)^{\beta-1} \exp\left(\frac{x-\delta}{\theta}\right)^{\beta} \quad x > \delta$$

It can be parameterized to fit a wide variety of observed distributions because of the shape parameter β, which can take a value between 0 and 20. θ is the scale parameter and δ is the location parameter (Fig. 8.3) that can be used to represent lag time in gastric emptying, for example, when x is time. Once the statistical distribution is specified and defined quantitatively with appropriate values for each of the biological parameters, the Monte Carlo (MC) approach involves[84,85] multiple sampling from each distribution using simple random or

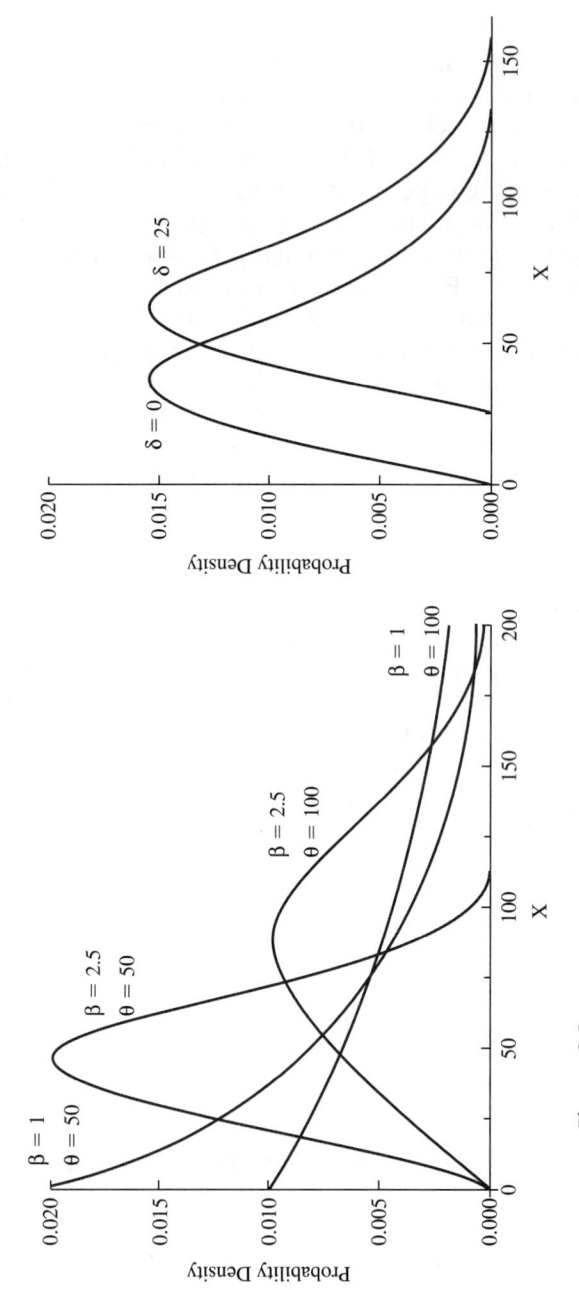

Figure 8.3. Weibull distribution: effects of the shape parameter, β, the scale parameter, θ, and the location parameter, δ, on the distribution.

8.4 MODELING UNCERTAINTY AND POPULATION VARIABILITY

Latin hypercube method[86,87] and computation of model outputs, resulting in a set of output values that can be statistically treated to obtain 90 or 95% confidence intervals for each of the PK parameters such as clearance, volume of distribution, and the like. Bayesian methods[88,89] use a set of known prior distributions that may be specified independent of each other to arrive at a posterior probability distribution, using newly available data. Bayesian methods are useful to update or modify prior probability distributions that were built from sparse data, as more population data/information becomes available. True determinants of variability from population data are obtained after removing the contributions from uncertainty. It is often difficult to obtain an analytical expression for the posterior probability distribution, which is needed for an MC simulation. Fortunately, with the advent of Markov chain Monte Carlo methods (MCMC),[90] sampling from posterior distributions lacking analytical expression has become possible.

Many physiological parameters vary in proportion to each other (correlated covariates). For example, cardiac output and breathing rate vary in proportion to one another. Similarly, the correlations, if any, in enzyme/transporter abundances should be captured. Blood flow rates to various organs are correlated, as the total blood flow to all organs should always add up to the cardiac output. A variability analysis that neglects these nested and universal interdependencies and constraints will overestimate the expected variability in a subpopulation. Thus, when sampling different physiological and anatomical parameters with MC, the covariance/interdependence of parameters should be addressed. One way to address interdependencies is to link physiological parameters to lean body mass or to other body parameters via fixed scaling coefficients. For example, a complete random sampling of cardiac output and body weight may yield an unrealistic combination of these two correlated parameters. A lean person cannot have the highest cardiac output. In order to avoid such a situation, deterministic allometric functions (see references in Appendices) identified from experimental data sets are employed, and these serve to reduce the extent of variability of cardiac output (CO).

$$CO = 187 \times BW^{0.81} \qquad (8.1)$$

This equation[91] quotes a coefficient of variation (CV) of 31% for the coefficient preceding the body weight (BW) and 1.3% on the exponent. Sampling body weight from a truncated normal distribution (to avoid unrealistically low or high body weights within an age group) thus provides variable CO for each BW. The cardiac output in turn determines the blood flow rates to individual organs. To ensure that the sum of all blood flow rates add up to CO, the blood flow rate to any organ is taken to be a proportion of CO. The proportions to different organs are constrained to vary within 1, by sampling from a multivariate Dirichlet distribution[4] in which parameters are represented as proportions rather than independent values. A further nested level of physiological variability in blood flow rates arises from environmental (stress), pharmacodynamic (anesthesia), and diurnal fluctuations.[92]

8.5 SENSITIVITY ANALYSIS

A number of compound-dependent variables and physiological parameters determine the final pharmacokinetic properties such as clearance, volume of distribution, half-life, elimination, and the like. A sensitivity analysis provides a measure of how sensitive a desired PK property is, to small changes in an influencing compound- or system-dependent parameter.

A general sensitivity strategy has been proposed by Ginot et al.[93] A univariate, local sensitivity coefficient, S_{ij} of an output parameter j to an input parameter i at their local, baseline values O and I is given by

$$S_{ij} = \frac{\partial O_j}{\partial I_i} \qquad (8.2)$$

The right-hand term represents the rate of change of O_j with respect to I_i, while keeping all other variables constant. A more valuable measure of sensitivity is the normalized sensitivity coefficient or the relative sensitivity coefficient, RS_{ij}, described as the partial derivative of each output variable of interest with respect to each model input parameter, normalized by both the output variable and model input parameter. It is given by

$$RS_{ij} = \frac{\partial O_j}{\partial I_i} \times \frac{I_i}{O_j} \qquad (8.3)$$

With this relative measure, it is possible to compare the sensitivity of a single output parameter with respect to several input parameters. Instead of using a simple difference as shown in equation 8.3, the derivative can be evaluated using a central difference[94] to consider the difference on either side of a central value for the model input parameter as shown in Figure 8.4. A 10% proportional change in a model output variable with a 10% change in the input parameter from its original value would yield a sensitivity value of 1 or represent a 1 : 1 sensitivity with the parameter of interest. The influence of a model input parameter on an output is considered insignificant if the corresponding normalized sensitivity coefficient is less than 0.1.

With local sensitivity analysis, covariances or interactions between the parameters cannot be assessed. For example, if the influence of intestinal blood flow on bioavailability is being analyzed, corresponding changes to other blood flow rates should also be taken into account, as blood flow rate changes are not independent of each other. A multivariate, global sensitivity analyses[86,95–98] that spans the entire parameter space and examines the impact simultaneously

8.5 SENSITIVITY ANALYSIS

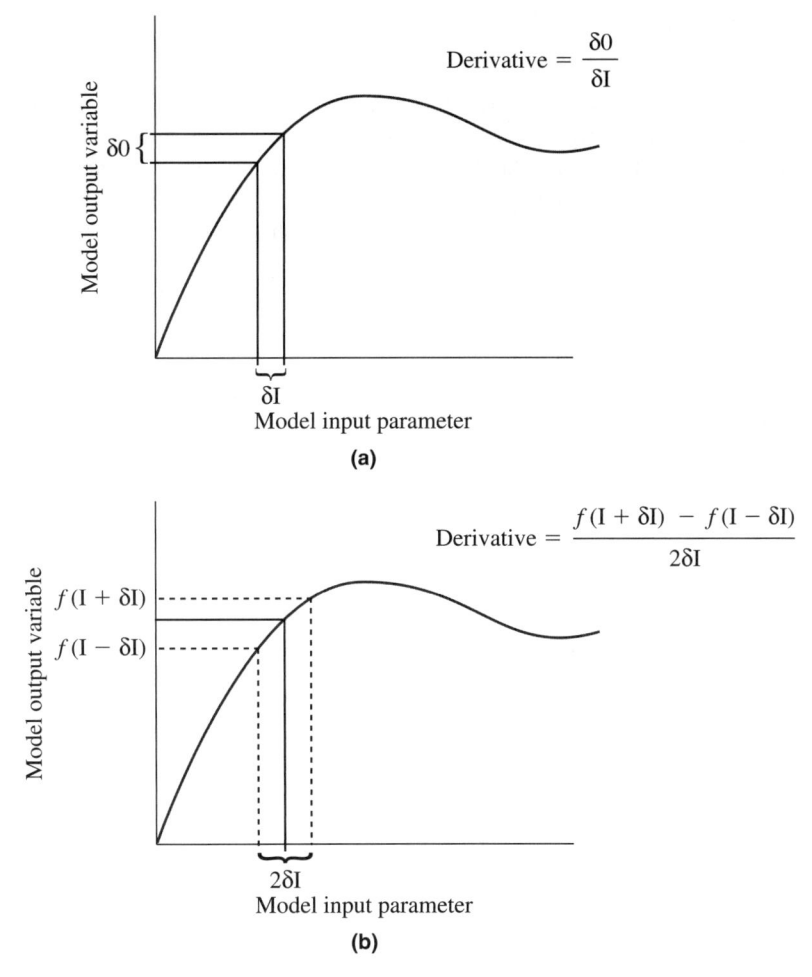

Figure 8.4. (a) Finite difference and (b) central difference methods for sensitivity derivative.

varying all parameters through their full range of values is needed for a better characterization. A workflow for sensitivity analysis has been proposed.[99]

In the design stage of drug discovery, sensitivity analysis can be very valuable in focusing on the modulation of the right properties (those that have maximum sensitivity) in order to achieve a desired change in a PK property. Sensitive input parameters affect an output to a large extent, and, therefore, any small error in their measurement will have a large impact on the accuracy of a predicted output parameter. Blood flow rates generally emerge as the most influential of the physiological parameters.

8.6 CONCLUSIONS

While uncertainty and sensitivity analyses are valuable tools in the drug discovery phase, variability analysis is extremely valuable during the clinical development of a drug. Population analysis seeks to characterize the variability in the time profile of plasma concentrations observed in different populations in terms of inter- and intraindividual variability. Such an analysis is critical in the assessment of a need for a minimum toxic dose (MTD) adjustment, especially for drugs having narrow therapeutic window. A single covariate may not lead to substantial changes to warrant dose adjustments—but a combination of several covariates, for example, an elderly population that happens to be poor or extensive metabolizers of a drug, are on several concurrent medications, and suffering from age-related diseases such as renal/hepatic impairment may need an adjusted dose to have the required efficacy or to stay below exposure levels associated with toxicity. Simulation of pharmacokinetic variability through a priori modeling of inter- and intraindividual variability of biological parameters in a virtual population enable an understanding of the sources of variability, which would then make it possible to predict the pharmacokinetics of a drug in a wider population, saving valuable time and money. Examples of how it can be achieved are detailed in Chapter 14.

KEYWORDS

A Posteriori: A posteriori knowledge is dependent on experience or empirical evidence.

A Priori: A priori knowledge is independent of experience.

Genetic Polymorphism: Gene variants that occur in at least 1% of the population, resulting in the expression of proteins with higher, lower, or no activity or no expression at all.

Latin Hypercube Method: Form of stratified sampling in which members of a population are grouped into relatively homogeneous mutually exclusive subgroups (such as male, female) prior to sampling. Then random or systematic sampling is applied within each stratum. It is an improvement over the classical random sampling technique in Monte Carlo methods, as convergence to a reasonably accurate random distribution is achieved in a lesser number or runs than necessary for a Monte Carlo simulation. This often improves the representativeness of the sample by reducing sampling error. It can produce a weighted mean that has less variability than the arithmetic mean of a simple random sample of the population.

Log-Normal Distribution: Parameters that are associated with low, positive mean values and large variances have skewed distributions that fit a log-normal distribution. It is characterized by a geometric mean and a multiplicative standard deviation.

Normal Distribution: Gaussian or bell-shaped function that describes the random variation that occurs in scientific data. It is characterized by an arithmetic mean and an additive standard deviation

Sensitivity Analysis: Sensitivity analysis provides an understanding of how a model responds to a parametric or structural change. In a global sensitivity analysis all the parameters are varied simultaneously and the sensitivity is measured over the entire range of each input parameter.

Uncertainty: Arises from lack of knowledge or errors in assumptions, experiments, and data processing techniques and could be reduced by increasing the size of the sample studied.

Variability: System-specific property that cannot be altered. The system can, for example, represent the human physiology of a population. The greater the sample size, the greater is the observed variability. The distribution of population variability can be modeled by a multivariate probability distribution function.

Weibull Distribution: Generic distribution function that reduces to exponential function when the shape parameter is 1, to Rayleigh distribution when the shape parameter is 2, and to **normal distribution** when the shape parameter is 3. It approximates the **log-normal distribution** for several values of the shape parameter.

REFERENCES

1. Nestorov I. Modeling and simulation of variability and uncertainty in toxicokinetics and pharmacokinetics. *Toxicol Lett.* 2001;120(1–3):411–420.
2. Woodruff TJ, Bois FY, Auslander D, Spear RC. Structure and parameterization of pharmacokinetic models: Their impact on model predictions. *Risk Anal.* 1992;12(2):189–201.
3. Farrar D, Allen B, Crump K, Shipp A. Evaluation of uncertainty in input parameters to pharmacokinetic models and the resulting uncertainty in output. *Toxicol Lett.* 1989;49(2–3):371–385.
4. Krewski D, Wang Y, Bartlett S, Krishnan K. Uncertainty, variability, and sensitivity analysis in physiological pharmacokinetic models. *J Biopharm Stat.* 1995;5(3):245–271.
5. Seng KY, Nestorov I, Vicini P. Physiologically-based pharmacokinetic modeling of drug disposition in rat and human: A fuzzy arithmetic approach. *Pharm Res.* 2008;25(8):1771–1781.
6. Seng KY, Vicini P, Nestorov IA. A fuzzy physiologically-based pharmacokinetic modeling framework to predict drug disposition in humans. *Conf Proc IEEE Eng Med Biol Soc.* 2006;Suppl:6485–6488.
7. Bois FY. Applications of population approaches in toxicology. *Toxicol Lett.* 2001;120(1–3):385–394.

8. Bernillon P, Bois FY. Statistical issues in toxicokinetic modeling: A Bayesian perspective. *Environ Health Perspect.* 2000;108 (Suppl 5):883–893.
9. Willmann S, et al. Development of a physiology-based whole-body population model for assessing the influence of individual variability on the pharmacokinetics of drugs. *J Pharmacokinet Pharmacodyn.* 2007;34(3):401–431.
10. Jamei M, Dickinson GL, Rostami-Hodjegan A. A framework for assessing interindividual variability in pharmacokinetics using virtual human populations and integrating general knowledge of physical chemistry, biology, anatomy, physiology and genetics: A tale of "bottom-up" vs "top-down" recognition of covariates. *Drug Metab Pharmacokinet.* 2009;24(1):53–75.
11. Bois FY, Jamei M, Clewell HJ. PBPK modeling of inter-individual variability in the pharmacokinetics of environmental chemicals. *Toxicology.* 2010.
12. Barton HA, et al. Characterizing uncertainty and variability in physiologically-based pharmacokinetic models: State of the science and needs for research and implementation. *Toxicol Sci.* 2007;99(2):395–402.
13. Bjornsson TD, et al. A review and assessment of potential sources of ethnic differences in drug responsiveness. *J Clin Pharmacol.* 2003;43(9):943–967.
14. Dorne JL, Walton K, Renwick AG. Human variability for metabolic pathways with limited data (CYP2A6, CYP2C9, CYP2E1, ADH, esterases, glycine and sulphate conjugation). *Food Chem Toxicol.* 2004;42(3):397–421.
15. Dorne JL, Walton K, Renwick AG. Human variability in the renal elimination of foreign compounds and renal excretion-related uncertainty factors for risk assessment. *Food Chem Toxicol.* 2004;42(2):275–298.
16. Haber LT, Maier A, Gentry PR, Clewell HJ, Dourson ML. Genetic polymorphisms in assessing interindividual variability in delivered dose. *Regul Toxicol Pharmacol.* 2002;35(2 Pt 1):177–197.
17. Ginsberg G, Guyton K, Johns D, Schimek J, Angle K, Sonawane B. Genetic polymorphism in metabolism and host defense enzymes: Implications for human health risk assessment. *Crit Rev Toxicol.* 2010;40(7):575–619.
18. Ginsberg G, Smolenski S, Hattis D, Guyton KZ, Johns DO, Sonawane B. Genetic polymorphism in glutathione transferases (GST): Population distribution of GSTM1, T1, and P1 conjugating activity. *J Toxicol Environ Health B Crit Rev.* 2009;12(5–6):389–439.
19. Neafsey P, Ginsberg G, Hattis D, Johns DO, Guyton KZ, Sonawane B. Genetic polymorphism in CYP2E1: Population distribution of CYP2E1 activity. *J Toxicol Environ Health B Crit Rev.* 2009;12(5–6):362–388.
20. Neafsey P, Ginsberg G, Hattis D, Sonawane B. Genetic polymorphism in cytochrome P450 2D6 (CYP2D6): Population distribution of CYP2D6 activity. *J Toxicol Environ Health B Crit Rev.* 2009;12(5–6):334–361.
21. Walker K, Ginsberg G, Hattis D, Johns DO, Guyton KZ, Sonawane B. Genetic polymorphism in N-acetyltransferase (NAT): Population distribution of NAT1 and NAT2 activity. *J Toxicol Environ Health B Crit Rev.* 2009;12(5–6):440–472.
22. Iwai M, Suzuki H, Ieiri I, Otsubo K, Sugiyama Y. Functional analysis of single nucleotide polymorphisms of hepatic organic anion transporter OATP1B1 (OATP-C). *Pharmacogenetics.* 2004;14(11):749–757.

23. Nishizato Y, et al. Polymorphisms of OATP-C (SLC21A6) and OAT3 (SLC22A8) genes: Consequences for pravastatin pharmacokinetics. *Clin Pharmacol Ther.* 2003;73(6):554–565.
24. Yang F, Tong X, McCarver DG, Hines RN, Beard DA. Population-based analysis of methadone distribution and metabolism using an age-dependent physiologically-based pharmacokinetic model. *J Pharmacokinet Pharmacodyn.* 2006;33(4):485–518.
25. Dorne JL, Ragas AM, Frampton GK, Spurgeon DS, Lewis DF. Trends in human risk assessment of pharmaceuticals. *Anal Bioanal Chem.* 2007;387(4):1167–1172.
26. Dorne JL, Skinner L, Frampton GK, Spurgeon DJ, Ragas AM. Human and environmental risk assessment of pharmaceuticals: Differences, similarities, lessons from toxicology. *Anal Bioanal Chem.* 2007;387(4):1259–1268.
27. Dorne JL, Renwick AG. The refinement of uncertainty/safety factors in risk assessment by the incorporation of data on toxicokinetic variability in humans. *Toxicol Sci.* 2005;86(1):20–26.
28. Dorne JL, Walton K, Renwick AG. Human variability in xenobiotic metabolism and pathway-related uncertainty factors for chemical risk assessment: A review. *Food Chem Toxicol.* 2005;43(2):203–216.
29. Renwick AG, Dorne JL, Walton K. An analysis of the need for an additional uncertainty factor for infants and children. *Regul Toxicol Pharmacol.* 2000;31(3):286–296.
30. Edginton AN, Schmitt W, Willmann S. Development and evaluation of a generic physiologically-based pharmacokinetic model for children. *Clin Pharmacokinet.* 2006;45(10):1013–1034.
31. Bouzom F, Walther B. Pharmacokinetic predictions in children by using the physiologically-based pharmacokinetic modeling. *Fundam Clin Pharmacol.* 2008;22(6):579–587.
32. Johnson TN, Rostami-Hodjegan A. Resurgence in the use of physiologically-based pharmacokinetic models in pediatric clinical pharmacology: Parallel shift in incorporating the knowledge of biological elements and increased applicability to drug development and clinical practice. *Paediatr Anaesth.* 2011;21(3):291–301.
33. Johnson TN, Boussery K, Rowland-Yeo K, Tucker GT, Rostami-Hodjegan A. A semimechanistic model to predict the effects of liver cirrhosis on drug clearance. *Clin Pharmacokinet.* 2010;49(3):189–206.
34. Du Bois D, Du Bois EF. A formula to estimate the approximate surface area if height and weight be known. 1916. *Nutrition.* 1989;5(5):303–11; discussion 312–313.
35. Verbraecken J, Van de Heyning P, De Backer W, Van Gaal L. Body surface area in normal-weight, overweight, and obese adults. A comparison study. *Metabolism.* 2006;55(4):515–524.
36. Mosteller RD. Simplified calculation of body-surface area. *N Engl J Med.* 1987;317(17):1098.
37. Haycock GB, Schwartz GJ, Wisotsky DH. Geometric method for measuring body surface area: A height-weight formula validated in infants, children, and adults. *J Pediatr.* 1978;93(1):62–66.
38. Gehan EA, George SL. Estimation of human body surface area from height and weight. *Cancer Chemother Rep.* 1970;54(4):225–235.

39. Schmitt W, Willmann S. Physiology-based pharmacokinetic modeling: Ready to be used. *Drug Discovery Today: Technologies*. 2004;1(4):449.
40. Cheymol G. Effects of obesity on pharmacokinetics implications for drug therapy. *Clin Pharmacokinet*. 2000;39(3):215–231.
41. Hanley MJ, Abernethy DR, Greenblatt DJ. Effect of obesity on the pharmacokinetics of drugs in humans. *Clin Pharmacokinet*. 2010;49(2):71–87.
42. Blouin RA, Warren GW. Pharmacokinetic considerations in obesity. *J Pharm Sci*. 1999;88(1):1–7.
43. Thompson CM, et al. Database for physiologically-based pharmacokinetic (PBPK) modeling: Physiological data for healthy and health-impaired elderly. *J Toxicol Environ Health B Crit Rev*. 2009;12(1):1–24.
44. Hense HW, et al. The associations of body size and body composition with left ventricular mass: Impacts for indexation in adults. *J Am Coll Cardiol*. 1998;32 (2):451–457.
45. Lauer MS, Anderson KM, Larson MG, Levy D. A new method for indexing left ventricular mass for differences in body size. *Am J Cardiol*. 1994;74(5):487–491.
46. Stanforth PR, et al. Generalized abdominal visceral fat prediction models for black and white adults aged 17–65 y: The HERITAGE family study. *Int J Obes Relat Metab Disord*. 2004;28(7):925–932.
47. Maroun LL, Graem N. Autopsy standards of body parameters and fresh organ weights in nonmacerated and macerated human fetuses. *Pediatr Dev Pathol*. 2005;8 (2):204–217.
48. Frankenfield DC, Rowe WA, Cooney RN, Smith JS, Becker D. Limits of body mass index to detect obesity and predict body composition. *Nutrition*. 2001;17(1):26–30.
49. Center for Disease Control and Prevention. Third national health and nutrition examination survey (NHANES III). Available at: http://www.cdc.gov/nchs/nhanes.htm.
50. International Commission on Radiological Protection (ICRP). *Basic Anatomical and Physiological Data for Use in Radiological Protection: Reference Values*. Amsterdam: ICRP publication 89; Elsevier Science, 2002.
51. Jamei M, et al. Population-based mechanistic prediction of oral drug absorption. *AAPS J*. 2009;11(2):225–237.
52. Kelly K, et al. Comparison of the rates of disintegration, gastric emptying, and drug absorption following administration of a new and a conventional paracetamol formulation, using gamma scintigraphy. *Pharm Res*. 2003;20(10):1668–1673.
53. Dressman JB, Fleisher D. Mixing-tank model for predicting dissolution rate control or oral absorption. *J Pharm Sci*. 1986;75(2):109–116.
54. Davis SS, Hardy JG, Fara JW. Transit of pharmaceutical dosage forms through the small intestine. *Gut*. 1986;27(8):886–892.
55. Olsson B, Szamosi J. Multiple dose pharmacokinetics of a new once daily extended release tolterodine formulation versus immediate release tolterodine. *Clin Pharmacokinet*. 2001;40(3):227–235.
56. Yu LX, Amidon GL. Characterization of small intestinal transit time distribution in humans. *Int. J. Pharm*. 1998;171:157.
57. Fallingborg J, Christensen LA, Ingeman-Nielsen M, Jacobsen BA, Abildgaard K, Rasmussen HH. pH-profile and regional transit times of the normal gut measured by a radiotelemetry device. *Aliment Pharmacol Ther*. 1989;3(6):605–613.

REFERENCES

58. Dressman JB, et al. Upper gastrointestinal (GI) pH in young, healthy men and women. *Pharm Res.* 1990;7(7):756–761.
59. Ibekwe VC, Fadda HM, McConnell EL, Khela MK, Evans DF, Basit AW. Interplay between intestinal pH, transit time and feed status on the in vivo performance of pH responsive ileo-colonic release systems. *Pharm Res.* 2008;25(8):1828–1835.
60. Evans DF, Pye G, Bramley R, Clark AG, Dyson TJ, Hardcastle JD. Measurement of gastrointestinal pH profiles in normal ambulant human subjects. *Gut.* 1988;29(8):1035–1041.
61. Russell TL, et al. Upper gastrointestinal pH in seventy-nine healthy, elderly, North American men and women. *Pharm Res.* 1993;10(2):187–196.
62. Arnold R. Diagnosis and differential diagnosis of hypergastrinemia. *Wien Klin Wochenschr.* 2007;119(19–20):564–569.
63. Lake-Bakaar G, et al. Gastric secretory failure in patients with the acquired immunodeficiency syndrome (AIDS). *Ann Intern Med.* 1988;109(6):502–504.
64. Lake-Bakaar G, et al. Gastropathy and ketoconazole malabsorption in the acquired immunodeficiency syndrome (AIDS). *Ann Intern Med.* 1988;109(6):471–473.
65. Fadda HM, McConnell EL, Short MD, Basit AW. Meal-induced acceleration of tablet transit through the human small intestine. *Pharm Res.* 2009;26(2):356–360.
66. Zhang QY, Dunbar D, Ostrowska A, Zeisloft S, Yang J, Kaminsky LS. Characterization of human small intestinal cytochromes P-450. *Drug Metab Dispos.* 1999;27(7):804–809.
67. Kolars JC, Schmiedlin-Ren P, Schuetz JD, Fang C, Watkins PB. Identification of rifampin-inducible P450IIIA4 (CYP3A4) in human small bowel enterocytes. *J Clin Invest.* 1992;90(5):1871–1878.
68. Watkins PB. Drug metabolism by cytochromes P450 in the liver and small bowel. *Gastroenterol Clin North Am.* 1992;21(3):511–526.
69. Paine MF, Hart HL, Ludington SS, Haining RL, Rettie AE, Zeldin DC. The human intestinal cytochrome P450 "pie". *Drug Metab Dispos.* 2006;34(5):880–886.
70. Paine MF, et al. Characterization of interintestinal and intraintestinal variations in human CYP3A-dependent metabolism. *J Pharmacol Exp Ther.* 1997;283(3):1552–1562.
71. Fisher MB, Paine MF, Strelevitz TJ, Wrighton SA. The role of hepatic and extrahepatic UDP-glucuronosyltransferases in human drug metabolism. *Drug Metab Rev.* 2001;33(3–4):273–297.
72. Ritter JK. Intestinal UGTs as potential modifiers of pharmacokinetics and biological responses to drugs and xenobiotics. *Expert Opin Drug Metab Toxicol.* 2007;3(1):93–107.
73. Johnson TN, Tanner MS, Taylor CJ, Tucker GT. Enterocytic CYP3A4 in a paediatric population: Developmental changes and the effect of coeliac disease and cystic fibrosis. *Br J Clin Pharmacol.* 2001;51(5):451–460.
74. Mouly S, Paine MF. P-glycoprotein increases from proximal to distal regions of human small intestine. *Pharm Res.* 2003;20(10):1595–1599.
75. Bruyere A, et al. Effect of variations in the amounts of P-glycoprotein (ABCB1), BCRP (ABCG2) and CYP3A4 along the human small intestine on PBPK models for predicting intestinal first-pass. *Mol Pharm.* 2010;7:1596.
76. Yamashiro W, Maeda K, Hirouchi M, Adachi Y, Hu Z, Sugiyama Y. Involvement of transporters in the hepatic uptake and biliary excretion of valsartan, a selective

antagonist of the angiotensin II AT1-receptor, in humans. *Drug Metab Dispos.* 2006;34(7):1247–1254.
77. Koukouritaki SB, et al. Developmental expression of human hepatic CYP2C9 and CYP2C19. *J Pharmacol Exp Ther.* 2004;308(3):965–974.
78. Lin YS, et al. Co-regulation of CYP3A4 and CYP3A5 and contribution to hepatic and intestinal midazolam metabolism. *Mol Pharmacol.* 2002;62(1):162–172.
79. Tateishi T, Watanabe M, Moriya H, Yamaguchi S, Sato T, Kobayashi S. No ethnic difference between Caucasian and Japanese hepatic samples in the expression frequency of CYP3A5 and CYP3A7 proteins. *Biochem Pharmacol.* 1999;57(8):935–939.
80. Ette EI, Williams PJ. Population pharmacokinetics II: Estimation methods. *Ann Pharmacother.* 2004;38(11):1907–1915.
81. Ette EI, Williams PJ. Population pharmacokinetics I: Background, concepts, and models. *Ann Pharmacother.* 2004;38(10):1702–1706.
82. Ette EI, Williams PJ, Lane JR. Population pharmacokinetics III: Design, analysis, and application of population pharmacokinetic studies. *Ann Pharmacother.* 2004;38(12):2136–2144.
83. Sheiner LB, Steimer JL. Pharmacokinetic/pharmacodynamic modeling in drug development. *Annu Rev Pharmacol Toxicol.* 2000;40:67–95.
84. Sheiner LB, Ludden TM. Population pharmacokinetics/dynamics. *Annu Rev Pharmacol Toxicol.* 1992;32:185–209.
85. Price PS, et al. Modeling interindividual variation in physiological factors used in PBPK models of humans. *Crit Rev Toxicol.* 2003;33(5):469–503.
86. Iman R, Helton J. An investigation of uncertainty and sensitivity analysis techniques for computer models. *Risk Anal.* 1988;8:71.
87. McKay M, Conover W, Beckman R. A comparison of three methods for selecting variables of input variables in the analysis of output of computer code. *Technometrics.* 1979;21:239.
88. Bennett JE, Wakefield JC. A comparison of a Bayesian population method with two methods as implemented in commercially available software. *J Pharmacokinet Biopharm.* 1996;24(4):403–432.
89. Wakefield J, Bennett J. The Bayesian modeling of covariates for population pharmacokinetic models. *J Am Stat Assoc.* 1996;91:917.
90. Gelman A, Rubin DB. Markov chain Monte Carlo methods in biostatistics. *Stat Methods Med Res.* 1996;5(4):339–355.
91. White L, Haines H, Adams T. Cardiac output related to body weight in small mammals. *Comp Biochem Physiol.* 1968;27:884–889.
92. Wada DR, Bjorkman S, Ebling WF, Harashima H, Harapat SR, Stanski DR. Computer simulation of the effects of alterations in blood flows and body composition on thiopental pharmacokinetics in humans. *Anesthesiology.* 1997;87(4):884–899.
93. Ginot V, Gaba S, Beaudouin R, Aries F, Monod H. Combined use of local and ANOVA-based global sensitivity analysis for the investigation of a stochastic dynamic model of a fish population. *Ecol Model.* 2006;193:479.
94. Campbell A. Development of PBPK model of molinate and molinate sulfoxide in rats and humans. *Regul Toxicol Pharmacol.* 2009;53(3):195–204.

REFERENCES

95. Saltelli A, Tarantola S, Campolongo F. Sensitivity analysis as an ingredient of modeling. *Stat Sci.* 2000;15:377.
96. Gueorguieva II, Nestorov IA, Rowland M. Fuzzy simulation of pharmacokinetic models: Case study of whole body physiologically-based model of diazepam. *J Pharmacokinet Pharmacodyn.* 2004;31(3):185–213.
97. Brochot C, Smith TJ, Bois FY. Development of a physiologically-based toxicokinetic model for butadiene and four major metabolites in humans: Global sensitivity analysis for experimental design issues. *Chem Biol Interact.* 2007;167(3):168–183.
98. Gueorguieva I, Nestorov IA, Rowland M. Reducing whole body physiologically-based pharmacokinetic models using global sensitivity analysis: Diazepam case study. *J Pharmacokinet Pharmacodyn.* 2006;33(1):1–27.
99. McNally K, Cotton R, Loizou GD. A workflow for global sensitivity analysis of PBPK models. *Front Pharmacol.* 2011;2:31.

9

EVALUATION OF DRUG–DRUG INTERACTION RISK WITH PBPK MODELS

CONTENTS

9.1 Introduction 184
9.2 Factors Affecting Drug–Drug Interactions.................... 186
9.3 *In Vitro* Methods to Evaluate Drug–Drug Interactions........... 190
    9.3.1 Candidate Drug as a Potential Inhibitor 190
    9.3.2 Candidate Drug as a Potential Victim of Inhibition...... 192
9.4 Static Models to Evaluate Drug–Drug Interactions 193
9.5 PBPK Models to Evaluate Drug–Drug Interactions 195
    9.5.1 Intrinsic Clearance of Victim (V) in the Absence of Inhibitor or Inducer.......................... 195
    9.5.2 Intrinsic Clearance of Victim (V) in the Presence of Inhibitor.................................... 196
    9.5.3 Time-Dependent Changes in the Abundance of an Enzyme Isoform Inhibited by an MBI 197
    9.5.4 Intrinsic Clearance of Victim (V) in the Presence of Inducer 197
9.6 Comparison of PBPK Models and Static Models for the Evaluation of Drug–Drug Interactions................... 198
    Keywords 201
    References.. 202

Physiologically-Based Pharmacokinetic (PBPK) Modeling and Simulations: Principles, Methods, and Applications in the Pharmaceutical Industry, First Edition. Sheila Annie Peters.
© 2012 John Wiley & Sons, Inc. Published 2012 by John Wiley & Sons, Inc.

9.1 INTRODUCTION

Concomitant administration of drugs can lead to drug–drug interactions (DDI) if one of them (often referred to as perpetrator of DDI) has the potential to inhibit or induce an enzyme that is critical to the distribution and elimination of another (victim of DDI). As a result of the dynamic interplay between transporters and metabolic enzymes in determining the disposition of a drug, inhibition or induction of transporters will also affect the exposure of a victim drug, depending on whether the transporter involved in the interaction is an uptake or efflux transporter. DDIs can be caused either by a reversible competitive, noncompetitive, or uncompetitive inhibition or by an irreversible **time-dependent inhibition (TDI)**.[1] A time-dependent inhibitor causes an inactivation of the drug-metabolizing enzyme (DME), permanently removing the DME from the pool of enzymes, until restored by the natural turnover of the enzyme. The inactivation is generally through a **mechanism-based inhibition (MBI)** in which an inhibitor gets activated by an enzyme and reversibly alters it. Alternatively, the inactivation can be through the formation of a metabolic intermediate complex (MIC) in which a metabolic product or intermediate of a drug coordinates tightly to the prosthetic heme[2] or covalently binds to the apoprotein. An example of a metabolic intermediate is the nitroso group formed from primary amines. Functional groups with likelihood for MIC formation have been reviewed.[3-5] Interactions related to plasma protein binding in which transient changes in unbound concentrations can occur due to disease state or due to concomitant drugs are not generally clinically significant, unless the victim drug is highly bound, has high extraction ratio, has a sharp response (high sensitivity to concentration changes), has a short equilibration time for efficacious or toxic effects compared to the time taken to regain equilibrium or has narrow therapeutic index, making it sensitive to such an interaction. Different mechanisms leading to DDI are summarized in Figure 9.1.

Drug–drug interactions can impact clinical pharmacokinetics in patients undergoing polytherapy, exposing them to a risk for reduced safety or efficacy, as victim concentrations exceed maximum tolerance concentration or fall below therapeutic concentration. The efficacy of a prodrug is affected if the metabolic route responsible for its conversion to active drug is inhibited. DDIs have been responsible for the early termination of development, refusal of regulatory approval, and market withdrawals. For example, in 1997, the FDA recalled the first nonsedating antihistamine, terfenadine (Seldane) based on its potential to reach lethal blood levels when coadministered with an antibiotic such as erythromycin. Both drugs are primarily metabolized by the cytochrome P450 (CYP) isofom, CYP3A4. The clinical result was an increase in the blood levels of Seldane to toxic levels resulting in lethal arrhythmias. Other examples of market withdrawals primarily due to DDIs include victim drugs such as cerivastatin (CYP2C8), and perpetrator drugs such as mibefradil (CYP3A4). The CYP3A and CYP2C subfamilies are the ones commonly implicated in DDI. In order to mitigate the risk of a costly developmental failure, DDI risk assessments

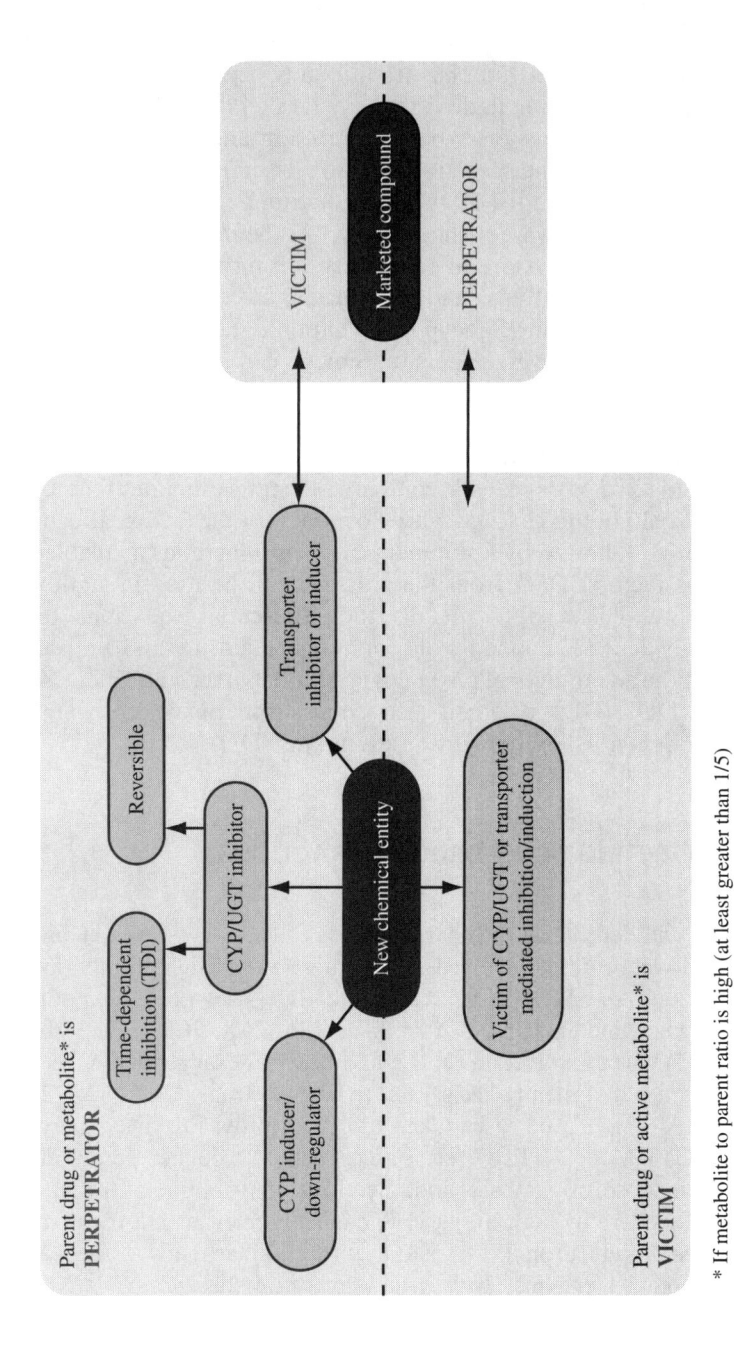

Figure 9.1. Potential sources of DDI risks for a new chemical entity. CYP: cytochrome P450; UGT: uridine 5′-diphospho-glucuronosyl transferase.

* If metabolite to parent ratio is high (at least greater than 1/5)

185

have been integrated into decision-making processes even in the discovery stage. Regulatory guidelines from FDA[6] and EMA[7] for DDI evaluation recommend *in vivo* interaction studies if *in vitro* studies indicate a high risk for DDI. DDI risk evaluation from *in vitro* data has traditionally been carried out using static equations that estimate area under the curve (AUC) ratios for reversible and irreversible, CYP, and transporter-based inhibition and induction in the presence (AUC^i) or absence of the perpetrator. However, in recent years, PBPK approaches have been developed for DDI risk evaluation.[8–11]

Commercially available software such as Simcyp[12] is being increasingly used for DDI risk assessment. A measure of variability arising from genetic polymorphisms, demographic variations, and populations of different ethnicities and disease states is possible using Simcyp's population databases. Thus, apart from predicting the risk in an average person, Simcyp can identify individuals who are at the extremes of population and are more vulnerable to DDI. All PBPK approaches have another advantage over static models. Estimates of the AUC ratio are based on dynamic variation of inhibitor concentrations and can therefore be expected to be more realistic compared to static estimates that tend to overpredict DDI risk. In this chapter, a brief overview of the factors affecting DDI will be presented, followed by a summary of *in vitro* methods available to capture the various factors. Data from these assays can be used in static or physiological models. This chapter details the differential equations that describe the mutually dependent kinetics of coadministered drugs and wraps up with a discussion of the advantages of physiological models over static models in the evaluation of DDI. DDI risk can also come from pharmacodynamic mechanisms. This chapter will only focus on PK-related DDI.

9.2 FACTORS AFFECTING DRUG–DRUG INTERACTIONS

The extent of DDI will depend on the characteristics of the enzyme as well as of the perpetrator and victim drugs involved in the interaction. High-affinity, low-capacity enzymes such as CYPs 2C9, 2C19, and 2D6 are more susceptible for inhibition. Polymorphic enzymes (CYPs 3A5, 2B6, 2C8, 2C9, 2C19, 2D6; UGTs 1A1 and 2B7) when involved in DDI could have an exaggerated effect in some patients. Enzymes that metabolize several marketed drugs (CYPs 3A4, 2C, 1A2; UGTs 1A1, 1A4, 2B7) need to be taken more seriously. Enzymes that are expressed in gut (CYP3A4, UGT2B7, etc.) are exposed to higher perpetrator concentrations during absorption phase, and, therefore, contribution of the gut to the overall risk is considerably high. In general, clinically relevant DDIs are rare for uridine 5′-diphospho-glucuronosyl transferases (UGTs), as many of them are low-affinity, high-capacity enzymes with broad substrate specificity.[13–15] Most interactions involving UGTs are associated with low AUC ratios and mostly confined to UGT2B7, a polymorphic enzyme that is involved in the metabolism of many marketed drugs. Both the EMA[7] and FDA[6] recommend that CYPs 1A2,

9.2 FACTORS AFFECTING DRUG–DRUG INTERACTIONS

3A4 (and 3A5), 2B6, 2C8, 2C9, 2C19, and 2D6 and UGTs 1A1 and 2B7 are investigated for evaluating the DDI risk. Among transporters, the uptake transporters OATP1B1, OATP1B3, OCT1, OCT2, OAT1, and OAT3 and efflux transporters P-gp, BCRP, and optionally BSEP are recommended for investigation by EMA.[7] OCT1 and BSEP are not mentioned by the FDA.[16] Among the recommended transporters, OATP1B1, OCT1, OCT2, BCRP, and BSEP are polymorphic. Statins are a class of drugs used in the treatment of hypercholesterolaemia, as they inhibit 3-hydroxy-3-methylglutaryl-coenzyme A (HMG-CoA) reductase, an important enzyme in the synthesis of cholesterol. Due to their low membrane permeability, many statins are dependent on OATP1B1 transporter for uptake into hepatocytes and, therefore, for their clearance. Statins are thus potential victims of DDI when coadministered with drugs that inhibit OATP1B1.[17] If an inhibitor is dependent on an inhibited uptake transporter for access to drug-metabolizing enzymes, then the extent of interaction is likely to be reduced because of reduced inhibitor concentration. Among the efflux transporters, interactions with P-glycoprotein (P-gp) are the most studied.[18,19] Due to the overlapping substrate specificity of P-gp and CYP3A, an inhibition of P-gp in the gut and the liver can impact exposures of CYP3A substrates. Inhibition of P-gp in the gut, increases the chances for a CYP3A substrate to get absorbed, while inhibition of P-gp in the bile canaliculi of the liver is likely to increase the cellular concentrations of a CYP3A substrate, thereby increasing the chances of its metabolism especially if the compound happens to have a low affinity to CYP3A. Inhibition of P-gp may not result in clinically significant differences in the exposure of the victim drug, but it can attenuate the efficacy of drugs targeting barrier tissues such as brain, lymphocytes, and tumor.[20]

With respect to inhibitor characteristics, the greater the inhibitor concentration in relation to its potency, the greater is the extent of inhibition. Whether to consider total or unbound inhibitor concentration has been debated. The FDA recommends the use of total inhibitor concentration,[6] while the EMA recommends the use of unbound inhibitor concentration[7] in the evaluation of DDI. The use of total concentration can result in an overprediction of risk. For example, montelukast, a potent CYP2C8 inhibitor that is highly bound to plasma proteins, does not affect the pharmacokinetics of the coadministered antidiabetic drug repaglinide.[21] The use of total concentrations in the evaluation of DDI using *in vitro* data would have resulted in categorizing this compound as risky with respect to DDI.

For a victim of inhibition, the greater its dependence on the inhibited route for its elimination, the greater is the potential for an altered pharmacokinetics of the drug. If more than 70% of a drug is eliminated using the inhibited pathway, the risk for interaction is high.[22] Thus, it is important to balance the clearance of a victim drug between hepatic, renal, and biliary routes (Fig. 9.2) during the design stage in order to minimize risk from drug–drug interactions. For a high-clearance victim drug, inhibition because of high first pass extraction, can result in a huge increase in the *AUC* in the presence of an inhibitor because of high first pass extraction, making it more sensitive to DDI compared

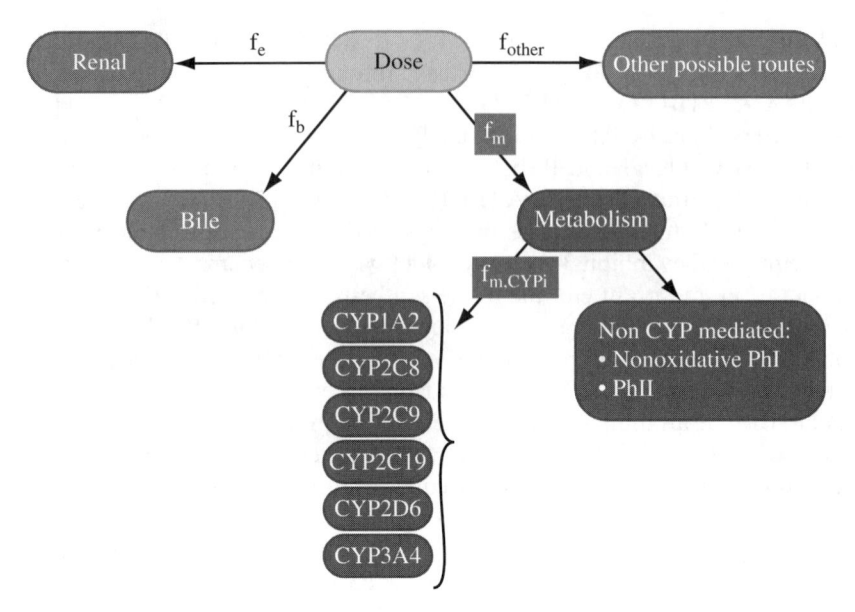

Figure 9.2. Parallel routes of elimination available to a compound. f_e and f_b are the fractions of compound eliminated in the urine and bile, respectively. f_m is the fraction of a compound metabolized. $f_{m,CYP}$ is the fraction of the metabolized compound that is dependent on a particular CYP. The smaller the product $f_m \times f_{m,CYP}$ compared to 1, the lower is the risk of interaction with a co-administered inhibitor.

to a low-clearance drug. Victim drugs associated with a narrow therapeutic window (e.g., phenytoin, carbamazepine, theophylline, digoxin, cyclosporin, phenobarbitone, warfarin, etc.) are more sensitive to DDI. Factors affecting DDI are summarized in Figure 9.3.

Inhibitor concentrations at any point in time will depend on the dose administered and the organ where the affected enzyme or transporter is located. Apart from the DMEs and transporters in the intestine and the liver, the transporters across the blood–brain barrier and those controlling renal tubular uptake and efflux can also be involved in DDI. DDI involving the blood–brain barrier has been reviewed.[23] Renal DDIs are rare and associated with significantly lower AUC ratios compared to hepatic DDIs. Victims of renal DDIs are generally compounds whose eliminations are largely dependent on the renal route. Perpetrators of renal DDIs are usually acids or bases with high therapeutic doses and inhibit the renal uptake transporters OATs 1 and 3 or OCT2. Probenecid inhibits the renal OATs and causes DDIs when coadministered with hydrophilic acids that are predominantly cleared by the kidney, such as penicillin, famotidine, and chlorthiazide.[24–27] Cimetidine, an OCT2 inhibitor, significantly reduces the renal clearance of bisoprolol, metformin, nicotine, and procainamide.[24–26,28–30] A summary of clinically observed renal DDIs is available in the literature.[31,32] Uptake transporters act in concert with efflux

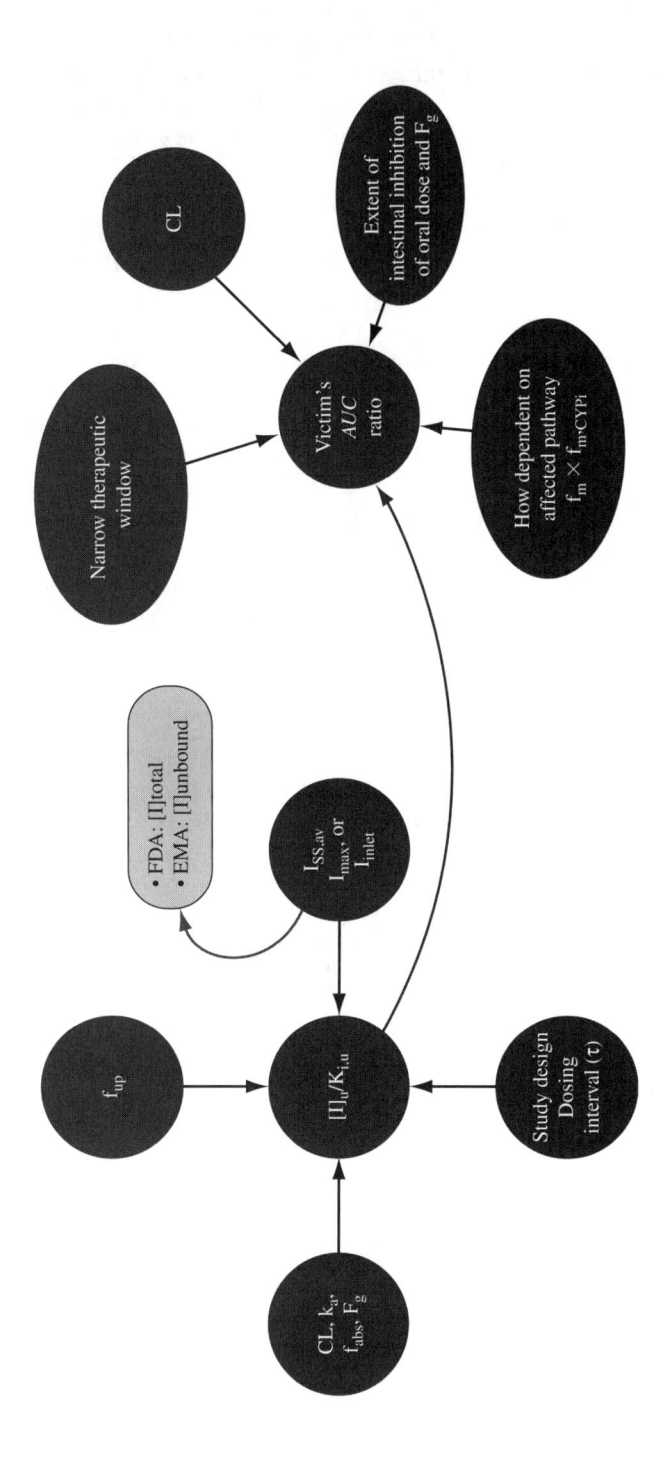

Figure 9.3. Factors affecting the extent of drug–drug interaction risk. AUC = Area under the curve; k_a: absorption rate constant; f_{abs}: fraction of drug absorbed; f_m: fraction of total elimination due to hepatic metabolism; $f_{m,CYP}$: fraction of total hepatic metabolism due to a specific cytochrome P450 isoform; F_g: gut bioavailability; $I_{ss,av}$: average steady state systemic inhibitor concentration; I_{max}: maximum inhibitor concentration; I_{inlet}: Sum of inhibitor concentrations from hepatic artery and hepatic portal vein; [I]: inhibitor concentration; CL: clearance; FDA: Food and Drug Administration; EMA: European Medical Agency.

transporters to eliminate toxins (OCT2, MATE1, and MATE2-K). Therefore, for secreted drugs that are reliant on transporters for overcoming the membrane barrier, the inhibition of efflux transporters could result in an accumulation of drugs in the cytoplasm of proximal tubular cells, leading to toxic effects in the kidney (nephrotoxicity). Decreases in renal function with age and disease need to be factored in the evaluation of DDI. Drugs that are predominantly cleared by the kidney cannot be high extraction drugs. Therefore, even if these drugs are highly bound to plasma proteins, DDI related to plasma protein binding cannot exist.

When multiple inhibitors inhibit multiple enzymes, the overall effect is additive, but if they inhibit the same enzyme, then the net effect defaults to the most potent inhibitor. Simultaneous inhibition and induction tend to reduce the net interaction. When a transporter and an enzyme are involved in DDI, then the net effect depends on whether the inhibitor or the substrate of the enzyme is dependent on the transporter for access to the enzyme. Rifampicin through its activation of the nuclear receptor, PXR, induces CYP3A4. Additionally, it also inhibits OATP1B1 uptake transporter. Thus, a single dose of rifampicin caused an eightfold increase in the exposure of atorvastatin.[33] However, multiple dose of rifampicin caused a decrease in the exposure of atorvastatin, due to induction.[34]

9.3 IN VITRO METHODS TO EVALUATE DRUG–DRUG INTERACTIONS

A candidate drug has to be evaluated for its potential to be a perpetrator or victim of DDI with likely coadministered drugs. *In vitro* methods that aid this assessment focus on reversible/irreversible inhibition or induction mediated by an enzyme or a transporter.

9.3.1 Candidate Drug as a Potential Inhibitor

9.3.1.1 Determination of IC_{50} and K_i for Reversible Enzyme- or Transporter-Mediated Inhibition. A commonly used measure of inhibitory potential of a drug is the inhibitor concentration required to inhibit 50% of the metabolic rate of a probe substrate (IC_{50}) *in vitro*. IC_{50} can be converted to absolute inhibition constant (K_i) using the Cheng–Prusoff equation.[35] For a competitive enzyme inhibition, K_i is given as $IC_{50}/[1 + (S/K_m)]$, where S is the substrate concentration and K_m is the Michaelis–Menten constant for the substrate. Under conditions when $S = K_m$, $K_i = IC_{50}/2$. When measurement of IC_{50} is done at substrate concentrations that are less than or approaching the K_m, IC_{50} is less than K_i for a reversible enzyme inhibition. When the substrate concentration exceeds its K_m, IC_{50} will overestimate K_i. For noncompetitive and uncompetitive inhibition, $IC_{50} = K_i$. Assumption of the mechanism of inhibition is thus necessary to get K_i from IC_{50}. The *in vitro* models employed to study enzyme inhibition are hepatocytes or microsomes, depending on the role of

9.3 IN VITRO METHODS TO EVALUATE DRUG–DRUG INTERACTIONS

phase II enzymes in the metabolism of the substrate. The advantage of these systems is the presence of enzymes in proportions found *in vivo*. A disadvantage of these systems is that only partial inhibition can be observed for substrates lacking specificity. In these cases, recombinant systems can be useful. Partial inhibition is also attributable to poor solubility of the inhibitor. The preferred *in vitro* systems for quantifying inhibitory potency of transporter inhibitors are Caco-2 cells (do not distinguish between transporters), HEK or MDCK transfected cells for uptake transporters, and membrane vesicles for efflux transporters. The inhibitor and substrate concentrations used in the *in vitro* assays should be clinically relevant. If the inhibitor concentrations in the assay are in excess of the clinical concentrations, the inhibitory potential determined would overestimate the risk *in vivo*. If the substrate concentration is different from clinically relevant plasma concentrations, then the relative contribution of different enzymes involved in the substrate metabolism could be different from that *in vivo*, making the *in vitro* IC_{50} irrelevant.

9.3.1.2 Determination of IC$_{50}$ Shift, Maximum Inactivation Rate (k_{inact}) and Inactivator Concentration at Half-Maximal k_{inact} (K_I) for TDI.

The clinical value of *in vitro* models that address TDI of drug-metabolizing enzymes has been reviewed.[36,37] The IC_{50} shift[38] and projected IC_{50}[39] assays are generally the first screening methods to assess TDI. Determination of the kinetic parameters k_{inact} and K_I[38,40] is often carried out for promising drug candidates. The test compound is preincubated with an enzyme source (usually pooled human liver microsomes) in at least five different inhibitor concentrations to get the pseudo-first-order rate constant (k_{obs}), which is related to k_{inact} and K_I by the Michaelis–Menten equation:

$$k_{obs} = \frac{k_{inact} \times [I]}{K_I + [I]} \tag{9.1}$$

where k_{inact} and K_I can be obtained either by a nonlinear regression of equation 9.1 or by plotting the Kitz–Wilson plot ($1/k_{obs}$ vs. $1/[I]$). Different *in vitro* approaches to evaluate MBI have been reviewed.[41] The quasi-irreversible MIC generated by coordinate bond formation to iron in the heme is identifiable by the absorbance of the Soret peak at 450–455 nm by difference spectra scanning[42] in a dual-beam spectrophotometer. Human liver microsomes or recombinant enzyme can be incubated with the test compound. Nitrogen-based MICs are further characterized by the reduction of the Soret absorbance on oxidation by potassium ferricyanide.[43,44] A key parameter that is necessary for the evaluation of TDI is the CYP degradation rate constant (k_{deg}) for the isoform that is inactivated. However, measured values vary widely.[45] Intestinal inhibition needs to be considered for the TDI observed for the CYP3A4 enzyme.[46]

9.3.1.3 Determination of E_{max} and IC$_{50}$ for CYP Induction.

Any *in vitro* system employed to evaluate induction potential of a candidate drug should retain activity for a considerable length of time. *In vitro* systems employed to study enzyme induction are cultured human hepatocytes,[47–49] immortalized hepatocytes,[50] minimally derived cell lines[51] such as HepaRG cells,[52] reporter gene assays,[53–55] and ligand binding assays. The classic method that is widely used in the pharmaceutical industry and accepted by the FDA[6] to measure induction is incubation of test compound with cultured human hepatocytes in which changes in messenger RNA (mRNA) expression and/or enzyme activity of target genes are measured and compared to untreated control hepatocytes.[56] Advantages of hepatocytes are many. They contain native receptors and transporters, and target genes are in their native context with full complement of regulatory elements. The disadvantage comes from the interdonor variability in CYP levels, which is difficult to distinguish from actual variability in induction response across the human population. Methods using *in vitro* data to predict induction potential have been reviewed.[57]

9.3.2 Candidate Drug as a Potential Victim of Inhibition

To evaluate a compound's potential to be a victim of DDI, it is necessary to identify the specific enzymes involved in its metabolism. A common experimental approach to **reaction phenotyping**[58,59] is the use of cDNA-expressed recombinant enzyme systems in which the test compound is incubated with a panel of individually expressed human recombinant enzymes. A typical panel of CYP enzymes includes CYPs 1A2, 2A6, 2B6, 2C8, 2C9, 2C19, 2D6, 2E1, 3A4, and 3A5. Using the known CYP abundances,[60,61] the percentage contribution of individual CYPs to the overall oxidative metabolism of a drug candidate ($f_{m,\mathrm{CYP}}$) can be estimated:

Percent contribution of an isoform i

$$= \frac{CL_{\mathrm{int},i}(\mu L/\min/\mathrm{pmolCYP}_i) \times \text{abundance of } \mathrm{CYP}_i(\mathrm{pmolCYP}_i/\mathrm{mg\ protein})}{\sum[CL_{\mathrm{int},i}(\mu L/\min/\mathrm{pmolCYP}_i) \times \text{abundance of } \mathrm{CYP}_i(\mathrm{pmolCYP}_i/\mathrm{mg\ protein})]}$$

(9.2)

where $CL_{\mathrm{int},i}$ is the intrinsic clearance of the ith CYP isoform in individually expressed a recombinant enzyme. The application of this method to estimate the relative contributions of the individual UGT enzymes to the overall glucuronidation rate is not so feasible today as the relative abundances of the various UGT enzyme activities in human liver are not well established. A second approach to **reaction phenotyping** involves the incubation of the test compound in hepatocytes, microsomes, or some other *in vitro* preparation, using normal tissues as the enzyme source with inclusion of selective chemical or immunoinhibitors of specific enzymatic pathways. By performing a series of incubations with various inhibitors, and comparing the relative rates of metabolism, one can identify which inhibitor reduces the overall metabolism to the greatest extent and thereby uncover the metabolic pathway that contributes the most to the clearance of a compound. The use of hepatocytes guarantees the full range of

9.4 STATIC MODELS TO EVALUATE DRUG–DRUG INTERACTIONS

phase I and phase II enzymes but is limited by the availability of specific inhibitors to some enzymes (e.g., UGT enzymes). The percentage contribution of an enzyme isoform to the metabolism of a victim drug provides a measure of the extent of its dependence on that isoform ($f_{m,\text{CYP}}$). In addition, a good assessment of victim potential requires knowledge of the fractions eliminated in bile and urine in order to get the fraction of compound metabolized (f_m).

9.4 STATIC MODELS TO EVALUATE DRUG–DRUG INTERACTIONS

Interaction data generated from the various *in vitro* approaches are employed in static or dynamic PBPK models in order to assess the extent of drug interactions. Static models for reversible, mechanism-based, and induction-based interactions, respectively[40,48,62–67] are shown in equations 9.3–9.5. AUC_i/AUC is the ratio of the area under the plasma concentration–time profile of the victim drug in the presence and absence of interacting agent, f^i_{gut} and f_{gut} are the fractions of an orally administered victim drug escaping gut metabolism in the presence and absence of the interacting agent, respectively; f_m is the fraction of total elimination that is due to hepatic metabolism, $f_{m,\text{CYP}}$ is the fraction of the total hepatic metabolism mediated through a specific CYP enzyme; I_u is the unbound concentration of the interacting drug. In equation 9.3 for reversible CYP inhibition, $K_{i,u}$ is the unbound inhibition constant:

$$\frac{AUC_i}{AUC} = \frac{f^i_{\text{gut}}}{f_{\text{gut}}} \times \frac{1}{\dfrac{\sum f_m \times f_{m,\text{CYP}}}{1 + \sum \dfrac{I_u}{K_{i,u}}} + \left(1 - \sum f_m \times f_{m,\text{CYP}}\right)} \quad (9.3)$$

In equation 9.4 for mechanism-based CYP inhibition k_{inact} is the maximal enzyme inactivation rate constant, k_{deg} is the first-order rate constant for the endogenous degradation of the CYP enzyme, and $K_{i,u}$ is the unbound inhibitor concentration at the half-maximal value of k_{inact}:

$$\frac{AUC_i}{AUC} = \frac{f^i_{\text{gut}}}{f_{\text{gut}}} \times \frac{1}{\dfrac{\sum f_m \times f_{m,\text{CYP}}}{1 + \sum \dfrac{k_{\text{inact}} \times I_u}{k_{\text{deg}} \times (K_{I,u} + I_u)}} + \left(1 - \sum f_m \times f_{m,\text{CYP}}\right)} \quad (9.4)$$

In equation 9.5 for induction-based interactions, E_{max} is the maximal activation observed in the *in vitro* hepatocyte incubations and EC_{50} is the concentration of the inducer that elicits half-maximal response:

$$\frac{AUC_i}{AUC} = \frac{f^i_{\text{gut}}}{f_{\text{gut}}} \times \frac{1}{\sum f_m \times f_{m,\text{CYP}} \times \left(1 + \sum \dfrac{E_{\text{max}} \times I_u}{EC_{50} + I_u}\right) + \left(1 - \sum f_m \times f_{m,\text{CYP}}\right)} \quad (9.5)$$

The ratio, $f^i_{\text{gut}}/f_{\text{gut}}$ is given by[68]

$$\frac{f^i_{\text{gut}}}{f_{\text{gut}}} = \frac{1}{f_{\text{gut}} + (1 - f_{m,\text{gut}}) \times E_{\text{gut}} + f_{m,\text{gut}} \times E_{\text{gut}} \times \left(1 + \sum \frac{E_{\max} \times I_u}{EC_{50} + I_u}\right)} \quad (9.6)$$

where $f_{m,\text{gut}}$ is the proportion of gut metabolism mediated by the inhibited enzyme; E_{gut} is the extraction in the gut.

A combined mathematical model for predicting DDI, considering simultaneous mechanisms—reversible and irreversible inhibition—as well as induction of CYP3A4, has been proposed by Fahmi et al.[69] by first defining the intrinsic clearance of the victim drug in the liver ($CL_{\text{int}, V, \text{LI}}$) in terms of basal abundance of an isoform, $E_{\text{LI}, 0, \text{isoform}}$:

$$CL_{\text{int}, V, \text{LI}} = \frac{V_{\max, \text{LI}}}{K_m} = \frac{k_{\text{cat}, \text{LI}} \times E_{\text{LI}, 0, \text{isoform}}}{K_m} = \frac{k_{\text{cat}, \text{LI}} \times \left(\frac{R_{\text{syn}, \text{LI}}}{k_{\text{deg}, \text{LI}}}\right)}{K_m} \quad (9.7)$$

where $V_{\max, \text{LI}}$ and K_M are the maximal velocity and Michaelis–Menten constant for the victim; $k_{\text{cat}, \text{LI}}$ is the first-order rate constant of the enzymatic/catalytic reaction. The steady-state abundance $E_{\text{LI}, 0, \text{isoform}}$ is simply the ratio of the zero-order synthesis rate $R_{\text{syn}, \text{LI}}$ and the first-order degradation rate constant $k_{\text{deg}, \text{LI}}$. Recognizing that K_m, $k_{\text{deg}, \text{LI}}$, and $R_{\text{syn}, \text{LI}}$ impact competitive inhibition, TDI, and enzyme induction, respectively, the following equations apply to these parameters in the presence of a perpetrator (superscripted with i):

$$K^i_m = K_m \left(1 + \frac{I_{u,\text{LI}}}{K_m}\right) \quad (9.8)$$

$$k^i_{\text{deg}, \text{LI}} = k_{\text{deg}, \text{LI}} + \frac{I_{u,\text{LI}} \times k_{\text{inact}}}{I_{u,\text{LI}} + K_I} \quad (9.9)$$

$$R^i_{\text{syn}, \text{LI}} = R_{\text{syn}, \text{LI}} + \frac{I_{u,\text{LI}} \times d \times E_{\max}}{I_{u,\text{LI}} + EC_{50}} \quad (9.10)$$

where $I_{u,\text{LI}}$ is the unbound concentration of the perpetrator in the liver; $E_{\max}$ and EC_{50} are the maximum induction and effective concentration of the inducer at half-maximal induction, respectively; and d is an *in vitro* to *in vivo* scaling factor, whose value depends on the *in vitro* system used and the method of quantification. Combining equations 9.8–9.10,

$$CL^i_{\text{int}, V, \text{LI}} = \frac{k_{\text{cat}, \text{LI}} \times \left(R_{\text{syn}, \text{LI}} + \frac{I_{u,\text{LI}} \times d \times E_{\max}}{I_{u,\text{LI}} + EC_{50}}\right)}{K_m \times \left(1 + \frac{I_{u,\text{LI}}}{K_m}\right) \times \left(k_{\text{deg}, \text{LI}} + \frac{I_{u,\text{LI}} \times k_{\text{inact}}}{I_{u,\text{LI}} + K_I}\right)} \quad (9.11)$$

Considering that the inhibited pathway is not the only elimination route for the victim compound,

9.5 PBPK MODELS TO EVALUATE DRUG–DRUG INTERACTIONS

$$CL^i_{\text{int},V,\text{LI}} = \left[\frac{k_{\text{cat,LI}} \times \left(R_{\text{syn,LI}} + \dfrac{I_{u,\text{LI}} \times d \times E_{\max}}{I_{u,\text{LI}} + EC_{50}}\right)}{K_m \times \left(1 + \dfrac{I_{u,\text{LI}}}{K_m}\right) \times \left(k_{\text{deg,LI}} + \dfrac{I_{u,\text{LI}} \times k_{\text{inact}}}{I_{u,\text{LI}} + K_I}\right)}\right] \quad (9.12)$$
$$\times f_m \times f_{\text{CYP}} + CL_{\text{int},V,\text{LI}} \times (1 - f_m \times f_{\text{CYP}})$$

Such an equation allows for the simultaneous treatment of inhibition and induction.

An uptake transporter-mediated DDI can be described by the following equation:[70]

$$\frac{AUC_i}{AUC} = \frac{1}{\dfrac{f_{\text{uptake}}}{\left(1 + \dfrac{I_u}{K_{i,u}}\right)} + (1 - f_{\text{uptake}})} \quad (9.13)$$

where f_{uptake} is the fraction of clearance of the substrate (victim) caused by a hepatic uptake transporter. When an enzyme inhibitor is a substrate to an uptake transporter, the unbound intracellular concentration available at the target enzyme is different from the unbound plasma concentration by a factor of θ.[70]

$$\frac{AUC_i}{AUC} = \frac{1}{\dfrac{f_m \times f_{m,\text{CYP}}}{\left(1 + \dfrac{\theta \times I_u}{K_{i,u}}\right)} + (1 - f_m \times f_{m,\text{CYP}})} \quad (9.14)$$

9.5 PBPK MODELS TO EVALUATE DRUG–DRUG INTERACTIONS

A PBPK model that is meant for evaluating DDI risk would need to simultaneously solve the differential equations of victim and perpetrator drugs. The clearance of a victim drug is reduced or increased due to loss or gain of active enzyme arising from reversible inhibition, TDI, or induction. In reversible inhibition, the reduction in clearance of a victim compound can be expressed in terms of inhibitor concentration relative to its potency. TDI, however, requires a proper consideration of the time-dependent removal of the isoform from the enzyme pool and the turnover rate of the isoform. Thus, consideration of TDI in a PBPK model requires that the intrinsic clearance of a substrate is expressed in terms of the abundance (E) of the enzyme isoforms that metabolizes it.

9.5.1 Intrinsic Clearance of Victim (*V*) in the Absence of Inhibitor or Inducer

The unbound hepatic intrinsic clearance of a victim drug ($CL_{\text{int},u,V,\text{LI}}$) in the absence of an inhibitor or inducer can be given in terms of enzyme abundance by equation 9.15:

$$CL_{\text{int},u,V,\text{LI}}(\text{mL/min}) = \sum CL_{\text{int},u,V,\text{LI}}(\text{mL/min/pmol isoform}) \times E_{\text{LI},0,\text{isoform}} \tag{9.15}$$

where $E_{\text{LI},0,\text{isoform}}$ is the basal abundance of an enzyme isoform in the absence of an inhibitor or inducer. $CL_{\text{int},u,V,\text{LI}}$ (mL/min/pmol isoform) is the unbound intrinsic clearance of the victim drug in the liver per pmol of the isoform. The summation goes over each of the enzyme isoforms that contributes to the overall clearance of the victim drug, if more than one isoform is involved. Applying the well-stirred model, $CL_{\text{int},u,V,\text{LI}}$ (mL/min/pmol isoform) is extracted from *in vivo* human clearance (from first-in-man PK study) or estimated human clearance (in the discovery phase) using the percentage contribution of each of the isoforms. Percentage contribution of an enzyme isoform to the overall clearance of the compound can be obtained from recombinant enzymes using the enzyme abundance of the isoform. Direct use of *in vitro* CL_{int} from recombinant CYPs for $CL_{\text{int},u,V,\text{LI}}$ (mL/min/pmol isoform) is likely to underestimate the extent of DDI predicted, as *in vitro* CL_{int} are generally lower than their corresponding values *in vivo* and the extent of DDI depends on the clearance of the victim drug.

9.5.2 Intrinsic Clearance of Victim (V) in the Presence of Inhibitor

The unbound hepatic intrinsic clearance of a victim drug ($CL^i_{\text{int},u,V,\text{LI}}$) in the presence of a competitive inhibitor can be given in terms of enzyme abundance by equation 9.16:

$$CL^i_{\text{int},u,V,\text{LI}}(\text{mL/min}) = \frac{CL_{\text{int},u,V,\text{LI}}(\text{mL/min/pmol isoform}) \times E_{\text{LI,active},t,\text{isoform}}}{\left(1 + \sum \frac{I_{u,\text{LI},t}}{K_i}\right)}$$

$$+ \sum CL_{\text{int},u,V,\text{LI}}(\text{mL/min/pmol isoform}) \times E_{\text{LI},0.\text{other}} \tag{9.16}$$

where $I_{u,\text{LI},t}$ and K_i are the unbound hepatic inhibitor concentration at time t and inhibition constant of a competitive inhibitor of an enzyme isoform, respectively. The first term in equation 9.16 represents the contribution to intrinsic clearance from the inhibited isoform and the second term represents contribution from all other isoforms to the total clearance of the victim. Differential equations representing the dynamic changes in the concentration of the inhibitor in the liver, $I_{u,\text{LI},t}$, can be relatively simple or elaborate depending on the availability of required data. $E_{\text{LI,active},t,\text{isoform}}$ is the active level in the liver at any time t of an enzyme isoform that is additionally inhibited by a time-dependent inhibitor. $E_{\text{LI},0,\text{other}}$ is the basal abundance of isoforms other than the one inhibited. The summation goes over each of the uninhibited isoforms that are involved in the clearance of the victim compound.

9.5.3 Time-Dependent Changes in the Abundance of an Enzyme Isoform Inhibited by an MBI

The rate of change of active levels of an enzyme isoform in the presence of a time-dependent inhibitor, $E_{\text{LI, active, }t,\text{ isoform}}$, is given by equation 9.17:

$$\frac{dE_{\text{LI, active, }t,\text{ isoform}}}{dt} = k_{\text{deg, LI, isoform}} \times (E_{\text{LI, 0, isoform}} - E_{\text{LI, active, }t,\text{ isoform}})$$
$$- \sum \left(\frac{k_{\text{inact}} \times \text{TDI}_{u,\text{LI}}}{K_I} \times E_{\text{LI, active, }t,\text{ isoform}} \right) \quad (9.17)$$

where $k_{\text{deg, LI, isoform}}$ is the first-order rate constant of enzyme degradation that when multiplied by the net abundance at time t, $E_{\text{LI, 0, isoform}} - E_{\text{LI, active, }t,\text{ isoform}}$, is the rate of synthesis of the enzyme isoform under equilibrium conditions. The last term is the rate of removal of the isoform from the enzyme pool, which depends on the first-order maximum rate of inactivation (k_{inact}), the unbound concentration of the time-dependent inhibitor in the liver ($TDI_{u,\text{LI}}$) and on the apparent dissociation constant, K_I, describing interaction between the enzyme and inhibitor. The summation runs over each of the isoforms.

Equations 9.15–9.17 can be applied to intestinal enzymes by substituting intestine-specific parameters. For self-inhibition, the inhibitor's intrinsic clearance is used in equations 9.15–9.17. In case of nonlinearity, V_{max}, K_m can substitute CL_{int}.

9.5.4 Intrinsic Clearance of Victim (*V*) in the Presence of Inducer

Inducible CYP isoforms include 1A1, 1A2, 1B1, 2A6, 2B6, and 3A4. Induction can be brought about either by increasing the zero-order rate of synthesis of an enzyme isoform (R_{syn}) or by a stabilization of the protein, by reducing its first-order enzyme degradation rate constant (k_{deg}). With the exception of CYP2E1, which is stabilized by inducers,[71] CYPs are primarily induced by the former mechanism.[72] Regulation of CYP transcription is controlled by nuclear receptors such as the arylhydrocarbon receptor (AHR), the pregnane X receptor (PXR), and the constitutive active receptor (CAR). An inducer binds to one of these nuclear receptors, thereby increasing the rate of transcription of the enzyme. A corresponding increase in the rate of translation of the induced enzyme follows. The mRNA turnover rates control the lag time that reflect the maximum rate of enzyme synthesis.[73] However, the time to reach maximum enzyme activity is determined by k_{deg}. The mRNA turnover and the k_{deg} thus determine the rate at which the new induced equilibrium is reached. Until the equilibrium is reached, differences in mRNA turnover and k_{deg} will influence the steady-state levels of the isoforms and, therefore, the extent of induction-mediated DDI. However, once equilibrium is reached, these two factors do not influence the extent of induction-mediated DDI.[68] A lack of consensus on the *in vivo* turnover half-lives of CYPs can potentially limit the success of predicting extent of DDI mediated by induction and TDI.[45] The isoform's basal abundance in the liver, $E_{LI,0,isoform}$, also determines the extent of induction-mediated DDI[68,74–76] through the maximal induction E_{max}, which is given by

$$E_{max} = \frac{E_{ind} - E_{LI,0,isoform}}{E_{LI,0,isoform}} \qquad (9.18)$$

where E_{ind} is the amount of enzyme at maximal induction. If $EC_{u,50}$ is the unbound concentration of the inducer associated with half-maximal induction (potency) corrected for any nonspecific binding, then the unbound intrinsic clearance of the victim drug in the presence of the inducer (CL'_{int}) in the liver is given as[71]

$$\begin{aligned}&CL^i_{int,u,V,LI}(mL/min) \\&= CL_{int,u,V,LI}(mL/min/pmol\ isoform) \times E_{LI,0,isoform} \times \left(\frac{E_{max} \times I_{u,LI,t}}{EC_{u,50} + I_{u,LI,t}}\right) \\&+ \sum CL_{int,u,V,LI}(mL/min/pmol\ isoform) \times E_{LI,0.isoform}\end{aligned}$$
$$(9.19)$$

In equation 9.15, $I_{u,LI,t}$ is the unbound inducer concentration in the liver at time t. A similar equation applies to induction of gut enzymes such as CYP3A4. The relative contribution of the gut with respect to the liver can be very different due to differences in inducer concentrations in the two organs and the differential expression of nuclear receptors[77] or regulatory promoters. Both liver[78] and gut[79] have been implicated as major contributors to induction-mediated DDI.

The intrinsic clearance of the victim in the presence and absence of the perpetrator are used in a PBPK model for the victim drug to get the concentration–time profiles of the victim drug in the two conditions commonly used as a useful way of monitoring conditions of a whole system. The AUC and C_{max} ratios can then be obtained. To estimate the physiological variability that might accompany the AUC ratio, all known measures of variability for the factors affecting DDI, namely, basal enzyme abundance, k_{deg}, mRNA turnover rates, and the like can be incorporated.

Examples of the use of PBPK models for the evaluation of DDI are available in the literature.[80–84] CYP-based mechanistic modeling to predict complex drug–drug interactions involving simultaneous competitive and time-dependent enzyme inhibition by parent compound and its metabolite in both liver and gut has been described.[85]

9.6 COMPARISON OF PBPK MODELS AND STATIC MODELS FOR THE EVALUATION OF DRUG–DRUG INTERACTIONS

When an inhibitor and substrate of an enzyme expressed in liver and gut are orally administered, the relevant inhibitor concentrations to consider for the

9.6 COMPARISON OF PBPK MODELS AND STATIC MODELS

inhibition of the substrate drug during intestinal and hepatic first pass and during systemic circulation are the enterocytic, portal vein, and the steady state systemic inhibitor concentrations, respectively. If the inhibited enzyme is present in the liver but not in the gut, only the portal vein and the systemic inhibitor concentrations need to be considered. One of the main advantages of PBPK over static equations in the evaluation of DDI is the possibility to consider dynamic changes in inhibitor concentrations mimicking the *in vivo* situation. In static equations, the inhibitor concentration is set to either the portal vein inhibitor concentration or maximum/average steady-state inhibitor concentration. The use of average systemic steady-state inhibitor concentrations in the evaluation of DDI gives the least prediction error. The differences in inhibitor concentrations in dynamic and static models are illustrated in Figure 9.4. The use of I_{gut} as dose/250 mL can overestimate the gut interaction

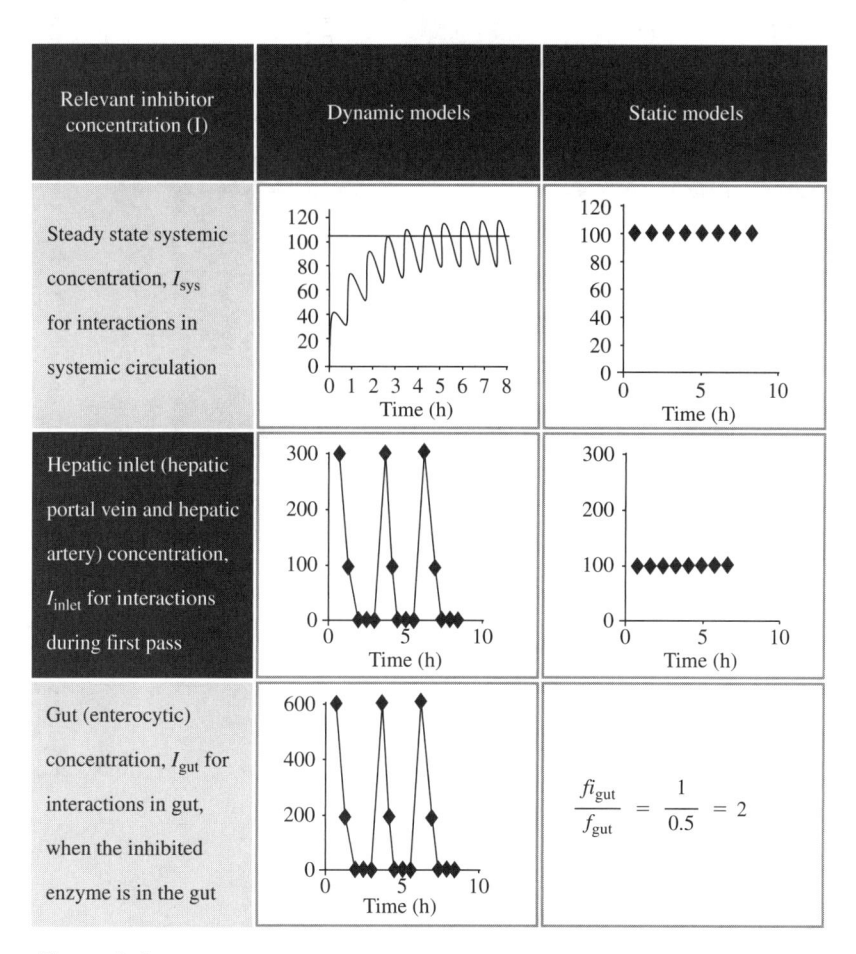

Relevant inhibitor concentration (I)	Dynamic models	Static models
Steady state systemic concentration, I_{sys} for interactions in systemic circulation	(oscillating curve approaching ~100, Time 0–8 h)	(constant at ~100, Time 0–10 h)
Hepatic inlet (hepatic portal vein and hepatic artery) concentration, I_{inlet} for interactions during first pass	(spikes to 300, Time 0–10 h)	(constant at ~100, Time 0–10 h)
Gut (enterocytic) concentration, I_{gut} for interactions in gut, when the inhibited enzyme is in the gut	(spikes to 600, Time 0–10 h)	$\dfrac{fi_{gut}}{f_{gut}} = \dfrac{1}{0.5} = 2$

Figure 9.4. Relevant inhibitor concentrations in dynamic and static models.

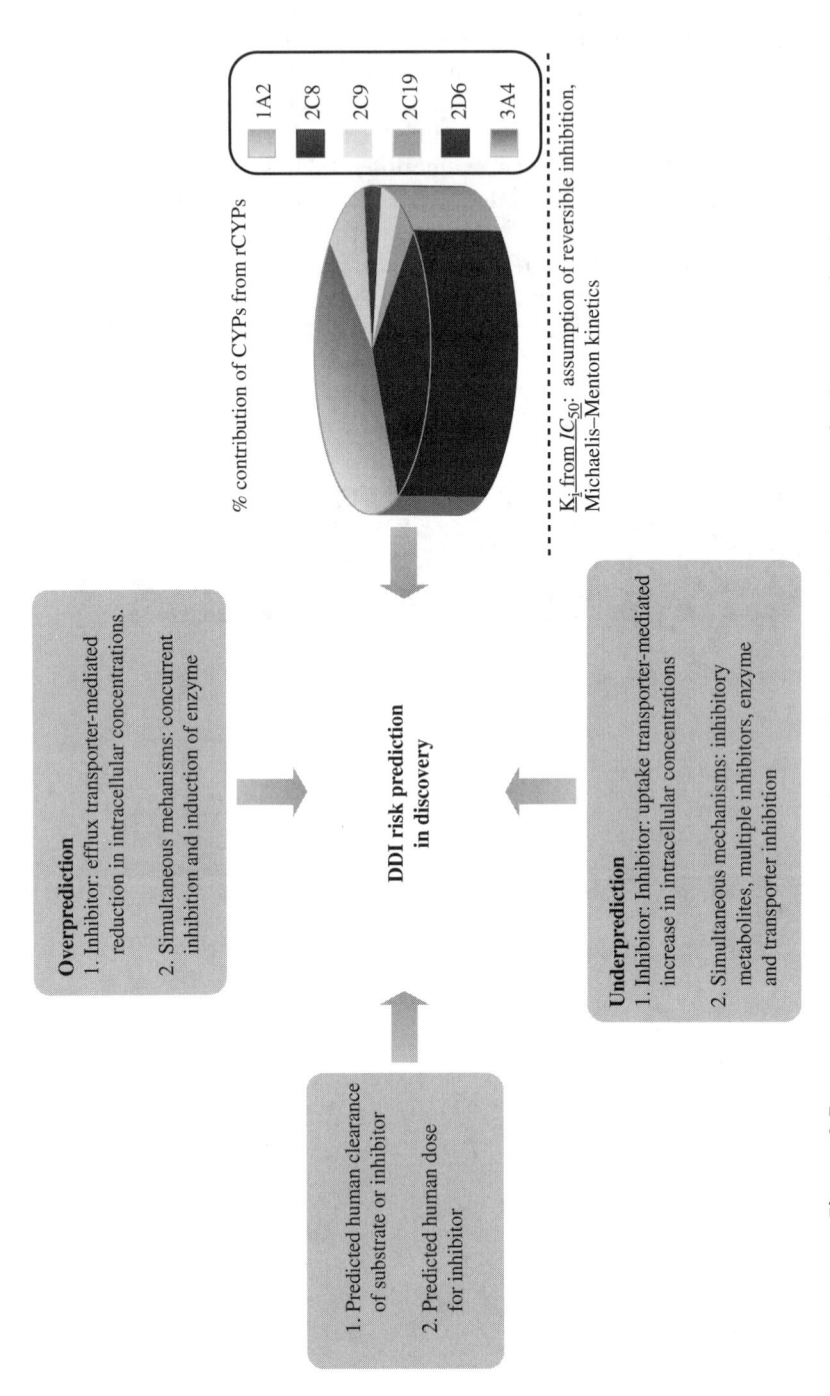

Figure 9.5. Uncertainties in input and limitations in the understanding of relevant mechanisms *in vivo* in human could limit the benefits of PBPK approach to DDI predictions in drug discovery and preclinical development phases.

as it represents the inhibitor concentrations in the small intestinal lumen. A more appropriate measure of inhibitor concentrations, accessible to enzymes is given by $f_{abs} \times k_a \times Q_{villi}$ where f_{abs} is the fraction of inhibitor dose absorbed, k_a is the absorption rate constant and Q_{villi} is the blood flow rate to the villi. Another way to consider gut interaction is to take $F_{gut} = 1$, assuming complete inhibition in the gut. Since competition from permeability is likely to restrict the metabolism of the victim drug, the gut bioavailability in the absence of inhibitor can be set to a mid value of 0.5. This assumption is quite relevant for many drugs that are extracted in the gut by CYP3A4. The use of average steady-state inhibitor concentration in static equations is likely to underestimate the extent of interaction for substrates with high hepatic extraction. PBPK approaches consider temporal changes in inhibitor and substrate concentrations, complex dosing regimens of inhibitor and substrate, as well as time-dependent changes in enzyme abundance (important for TDI and induction), thereby enabling a more realistic assessment of DDI. They can also take into consideration DDI dependence on substrate clearance and interindividual variability in enzyme expression levels arising from genetic, anatomical, demographic, and pathophysiological differences. However, despite these advantages of PBPK approaches, uncertainties in the input data generated from *in vitro* experiments and limitations in the understanding of compensatory and simultaneous mechanisms *in vivo* (Fig. 9.5) are so high that *AUC* ratios obtained from PBPK approaches are similar to those from static models.[86–88] Uncertainty in k_{deg} can lead to inaccuracy in the prediction of induction and TDI.[45] When the mean *AUC* ratios are not predicted well, it does not make sense to rely on the absolute values of *AUC* ratios of the extreme individuals identified by PBPK approaches. Still, when clinical PK data becomes available, a knowledge of the genetic, anatomical, demographic, and pathophysiological profile of these extreme individuals who are at a higher risk for adverse reactions can be valuable in designing the next clinical trials.

KEYWORDS

Mechanism-Based Inhibition (MBI): A potential mechanism for time-dependent inhibition (TDI), where a more inhibitory metabolite relative to parent causes the inactivation of a CYP by a protein or heme adduct formation.[41]

Reaction Phenotyping: The estimation of the relative contributions of specific enzymes to the metabolism of a test compound.

Time-Dependent Inhibition (TDI): A collective term that refers to the increase in the extent of inhibition of the substrate, when the inhibitor is incubated with the enzyme prior to the addition of the substrate[37,41] *in vitro* or during the dosing period *in vivo*.

REFERENCES

1. Venkatakrishnan K, Obach RS. Drug-drug interactions via mechanism-based cytochrome P450 inactivation: Points to consider for risk assessment from in vitro data and clinical pharmacologic evaluation. *Curr Drug Metab*. 2007;8(5):449–462.
2. Silverman RB, Hiebert CK. Inactivation of monoamine oxidase A by the monoamine oxidase B inactivators 1-phenylcyclopropylamine, 1-benzylcyclopropylamine, and N-cyclopropyl-alpha-methylbenzylamine. *Biochemistry*. 1988;27(22):8448–8453.
3. Kalgutkar AS, Obach RS, Maurer TS. Mechanism-based inactivation of cytochrome P450 enzymes: Chemical mechanisms, structure-activity relationships and relationship to clinical drug-drug interactions and idiosyncratic adverse drug reactions. *Curr Drug Metab*. 2007;8(5):407–447.
4. Riley RJ, Grime K, Weaver R. Time-dependent CYP inhibition. *Expert Opin Drug Metab Toxicol*. 2007;3(1):51–66.
5. Hollenberg PF, Kent UM, Bumpus NN. Mechanism-based inactivation of human cytochromes p450s: Experimental characterization, reactive intermediates, and clinical implications. *Chem Res Toxicol*. 2008;21(1):189–205.
6. FDA Center for Drug Evaluation and Research. Drug interaction studies—study design, data analysis and implications for dosing and labelling. 2006.
7. European Medicines Agency. Available at: http://www.ema.europa.eu/docs/en_GB/document_library/Scientific_guideline/2010/05/WC500090112.pdf.
8. Kanamitsu S, Ito K, Green CE, Tyson CA, Shimada N, Sugiyama Y. Prediction of in vivo interaction between triazolam and erythromycin based on in vitro studies using human liver microsomes and recombinant human CYP3A4. *Pharm Res*. 2000;17(4):419–426.
9. Zhang X, Quinney SK, Gorski JC, Jones DR, Hall SD. Semi-physiologically-based pharmacokinetic models for the inhibition of midazolam clearance by diltiazem and its major metabolite. *Drug Metab Dispos*. 2009;37(8):1587–1597.
10. Fenneteau F, Poulin P, Nekka F. Physiologically-based predictions of the impact of inhibition of intestinal and hepatic metabolism on human pharmacokinetics of CYP3A substrates. *J Pharm Sci*. 2010;99(1):486–514.
11. Bois FY. Physiologically-based modelling and prediction of drug interactions. *Basic Clin Pharmacol Toxicol*. 2010;106(3):154–161.
12. Jamei M, Marciniak S, Feng K, Barnett A, Tucker G, Rostami-Hodjegan A. The simcyp population-based ADME simulator. *Expert Opin Drug Metab Toxicol*. 2009;5(2):211–223.
13. Williams JA, et al. Drug-drug interactions for UDP-glucuronosyltransferase substrates: A pharmacokinetic explanation for typically observed low exposure (AUC_i/AUC) ratios. *Drug Metab Dispos*. 2004;32(11):1201–1208.
14. Kiang TK, Ensom MH, Chang TK. UDP-glucuronosyltransferases and clinical drug-drug interactions. *Pharmacol Ther*. 2005;106(1):97–132.
15. Miners JO, Mackenzie PI, Knights KM. The prediction of drug-glucuronidation parameters in humans: UDP-glucuronosyltransferase enzyme-selective substrate and inhibitor probes for reaction phenotyping and in vitro-in vivo extrapolation of drug clearance and drug-drug interaction potential. *Drug Metab Rev*. 2010;42(1):189–201.

REFERENCES

16. International Transporter Consortium, Giacomini KM, Huang SM, et al. Membrane transporters in drug development. *Nat Rev Drug Discov.* 2010;9(3):215–236.
17. Igel M, Sudhop T, von Bergmann K. Metabolism and drug interactions of 3-hydroxy-3-methylglutaryl coenzyme A-reductase inhibitors (statins). *Eur J Clin Pharmacol.* 2001;57(5):357–364.
18. Fenner KS, et al. Drug-drug interactions mediated through P-glycoprotein: Clinical relevance and in vitro-in vivo correlation using digoxin as a probe drug. *Clin Pharmacol Ther.* 2009;85(2):173–181.
19. Tachibana T, Kato M, Watanabe T, Mitsui T, Sugiyama Y. Method for predicting the risk of drug-drug interactions involving inhibition of intestinal CYP3A4 and P-glycoprotein. *Xenobiotica.* 2009;39(6):430–443.
20. Lee CA, Cook JA, Reyner EL, Smith DA. P-glycoprotein related drug interactions: Clinical importance and a consideration of disease states. *Expert Opin Drug Metab Toxicol.* 2010;6(5):603–619.
21. Kajosaari LI, Niemi M, Backman JT, Neuvonen PJ. Telithromycin, but not montelukast, increases the plasma concentrations and effects of the cytochrome P450 3A4 and 2C8 substrate repaglinide. *Clin Pharmacol Ther.* 2006;79(3):231–242.
22. Rodrigues AD, Winchell GA, Dobrinska MR. Use of in vitro drug metabolism data to evaluate metabolic drug-drug interactions in man: The need for quantitative databases. *J Clin Pharmacol.* 2001;41(4):368–373.
23. Eyal S, Hsiao P, Unadkat JD. Drug interactions at the blood-brain barrier: Fact or fantasy? *Pharmacol Ther.* 2009;123(1):80–104.
24. Shitara Y, Sato H, Sugiyama Y. Evaluation of drug-drug interaction in the hepatobiliary and renal transport of drugs. *Annu Rev Pharmacol Toxicol.* 2005;45:689–723.
25. Ho RH, Kim RB. Transporters and drug therapy: Implications for drug disposition and disease. *Clin Pharmacol Ther.* 2005;78(3):260–277.
26. Ayrton A, Morgan P. Role of transport proteins in drug absorption, distribution and excretion. *Xenobiotica.* 2001;31(8–9):469–497.
27. Inotsume N, Nishimura M, Nakano M, Fujiyama S, Sato T. The inhibitory effect of probenecid on renal excretion of famotidine in young, healthy volunteers. *J Clin Pharmacol.* 1990;30(1):50–56.
28. Bendayan R, Sullivan JT, Shaw C, Frecker RC, Sellers EM. Effect of cimetidine and ranitidine on the hepatic and renal elimination of nicotine in humans. *Eur J Clin Pharmacol.* 1990;38(2):165–169.
29. Kirch W, Rose I, Demers HG, Leopold G, Pabst J, Ohnhaus EE. Pharmacokinetics of bisoprolol during repeated oral administration to healthy volunteers and patients with kidney or liver disease. *Clin Pharmacokinet.* 1987;13(2):110–117.
30. Somogyi A, Stockley C, Keal J, Rolan P, Bochner F. Reduction of metformin renal tubular secretion by cimetidine in man. *Br J Clin Pharmacol.* 1987;23(5):545–551.
31. Masereeuw R, Russel FG. Mechanisms and clinical implications of renal drug excretion. *Drug Metab Rev.* 2001;33(3–4):299–351.
32. Masereeuw R, Russel FG. Therapeutic implications of renal anionic drug transporters. *Pharmacol Ther.* 2010;126(2):200–216.

33. Lau YY, Huang Y, Frassetto L, Benet LZ. Effect of OATP1B transporter inhibition on the pharmacokinetics of atorvastatin in healthy volunteers. *Clin Pharmacol Ther.* 2007;81(2):194–204.
34. Backman JT, Luurila H, Neuvonen M, Neuvonen PJ. Rifampin markedly decreases and gemfibrozil increases the plasma concentrations of atorvastatin and its metabolites. *Clin Pharmacol Ther.* 2005;78(2):154–167.
35. Cheng Y, Prusoff WH. Relationship between the inhibition constant (K_1) and the concentration of inhibitor which causes 50 percent inhibition (I_{50}) of an enzymatic reaction. *Biochem Pharmacol.* 1973;22(23):3099–3108.
36. Venkatakrishnan K, von Moltke LL, Obach RS, Greenblatt DJ. Drug metabolism and drug interactions: Application and clinical value of in vitro models. *Curr Drug Metab.* 2003;4(5):423–459.
37. Grimm SW, et al. The conduct of in vitro studies to address time-dependent inhibition of drug-metabolizing enzymes: A perspective of the Pharmaceutical Research and Manufacturers of America. *Drug Metab Dispos.* 2009;37(7):1355–1370.
38. Obach RS, Walsky RL, Venkatakrishnan K. Mechanism-based inactivation of human cytochrome p450 enzymes and the prediction of drug-drug interactions. *Drug Metab Dispos.* 2007;35(2):246–255.
39. Atkinson A, Kenny JR, Grime K. Automated assessment of time-dependent inhibition of human cytochrome P450 enzymes using liquid chromatography-tandem mass spectrometry analysis. *Drug Metab Dispos.* 2005;33(11):1637–1647.
40. Mayhew BS, Jones DR, Hall SD. An in vitro model for predicting in vivo inhibition of cytochrome P450 3A4 by metabolic intermediate complex formation. *Drug Metab Dispos.* 2000;28(9):1031–1037.
41. Grime KH, Bird J, Ferguson D, Riley RJ. Mechanism-based inhibition of cytochrome P450 enzymes: An evaluation of early decision making in vitro approaches and drug-drug interaction prediction methods. *Eur J Pharm Sci.* 2009;36(2–3):175–191.
42. Franklin MR. Cytochrome P450 metabolic intermediate complexes from macrolide antibiotics and related compounds. *Methods Enzymol.* 1991(206):559–573.
43. Watanabe A, Nakamura K, Okudaira N, Okazaki O, Sudo K. Risk assessment for drug-drug interaction caused by metabolism-based inhibition of CYP3A using automated in vitro assay systems and its application in the early drug discovery process. *Drug Metab Dispos.* 2007;35(7):1232–1238.
44. Lim HK, Duczak N, Jr, Brougham L, Elliot M, Patel K, Chan K. Automated screening with confirmation of mechanism-based inactivation of CYP3A4, CYP2C9, CYP2C19, CYP2D6, and CYP1A2 in pooled human liver microsomes. *Drug Metab Dispos.* 2005;33(8):1211–1219.
45. Yang J, et al. Cytochrome p450 turnover: Regulation of synthesis and degradation, methods for determining rates, and implications for the prediction of drug interactions. *Curr Drug Metab.* 2008;9(5):384–394.
46. Galetin A, Burt H, Gibbons L, Houston JB. Prediction of time-dependent CYP3A4 drug-drug interactions: Impact of enzyme degradation, parallel elimination pathways, and intestinal inhibition. *Drug Metab Dispos.* 2006;34(1):166–175.

47. LeCluyse E, Madan A, Hamilton G, Carroll K, DeHaan R, Parkinson A. Expression and regulation of cytochrome P450 enzymes in primary cultures of human hepatocytes. *J Biochem Mol Toxicol.* 2000;14(4):177–188.
48. Kato M, Chiba K, Horikawa M, Sugiyama Y. The quantitative prediction of in vivo enzyme-induction caused by drug exposure from in vitro information on human hepatocytes. *Drug Metab Pharmacokinet.* 2005;20(4):236–243.
49. Luo G, Guenthner T, Gan LS, Humphreys WG. CYP3A4 induction by xenobiotics: Biochemistry, experimental methods and impact on drug discovery and development. *Curr Drug Metab.* 2004;5(6):483–505.
50. Mills JB, Rose KA, Sadagopan N, Sahi J, de Morais SM. Induction of drug metabolism enzymes and MDR1 using a novel human hepatocyte cell line. *J Pharmacol Exp Ther.* 2004;309(1):303–309.
51. Aninat C, et al. Expression of cytochromes P450, conjugating enzymes and nuclear receptors in human hepatoma HepaRG cells. *Drug Metab Dispos.* 2006;34(1):75–83.
52. Kanebratt KP, Andersson TB. HepaRG cells as an in vitro model for evaluation of cytochrome P450 induction in humans. *Drug Metab Dispos.* 2008;36(1):137–145.
53. Sinz M, et al. Evaluation of 170 xenobiotics as transactivators of human pregnane X receptor (hPXR) and correlation to known CYP3A4 drug interactions. *Curr Drug Metab.* 2006;7(4):375–388.
54. El-Sankary W, Gibson GG, Ayrton A, Plant N. Use of a reporter gene assay to predict and rank the potency and efficacy of CYP3A4 inducers. *Drug Metab Dispos.* 2001;29(11):1499–1504.
55. Persson KP, et al. Evaluation of human liver slices and reporter gene assays as systems for predicting the cytochrome p450 induction potential of drugs in vivo in humans. *Pharm Res.* 2006;23(1):56–69.
56. Chu V, et al. *In vitro* and *in vivo* induction of cytochrome p450: A survey of the current practices and recommendations: A Pharmaceutical Research and Manufacturers of America perspective. *Drug Metab Dispos.* 2009;37(7):1339–1354.
57. Fahmi OA, Ripp SL. Evaluation of models for predicting drug-drug interactions due to induction. *Expert Opin Drug Metab Toxicol.* 2010;6(11):1399–1416.
58. Zhang H, Davis CD, Sinz MW, Rodrigues AD. Cytochrome P450 reaction-phenotyping: An industrial perspective. *Expert Opin Drug Metab Toxicol.* 2007;3(5):667–687.
59. Harper TW, Brassil PJ. Reaction phenotyping: Current industry efforts to identify enzymes responsible for metabolizing drug candidates. *AAPS J.* 2008;10(1):200–207.
60. Rowland-Yeo K, Rostami-Hodjegan A, Tucker GT. Abundance of cytochromes P450 in human liver: A meta-analysis. *Br. J. Clin. Pharmacol.* 2004;57:687.
61. Rodrigues AD. Integrated cytochrome P450 reaction phenotyping: Attempting to bridge the gap between cDNA-expressed cytochromes P450 and native human liver microsomes. *Biochem Pharmacol.* 1999;57(5):465–480.
62. Weaver RJ. Assessment of drug-drug interactions: Concepts and approaches. *Xenobiotica.* 2001;31(8–9):499–538.
63. Wang YH, Jones DR, Hall SD. Prediction of cytochrome P450 3A inhibition by verapamil enantiomers and their metabolites. *Drug Metab Dispos.* 2004;32(2):259–266.

64. Fahmi OA, et al. Comparison of different algorithms for predicting clinical drug-drug interactions, based on the use of CYP3A4 in vitro data: Predictions of compounds as precipitants of interaction. *Drug Metab Dispos.* 2009;37(8):1658–1666.
65. Kozawa M, Honma M, Suzuki H. Quantitative prediction of in vivo profiles of CYP3A4 induction in humans from in vitro results with a reporter gene assay. *Drug Metab Dispos.* 2009;37(6):1234–1241.
66. Ohno Y, Hisaka A, Ueno M, Suzuki H. General framework for the prediction of oral drug interactions caused by CYP3A4 induction from in vivo information. *Clin Pharmacokinet.* 2008;47(10):669–680.
67. Shou M, et al. Modeling, prediction, and in vitro in vivo correlation of CYP3A4 induction. *Drug Metab Dispos.* 2008;36(11):2355–2370.
68. Almond LM, Yang J, Jamei M, Tucker GT, Rostami-Hodjegan A. Towards a quantitative framework for the prediction of DDIs arising from cytochrome P450 induction. *Curr Drug Metab.* 2009;10(4):420–432.
69. Fahmi OA, Boldt S, Kish M, Obach RS, Tremaine LM. Prediction of drug-drug interactions from in vitro induction data: Application of the relative induction score approach using cryopreserved human hepatocytes. *Drug Metab Dispos.* 2008;36 (9):1971–1974.
70. Obach RS. Predicting drug-drug interactions from in vitro drug metabolism data: Challenges and recent advances. *Curr Opin Drug Discov Devel.* 2009;12(1):81–89.
71. Roberts BJ, Song BJ, Soh Y, Park SS, Shoaf SE. Ethanol induces CYP2E1 by protein stabilization. Role of ubiquitin conjugation in the rapid degradation of CYP2E1. *J Biol Chem.* 1995;270(50):29632–29635.
72. Jana S, Paliwal J. Molecular mechanisms of cytochrome p450 induction: Potential for drug-drug interactions. *Curr Protein Pept Sci.* 2007;8(6):619–628.
73. Gordi T, Xie R, Huong NV, Huong DX, Karlsson MO, Ashton M. A semiphysiological pharmacokinetic model for artemisinin in healthy subjects incorporating autoinduction of metabolism and saturable first-pass hepatic extraction. *Br J Clin Pharmacol.* 2005;59(2):189–198.
74. McCune JS, et al. *In vivo* and *in vitro* induction of human cytochrome P4503A4 by dexamethasone. *Clin Pharmacol Ther.* 2000;68(4):356–366.
75. Gorski JC, et al. The effect of Echinacea (Echinacea purpurea root) on cytochrome P450 activity *in vivo*. *Clin Pharmacol Ther.* 2004;75(1):89–100.
76. van Heeswijk RP, et al. Effect of high-dose vitamin C on hepatic cytochrome P450 3A4 activity. *Pharmacotherapy.* 2005;25(12):1725–1728.
77. Xu Y, Iwanaga K, Zhou C, Cheesman MJ, Farin F, Thummel KE. Selective induction of intestinal CYP3A23 by 1alpha,25-dihydroxyvitamin D3 in rats. *Biochem Pharmacol.* 2006;72(3):385–392.
78. Mouly S, et al. Hepatic but not intestinal CYP3A4 displays dose-dependent induction by efavirenz in humans. *Clin Pharmacol Ther.* 2002;72(1):1–9.
79. Fromm MF, Busse D, Kroemer HK, Eichelbaum M. Differential induction of prehepatic and hepatic metabolism of verapamil by rifampin. *Hepatology.* 1996;24 (4):796–801.
80. Watanabe T, Kusuhara H, Maeda K, Shitara Y, Sugiyama Y. Physiologically-based pharmacokinetic modeling to predict transporter-mediated clearance and distribution of pravastatin in humans. *J Pharmacol Exp Ther.* 2009;328(2):652–662.

81. Chien JY, Mohutsky MA, Wrighton SA. Physiological approaches to the prediction of drug-drug interactions in study populations. *Curr Drug Metab*. 2003;4(5):347–356.
82. Yang J, et al. Implications of mechanism-based inhibition of CYP2D6 for the pharmacokinetics and toxicity of MDMA. *J Psychopharmacol*. 2006;20(6):842–849.
83. Ito K, Ogihara K, Kanamitsu S, Itoh T. Prediction of the in vivo interaction between midazolam and macrolides based on in vitro studies using human liver microsomes. *Drug Metab Dispos*. 2003;31(7):945–954.
84. Perdaems N, et al. Predictions of metabolic drug-drug interactions using physiologically-based modelling: Two cytochrome P450 3A4 substrates coadministered with ketoconazole or verapamil. *Clin Pharmacokinet*. 2010;49(4):239–258.
85. Rowland Yeo K, Jamei M, Yang J, Tucker GT, Rostami-Hodjegan A. Physiologically-based mechanistic modelling to predict complex drug-drug interactions involving simultaneous competitive and time-dependent enzyme inhibition by parent compound and its metabolite in both liver and gut—the effect of diltiazem on the time-course of exposure to triazolam. *Eur J Pharm Sci*. 2010;39(5):298–309.
86. Wang YH. Confidence assessment of the simcyp time-based approach and a static mathematical model in predicting clinical drug-drug interactions for mechanism-based CYP3A inhibitors. *Drug Metab Dispos*. 2010;38(7):1094–1104.
87. Einolf HJ. Comparison of different approaches to predict metabolic drug-drug interactions. *Xenobiotica*. 2007;37(10–11):1257–1294.
88. Guest EJ, Rowland-Yeo K, Rostami-Hodjegan A, Tucker GT, Houston JB, Galetin A. Assessment of algorithms for predicting drug-drug interactions via inhibition mechanisms: Comparison of dynamic and static models. *Br J Clin Pharmacol*. 2011;71(1):72–87.

10

PHYSIOLOGICALLY-BASED PHARMACOKINETICS OF BIOTHERAPEUTICS

CONTENTS

10.1	Introduction.	210
10.2	Therapeutic Proteins.	210
	10.2.1 Peptides and Proteins.	210
	10.2.2 Monoclonal Antibodies	212
10.3	Pharmacokinetics of Therapeutic Proteins	214
	10.3.1 Peptides and Proteins.	215
	10.3.2 Monoclonal Antibodies	224
10.4	PBPK/PD Modeling for Therapeutic Proteins	230
	10.4.1 Need for PBPK Modeling for Therapeutic Proteins	230
	10.4.2 PBPK Modeling for Therapeutic Proteins	231
	10.4.3 Pharmacokinetic Scaling.	239
	10.4.4 Applications of PBPK Models of Therapeutic Proteins	242
	10.4.5 PBPK Integration with Pharmacodynamics	244
10.5	Antisense Oligonucleotides and RNA Interferance	245
	10.5.1 Antisense Oligonucleotides (ASOs)	245
	10.5.2 Ribonucleic Acid Interference (RNAi).	245
	10.5.3 Pharmacokinetics of ASOs[50] and Double-Stranded RNAs	247

Physiologically-Based Pharmacokinetic (PBPK) Modeling and Simulations: Principles, Methods, and Applications in the Pharmaceutical Industry, First Edition. Sheila Annie Peters.
© 2012 John Wiley & Sons, Inc. Published 2012 by John Wiley & Sons, Inc.

10.5.4 Design and Modifications of ASOs to Improve Target Affinity and PD: the First, Second, and Third Generation ASOs. 249
10.5.5 Integration of PK/PBPK and PD Modeling 253
Keywords 254
References. 256

10.1 INTRODUCTION

Biologics, also called biotech drugs or biopharmaceuticals, are biotechnology-derived products, typically of high molecular weights (>1 kDa), representing the cutting edge of biomedical research (Fig. 10.1). All biologicals are derived from living materials—human, plant, animal, or microorganism. Unlike small molecules that are defined by their chemical structure and purity, biologics are characterized by their biological production processes. Today, there are a number of macromolecules in the market for treating a variety of diseases, mainly cancers, infections, diabetes, autoimmune disorders, and cardiovascular diseases.

According to a PhRMA report,[1] the U.S. FDA has already approved 125 biotech drugs, while over 400 drugs are in development for more than 100 diseases. It is widely believed that in the future, they will offer the most effective means of treatment for a variety of medical conditions that have no other treatment options. Future growth, however, depends largely on the industry overcoming a number of hurdles, some of them related to the pharmacokinetics (PK) of these molecules, including delivery of the drug to the site of action and the high cost of material production due to dose requirements. Therefore, it is necessary to understand the whole-body disposition characteristics of these compounds in order to design therapeutics with optimal pharmacokinetic and pharmacodynamic (PD) characteristics. This chapter will describe the pharmacokinetics of biotherapeutics and highlight the need to apply physiologically-based pharmacokinetic modeling concepts to model their PK and to link with PD.

10.2 THERAPEUTIC PROTEINS

10.2.1 Peptides and Proteins

Diseases caused by defective genes or disease processes that do not produce sufficient levels of proteins required by the body can be treated by protein replacement therapies. Examples include factor VIII, a protein involved in the blood-clotting process, lacking in hemophiliacs, and insulin, a protein hormone that regulates blood glucose levels and lacking in diabetics. Leading classes

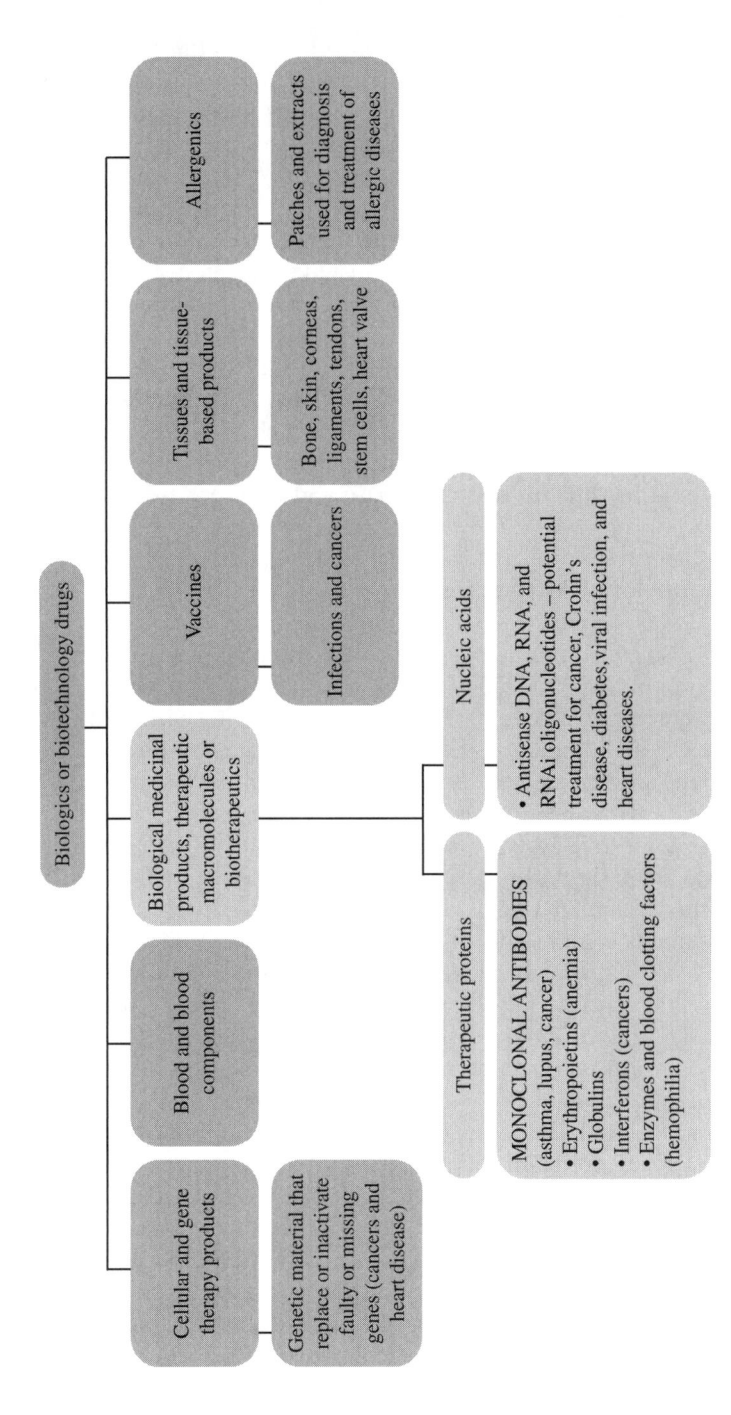

Figure 10.1. Biologics or biotechnology drugs categorized.

of proteins and peptides include enzymes, growth factors [erythropoietins, granulocyte-colony stimulation factor (G-CSF), thrombopoietin, etc.], inflammatory mediators (cytokines, interferons), and blood factors. A number of therapeutic peptides contain 8–20 amino acids and are analogs of endogenous hormones including oxytocin, arginine vasopressin (also called antidiuretic hormone, ADH), somatostatin, gonadotropin-releasing hormone, and luteinizing hormone. Because of their small size, these peptides have the possibility to be synthesized chemically. The peptides of intermediate complexity contain 20–50 amino acids and include insulin, glucagon, teriparatide, nesiritide, enfuvirtide, refludan, and sermorelin. They are mimics of endogenous proteins. Many proteins are produced by splicing genes into bacteria by recombinant DNA (rDNA) technology or by cell cloning. Recombinant protein-based therapeutics include interleukins, deoxyribonuclease, replacement enzymes for metabolic disease, growth factors such as G-CSF, platelet-derived growth factor, and the like. Some of these undergo post-translation modifications that could be difficult to characterize and replicate.

10.2.2 Monoclonal Antibodies

Antibodies, also called immunoglobulins (Ig), are endogenous macromolecules [molecular weight (MW) ∼150 kDa] that are a key component of the humoral immune system responsible for neutralization of invading pathogens. Antibodies have two functional domains—the fragment of antigen binding (Fab), which is responsible for antigen recognition, and the fragment crystallizable (Fc) region, which interacts with the surface Fc receptors of effector cells (Fig. 10.2). In humans, there are five isotypes of antibodies (Table 10.1)—immunoglobulin alpha (IgA), delta (IgD), epsilon (IgE), gamma (IgG), and mu (IgM)—with differences in molecular weight and antigen-binding sites. The most prevalent isotype in humans is IgG (85% of Ig in serum), which also has a relatively long half-life of 20–21 days for IgG1, IgG2, and IgG4 and 7 days for IgG3.

Therefore, a majority of therapeutic antibodies are of the IgG isotype. Endogenously, antibodies to a particular target are generally polyclonal—that is, antibodies produced by different B-cell lineages, hence with differing affinity to the target antigen and also binding different parts of the antigen (epitopes). However, monoclonal antibodies (mAbs) are identical monospecific antibodies produced by a single type of immune cell that are all clones of a single parent cell (Fig. 10.3). Rodent antibodies are more easily generated compared to that of humans, but they could trigger a rapid immune response in humans. So, several strategies were devised to humanize rodent antibodies. Chimeric antibodies with human constant regions and rodent variable regions, humanized or CDR-grafted monoclonal antibodies with human IgG variable sequence except for the complementarity-determining regions (CDRs), and fully human antibodies have been developed as therapeutic drugs. The names of chimerics contain the affix -xi- (e.g., infliximab), humanized antibodies have the infix-

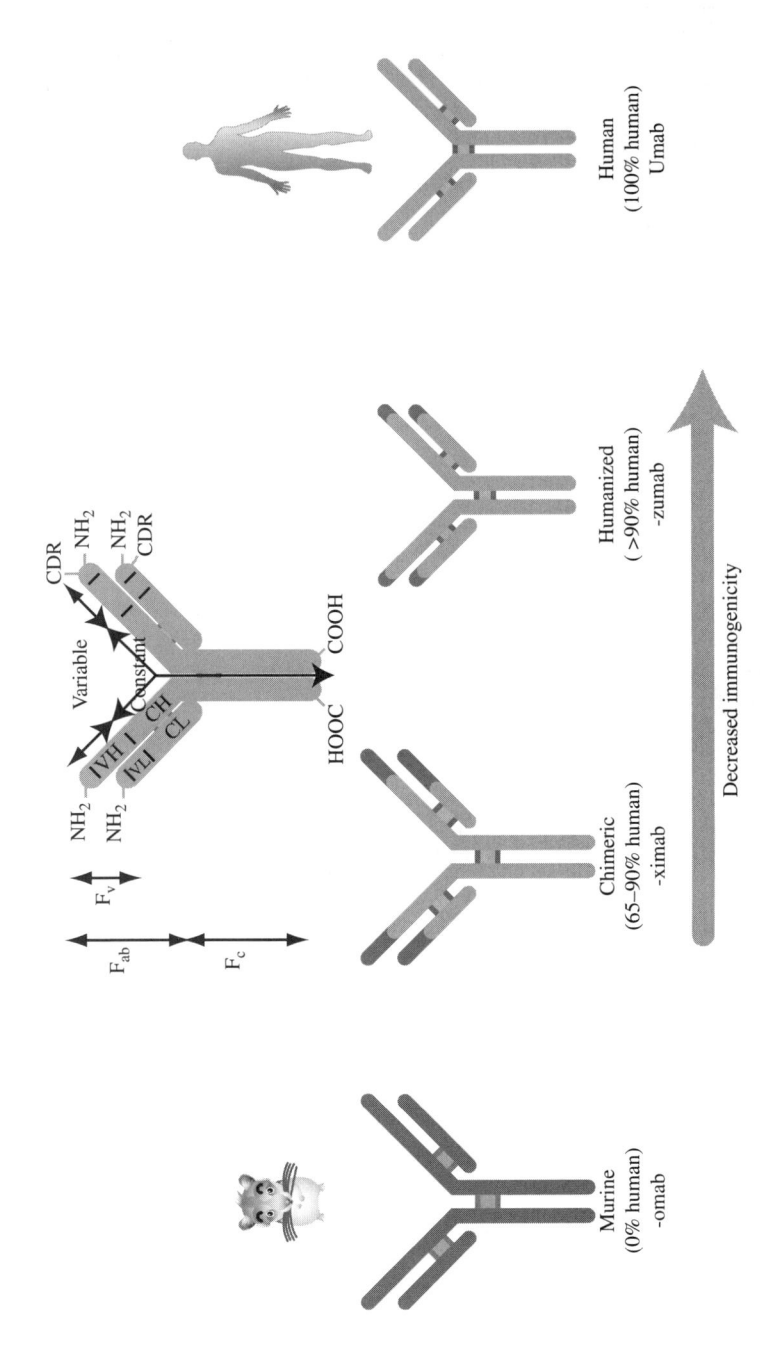

Figure 10.2. Complementarity-determining regions (CDRs) are found in the variable fragment (Fv) portion of the antigen-binding fragment (Fab). Chimeric mAbs are constructed with variable regions on the light and heavy chains (VL and VH) derived from a murine source and constant regions of light and heavy chains (CL and CH) derived from a human source. Humanized therapeutic mAbs are predominantly derived from a human source except for the CDRs, which are murine.

TABLE 10.1. Human Immunoglobulins

Class	Function	MW (kDa)	Half-life (days)
IgG (1, 2, 3, and 4)	Main Ig in blood. Coats microbes and toxins to enhance their uptake by other immune system cells.	150	21 for IgG1, IgG2, and IgG4; 7 for IgG3
IgA (1 and 2)	Concentrates in body fluids, guarding entrances to body.	160	6
IgD	Remains attached to B cells. Plays a key role in initiating early B-cell responses. Mainly in humans.	180	3
IgM	Effective in killing bacteria by favoring agglutination.	970	10
IgE	Protects against parasitic infections. Responsible for symptoms of allergy.	190	2

zu- (e.g., bevacizumab), and antibodies originating in humans use -u- (e.g., adalimumab). The human immune system reacts to murine, chimeric, and humanized mAbs, by producing human antimurine, antichimeric, and antihumanized antibodies (**HAMA, HACA** and **HAHA,** respectively).

It is possible to create monoclonal antibodies that specifically bind to any antigen. In cancer therapy, for example, mAb can target a tumor-specific antigen, which is selective to the tumor or at least overexpressed in the tumor. The greater the selectivity to the target, the smaller is the risk for adverse events. Developing mAbs as drugs is an iterative design process that involves the generation and optimization of antibodies to improve their clinical potential. Some antibodies are loaded with cytotoxins, including radionucleotides or bacterial toxins, and mediate targeted drug delivery (antibody-directed drug delivery). Others are designed to function naturally, depending on the therapeutic application. Antibodies are used to neutralize toxins (immunotoxicotherapy), mediate the destruction of cells, or alter cellular function. The antibody marked cancer cell is recognized by immune system as invaders and destroyed. Immune mechanisms (Fig. 10.4) explaining antibody action includes **complement-dependent cytotoxicity (CDC), complement-dependent cell-mediated cytotoxicity (CDCC),** and **antibody dependent cellular cytotoxicity (ADCC).**

10.3 PHARMACOKINETICS OF THERAPEUTIC PROTEINS

This section provides a brief overview of the pharmacokinetics of therapeutic proteins. A very detailed and comprehensive discussion of the subject is available elsewhere.[2–4]

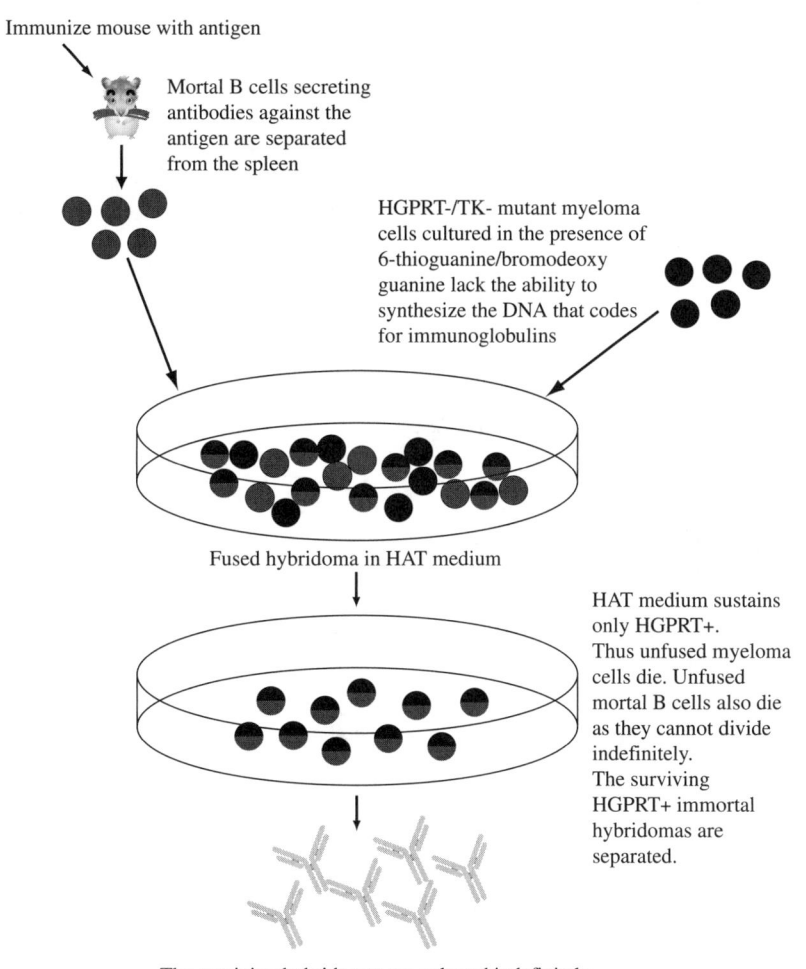

Figure 10.3. Production of monoclonal antibodies by hybridoma technology. HGPRT: hypoxanthine–guanine phosphoribosyl transferase; HAT: hypoxanthine aminopterin thymine; TK: thymidine kinase.

10.3.1 Peptides and Proteins

10.3.1.1 Absorption. Oral absorption of therapeutic proteins is limited by their size, charge, denaturation in gastric pH, and proteolysis by the exo- and endo-peptidases secreted by the pancreas into the lumen of the stomach and intestine, as well as expressed glycoproteins on the brush border membrane. Consequently, therapeutic proteins have to be administered as an intravenous (bolus injection or constant rate infusion), subcutaneous (SC) or intramuscular

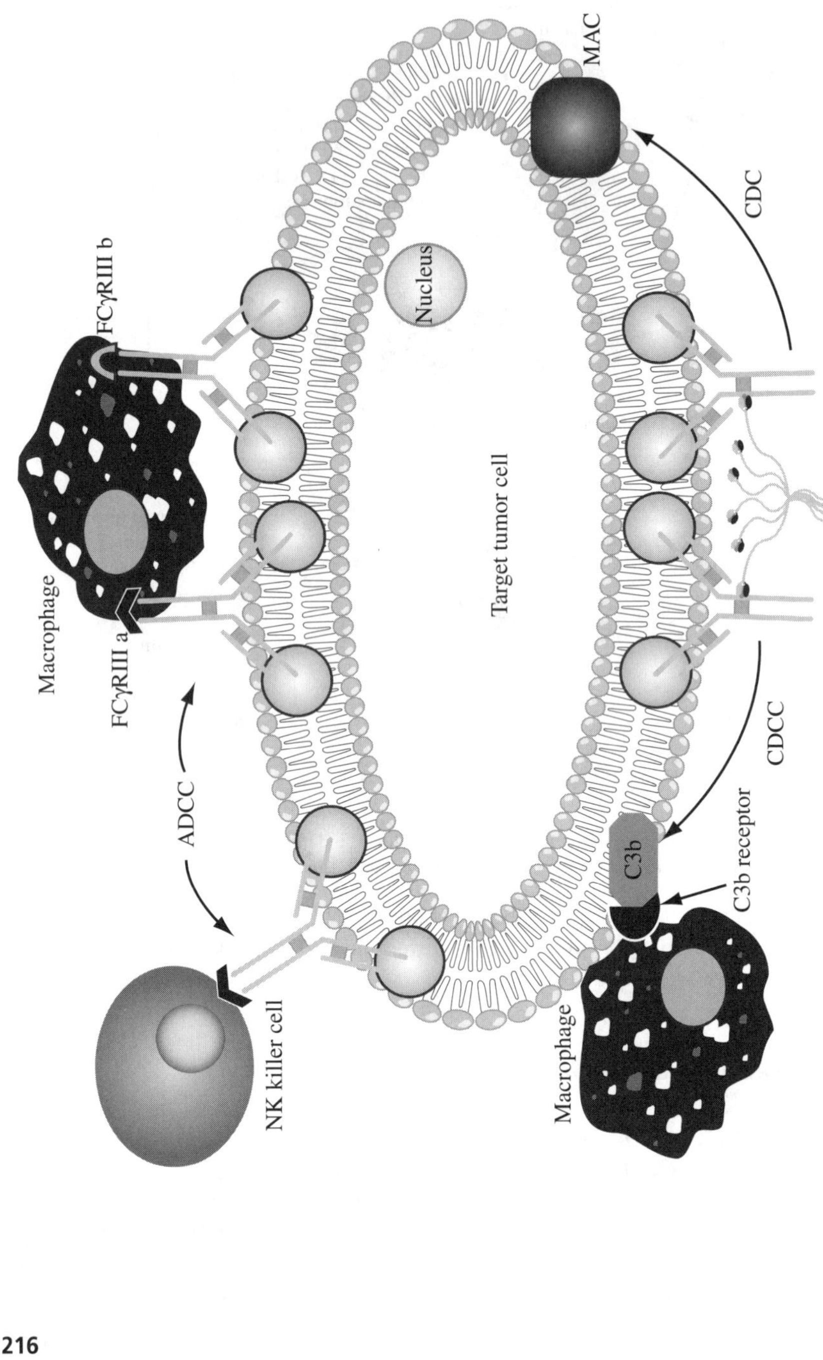

Macrophage
FCγRIII a
FCγRIII b
ADCC
NK killer cell
Nucleus
Target tumor cell
MAC
CDC
C1q
CDCC
C3b
C3b receptor
Macrophage

10.3 PHARMACOKINETICS OF THERAPEUTIC PROTEINS 217

Figure 10.4. mAb therapeutic action by immune mechanisms. The binding of mAb to its target antigen (eg., a protein specific to the tumor cell) triggers the recruitment of immune effector cells (macrophages, monocytes, cytotoxic T cells, and natural killer cells) to the area and the activation of the classical complement pathway. This is followed by an immune response that ultimately results in cell lysis and release of cytokines into circulation. The complement system comprises 30 proteins circulating in blood plasma. One of the complement proteins, C1q interacts with the CH2 constant region of the mAb, which leads to the activation of a proteolytic cascade and consequently induces the formation of a membrane-attack complex (MAC) for the lysis of tumor cells; this effect is termed complement-dependent cytotoxicity (CDC). C3b, which is generated during this cascade reaction, functions as an opsonin to facilitate phagocytosis and cytolysis through its interaction with the C3b receptor (C3bR) on a macrophage or natural killer (NK) cell; this activity is termed complement-dependent cell-mediated cytotoxicity (CDCC). In addition, mAb binding to tumor cells induces antibody-dependent cellular cytotoxicity (ADCC); immune-effector cells such as macrophages and NK cells that are recruited to the area interact with the CH3 region of the mAbs through FcγRIIIa expressed by both effector cells. Then, mAb-coated tumor cells are phagocytosed by macrophages or undergo cytolysis by NK cells. On the other hand, there is a negative regulation to modulate the cytotoxic response against tumors through FcγRIIb, which is expressed on the cell surface of macrophages. Immunoglobulin G1 (IgG1) and IgG3 can activate the classical complement pathway and interact with Fcγ receptors more potently than IgG2 or IgG4. In particular, IgG4 cannot activate the classical complement pathway.

◄──

(IM) injection. Several strategies for oral delivery are emerging, notable among them being the use of absorption enhancers, bile salts and surfactants, coadministration of protease inhibitors, and amino acid backbone modification to make them less susceptible to degradation, uptake of encapsulated, biodegradable, and biocompatible micro or nanoparticles via transcytosis or delivery through Peyer's patches (a collection of large oval lymph tissues located in the mucous secreting lining of the small intestine). The use of nanoparticulate delivery systems, particularly mucoadhesive nanoparticulate carrier systems, appear to be promising as they enhance membrane permeation as well as simultaneously protecting the therapeutic protein from enzyme degradation. Modulation of intestinal tight junctions to increase paracellular transport can increase absorption of therapeutic proteins but is associated with the potential risk of transporting toxic and unwanted pathogens too. Uptake of therapeutic proteins into systemic circulation following SC administration is directly via blood capillaries and through the more porous lymphatic vessels, depending on their size. Peptides and small proteins enter blood circulation while larger proteins predominantly enter the lymphatic system and are unidirectionally transported into the veins due to semilunar valves. Lymph fluid consists of excess interstitial water, electrolytes, and proteins that drain from interstitial space into the lymphatic capillaries. The source of this excess interstitial fluid is a net leakage across endothelium of local blood capillaries in tissues. The lymph fluid drains from lymphatic capillaries to the collecting vessels, which pass through lymph nodes that act as filters to trap the foreign particles. These vessels then drain into larger trunks that in turn lead into the ducts. The lymphatic system thus plays an important role in maintaining homeostasis of body water, by constantly moving excess fluids, lipids, and immune-system-related products around the body. Due to a lack of a pumping system, the flow rate of lymph is 100–500 times slower than the blood flow rate. Consequently, the rate and extent of absorption of large proteins administered by the SC route is limited. Several studies have shown that there is minimal enzymatic degradation or leakage of proteins during lymph transport. The site of SC injection might affect the rate of lymphatic absorption. Presystemic metabolism after IM or SC is possible, causing a reduction in the bioavailability of drugs administered in these routes. Thus,

$$F = \frac{k_a}{k_a + k_{\text{deg}}} = \frac{k_a}{k_{\text{app}}} \qquad (10.1)$$

where F is the systemic availability following IM or SC, and k_a, k_{deg}, and k_{app} are the rate constants of absorption, metabolic degradation, and apparent absorption from either route. The small injection volumes that are needed to avoid pain in SC (< 2.5 mL) and IM (< 5 mL) are not always possible due to the solubility of mAbs (usually in the range of 100 mg/mL). Other parenteral routes of delivering therapeutic proteins include nasal, buccal, rectal, vaginal, transdermal, ocular, and pulmonary, although clinical examples of these are rare.

10.3.1.2 Distribution. Once in the bloodstream, therapeutic proteins must be transported from the blood circulation to the site of action in target tissues in order to exert their pharmacological action. This happens by extravasation from vascular space to the interstitium and diffusion of the drugs through the extracellular matrix to a target at the cell surface. If the target is intracellular (cytoplasmic or nuclear), the plasma membrane poses an additional barrier to the therapeutic protein reaching its target. According to the two-pore formalism,[5] the vascular wall has a large number of small pores of radii 4–10 nm and a much smaller number of large pores of radii 40–60 nm per unit area of the wall. Extravasation of therapeutic proteins through either of these pores into interstitial fluid is feasible as a result of higher hydrostatic pressure on the arterial end of the capillaries compared to the colloid osmotic pressure of the surrounding tissues. On the venous end of the capillary, the colloid osmotic pressure exceeds the blood pressure, resulting in the interstitial fluid being drawn into the venule. The extravasation across the vascular wall caused by a pressure gradient is called convective transport. Extravasation rates depend on the regional differences in capillary structures, disease state of the tissue, and molecular size, shape, charge, and polarity of the macromolecules. For example, inflammation and angiogenesis cause the local endothelium to be hyperpermeable to macromolecules. Some exogenous and endogenous proteins such as histamine and thrombin can increase the endothelial permeability. In addition to convective transport through the pores of the endothelium, the process of transcytosis via membrane vesicles also plays an important role in the transfer of therapeutic proteins from the vascular space to the interstitial space. Following convective or transcytotic extravasation, protein drugs distribute into the interstitium by diffusion. Distribution occurs by the simultaneous diffusion and convective transport. Based on molecule size and the three-dimensional structure, one process is more prominent than the other. Only part of the total interstitial fluid space (V) is available to the macromolecules. This available volume is denoted by V_a. Mutual exclusion between protein therapeutics and the structural molecules of the extracellular matrix such as the negatively charged glycosaminoglycans and collagens results in exclusion volume (V_e), which is simply $V - V_a$ and dependent on the molecular weight of the macromolecule. The cationic IgG1 has a significantly greater available volume than the anionic IgG4. The greater the exclusion volume, the greater is the rate of lymphatic elimination, and the lower is the interstitial concentration. Part of the drug that enters the interstitium drains into the lymphatic capillary facilitated by the lymph flows (Fig. 10.5) and ultimately returned to systemic circulation.

Unlike small molecules that have good nonspecific distribution by passive diffusion, proteins and antibodies are limited to vascular space due to permeability limitation arising from size. However, extensive binding of a high-affinity drug to its pharmacological target can lead to an increased tissue uptake and clearance. This is called **target-mediated drug disposition (TMDD)**.[6–8] Although TMDD can apply to any drug, therapeutic proteins are more prone to TMDD because of their generally low doses, which guarantee nonsaturation of the target. TMDD is easily observed in the PK profiles of proteins and antibodies as it leads to a rapid initial distribution and long terminal elimination phase for low doses (Fig. 10.6). At high doses, the capacity limitation of the target results in volumes of distribution that approximately equal the extracellular space (vascular and interstitial space). Thus, TMDD leads to dose dependence in distribution and clearance. A relatively modern concept that should be included within TMDD is **receptor-mediated endocytosis (RME)** (Fig. 10.7) in which target binding initiates internalization of the drug–receptor complex followed by intracellular drug elimination. If the number of pharmacological target receptors is of the order of the number of drug molecules and the turnover rates of the bound target relative to the elimination rate of the naked mAb is high, RME can be a major contributor to the elimination of the drug. In this case, PK and PD become inseparable and bidirectionally interdependent.

10.3.1.3 Elimination. While dietary proteins are completely broken down in the gut, the degradation of systemic functional and regulatory proteins, as well as exogenous proteins, is a highly complex and temporally controlled process.

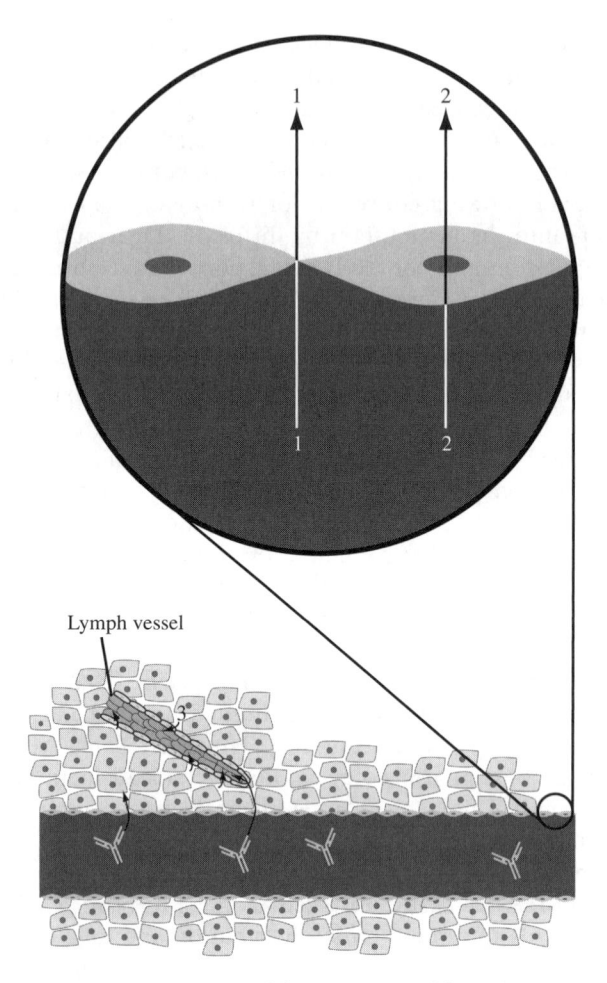

Figure 10.5. Convective extravasation (1), transcytosis (2), and lymphatic drainage (3) of therapeutic proteins.

There are two major mechanisms of protein elimination—the lysosomal and ubiquitin-mediated pathway. Within cells, the lysosome is a vacuolar structure that is produced by the Golgi apparatus, whose function is to destroy microorganelles and damaged or defective intra- and extracellular proteins, thereby allowing for replenishment. Such an elimination involving the cells own components is called autophagocytosis. Exogenous proteins and antibodies are internalized by specific RME (the receptor could be the pharmacological target or others) or by nonspecific pinocytosis. The inward budding of vesicles from plasma membranes are processed in the endosomal compartment and delivered to the lumen of lysosome for proteolytic degradation. The lysosomal compartment with an acidic pH of around 5 compared to the cytosolic pH of 7.2 is ideal

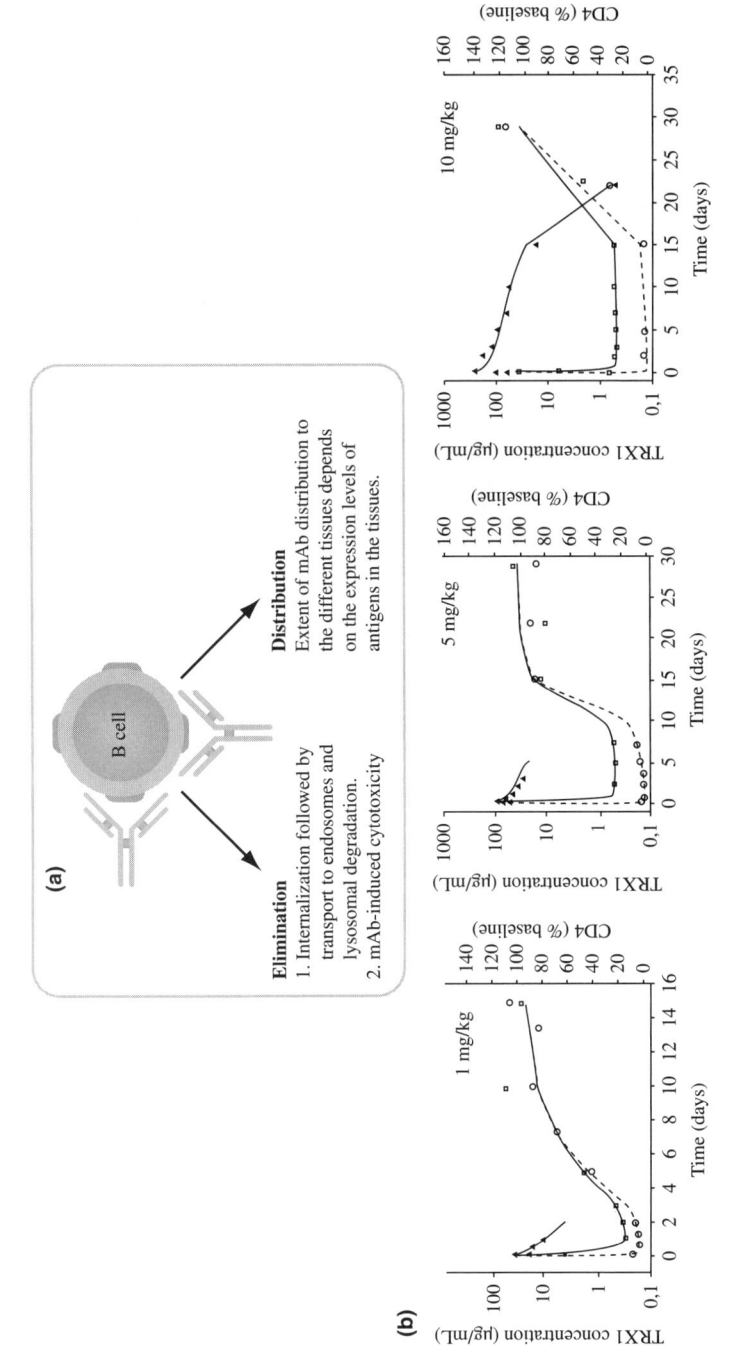

Figure 10.6. Target-mediated drug disposition. (a) Target binding of mAb influences both distribution and clearance. (b) Interdependence of PK (closed triangles) and PD of anti-CD4 monoclonal antibody (TRX1) due to TMDD. The cellular binding and internalization of TRX1 leads to a rapid loss of the mAb, which in turn causes CD4 levels to return to the baseline. Demonstrates the saturable nature of TMDD arising from capacity limitation. [Reprinted from "Pharmacokinetics/pharmacodynamics of non-depleting anti-CD4 monoclonal antibody (TRX1) in healthy human volunteers." *Pharm Res* **23**:95–103. Copyright (2006). With permission from Springer.]

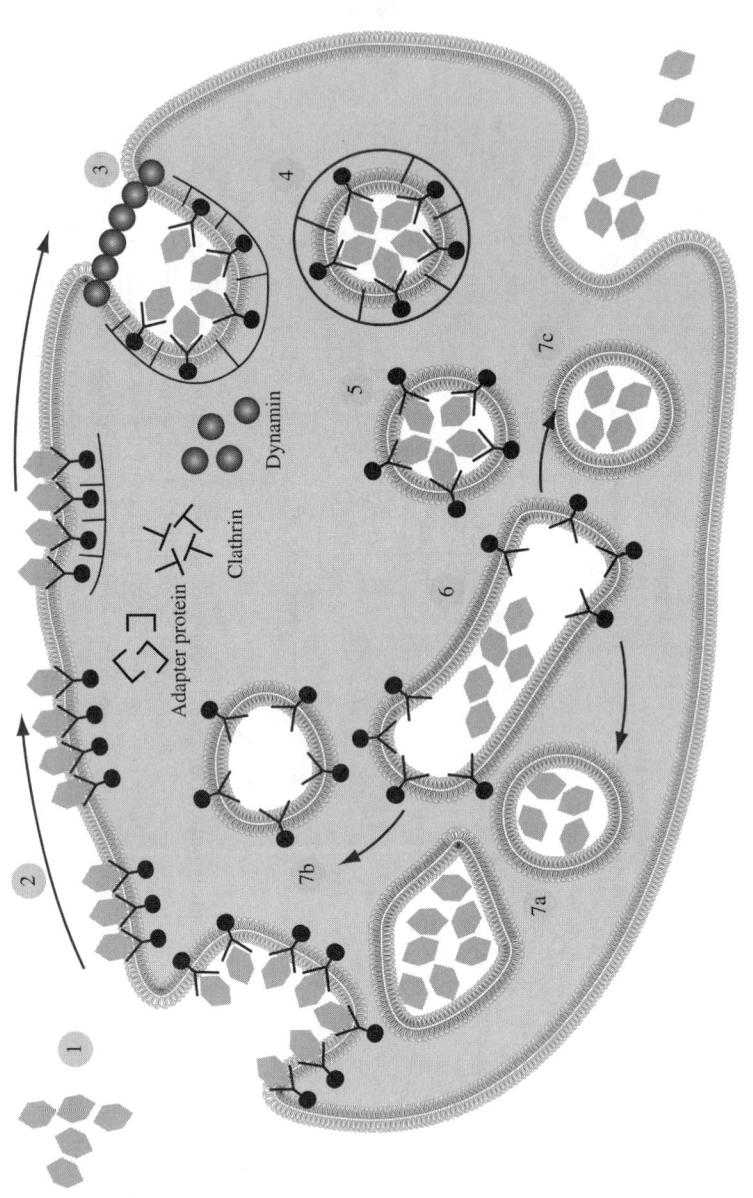

Figure 10.7. Receptor-mediated endocytosis. Internalization and subsequent lysosomal digestion in hepatocytes and other cells. (1) The molecules that will be internalized bind to specific receptors on the surface of the plasma membrane. (2) Receptor–ligand complexes accumulate in coated pits, where (3) invagination is facilitated by adaptor protein, clathrin, and dynamin on the cytosolic surface of the membrane. (4) An internalized coated vesicle that (5) quickly loses its clathrin coat. The uncoated vesicle is now free to (6) fuse with other intracellular membranes, usually a membrane surrounding an early endosome, where internalized material is sorted. The fate of the receptors and the ingested molecules depends on the nature of the material. Transport vesicles often carry the internalized material to a late endosome for digestion (7a). Alternative pathways include recycling to the plasma membrane (7b) or transport to another region of the plasma membrane and exocytosis (7c).

for denaturing the proteins. The denatured proteins are then destroyed by the endo- and exopeptidases while glycoproteins are acted on by glycosidases such as sialidase and β-galactosidase first. The survival of lysosomal proteases in the acidic environment is attributed to their association with cathepsin-A, a lysosomal protective protein, whose deficiency can cause severe metabolic disorders. A second metabolic pathway for protein degradation is the ubiquitin–proteasome pathway.[9] Degradation of a protein by the ubiquitin system involves two distinct and successive steps: (i) covalent attachment of multiple ubiquitin molecules to the target protein and (ii) degradation of the tagged protein by the 26S proteasome (an ATP-dependent, multifunctional proteolytic complex, consisting of a proteolytic core, 20S proteasome, and sandwiched between 19S regulatory complexes) or, in certain cases, by the lysosomes/vacuole.

Hepatic proteolysis starts with the endopeptidases attacking the middle of the protein and further degradation by exopeptidases. Proteolytic enzymes being ubiquitous, elimination of proteins occur in almost every tissue in the body but predominantly in blood and hepatic and reticuloendothelial cells. As RME can also be important in organs where the target abounds, elimination of therapeutic proteins are not necessarily restricted to the central compartment comprising the vascular space and fast perfused organs. Target-mediated effects (distribution and elimination) are modulated by receptor up- or down-regulation. Recombinant human G-CSF is a good example of RME. A good correlation between G-CSF and percentage of G-CSF-positive neutrophils have been demonstrated in cancer patients.[10] Administration of G-CSF increases the neutrophil levels and causes an up-regulation of G-CSF receptors, thus modulating its own elimination. Another example is the specific and saturable internalization of erythropoietin in erythroid progenitor cells.[11]

Biliary and renal excretion are generally rare. Renal elimination depends on size and is only important for peptides and proteins whose molecular weight is <30 kDa. For small peptides, however, renal elimination need not be the major route of elimination. Assuming that the glomerular filtration rate (GFR) in human is 100 mL/min in a 70-kg man, it should take about 60 min to completely eliminate a compound from blood circulation, once it is absorbed. However, the half-life of many peptides is much shorter, suggesting that other catabolic clearance pathways may play a quantitatively more important role. Renal clearance becomes significant for proteins whose diameters are <800–1000 nm—the effective pore size of the glomerular filtration. Filgrastim, a recombinant methionyl G-CSF with a diameter of 250 nm and MW of 18.8 Da, is eliminated mainly by renal clearance.

Chronic administration of therapeutic proteins can trigger antibody formation, especially for those derived from animal proteins, and this is termed as **immunogenicity**. The antibody–therapeutic protein drug complex serves as a depot for the therapeutic protein and could be beneficial in extending the therapeutic activity of the drug, if complex formation does not adversely affect its activity. However, immunogenicity can also be detrimental if the antibodies raised against therapeutic proteins cross-react with endogenous proteins. For example, treatment of recombinant human erythropoietin (rhEPO) can sometimes lead to near total shutdown of red blood cell production, a situation termed pure red cell aplasia, which developed 3–67 months after treatment. Loss of efficacy due to irreversible elimination is another consequence of immunogenicity. Many factors can affect the immunogenicity of therapeutic proteins—sequence variation, glycosylation, genetic background of patients, type of disease, length of treatment, dosage regimen, route of administration, formulation, impurity and contaminants of the therapeutic protein, and other unknown factors. Protein aggregate and precipitate formation that can accompany extravascular dosing make the proteins administered in those routes more immunogenic compared to the IV route. SC injections have been shown to be more immunogenic compared to IM. Protein conjugation with polyethylene glycol (PEG) has been shown to shield (by steric hindrance) antigenic determinants on the protein from detection by the immune system. For example, **PEGylation** of pegaspargase protects it against allergic reactions towards L-asparaginase. Other advantages of PEGylation include improved PK and physicochemical properties. Pegfilgrastim, the PEGylated version of filgrastim, a G-CSF used for the management of chemotherapy-induced neutropenia, has a reduced renal elimination. The alternate pathway of neutrophil-mediated clearance is not very effective due to the low availability of neutrophils, resulting in a prolonged half-life and sustained action. The increase in the hydrodynamic volume of the protein afforded by PEGylation causes reduced renal elimination, restricted biodistribution, prolonged residence time, and, therefore, more sustained duration of action. It also increases solubility and protects against proteolytic degradation.

10.3.2 Monoclonal Antibodies

Successful preclinical and clinical development, approval and therapeutic application of mAbs rely on a thorough understanding of their PK characteristics. The PK mechanisms described for proteins are also applicable for mAbs. The clinical PK of mAbs has been recently reviewed.[12] Convective transport, transcytosis, lymphatic elimination, TMDD, internalization by RME, or pinocytosis and lysosomal catabolic proteolysis are key mechanisms underlying the disposition of mAbs, causing their pharmacokinetics (Table 10.2) to be nonlinear and complex. Given that therapeutic antibodies are often developed for high-affinity binding to cellular antigens, TMDD often plays an important role in determining the distribution volume as well as clearance of mAbs. Interaction of mAbs with cellular antigens could impact its tissue distribution and elimination. Receptor-mediated internalization followed by lysosomal degradation and mAb-induced cytotoxicity (Fig. 10.4) contribute to target-mediated elimination. Due to high target affinity, these distribution and elimination routes are saturable and limited by the capacity of the target. For

TABLE 10.2. PK of Therapeutic Proteins

Therapeutic Protein	Function	Half-life (h)	Disposition: Absorption, Metabolism, Elimination, and Distribution[a]	Examples
Peptides	Hormones	Short (<1 h)	**A:** Diffusion, carrier-mediated uptake. **M:** CYP-mediated hepatic metabolism and proteolysis by peptidases and proteases. **E:** Renal elimination.	Insulin proteolysis by the aspartic acid protease, cathepsin D.
Proteins	Transport	Long	**D:** Due to size, largely confined to vascular space. Extravasation from vascular space occurs paracellularly via convection (due to osmotic pressure and sieve effect from hydrostatic pressure—appreciable for MW<150 kDa), or transcellularly via pinocytosis (fluid-phase endocytosis). Convective elimination into lymphatic drainage. High-affinity, high-capacity interaction of a therapeutic protein with a pharmacological target leads to target-mediated drug disposition (TMDD). **Proteolysis:** Uptake into hepatocytes, therapeutic target cells, and other cells by receptor-mediated endocytosis (RME). Internalization by RME is followed by lysosomal or ubiquitin-mediated catabolic degradation by proteases.	Enzymes, erythropoietins, interferons, and blood factors. Internalization of erythropoietin in erythroid progenitor cells in the bone marrow is an example of RME. Filgrastim, MW 18.8 kDa and diameter of around 250 nm is predominantly cleared renally.

(Continued)

TABLE 10.2. (Continued)

Therapeutic Protein	Function	Half-life (h)	Disposition: Absorption, Metabolism, Elimination, and Distribution[a]	Examples
			E: Renal elimination, depending on size. Glomerular filtration is efficient for macromolecules with MW <30 kDa. Cation > neutral > anion. Elimination by **opsonization** and phagocytosis, for therapeutic proteins with MW > 200 kDa.	
mAbs	Long-term immunity	Long (1–50)	**D:** Extravasation from vascular space similar to proteins. Additionally, Fc receptor-mediated transcytosis—bidirectional transport to the blood and tissue interstitium. Convective elimination into lymphatic drainage, TMDD, RME, and catabolism—same as proteins. **E:** FcRn binding in tissue cells protects mAb from intracellular catabolism. Renal elimination, opsonization, and phagocytosis similar to proteins.	Nonlinear PK of efalizumab, a recombinant humanized IgG1, used in the treatment chronic psoriasis has been attributed to receptor-mediated mechanism.

[a] **A**, absorption; **D**, Distribution; **M**, metabolism, and **E**, elimination.

10.3 PHARMACOKINETICS OF THERAPEUTIC PROTEINS

most antibodies, the total tissue (vasculature + interstitial + cellular) : blood concentrations ratios are in the range of 0.1–0.5, and the expected distribution volume is 2–5 times the plasma volume. However, contrary to expectations, the steady-state volume of distribution, V_{ss}, reported in the literature for many mAbs are low, only about the plasma volume of the subject. This is because the statistical moment theory that is used for the calculation of V_{ss} makes the assumption of elimination only from the central compartment, which is not always applicable for mAbs. Immunogenicity can reduce the half-lives and efficacy of mAbs, apart from the adverse reactions they can cause. Therapeutic mAbs can have varying half-lives depending on their origin ranging from a few days for murine to a few weeks for human mAbs. Imunogenicity decreases as follows: fully rodent > rodent/human chimeric > CDR-grafted human > fully human. Following the administration of an immunogenic protein or antibody, a measurable antibody response is typically observed within 7–10 days (Fig. 10.8). It has been shown that the anti-infliximab antibodies cause an increased elimination of infliximab. In addition to PK mechanisms common with proteins, FcRn-mediated extravasation into tissue interstitium and protection against catabolism (Fig. 10.9) should also be considered for IgG antibodies. Factors influencing half-lives include target affinity and degree and nature of glycosylation of the antibody and FcRn-mediated protection against catabolism. The generally long half-lives of IgG antibodies have been attributed to neonatal Fc receptor or Brambell receptor (FcRn), a heterodimeric receptor composed of a nonclassical major histocompatibility complex (MHC) with class I α chain and $\beta 2$ microglobulin. It was identified in 1972 as the receptor that mediates the acquisition of maternal IgG in neonatal rat intestine

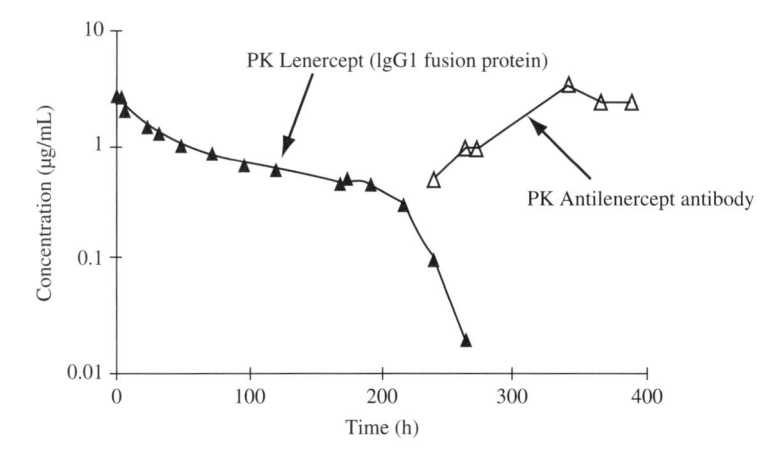

Figure 10.8. Effect of antilenercept antibody on the pharmacokinetic profile of lenercept. The anti-antibodies increase the clearance of the antibody with time. [Reprinted from "Antibody pharmacokinetics and pharmacodynamics." *J Pharm Sci* **93**:2645–2668. Copyright (2004). With permission from Wiley.]

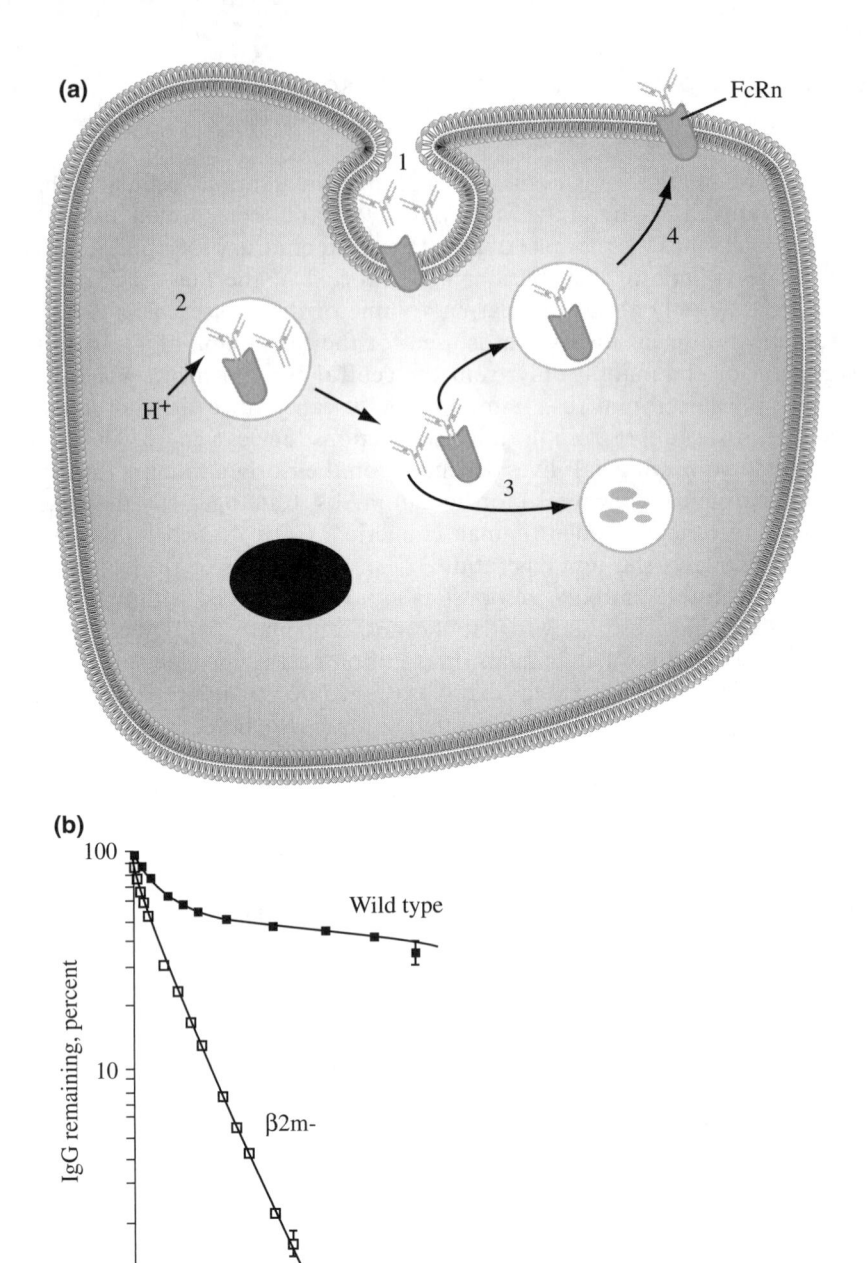

Figure 10.9. FcRn protection. (a) mAbs show a pH-dependent affinity to FcRn with a K_D of ~20 nM at pH 6.5. (1) Pinocytosis of IgG. (2) H$^+$ entry into endosome and subsequent FcRn–IgG binding. (3) Unbound IgG catabolyzed by lysosomal proteases. (4) Release of protected IgG into interstitial fluid or plasma on exposure to physiological pH. (b) IgG clearance in β2m knockout mice lacking expression of functional FcRn is ~10-fold that of wild type. [Reprinted from "The protection receptor for IgG catabolism is the beta2-microglobulin-containing neonatal intestinal transport receptor." *Proc Natl Acad Sci U S A* **93**:5512–5516. Copyright (1996) National Academy of Sciences, USA.]

and hence its name. FcRn binds only to the IgG class of antibodies in a pH-dependent manner, with a stronger binding at pH < 6 than at the physiological pH, 7.4. FcRn primarily resides in the endosomal vesicles within cells, with limited expression on the cell surface. Following IgG uptake into vascular endothelial cells through fluid-phase pinocytosis, the Fc part of the IgG antibodies bind to the Fc receptor in the acidic sorting endosomes. The unbound FcRn is catabolized by the proteases in the lysosomes, while the FcRn bound IgG are bidirectionally transported to the interstitial fluid or to systemic circulation (Fig. 10.9). Once outside the cell, the FcRn–IgG complex breaks up at the physiological pH. This FcRn-mediated protection from lysosomal degradation and transcytosis contributes to the distribution and long half-life of humanized and human IgG. Apart from the vascular endothelial cells where FcRn is highly expressed, it is also expressed in a wide variety of tissues and cells, where it protects the IgG from catabolism as described before. Cells expressing FcRn include renal glomerular epithelial cells, skin, bone marrow, muscle, lungs, hepatocytes, intestinal macrophages, B lymphocytes, platelets, peripheral blood monocytes, and dendridic cells. Ten times higher IgG clearance rates in FcRn-deficient mice (knockout of B2M gene) have demonstrated the protective role of FcRn binding. Due to the high level of expression in a variety of tissues, FcRn is not readily saturated, unless overloaded with exogenous IgG or albumin. IgG also binds to Fcγ receptors (FcγR) belonging to the immunoglobulin superfamily. They are the most important Fc receptors for inducing phagocytosis of opsonized microbes or tumor cells (see Fig. 10.4) and are expressed in macrophages, neutrophils, eosinophils, platelets, langerhans cells, B cells, mast cells, and dendridic cells. Apart from the nonspecific binding to FcRn and FcγR, IgG specifically binds to antigens via the Fab region of IgG, which is responsible for its therapeutic role. Specific binding of many mAbs with their antigens are almost irreversible with affinity constants in the range of 10^{10}–10^{11} M. Depending upon whether the antigen is in systemic circulation, located on the surface of a cell, or on the cell surface and shed into systemic circulation at a certain rate, the extent of competition from the endogenous ligand can vary. In general, the mAb preferentially binds to soluble antigen, which is in systemic circulation, while the endogenous ligand favors the antigen on the cell surface. Species differences in binding affinity and target density means that the choice of animal models for testing the drugs is crucial. Nonhuman primates are often the closest to humans and preferable over rats and dogs. Alternatively, transgenic animal models can be used.

Due to their size, physicochemical properties and similarity to endogenous nutrient and regulatory proteins, therapeutic proteins can pose bioanalytical challenges in pharmacokinetic and pharmacodynamic studies. Unlike **immunoassays** and MS-based methods, which are normally accurate and specific, **bioassays** for measuring and detecting mAbs and conjugated (PEGylated or glycosylated) proteins are generally variable due to matrix effects and reagent quality than a typical small-molecule bioassay. Despite these challenges, therapeutic proteins have an overall developmental success rate of 17% compared to 5% for small-molecule drugs because of therapeutic advantages such as high potency, long half-life, and limited off-target toxicity. Toxicity in biologicals is generally the result of an exaggerated pharmacological effect as in the case of TGN1412,[13] which means that the PD model and biomarker can be extended to safety as well. Drug–drug interactions are rare, as metabolism of therapeutic proteins is not CYP dependent. Catabolic degradation products are amino acids, which are recycled into the endogenous amino acid pool, excluding the possibility of toxic metabolite accumulation in tissues. Thus, routine drug interaction studies from *in vitro* CYP, UGT metabolism-based systems, and biotransformation studies are not needed.

10.4 PBPK/PD MODELING FOR THERAPEUTIC PROTEINS

10.4.1 Need for PBPK Modeling for Therapeutic Proteins

Compartmental modeling is sufficient for small molecules, for which linear PK is valid and elimination occurs in the central compartment comprising fast perfused organs (such as liver and kidney that are in rapid equilibrium with plasma). These simplifying assumptions may not be applicable for some therapeutic proteins. Due to size restrictions, the rate of their tissue distribution is by no means rapid. Therefore, noncompartmental determination of steady-state volume of distribution, using statistical moment theory and assuming first-order disposition and elimination from central compartment may not be valid for therapeutic proteins, for which substantial proteolysis can occur in peripheral tissues. For example, IgG elimination occurs in all tissues, including those that are not in rapid equilibrium with plasma. Convective transport is what drives the tissue distribution of therapeutic macromolecules. The saturable nature of several mechanisms of transport, distribution, and elimination that are unique to therapeutic proteins may result in nonlinear PK. These nonlinear processes include:

1. FcRn-mediated bidirectional transcytosis and protection from catabolism for mAbs
2. Target-mediated distribution and elimination

A physiologically-based PK model can accommodate several mechanisms that are important for therapeutic proteins such as transcapillary drug exchange, two-pore formalism for paracellular extravasation, TMDD arising from specific binding to target, convective lymphatic drainage, and return to circulation. It is, therefore, better suited to describe the PK of therapeutic proteins. In addition to the mechanisms considered for proteins, FcRn-mediated transport and protection from catabolism of IgG are important for mAbs. A mechanistic approach provides a means of taking into account the nonlinearity arising from saturation, if we know the abundance and turnover of the receptors involved. Extrapolation across species and site/route of administration also becomes feasible with mechanistic approaches. PBPK allows optimization of absorption efficiency to minimize cost of goods. It provides an understanding of what concentrations can be expected at the site of the target. It also facilitates the design of biologic derivatives such as, for example, single-chain variable fragment (scFv), with more optimal PK properties.

Physiologically-based PK modeling of macromolecules requires the subcompartmentalization of tissue compartments into vascular, intracellular, endosomal, and interstitial spaces to capture the permeability-limited distribution and the nonuniform distribution within a tissue arising from the different PK mechanisms described above. Detailed information on tissue interstitial and cellular drug concentrations needed for PBPK are obtainable from techniques such as fluorescence-activated cell sorting (FACS), microdialysis, and noninvasive *in vivo* imaging methods such as positron emission tomography (PET). These techniques are particularly valuable for drugs exhibiting TMDD. Antibody elimination in sites that are not in rapid equilibrium with plasma necessitates direct assessment of antibody concentrations in tissue via biopsy or necropsy. Measures of tissue distribution rates derived from these techniques can be validated in the PBPK model of the same species, before they can be employed in a human PBPK model to predict antibody levels in the target tissue. Time course of target tissue concentrations can then be combined with pharmacodynamic models (see Chapter 19). The structure of blood vessels in tumors is markedly different from those in normal tissues. The quick development of tumor cells is possible only if they stimulate the formation of new blood vessels for nutritional and oxygen supply. Thus, tumor blood vessels have an abnormal architecture, as their endothelial cells are poorly aligned, with large pore sizes, wide fenestrae, and lacking a smooth muscle layer and effective lymphatic drainage. Thus, macromolecules accumulate more in tumor tissues than in normal tissues. The enhanced permeability and retention (EPR) effect of tumor tissue justifies the need to treat it as a separate compartment. The vascular and interstitial barriers to delivery of therapeutic macromolecules in tumors[14] differ with the type and structure of tumor, making it necessary to consider case-based modeling. For mAbs targeting cancer, it is important to consider antibody-mediated antigen stripping from the cell surface, which results in the presence of solubilized or shed receptors in the circulating blood (referred to as antigenemia)[15] and in the loss of antibody bindability. Antigenia occurs with several membrane-bound–tumor-linked antigens such as the epidermal growth factor receptor in breast cancer, CA125 in ovarian cancer, and the like.

10.4.2 PBPK Modeling for Therapeutic Proteins

A PBPK model for an antibody was presented by Covell et al. as early as 1986.[16] This six-compartment model considered diffusion and convection-driven tissue uptake from plasma. Baxter et al.[17] further refined this model by introducing the additional tumor compartment and the two-pore formalism. Ferl et al.[18] introduced FcRn binding but only in skin and muscle tissue. Tissue uptake via convection and endocytosis as well as FcRn protection from catabolism in all tissues were the highlights of the PBPK model presented by Garg and Balthasar.[19] PBPK models describing the disposition of single-chain Fv or FvFc antibody fragments have also been reported in the literature.[20,21] A PBPK model that combines all of these is depicted in Figure 10.10.

According to the two-pore formalism, even under isogravimetric conditions, when net fluid flow across the capillary is zero, the filtration of macromolecules through the large pores is counterbalanced by an osmotic absorption of mainly protein-free fluid through the small pores. The recirculation of fluid defined by the term J_{iso} causes a continuous net flux of macromolecules (Fig. 10.11) across the membrane. The extravasation or the net flux of macromolecules in each tissue, T, is J_T and given by

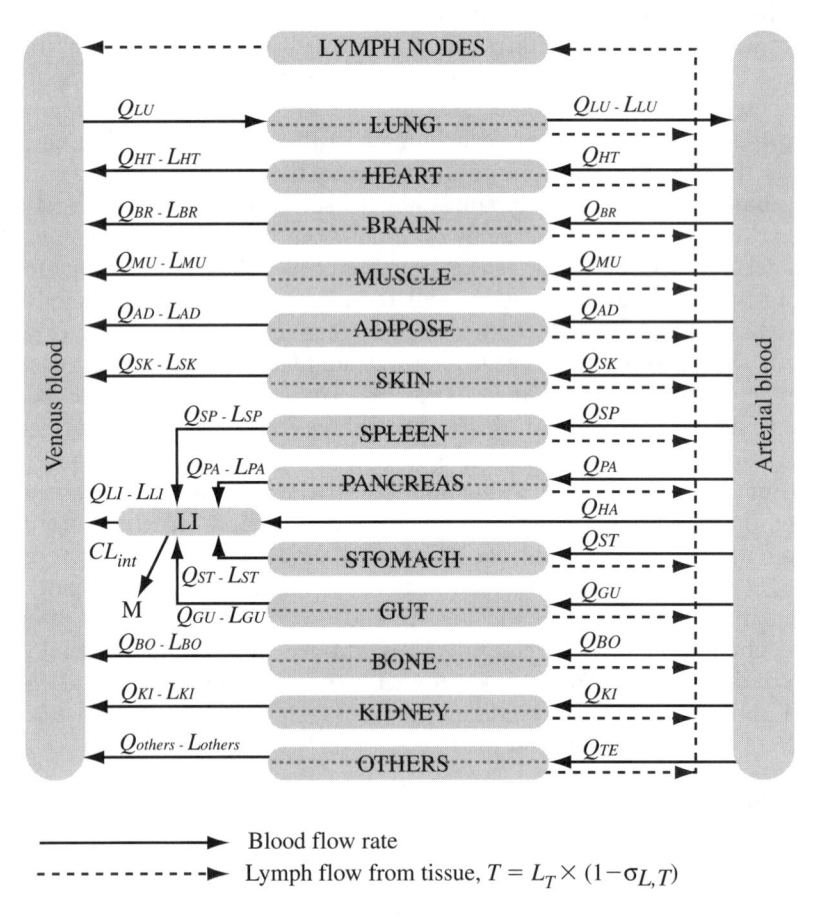

Figure 10.10. Schematic representation of a generic PBPK model for macromolecules.

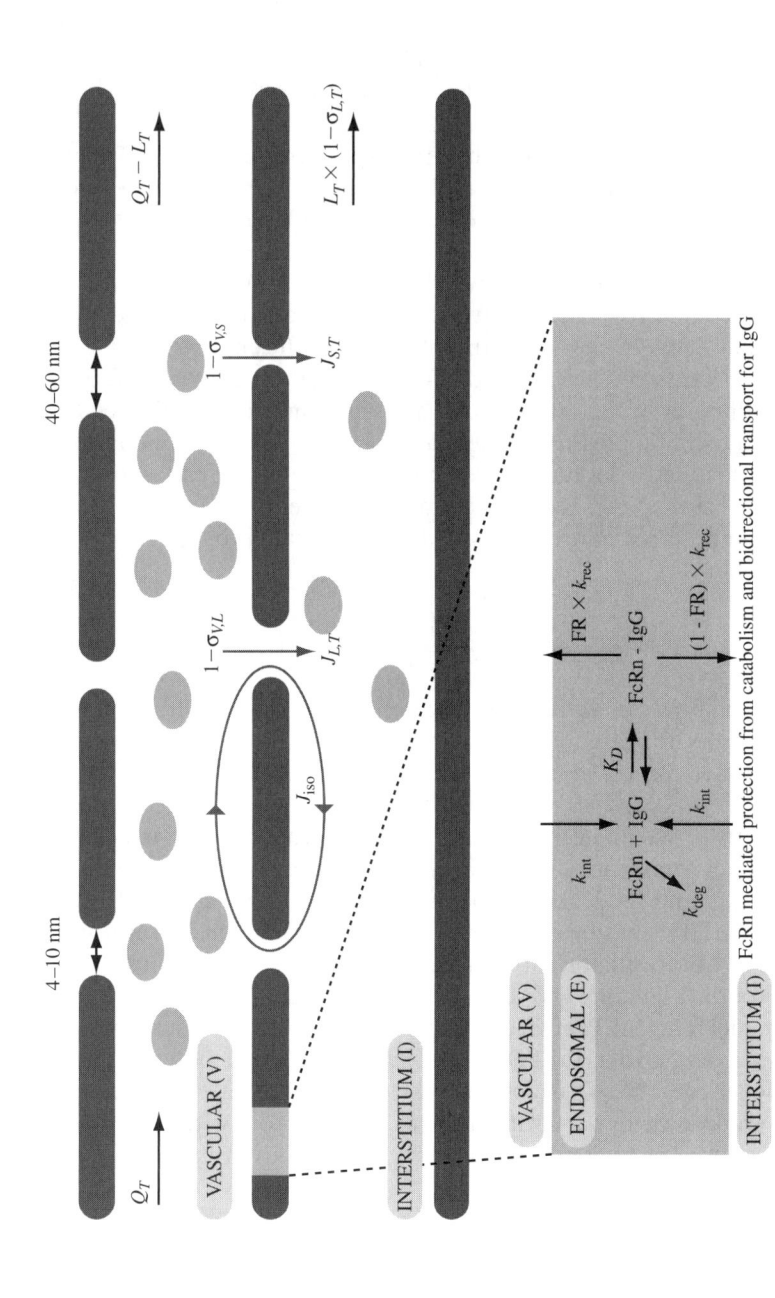

Figure 10.11. Modeling convective transport and FcRn-dependent PK of mAbs. $\sigma_{V,S}$, $\sigma_{V,L}$: Vascular reflection coefficients or the fractions that do not pass through the small and large pores of the capillary; $\sigma_{L,T}$: Lymph reflection coefficient in tissue T; L_T: Lymph flow rate in tissue T; Q_T: Blood flow rate in tissue T; $J_{S,T}$, $J_{L,T}$: Rates of fluid flow through the small pores in a tissue T; FR: Fraction of FcRn-bound IgG that is recycled into the vascular and interstitial spaces from the endosomal compartment at a rate of k_{rec}; K_D Dissociation constant of the FcRn-IgG complex; k_{int}: First order rate constant of internalization of IgG in vascular and interstitial space; k_{deg}: First order rate constant of degradation of FcRn; J_{iso}: Recirculation of fluid across vascular wall.

$$J_T = J_{L,T}(1 - \sigma_{V,L}) \times C_T^V + PS_{L,T} \times \left(C_T^V - \frac{C_T^I}{K_T}\right) \times \frac{Pe_L}{e^{Pe_L} - 1}$$
$$+ J_{S,T}(1 - \sigma_{V,S}) \times C_T^V + PS_{S,T} \times \left(C_T^V - \frac{C_T^I}{K_T}\right) \times \frac{Pe_S}{e^{Pe_S} - 1} \quad (10.2)$$

The transport of macromolecules across the capillary is determined by diffusion (second and fourth terms) and convection (first and third terms). The permeability (P), surface area (S) product (PS) describes the diffusion while the flow rates, J_S and J_L, describe the convection through small and large pores, respectively; C_T^V and C_T^I are the macromolecular concentrations in the vascular and interstitial space in a tissue, respectively; K_T is the partition coefficient of the macromolecules between the tissue and plasma. Coefficients $\sigma_{v,L}$ and $\sigma_{v,S}$ are the vascular reflection coefficients or the fractions that do not pass through the large and small pores of the capillary and are determined by the differences in hydrostatic and osmotic pressures. These are calculated using antibody size and pore size.[22] Pe is the Peclet number representing the ratio of convective to diffusive transport across the vascular membrane. The rates of fluid flow through the small and large pores ($J_{S,T}$ and $J_{L,T}$) are related to $J_{iso,T}$ by

$$J_{L,T} = J_{iso,T} + \alpha_L L_T \quad (10.3)$$

$$J_{S,T} = -J_{iso,T} + \alpha_S L_T \quad (10.4)$$

where L is the lymph flow rate through the organ; α_L and α_S are the fractions of hydraulic conductivity for large and small pores, respectively, describing the ease with which water can move through the pores.

It has been suggested that TMDD and RME followed by lysosomal degradation contribute to the systemic clearance of several therapeutic proteins including recombinant human erythropoietin (rHuEPO),[23] thrombopoietin (TPO),[24] interferon (IFN-α),[25] and the leukemia inhibitor factor.[26] A general model for drugs exhibiting TMDD[27] has been described and further simplified[28] for drugs that are in rapid or quasi-equilibrium with the drug–target complex, thus eliminating the need for drug-binding and dissociation rate constants such as k_{on} and k_{off}. A PK/PD model for rHuEPO that closely reflects the underlying mechanisms of disposition and dynamics of rHuEPO has also been described in the literature.[29]

A general model for TMDD is described below (Fig. 10.12). If the target concentrations are known to vary with time (as for receptors with functional adaptation), the rate of change of target concentration is given by

$$\frac{dR}{dt} = k_{syn,R} - k_{deg,R} \times R - k_{on} \times C \times R + k_{off} \times RD \quad (10.5)$$

where R is the target concentration, RD is the concentration of the drug-target complex, $k_{syn,R}$ is the zero-order rate constant for target synthesis, $k_{deg,R}$ is the rate constant for target degradation, k_{on} is a second-order rate constant for

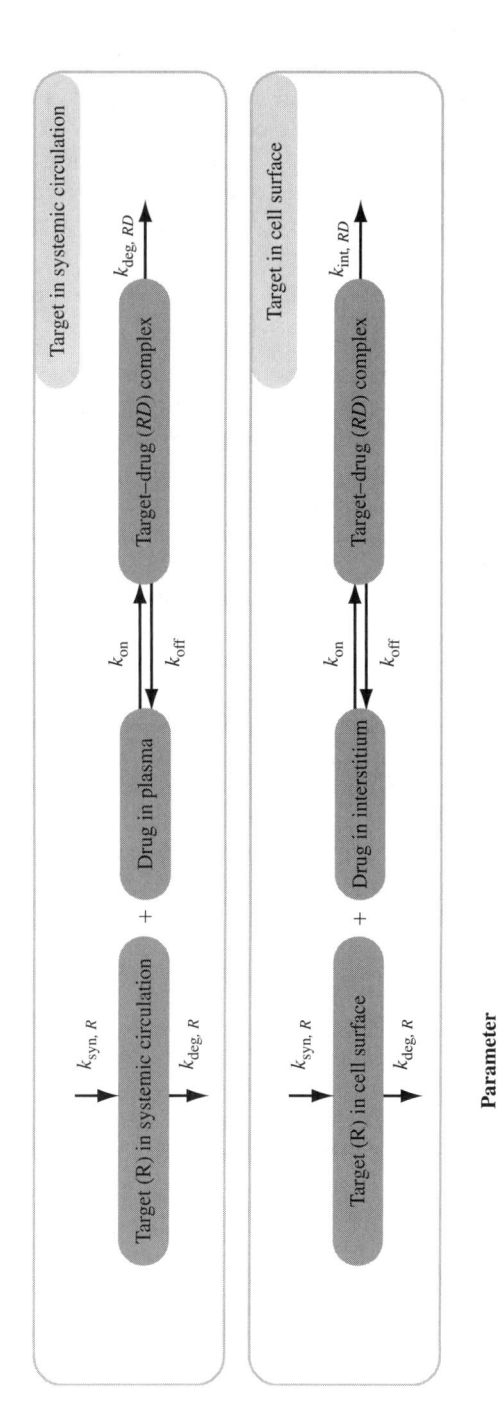

Figure 10.12. Modeling target-mediated drug disposition.

drug–target association, k_{off} is first-order rate constant for drug–target dissociation, and C is the concentration of the therapeutic protein in the plasma or the target tissue interstitium, depending on whether the target is in the systemic circulation or on the cell surface of the target tissue cells. The first two terms refer to turnover rate of the target and the last to the association and dissociation rates of the drug–target complex. If the total target concentration (R_{total}) can be assumed to be time invariant, the free concentration of the target R can then be defined as $R_{total}-RD$:

$$\frac{dRD}{dt} = k_{on} \times C \times R - k_{off} \times RD - k_{deg/int, RD} \times RD \quad (10.6)$$

where $k_{deg/int, RD}$ is the rate constant of degradation or internalization of the drug–target complex depending on whether the target is in the systemic circulation or on the surface of the target tissue cells. For a time-invariant total target concentration, R in equation 10.6 can be replaced with $R_{total}-RD$. At high drug concentrations, when RD approaches R_{total}, the first term, representing formation of the drug–target complex, approaches zero, implying saturation of the formation of the complex. Even the simpler approximations of TMDD models can be overparameterized if target and drug–target complex concentrations are not available. Therefore, parameter identifiability analysis should be an integral part of the TMDD modeling. Examples demonstrate[7] that use of the TMDD equations for the population PK and PD modeling is most successful when the concentrations of target, drug, and drug–target complex are known.

For IgG antibodies, the FcRn binding and the bidirectional transport and protection from catabolism that it mediates should be additionally considered. The extent of binding of an IgG to FcRn is a key parameter as it is only the bound IgG that is bidirectionally recycled and protected from degradation. $f_{u,T}$, the fraction of IgG that is not bound to FcRn in a tissue is derived from K_D, the dissociation constant of the FcRn–IgG complex[19] as

$$f_{u,T} = 1 - \frac{1}{2 \times C_T^E} \times [(K_D + C_{FcRn,T} + C_T^E) - \sqrt{(K_D + C_{FcRn,T} + C_T^E)^2 - 4 \times C_{FcRn,T} \times C_T^E}] \quad (10.7)$$

where $C_{FcRn,T}$ is the concentration of the FcRn in any organ. C_T^E is the endosomal concentration of FcRn in a tissue. For non-antibody proteins, $f_{u,T}$ will be 1.

10.4.2.1 Plasma (PL)

$$\frac{dC_{PL}}{dt} = \frac{1}{V_{PL}} \left\{ \begin{array}{l} (Q_{LU} - L_{LU}) \times C_{LU}^V - \left(L_{GI} + L_{SP} + \sum_{T_1} Q_{T_1}\right) \\ \times C_{PL} + \sum_T [(1 - \sigma_{L,T}) \times L_T \times C_T^I] \\ -k_{on} \times R \times C_{LU}^I + k_{off} \times RD \end{array} \right\} \quad (10.8)$$

10.4 PBPK/PD MODELING FOR THERAPEUTIC PROTEINS

where C is concentration of the macromolecule, Q and L are the blood and lymph flow rates, respectively, in organs indicated as subscripts: PL, plasma; LU, lung; SP, spleen; GI, gastrointestinal; T_I, set of organs comprising liver (LI), heart (HT), kidney (KI), skin (SK), muscle (MU), tumour (TU) and others (OT); T, T_I organs plus GI and SP. The superscripts V and I are vascular and interstitial sub-compartments of the organs. $\sigma_{L,T}$ is the lymph reflection coefficient. The first term represents inflow from the lung, the second term is the loss into vascular compartments of different organs, and the third term represents the returning lymph fluid from the different tissues (see Fig. 10.10). The last 2 terms describe changes in plasma drug concentrations arising from target binding, which needs to be included to account for TMDD only for a directly accessible pharmacological target in systemic circulation. Otherwise, k_{on} and k_{off} are simply taken as zero.

10.4.2.2 Lung (LU)

Vascular Space.

$$\frac{dC_{LU}^V}{dt} = \frac{1}{V_{LU}^V} \left\{ \begin{array}{l} \sum_{T_I}[(Q_{T_I} - L_{T_I}) \times C_{T_I}^V] - [(Q_{LU} - L_{LU}) \times C_{LU}^V] \\ -J_{LU} - (k_{int} \times V_{LU}^V \times C_{LU}^V) \\ +[FR \times k_{rec} \times (1 - f_{u,LU}) \times V_{LU}^E \times C_{LU}^E] \end{array} \right\} \quad (10.9)$$

where $\sigma_{L,LU}$ is the lymph reflection coefficient and J_{LU} is the net flux of macromolecules across the blood vessel in the lung shown in equation 10.2, and k_{int} is the rate constant of internalization. The first two terms represent the rates of drug entry and removal from vascular space, the third and fourth terms represent the net flux of drug leaving the vascular space either by convection or fluid-phase pinocytosis. For antibodies with FcRn-mediated protection against catabolism, the last term is necessary. FR is the fraction of FcRn-bound IgG that is recycled into the vascular and interstitial spaces from the endosomal compartment at a rate of k_{rec}; $f_{u,LU}$ is the fraction of IgG that is not bound to FcRn in the lung. For non-IgG proteins the last term in equation 10.9 will automatically vanish as the unbound fraction will then be 1:

Endosomal Space.

$$\frac{dC_{LU}^E}{dt} = \frac{1}{V_{LU}^E} \left\{ \begin{array}{l} (k_{int} \times V_{LU}^V \times C_{LU}^V) + (k_{int} \times V_{LU}^I \times C_{LU}^I) \\ -(k_{deg} \times f_{u,LU} \times V_{LU}^E \times C_{LU}^E) \\ -[k_{rec} \times (1 - f_{u,LU}) \times V_{LU}^E \times C_{LU}^E] \end{array} \right\} \quad (10.10)$$

where k_{int} is the rate constant of internalization of the antibody from the vascular and interstitial space. The first and second terms in equation 10.10 represent internalization, the third term represents degradation of the therapeutic protein. For non-IgG proteins, $f_{u,\text{LU}} = 1$, and so all of the protein can be degraded. However, only the unbound IgG are subject to degradation. The last term is the rate of recycling of the FcRn–bound IgG.

Interstitial Space.

$$\frac{dC_{\text{LU}}^{\text{I}}}{dt} = \frac{1}{V_{\text{LU}}^{\text{I}}} \left\{ \begin{array}{l} J_{\text{LU}} - (k_{\text{int}} \times V_{\text{LU}}^{\text{I}} \times C_{\text{LU}}^{\text{I}}) - (1 - \sigma_{L,\text{LU}}) \times L_{\text{LU}} \times C_{\text{LU}}^{\text{I}} \\ -k_{\text{on}} \times R \times C_{\text{LU}}^{\text{I}} + k_{\text{off}} \times RD + [(1 - \text{FR}) \times k_{\text{rec}} \\ \times (1 - f_{u,\text{LU}}) \times V_{\text{LU}}^{\text{E}} \times C_{\text{LU}}^{\text{E}}] \end{array} \right\}$$

(10.11)

where R is the target concentration and k_{on} is the rate constant for target binding for a target located on a cell surface. The first three terms represent the flux of convective flow into the tissue interstitium from vascular space, the internalization of the antibody, and the lymph drainage. The fourth and fifth terms describe changes in interstitial drug concentrations arising from target binding, which needs to be included to account for TMDD only if the organ expresses the pharmacological target. Otherwise, k_{on} and k_{off} are simply taken as zero. The last term is specific to antibodies and represents the recycling of bound IgG from the endosomal space.

10.4.2.3 Other Organs (OT) and Tumor (TU)

Vascular Space. Similar to equation 10.9, except that the first term representing the rate of drug entry into the tissue should be replaced by

$$(Q_{\text{GI}} - L_{\text{GI}}) \times C_{\text{GI}}^{\text{V}} + (Q_{\text{SP}} - L_{\text{SP}}) \times C_{\text{SP}}^{\text{V}}$$
$$+ (Q_{\text{LI}} - Q_{\text{GI}} - Q_{\text{SP}} + L_{\text{GI}} + L_{\text{SP}}) \times C_{\text{PL}}$$

for liver and by $Q_T \times C_T$ for all other tissues.

Endosomal Space. Similar to equation 10.10 for all tissues.

Interstitial Space. Similar to equation 10.11 for all tissues.

PBPK modeling of therapeutic proteins can provide a meaningful extrapolation to humans, once the model parameters are obtained by fitting the concentration–time profiles in preclinical species (see Section 10.4.3 on PK scaling). PBPK modeling facilitates a population-based prediction in which the influence of demographic (age, height, weight, and gender) and pathophysiological variability on the PK are captured in the modeling of healthy volunteers and subpopulations (children, obese, renally impaired, and elderly).

10.4.2.4 Model Parameters.
Physiological parameters such as blood flow rates, tissue volumes, target and FcRn turnover rates, and their concentrations

in different tissues are all obtained from the literature. Interstitial and vascular volumes in various tissues are obtained by scaling up from murine values (see Appendices). Lymphatic flow rates for visceral and nonvisceral organs are assumed to be 2 and 4 times plasma tissue flow rates, respectively.[19] Lymph flow rates and fluid circulation rates have been obtained by scaling up from murine values.[20] The k_{int}, k_{deg} and the nonspecific binding of antibodies to FcRn are likely to be similar for all antibodies. K_D is assumed to be 750 nM.[19] It may be expected that the human antibodies have a higher binding to human FcRn rather than to murine. Target-specific binding parameters (k_{on} and k_{off}) need to be measured. Permeability–surface area product (PS) of the antibody is related to its diffusion coefficient and the total surface area of the pores in the vasculature. The PS of an antibody in all tissues except spleen and liver is scaled from that of albumin by its diffusion coefficient in normal tissue.[30] The PS of an antibody in the spleen and liver are assumed to 10-fold higher than in other organs, while that in tumor tissue is obtained through curve fitting.[20] Reflection coefficients are dependent on the size of the macromolecules, and, since the size of all IgG are same, they can be assumed to be 0.2, 0.1, and 0.95 for the lymphatic, large, and small pores, respectively.[19] Tissue partition coefficients can be obtained from whole-body autoradiography or other imaging techniques. The rest of the parameters are obtained through fitting. Some of these fitted parameters are common to all IgG and could be obtained from the literature. Table 10.3 summarizes the source of different model parameters.

10.4.3 Pharmacokinetic Scaling

Empirical allometric scaling of PK parameters such as volume of distribution and clearance uses the equation

$$\text{PK parameter} = a(\text{BW})^b \qquad (10.12)$$

where a is the allometric constant, BW is the body weight, and b is the allometric scaling exponent. The constant a tends to vary depending on the compound and the PK parameter, while b is specific to the PK parameter. It is generally assumed to be 1 for volume of distribution, 0.75 for clearance, and 0.25 for half-life. Allometry may be expected to be more accurate for therapeutic proteins than for small molecules because of limited nonspecific distribution and the limited role of the highly species-variable hepatic enzymes. Examples of allometric scaling for proteins[29] and antibodies[31] can be found in the literature. However, the importance of TMDD for biologics requires that PK scaling takes into consideration the target abundance and dynamics across different species. Allometric scaling could be successful[32] when TMDD is not relevant, as for antibodies binding to soluble antigens such as cytokines. Other factors that need to be taken into account are species differences in antibody–antigen binding and the impact of antigen binding on antibody kinetics, as well

TABLE 10.3. Physiological and Compound-Dependent Parameters

	Species Dependent Physiological Parameter		Scaled (Sc); Literature (L); Fitting (F)	Drug Dependent or Biochemical Parameter		Measured (M); Lit (L); Fitting (F)
Convective extravasation	Net flux of macromolecules across the vascular endothelium in each tissue	$J_{iso,T}$	F	Lymph reflection coefficients	$\sigma_{L,T}$	L
	Rates of fluid flow through small and large pores	$J_{S,T}, J_{L,T}$	Calculated from J_{iso} (Eqs. 10.3 and 10.4)	Vascular reflection coefficients for small and large pores	$\sigma_{V,S}$ $\sigma_{V,L}$	Calculated
	Surface area	S	Sc	Permeability	P	Sc
	Fractions of hydraulic conductivity	α_S, α_L		Partition coefficients in different tissues	K_T	M
TMDD	Target turnover rates (synthesis and degradations rates) or	k_{syn}, k_{deg}	L	Target association and dissociation rate constants	k_{on}, k_{off}	M
	Target expression levels in different tissues	R_{total}	L			
	Rate constant of degradation or internalization of the drug–target complex	$k_{deg/int, RD}$	F			

FcRn binding (only for humanized or human IgG)	Concentration of FcRn in each tissue	$C_{FcRn,T}$	L
	Rate constant of internalization into endosome	k_{int}	F
	Rate constant of recycle	k_{rec}	F
	Rate constant of degradation	k_{deg}	F
	Fraction of IgG that is not bound to the FcRn	$f_{u,T}$	L
	Fraction of FcRn bound IgG that is recycled	FR	F/L
Others	Blood flow rates of different tissues	Q_T	L
	Lymph flow rates of different tissues	L_T	Sc
	Vascular, endosomal, and interstitial tissue volumes	V_T	L

as possible differences in binding to FcRn receptor between species and immunogenic potential of antibody in various species. Allometry cannot take these species differences into consideration as these are not related to body weight differences between the species. Allometric scaling is, therefore, used to describe the relationship between body weight and PK parameters but rarely used for prospective prediction of antibody PK in humans. Since species-specific parameters can readily be incorporated into the mechanism-based PBPK models, they are best suited for antibody scaling. Heiskanen et al.[33] developed a PBPK for a radiolabeled mAb, incorporating all the mechanisms that describe its biodistribution in mice. Parameters of the model were obtained with empirical correlations or by fitting the simulated curves to experimental time-series data. The model was then used to scale PK from mice to humans with tumors. A PBPK model for specific and nonspecific antibodies and fragments in normal tissues and a human colon carcinoma xenograft in nude mice was developed,[17] which could explain experimental data using as few adjustable parameters as possible. The authors could demonstrate that, for antibodies and fragments, the tumor itself had no significant influence on the pharmacokinetics in normal tissues. A sensitivity analysis showed that the lymph flow rate and transvascular fluid recirculation rate were important parameters for the uptake of antibodies, while for the retention of specific antibodies, extravascular binding was the key parameter. The model could be useful in scaling up antibody pharmacokinetics from mice to humans.

In general, a PBPK-based scaling to humans relies on relevant parameters being extracted from PBPK models built for one preclinical species and validated in another. Compound-dependent parameters that are measured, extracted from preclinical species, and scaled to humans are combined with human-specific physiological parameters (see Table 10.3) in a PBPK model to predict human PK for a therapeutic protein. A schematic representing the use of PBPK for human PK predictions is illustrated in Figure 10.13.

10.4.4 Applications of PBPK Models of Therapeutic Proteins

Kletting et al.[34] applied a PBPK model to determine the optimal preload for radioimmunotherapy (a method to selectively deliver radioactivity to cancer cells via specific antibodies) with the YAML568 anti-CD45 antibody. An optimal preloading with unlabelled mAbs aids in the selective targeting of cancer cells and a decrease in the toxic burden to normal cells. The PBPK model was fitted to PK data from 5 patients with acute myeloid leukemia. Model parameters were estimated from the fitting. Simulations of a 0–534 nmol preload of unlabeled antibody were done and the organ residence times of the antibody were estimated. The amount of antibody leading to the most favorable biodistribution for the therapy is determined. The approach was applied to other antibodies too.[35,36]

The mechanisms for tumor evasion of immune response are poorly understood. One viable explanation is that tumor-killing lymphocytes cannot reach

10.4 PBPK/PD MODELING FOR THERAPEUTIC PROTEINS

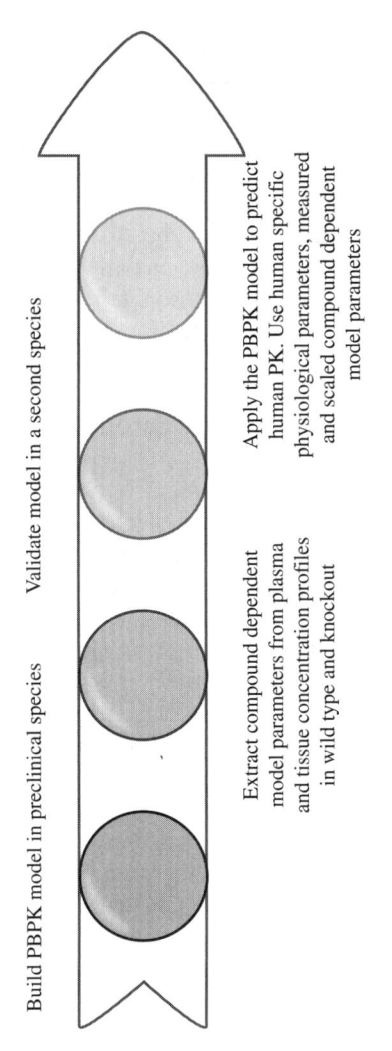

Figure 10.13. Schematic illustration of PBPK-based scaling of therapeutic protein PK to humans.

the tumor cells in sufficient quantity to keep the tumor in check. Recently, the use of bifunctional antibodies (BFAs) has been proposed as a way to direct immune cells to the tumor: One arm of the antibody is specific for a known tumor-associated antigen and the other for a lymphocyte marker such as CD3. Injecting this BFA should presumably result in cross-linking of lymphocytes (either endogenous or adoptively transferred) with tumor cells, thereby enhancing therapy. However, frequently no benefit has been observed by using the BFA. Using a physiologically-based whole-body pharmacokinetic model that accounts for interactions between all relevant species in the various organs and tumor, Friedrich et al.[37] suggested that the design of the BFA is critical and demonstrated the need for optimizing the binding constants of the antigen and lymphocyte-binding epitopes for successful therapy.

Antibody-directed enzyme prodrug therapy (ADEPT) is a 2-step process in which a drug-activating enzyme is targeted to the tumors by a tumor-targeting antibody. This is then followed by administration of a nontoxic prodrug that is converted to its active form by the tumor-localized enzyme. A PBPK model was used to understand the complexity of the ADEPT *in vivo* and to explore the optimal dosing interval between the enzyme and prodrug administration. A sensitivity analysis identified the parameters that were sensitive to the conversion of prodrug to active drug. The model was used to predict the optimal dosing regimen for ADEPT.

10.4.5 PBPK Integration with Pharmacodynamics

Integration of PK and PD of mAbs requires clear understanding of the interaction of antibody with the immune system, the interdependence of PK and PD, the physiology of the system being targeted, and the impact of disease on the physiology. Very few publications describing the PK/PD modeling of recombinant proteins and mAbs are available in the literature.[38-41] PD of most antibodies is characterized by indirect response models,[42] which account for the persistent response long after the antibody is completely cleared from the body and for the lag periods between drug action and measurable response. As a direct consequence of size and permeability-limited tissue distribution, the target concentrations of therapeutic proteins are much less than their plasma concentration, making the integration of plasma concentrations with PD irrelevant. The integration of tissue concentrations from PBPK models with PD components, mechanistic or otherwise, provide reliable results. For macromolecules, since target binding can trigger drug elimination (TMDD), physiological, pharmacological, and pathological factors modulating the target can cause transient changes in their PK, which in turn can affect their PD effects.[43] This mutual interdependence of PK and PD arising from TMDD means that the conventional approach of building an empirical PK model, which is then combined with a PD model, cannot describe the simultaneous changes. Instead, differential equations that describe the mechanisms underlying PK and target dynamics have to be solved simultaneously in order to best address the interdependence of PK and PD. As described in Section 10.4.1, TMDD also introduces nonlinearity in PK and consequently in the PD elicited by the drug, making conventional, empirical PK modeling unreliable. A general modeling approach of combining equations 10.5 and 10.6 with PBPK equations was described in Section 10.4.2 to show how PK and PD equations can be solved simultaneously. However, since drugs elicit pharmacological response through a wide range of mechanisms, PK/PD models should be built, tailor suited to the target and relevant rate-limiting steps in the biology of the system of interest. Thygesen et al.[44] have reviewed the advantages of PBPK/PD modeling of therapeutic macromolecules over conventional approaches.

10.5 ANTISENSE OLIGONUCLEOTIDES AND RNA INTERFERANCE

10.5.1 Antisense Oligonucleotides (ASOs)

When a gene is transcribed, the resulting messenger RNA (mRNA) contains the sense sequence, which is translated into a protein. **Antisense** oligonucletides (ASOs)[45] comprise 15–20 bases and are designed to hybridize to RNA through the Watson–Crick base pairing. Upon specific hybridization to its complement in the targeted mRNA, ASOs induce inhibition of gene expression. Knowing the partial sequence of the sense, it is easy to deduce its complementary antisense sequence. Because the gene structure itself would dictate the structure of this antisense drug, troublesome and time-consuming drug-screening programs can be avoided, enabling a quick transition from concept to humans.

The potential use of ASOs for a number of diseases have been reviewed.[46–48] There are many ways by which an ASO can achieve the desired effects in cells and in tissues. Antisense mechanisms fall into two broad classes—nondegradation (or noncleavage) based and degradation (or cleavage) based. Theoretically, an ASO could bind to a sense sequence, effectively blocking the protein translational machinery (Fig. 10.14), through the inhibition of translation initiation, exon inclusion, or polyadynylation. Alternatively, the mRNA is marked for degradation by endogenous enzymes such as RNAaseH, **RNA interference** (Ago 2), ribozymes, and double-stranded RNases (dsRNases). The only marketed ASO product (Vitravene, CibaVision, Isis Pharmaceuticals/Novartis) and the most clinically advanced ASO compounds exploit the RNaseH mechanism to block the production of proteins encoded by the targeted mRNA. RNaseH is a ubiquitous enzyme involved in the digestion of RNA in DNA–RNA hybrid molecules that originate when viruses infect cells. Thus, genes identified in disease pathogenesis are inhibited or shut down by short oligonucleotides. Antisense drugs appear to be effective even when the specific genes that they were supposed to target are knocked out, thus leading to the idea that the drug action could also involve nonspecific boosting of the immune system coming from the cytosine–guanine (CG) motifs that some early antisense drugs had in common. CG sequences are unique to viruses and the human immune system uses this as a marker.

Vitravene, a drug for treating cytomegalovirus (CMV)-induced retinitis in acquired immunodeficiency syndrome (AIDS) patients, was the first antisense drug to reach the market. Following this success, however, there were a series of failures that encouraged California-based ISIS Pharmaceuticals to acquire all the intellectual property rights at rock bottom prices from companies that hastily wound up operations in the field.

10.5.2 Ribonucleic Acid Interference (RNAi)

Regulation of transcription and translation in cells is facilitated by a diverse class of small noncoding, double-stranded RNAs called **micro-RNAs**

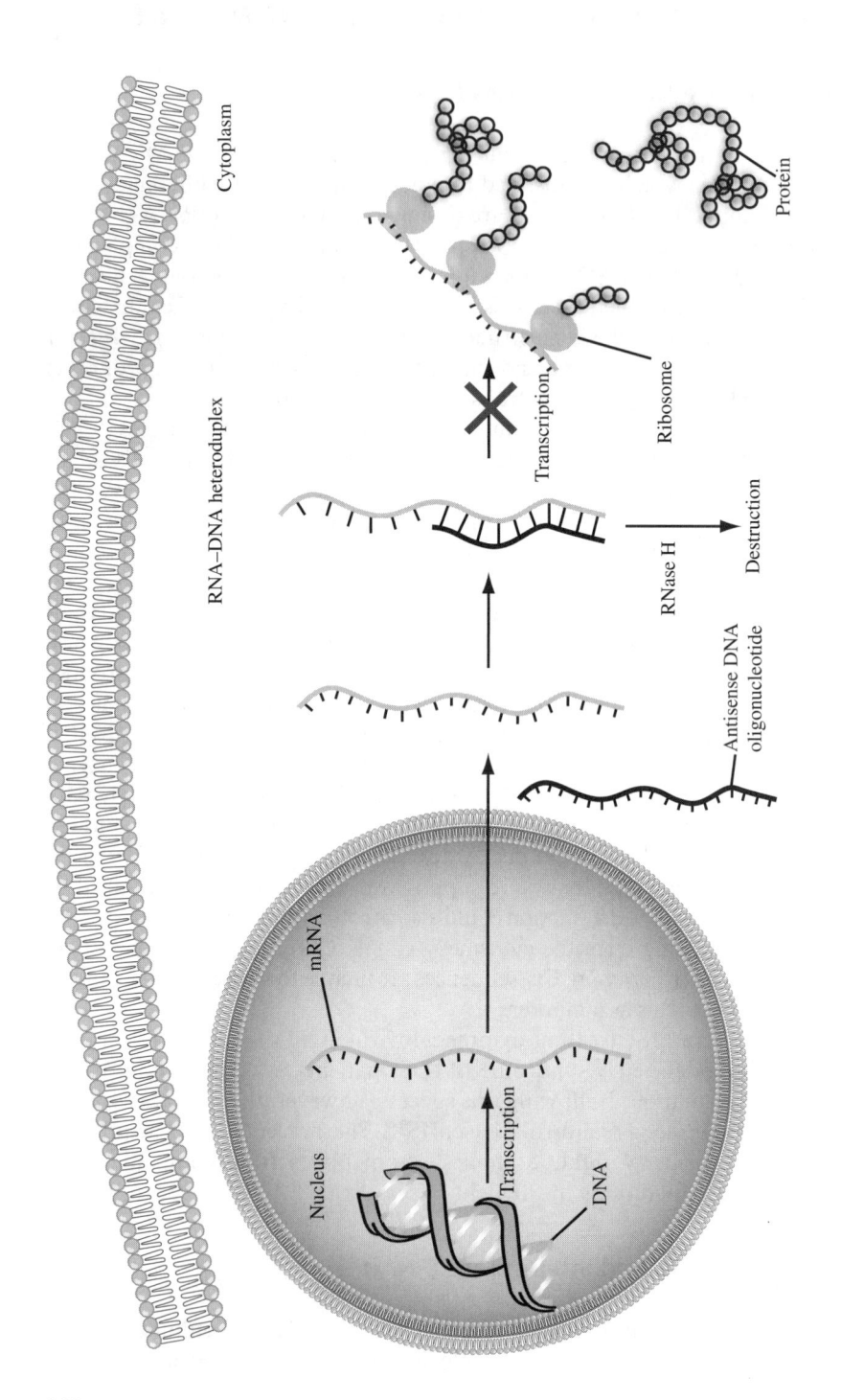

Figure 10.14. Antisense oligonucleotides as therapeutics.

10.5 ANTISENSE OLIGONUCLEOTIDES AND RNA INTERFERANCE

(miRNAs). The only RNA found in a cell should be single stranded. So a double-stranded RNA (dsRNA) is recognized as an abnormality and is chopped up by a dsRNA-specific endonuclease, called a dicer, into small fragments between 21–25 base pairs in length (Fig. 10.15). These short RNA fragments bind to an **RNA-induced silencing complex (RISC)**, which unwind them into single strands. The unwinding activates the RISC, and it binds in a sequence-specific manner to the corresponding mRNA using the sequence of the single RNA strand. The RISC complex contains an enzyme called the slicer, an argonaute protein (Ago-2), which has an RNase-H like domain (Piwi) that cleaves the mRNA. The cleaved mRNA is recognized as aberrant and is destroyed by the cell. Thus, translation into functional protein is prevented and the target gene is silenced (**epigenetics**). This RNAi pathway is found in many eukaryotic cells and plays a vital role in gene regulation and in antiviral defense. miRNAs have been found to be down-regulated in a number of cancers. They are involved in the pathogenesis of a variety of disease areas including cardiovascular, central nervous system, diabetes, liver, kidney, and infection. Thus, RNAi-based therapeutics have emerged as a new class of drugs.[49] Additionally, endogenous small RNAs may serve as novel biomarkers and therapeutic targets for the majority of diseases. Mimicking the mechanisms of the endogenous miRNAs, exogenous small interfering RNAs (siRNAs) target cellular genes that are implicated in human disease and cause sequence-specific gene suppression. This revolutionary technique was discovered in 1998. Unlike the endogenous miRNAs, double-stranded siRNAs are exact matches to the mRNAs they target and are thus highly specific. siRNAs are more specific and thorough compared to antisense. However, the risk for side effects arising from competition with endogenous miRNAs cannot be ruled out.

10.5.3 Pharmacokinetics of ASOs[50] and Double-Stranded RNAs

The ASOs and double-stranded siRNAs are intermediate in size (molecular weight between 4500 and 18,000 Da) between small-molecule drugs and protein-based biologicals. The phosphodiester backbone renders the oligos with negative charges. Oral administration is, therefore, not viable due to low permeability. Use of absorption enhancers to increase permeability of tight junctions is an option. However, due to high endonuclease activity in the intestine, bioavailability is generally less than 1%. As rapid intravenous (IV) injection may lead to hemodynamic toxicity, RNA-based drugs can only be administered as IV infusions. Nearly all the drug is absorbed from a subcutaneous administration. Once in systemic circulation, oligonucleotides are quickly degraded by RNase-H and endo- and exonuclease activities. Their size and negative charges make it difficult for them to cross the cell membrane barrier to reach their targets. The very low plasma protein binding of ASOs in combination with their hydrophilic nature makes them highly susceptible to renal elimination. The half-life of these compounds is, therefore, very short. Chemical modifications that aim to improve their

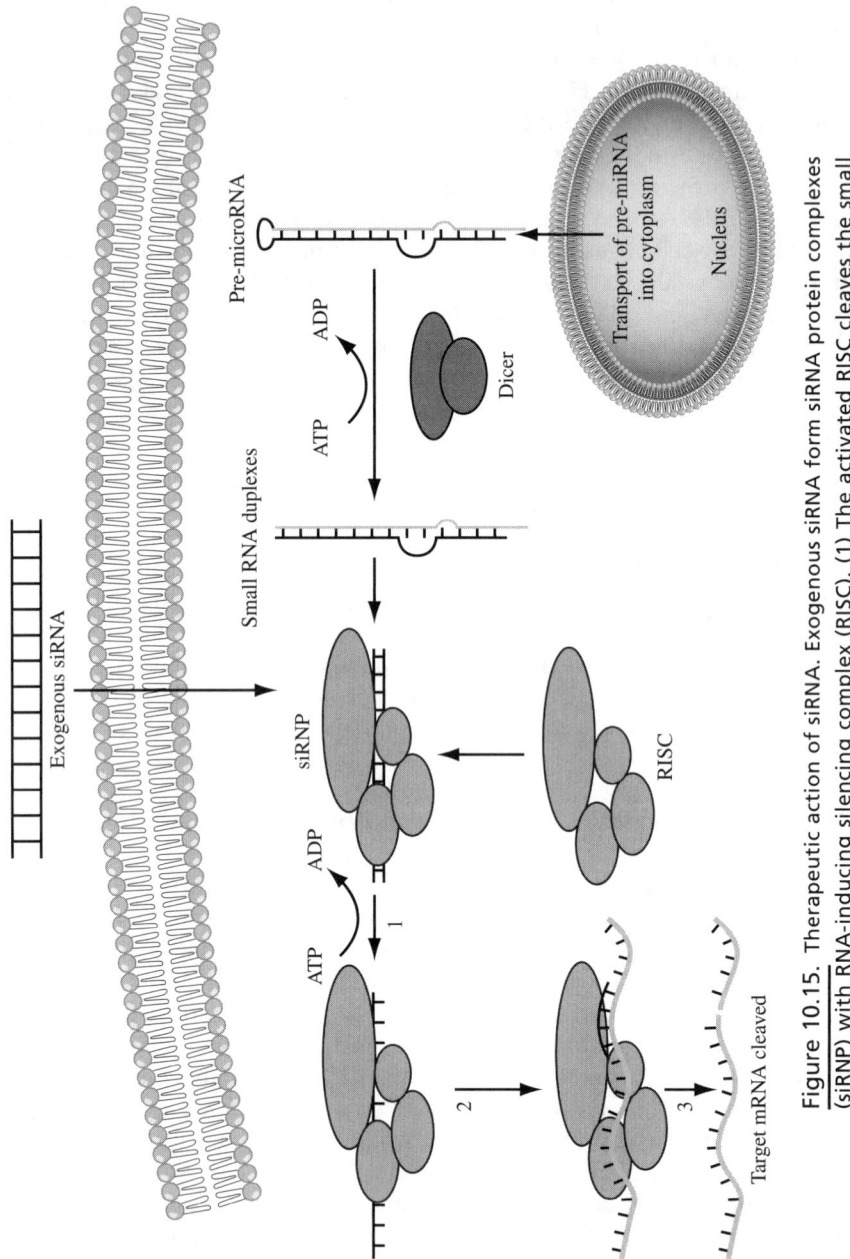

Figure 10.15. Therapeutic action of siRNA. Exogenous siRNA form siRNA protein complexes (siRNP) with RNA-inducing silencing complex (RISC). (1) The activated RISC cleaves the small RNA duplexes. (2) siRNA-mediated target recognition and association with target mRNA. (3) Cleavage of the target mRNA.

pharmacokinetics without compromising much on their efficacy and drug delivery strategies are vital to make them valuable as drugs.

10.5.4 Design and Modifications of ASOs to Improve Target Affinity and PD: the First, Second, and Third Generation ASOs

The stability of the RNA–DNA duplex in terms of hybridization and half-life is crucial to successful gene inhibition. Under physiological conditions, however, unmodified oligodeoxynucleotides (ODNs) with natural phosphodiester bonds are highly susceptible to rapid degradation by cellular nucleases, which hydrolyze the phosphodiester bond of the internucleotide backbone. Capping the 3′ and 5′ ends renders protection from exonucleases but not from endonuclease activity. In addition, they are only weekly bound to the plasma proteins, which mean that they are rapidly eliminated into urine. Various modifications (Fig. 10.16) that are compatible with Watson–Crick hybridization of oligonucleotide, conferring enhanced, high-affinity recognition of DNA/RNA targets and reduced susceptibility to nuclease degradation, were suggested and are discussed below.

10.5.4.1 Phosphorothioate Oligodeoxynucleotides (PS ODNs) or First Generation Oligos. The first-generation ASOs are modified oligonucleotides in which a nonbridging oxygen atom in the phosphate group is replaced with sulfur (phosphorothioate backbone modification). These are resistant to endonucleases but are still susceptible to RNaseH activity. Exonuclease activity is limited to successive removal of bases from the 3′ end in the plasma and from 3′ and 5′ ends in the tissues. They hybridize to the target sequences with lesser affinity than the phosphodiester counterpart. The sulfur-substituted oligonucleotides have a phosphorothioate linkage and are termed as phosphorothioates or simply as S-oligo. *In vitro*, liposomes or viral vectors are required to efficiently transfect cells with phosphorothioate ASOs, but transfection does not seem necessary *in vivo*, as the inhibition of the expression of the target can be accomplished with ASOs alone. Phosphorothioates readily distribute into tissues and do not require formulations to reach their targets. An increased binding to plasma proteins confer an enhanced systemic availability. However, the serum and tissue half-lives of phosphorothioate ASOs are less than 2 and 4 h, respectively, and minimal amounts of full-length ASOs can be found in tissues after 48 h. Consequently, continuous or frequent intravenous infusions have been used to administer conventional phosphorothioate ASOs in clinical trials. As a class, phosphorothioate ASOs are well tolerated, mostly with non-sequence-specific toxicity attributable to the phosphorothioate backbone and the polyanionic nature of these compounds.

10.5.4.2 Modifications in the 2′ Position of the Sugar–Second Generation Oligos. Considerable efforts have gone into generating the second generation ASOs that have improved stability and efficacy by chemically modifying the

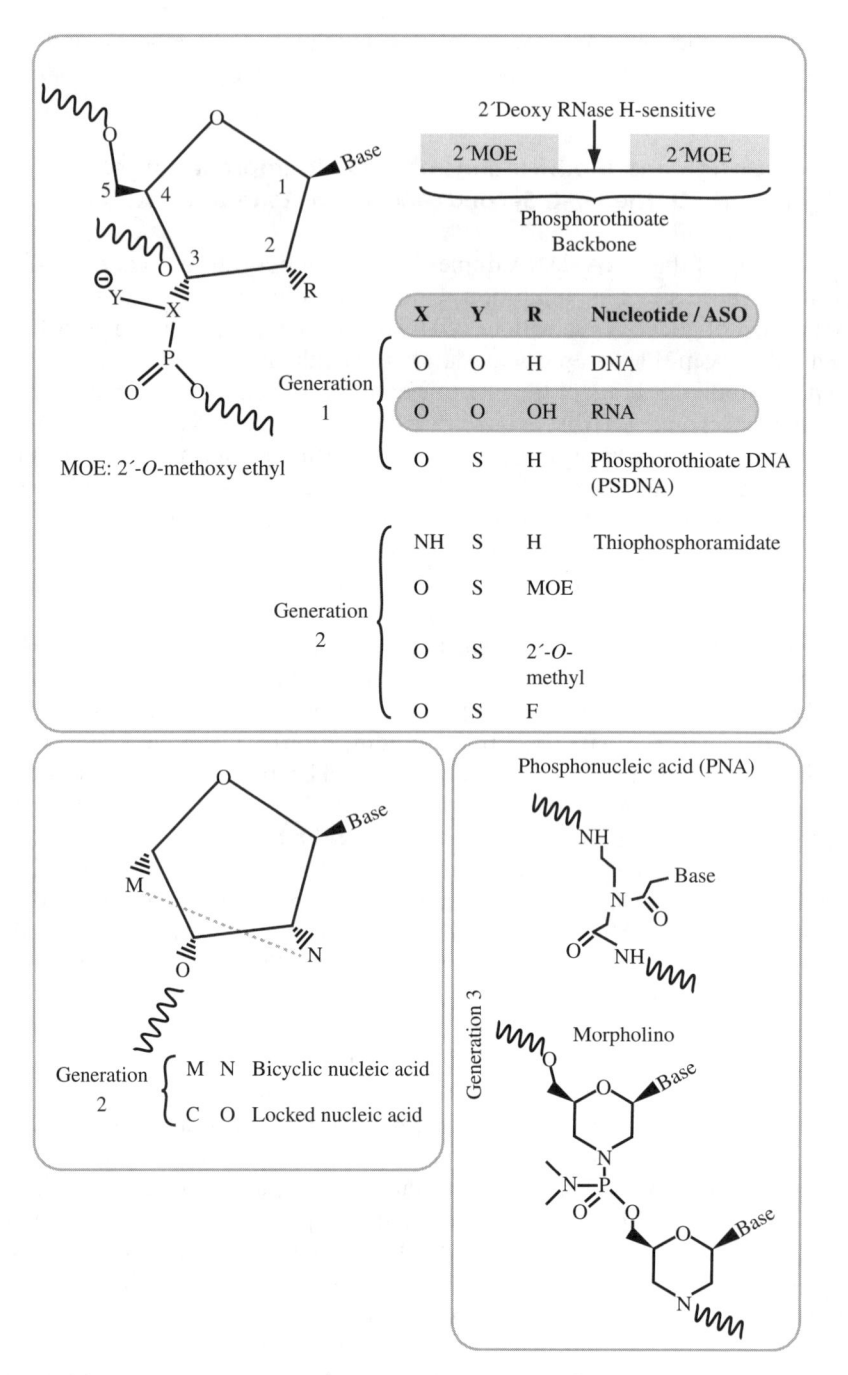

Figure 10.16. Chemical modifications of antisense oligonucleotides to improve efficacy and exposure.

ribose, and in particular, the 2′ position with electronegative substitutes such as 2′-O-methyl or the 2′-O-methoxyethyl (MOE) group. A 2′-O,4′-C-methylene bridge (locked nucleic acid, LNA) confer an RNA-like C3′-endo conformation to the oligonucleotide that greatly increases affinity, stability, safety, and tolerability *in vivo*. While the RNA-like conformation improves binding, it adversely affects the oligonucleotide's ability to activate RNase-H, a crucial aspect of ASO drug action. This dilemma has been addressed by generating chimeric oligonucleotides with 2′-modified nucleotides placed only at the ends of the oligonucleotide, thereby leaving a central RNase-compatible DNA gap. These advanced chemistry "gapmers" (e.g., OGX-011) possess favorable physicochemical, biochemical, and pharmacokinetic properties, with increased RNA affinity tissues and reduced nonspecific immune stimulation. Most importantly, improved resistance against nuclease-mediated metabolism resulting in significantly improved tissue and plasma half-life *in vivo* enables a longer duration of action and a more patient-friendly intermittent dosing schedule. It is generally believed that LNA-substituted oligos with phosphorothioate linkages presents the most stable hybridization with least susceptibility to nuclease degradation.

10.5.4.3 Morpholinos or Third Generation Oligos.

Third generation ASOs such as morpholinos and peptide nucleic acids (PNAs) promise further improvement in pharmacokinetic properties. Morpholinos are sometimes referred to as PMOs (phosphorodiamidate morpholino oligonucleotides). Structurally, the difference between morpholinos and DNA is that while morpholinos have standard nucleic acid bases, those bases are bound to morpholine rings instead of deoxyribose rings and linked through phosphorodiamidate groups instead of phosphates. Morpholinos do not degrade their target RNA molecules, unlike phosphorothioates. Instead, morpholinos act by "steric blocking," binding to a target sequence within an RNA and simply getting in the way of molecules that might otherwise interact with the RNA. Replacement of anionic phosphates with the uncharged phosphorodiamidate groups eliminates ionization in the usual physiological pH range, so morpholinos in organisms or cells are uncharged molecules. Morpholinos are most commonly used as single-stranded oligos, though heteroduplexes of a morpholino strand and a complementary DNA strand may be used in combination with cationic cytosolic delivery reagents. PNAs have a peptide substituted for the sugar–phosphate backbone. PNAs are highly resistant to nuclease and protease activity, while still retaining high affinity to RNA and DNA. PNAs are used as splicing modulators or translation inhibitors, as they do not activate RNase-H activity. PNAs have low tissue distribution, which is overcome by conjugation with short peptides. Table 10.4 summarizes the pharmacokinetics of ASOs and siRNA. The second-generation ASOs are by far the most researched oligos as they are safe and easily administered by SC injection. Their long half-lives allow for once/week administration. Further research should be directed toward understanding mechanisms underlying their cellular uptake.[50]

TABLE 10.4. Pharmacokinetics of Modified Oligos and Double-Stranded siRNA

PK Property	Phosphorothioates (PS ODNs)	2′ Sugar Modifications	PNAs, Morpholinos, and siRNA
Plasma protein binding	High binding to α2-macroglobulin and albumin. ≈90% and above	Dependent on the type of modification	Uncharged. Low plasma protein binding. Cholesterol conjugation of siRNAs to increase binding
Metabolism	Reduced nuclease activity. Clearance (CL) is species independent—1–3 mL/min/kg in rat, dog, monkey, and human. Higher CL in mice. Allometric scaling with exponent on body weight = 1 is successful	Increased resistance to exonucleases and therefore reduced rate of metabolism compared to PS ODNs	Highly stable
Volume of distribution	Distribution to all tissues except to brain. Improved over unmodified oligos. Species independent. Accumulation mainly in liver and kidney but also in spleen, bone marrow, adipocytes, and lymph nodes with 1–5 days. Saturable tissue uptake leading to nonlinear PK. Tissue uptake mainly via pinocytosis	Similar to PS ODNs. Less pronounced nonlinearity	Low. Conjugation of PNAs with short peptides and siRNAs with cholesterol to improve tissue distribution. Cholesterol-conjugated siRNA appears to be transported via the SID-1 transmembrane channel
Elimination	Negligible renal or biliary elimination. Renal excretion of shorter length metabolites is nonlinear	Renal elimination inversely proportional to the extent of plasma protein binding	Rapidly excreted in urine
Half-life	Long duration of action. Correlates with tissue residence time	Higher compared to the PS ODNs	Short

10.5.5 Integration of PK/PBPK and PD Modeling

An extensive PK/PD analysis of the phosphorothioate G3139 given in combination with carboplatin and paclitaxel, in a phase I trial to patients with advanced solid tumors, was done to define the activity of G3139 in tumors.[51] Bcl-2 is an antiapoptotic protein whose up-regulated expression in many types of cancers disturbs the balance between pro- and antiatoptotic regulatory signals, thus promoting the development of many solid tumors. By targeting Bcl-2, it is possible to restore normal regulatory mechanisms and thus enhance the effectiveness of chemotherapies. G3139 binds to the first 6 codons of the human Bcl-2 mRNA and down-regulates the Bcl-2. Steady-state plasma concentrations were reached 10 h after an IV infusion of the drug. The drug has a terminal plasma half-life of about 2 h. The authors showed that the PK of Bcl-2 was unaffected by the coadministration of carboplatin and paclitaxel. They showed that a reduction in the mRNA and protein expression levels in paired tumor biopsies and in peripheral blood mononuclear cells correlated well with intratumoral G3139 concentrations. In any PK/PD integration, it is the target concentrations that are relevant for combining with the PD data. In this clinical study, since G3139 was an anticancer agent, it was possible to get target (tumor) concentrations from biopsies. However, this may not be possible for many drugs. PBPK modeling can then be employed to extract tissue distribution parameters (tissue partition coefficients and product of permeability and surface area, or PS) from animal PK profiles in plasma and relevant tissues and to scale it to humans. As explained in the earlier section, target availability of the bulky oligos is limited by membrane permeability. To adequately model permeability rate-limited tissue distribution, each of the tissues need to be partitioned into vascular and extravascular compartments with a permeability barrier at the capillary wall membrane. A PBPK model describing the tissue distribution kinetics of ISIS 1082, a PS ODN, administered as IV bolus has been developed.[52] Radioactivity was measured at different time intervals up to 72 h. The concentrations of a parent compound 90 min postadministration in the blood and in various tissues were determined using capillary gel electrophoresis. In the liver and kidney, where the compound was degraded, the concentrations were measured at 24 h also. Nonlinear regression was employed to fit a whole-body PBPK model to the observed data in order to extract tissue distribution parameters. The authors observed that the tissue partition coefficients were not correlated with the PS. For all tissues, the ratio of effective PS to the blood flow rates was very low, suggesting that tissue distribution of the oligos is highly limited by permeability rates. Models such as these not only provide insights into the handling of oligos in the body of the species studied but also aid scaling to humans.

In conclusion, although size can be an excellent parameter to guide PK predictions of biotherapeutics, incorporating the known mechanisms affecting the PK and PD of biotherapeutics into PBPK/PD models allows for considering the interdependence of PK and PD, aids prospective predictions and extrapolations, and enables mechanistic insights. This chapter has laid the foundations for building PBPK models for antibodies, but the principles laid out can be applied to build more specific models.

KEYWORDS

Antisense: An antisense drug is the complementary of a small segment of messenger RNA (mRNA), the target that carries the "sense" sequence. The antisense drug readily binds to the mRNA strand inhibiting its further translation into a functional protein.

Antibody-Dependent Cellular Cytoxicity (ADCC): ADCC along with CDC and CDCC are immune mechanisms initiated by mAb against tumor cells. mAb binding to tumor cells induces antibody-dependent cellular cytotoxicity. Immune-effector cells such as macrophages and natural killer (NK) cells are recruited and they interact with the CH3 region of the mAbs through FcγRIIIa receptors expressed by both effector cells. Then, mAb-coated tumor cells are phagocytosed by macrophages or undergo cytolysis by NK cells.

Bioassay: Assay that measures biological activity of a substance based on its specific functional or biological response *in vivo* or quantifiable cellular response to stimulation of engineered cell lines *in vitro*. Results from *in vitro* assays may be perturbed by active metabolites, biological matrix, and environmental conditions but offer the advantage of low assay variability compared to *in vivo* methods. Examples of *in vitro* bioassays are reporter gene assay, kinase receptor activation assay, and surface plasmon resonance.

Complement-Dependent Cytotoxicity (CDC): CDC along with ADCC and CDCC are immune mechanisms initiated by mAb against tumor cells. The binding of monoclonal antibody to its target antigen on a tumor cell triggers the complement system comprising 30 proteins circulating in blood plasma, through the so-called classical pathway. One of the complement proteins, C1q, interacts with the CH2 constant region of the mAb, which leads to the activation of a proteolytic cascade and consequently induces the formation of a membrane-attack complex (MAC) for the lysis of tumor cells; this effect is termed complement-dependent cytotoxicity.

Complement-Dependent Cell-Mediated Cytotoxicity (CDCC): CDCC along with CDC and ADCC are immune mechanisms initiated by mAb against tumor cells. The binding of monoclonal antibody to its target antigen on a tumor cell triggers the complement system comprising 30 proteins circulating in blood plasma, through the so-called classical pathway. One of the complement proteins, C3b, functions as an opsonin to facilitate phagocytosis and cytolysis through its interaction with the C3b receptor (C3bR) on a macrophage or natural killer (NK) cell; this activity is termed complement-dependent cell-mediated cytotoxicity.

Epigenetics: Epigenetics is concerned with the mechanisms that control/regulate gene activity in the cell. These mechanisms switch individual genes and/or gene segments on or off without altering the DNA sequence and give rise to superordinate expression patterns that are not predefined in the gene sequence and can be passed on from cells to daughter cells and from parent generations to their offspring. DNA methylation, RNA interference, and histone modification are among the most important epigenetic regulatory mechanisms.

HAMA/HACA/HAHA: Immune system recognizes mAbs as foreign proteins and may react by generating natural antibodies—HAMA (human antimouse antibody), HACA (human antichimeric antibody), or HAHA (human antihumanized antibody). Typically, these adaptive immune responses are less common with chimeric, humanized, or fully human monoclonal antibodies than with murine antibodies. Since B cells are the immune cells responsible for generating the natural antibodies necessary for an adaptive immune response, mAbs targeting B cells often trigger lesser immune response.

Immunoassay: Frequently used analytical technique for therapeutic proteins in biological fluids. Advantages over bioassay include high sensitivity, specificity, precision, availability, and ease of performance. Examples of immunoassays include enzyme-linked immunoabsorbant assays (ELISA) and radioimmunoassay (RIA). ELISA does not discriminate between the active, inactive, and the metabolized forms or between the endogenous and exogenous compound leading to overestimation of concentration. Radio labeling in RIA may alter the PK of the native compound. Liquid chromatography mass spectrometry (LC/MS) methods have comparable sensitivity to immunoassays but higher selectivity.

Immunogenicity: Degree to which an antigen can stimulate the production of an immune response. It is also referred to as antigenicity.

Micro RNA (miRNAs): Short RNAs of about 20 nucleotides that bind to RISC and guide it to the 3′ UTR of the mRNA with the complementary sequence.

Monoclonal Antibodies (mAb): Large protein molecules produced by white blood cells that seek out and destroy harmful foreign substances.

Opsonization: Both phagocytic membranes and its target (antigens) are negatively charged, preventing their coming together. The alteration of bacteria by opsonins that then enables the phagocytes to engulf the target readily is called opsonization.

Pegylation: Refers to the covalent attachment of polyethylene glycol (PEG) to a therapeutic protein by incubation of a reactive derivative of PEG with the target macromolecule. PEGylation can mask the macromolecule from the host's immune system (reduced immunogenicity or antigenicity) and increase the hydrodynamic size (size in solution) of the agent that prolongs its circulatory time by reducing renal clearance. PEGylation can also provide water solubility to hydrophobic drugs and proteins.

RNA-Induced Silencing Complex (RISC): Multiprotein complex that uses a sequence in an miRNA or an siRNA to recognize the mRNA of a target gene meant for suppression.

RNA Interference (RNAi): A cellular pathway in many eukaryotic cells that plays an important role in gene regulation and antiviral defense.

Receptor-Mediated Endocytosis: Internalization of the ligand–target complex by endocytosis, triggered by the binding of the therapeutic protein to specific receptors on the surface of the plasma membrane.

Target-Mediated Disposition: Binding of therapeutic proteins to high-affinity pharmacological targets can trigger uptake by endocytosis (receptor-mediated endocytosis, except that the receptor is the pharmacological target) followed by proteolytic digestion in late endosome, thereby impacting their distribution volume and clearance. The capacity limitation of the receptor of interest, results in saturable and therefore dose-dependent uptake and clearance (complex, nonlinear disposition), that is further modulated with respect to time, due to ligand binding-induced receptor up- or down-regulation.

REFERENCES

1. Phrma. 418 biotechnology medicines in testing promise to bolster the arsenal to against disease. Available at: http://www.phrma.org/files/attachments/Biotech%202006.pdf.
2. Meibohm B, ed. *Pharmacokinetics and Pharmacodynamics of Biotech Drugs.* Weinheim, Germany: Wiley-VCH, 2006.
3. Tang L, Persky AM, Hochhaus G, Meibohm B. Pharmacokinetic aspects of biotechnology products. *J Pharm Sci.* 2004;93(9):2184–2204.
4. Lin JH. Pharmacokinetics of biotech drugs: Peptides, proteins and monoclonal antibodies. *Curr Drug Metab.* 2009;10(7):661–691.
5. Rippe B, Haraldsson B. Transport of macromolecules across microvascular walls: The two-pore theory. *Physiol Rev.* 1994;74(1):163–219.
6. Mager DE. Target-mediated drug disposition and dynamics. *Biochem Pharmacol.* 2006;72(1):1–10.
7. Gibiansky L, Gibiansky E. Target-mediated drug disposition model: Approximations, identifiability of model parameters and applications to the population pharmacokinetic-pharmacodynamic modeling of biologics. *Expert Opin Drug Metab Toxicol.* 2009;5(7):803–812.
8. Peletier LA, Gabrielsson J. Dynamics of target-mediated drug disposition. *Eur J Pharm Sci.* 2009;38(5):445–464.
9. Ciechanover A. The ubiquitin-proteasome pathway: On protein death and cell life. *EMBO J.* 1998;17(24):7151–7160.
10. Terashi K, et al. Close association between clearance of recombinant human granulocyte colony-stimulating factor (G-CSF) and G-CSF receptor on neutrophils in cancer patients. *Antimicrob Agents Chemother.* 1999;43(1):21–24.
11. Mufson RA, Gesner TG. Binding and internalization of recombinant human erythropoietin in murine erythroid precursor cells. *Blood.* 1987;69(5):1485–1490.
12. Keizer RJ, Huitema AD, Schellens JH, Beijnen JH. Clinical pharmacokinetics of therapeutic monoclonal antibodies. *Clin Pharmacokinet.* 2010;49(8):493–507.

13. Marshall E. Clinical research. Lessons from a failed drug trial. *Science*. 2006;313 (5789):901.
14. Jain RK. Physiological barriers to delivery of monoclonal antibodies and other macromolecules in tumors. *Cancer Res*. 1990;50(3 Suppl):814s–819s.
15. Junghans RP, Carrasquillo JA, Waldmann TA. Impact of antigenemia on the bioactivity of infused anti-tac antibody: Implications for dose selection in antibody immunotherapies. *Proc Natl Acad Sci U S A*. 1998;95(4):1752–1757.
16. Covell DG, Barbet J, Holton OD, Black CD, Parker RJ, Weinstein JN. Pharmacokinetics of monoclonal immunoglobulin G1, F(ab')2, and fab' in mice. *Cancer Res*. 1986;46(8):3969–3978.
17. Baxter LT, Zhu H, Mackensen DG, Jain RK. Physiologically-based pharmacokinetic model for specific and nonspecific monoclonal antibodies and fragments in normal tissues and human tumor xenografts in nude mice. *Cancer Res*. 1994;54(6):1517–1528.
18. Ferl GZ, Wu AM, DiStefano JJ, 3rd. A predictive model of therapeutic monoclonal antibody dynamics and regulation by the neonatal fc receptor (FcRn). *Ann Biomed Eng*. 2005;33(11):1640–1652.
19. Garg A, Balthasar JP. Physiologically-based pharmacokinetic (PBPK) model to predict IgG tissue kinetics in wild-type and FcRn-knockout mice. *J Pharmacokinet Pharmacodyn*. 2007;34(5):687–709.
20. Davda JP, Jain M, Batra SK, Gwilt PR, Robinson DH. A physiologically-based pharmacokinetic (PBPK) model to characterize and predict the disposition of monoclonal antibody CC49 and its single chain fv constructs. *Int Immunopharmacol*. 2008;8(3):401–413.
21. Ferl GZ, Kenanova V, Wu AM, DiStefano JJ, 3rd. A two-tiered physiologically-based model for dually labeled single-chain fv-fc antibody fragments. *Mol Cancer Ther*. 2006;5(6):1550–1558.
22. Taylor AE, Granger DN. *Handbook of Physiology*, Vol. 4. Bethesda MD: American Physiological Society, 1984.
23. Kato M, et al. The quantitative prediction of CYP-mediated drug interaction by physiologically-based pharmacokinetic modeling. *Pharm Res*. 2008;25(8):1891–1901.
24. Jin F, Krzyzanski W. Pharmacokinetic model of target-mediated disposition of thrombopoietin. *AAPS J*. 2004;6(1):86–93.
25. Mager DE, Wyska E, Jusko WJ. Diversity of mechanism-based pharmacodynamic models. *Drug Metab Dispos*. 2003;31(5):510–518.
26. Segrave AM, Mager DE, Charman SA, Edwards GA, Porter CJ. Pharmacokinetics of recombinant human leukemia inhibitory factor in sheep. *J Pharmacol Exp Ther*. 2004;309(3):1085–1092.
27. Mager DE, Jusko WJ. Pharmacodynamic modeling of time-dependent transduction systems. *Clin Pharmacol Ther*. 2001;70(3):210–216.
28. Mager DE, Krzyzanski W. Quasi-equilibrium pharmacokinetic model for drugs exhibiting target-mediated drug disposition. *Pharm Res*. 2005;22(10):1589–1596.
29. Woo S, Krzyzanski W, Jusko WJ. Target-mediated pharmacokinetic and pharmacodynamic model of recombinant human erythropoietin (rHuEPO). *J Pharmacokinet Pharmacodyn*. 2007;34(6):849–868.

30. Rippe B, Haraldsson B. Fluid and protein fluxes across small and large pores in the microvasculature application of two-pore equations. *Acta Physiol Scand.* 1987;131(3):411–428.
31. Grene-Lerouge NA, Bazin-Redureau MI, Debray M, Scherrmann JM. Interspecies scaling of clearance and volume of distribution for digoxin-specific fab. *Toxicol Appl Pharmacol.* 1996;138(1):84–89.
32. Vugmeyster Y, Szklut P, Tchistiakova L, Abraham W, Kasaian M, Xu X. Preclinical pharmacokinetics, interspecies scaling, and tissue distribution of humanized monoclonal anti-IL-13 antibodies with different IL-13 neutralization mechanisms. *Int Immunopharmacol.* 2008;8(3):477–483.
33. Heiskanen T, Heiskanen T, Kairemo K. Development of a PBPK model for monoclonal antibodies and simulation of human and mice PBPK of a radiolabelled monoclonal antibody. *Curr Pharm Des.* 2009;15(9):988–1007.
34. Kletting P, Bunjes D, Reske SN, Glatting G. Improving anti-CD45 antibody radioimmunotherapy using a physiologically-based pharmacokinetic model. *J Nucl Med.* 2009;50(2):296–302.
35. Kletting P, Reske S, Glatting G. Physiologically-based pharmacokinetic (PBPK) model for CD45-antibody radioimmunotherapy (RIT) with preloading. *J Nuclear Med.* 2008;49(1):320P.
36. Urva SR, Yang VC, Balthasar JP. Physiologically-based pharmacokinetic model for T84.66: A monoclonal anti-CEA antibody. *J Pharm Sci.* 2009;99(3):1582–1600.
37. Friedrich SW, Lin SC, Stoll BR, Baxter LT, Munn LL, Jain RK. Antibody-directed effector cell therapy of tumors: Analysis and optimization using a physiologically-based pharmacokinetic model. *Neoplasia.* 2002;4(5):449–463.
38. Meno-Tetang GM, Lowe PJ. On the prediction of the human response: A recycled mechanistic pharmacokinetic/pharmacodynamic approach. *Basic Clin Pharmacol Toxicol.* 2005;96(3):182–192.
39. Mager DE, Neuteboom B, Efthymiopoulos C, Munafo A, Jusko WJ. Receptor-mediated pharmacokinetics and pharmacodynamics of interferon-beta1a in monkeys. *J Pharmacol Exp Ther.* 2003;306(1):262–270.
40. Benincosa LJ, Chow FS, Tobia LP, Kwok DC, Davis CB, Jusko WJ. Pharmacokinetics and pharmacodynamics of a humanized monoclonal antibody to factor IX in cynomolgus monkeys. *J Pharmacol Exp Ther.* 2000;292(2):810–816.
41. Kamiya H, Akita H, Harashima H. Pharmacokinetic and pharmacodynamic considerations in gene therapy. *Drug Discov Today.* 2003;8(21):990–996.
42. Mould DR, Sweeney KR. The pharmacokinetics and pharmacodynamics of monoclonal antibodies--mechanistic modeling applied to drug development. *Curr Opin Drug Discov Devel.* 2007;10(1):84–96.
43. Agoram BM, Martin SW, van der Graaf PH. The role of mechanism-based pharmacokinetic-pharmacodynamic (PK-PD) modelling in translational research of biologics. *Drug Discov Today.* 2007;12(23–24):1018–1024.
44. Thygesen P, Macheras P, Van Peer A. Physiologically-based PK/PD modelling of therapeutic macromolecules. *Pharm Res.* 2009;26(12):2543–2550.
45. Bennett CF, Swayze EE. RNA targeting therapeutics: Molecular mechanisms of antisense oligonucleotides as a therapeutic platform. *Annu Rev Pharmacol Toxicol.* 2010;50:259–293.

REFERENCES

46. Gleave ME, Monia BP. Antisense therapy for cancer. *Nat Rev Cancer.* 2005;5(6): 468–479.
47. Jackson JK, Gleave ME, Gleave J, Burt HM. The inhibition of angiogenesis by antisense oligonucleotides to clusterin. *Angiogenesis.* 2005;8(3):229–238.
48. Miyake H, Hara I, Fujisaw M, Gleave ME. Antisense oligodeoxynucleotide therapy for bladder cancer: Recent advances and future prospects. *Expert Rev Anticancer Ther.* 2005;5(6):1001–1009.
49. Tiemann K, Rossi JJ. RNAi-based therapeutics—current status, challenges and prospects. *EMBO Mol Med.* 2009;1(3):142–151.
50. Geary RS. Antisense oligonucleotide pharmacokinetics and metabolism. *Expert Opin Drug Metab Toxicol.* 2009;5(4):381–391.
51. Liu G, et al. A phase I pharmacokinetic and pharmacodynamic correlative study of the antisense bcl-2 oligonucleotide g3139, in combination with carboplatin and paclitaxel, in patients with advanced solid tumors. *Clin Cancer Res.* 2008;14(9): 2732–2739.
52. Peng B, Andrews J, Nestorov I, Brennan B, Nicklin P, Rowland M. Tissue distribution and physiologically-based pharmacokinetics of antisense phosphorothioate oligonucleotide ISIS 1082 in rat. *Antisense Nucleic Acid Drug Dev.* 2001; 11(1):15–27.

SECTION II

APPLICATIONS IN THE PHARMACEUTICAL INDUSTRY

11

DATA INTEGRATION AND SENSITIVITY ANALYSIS

CONTENTS

11.1 Introduction...263
11.2 Examples of Data Integration with PBPK Modeling...........264
11.3 Examples of Sensitivity Analysis with PBPK Modeling........267
     References...271

11.1 INTRODUCTION

In the drug discovery phases of lead generation and lead optimization, efforts are focused on identifying a lead series and optimizing a lead compound with respect to potency, selectivity, physicochemical, and pharmacokinetic properties to meet a set of predefined criteria. Optimization of the pharmacokinetics (PK) of a lead compound is critical to ensuring its availability at the target sites. Prediction of the dynamic changes in the target site concentrations of a lead compound should consider the collective influence of a variety of compound-dependent physicochemical properties as well as compound- and

---

Physiologically-Based Pharmacokinetic (PBPK) Modeling and Simulations: Principles, Methods, and Applications in the Pharmaceutical Industry, First Edition. Sheila Annie Peters.
© 2012 John Wiley & Sons, Inc. Published 2012 by John Wiley & Sons, Inc.

system-dependent biochemical properties in a physiological context. An integration of all relevant measured and calculated data with due consideration of how they might change in the physiological environment is key to a good prediction of pharmacological and toxicologically relevant concentrations at the biological target sites. Physiologically-based pharmacokinetic modeling provides a framework for data integration within a physiological context, allowing for simultaneous consideration of multiple factors affecting the pharmacokinetics of a compound.

Sensitivity analysis aids an understanding of the impact of different model input parameters on the pharmacokinetics of a compund and can be valuable in designing compounds that meet the predefined criteria for PK. Data integration is essential for a sensitivity analysis, as it allows for the consideration of simultaneous changes in the PK, pharmacodynamic or toxicokinetic properties of interest in response to small changes in a set of impacting parameters. This chapter aims to provide some examples of PBPK-aided data integration and sensitivity analysis in early drug discovery.

11.2 EXAMPLES OF DATA INTEGRATION WITH PBPK MODELING

Chapters 4–6 detailed the physiological models for absorption, distribution, metabolism, and excretion (ADME). In each of these models, compound- and species-dependent data were employed along with physiological parameters to predict PK properties. Drug absorption, for example, is determined by the solubility and membrane permeability of a compound, both of which are compound- and species-dependent variables. Setting cutoffs for solubility and membrane permeability for compound selection in the discovery phase ignores the interdependence of these two in the dynamic, physiological environment *in vivo*. Figure 11.1 shows the absorption of marketed drugs with a wide range of solubility and permeability values. No cutoffs are obvious. The high absorption of some low solubility compounds can be explained by understanding that the dissolution rates of these compounds do not limit their absorption under the sink conditions *in vivo*, where dissolution, membrane permeation, and transit down the small intestine are happening simultaneously. PBPK models are best suited to address this dependence of PK properties on the physiology. In the absorption model discussed in Chapter 4, the gastric emptying rate, intestinal transit rates, and regional pHs in the different gut compartments, and the differences in these parameters in the fasted and fed states are combined with compound properties to predict drug absorption. The pH dependence of solubility using the pK_a of a compound is also considered. In addition, paracellular contribution to the overall absorption is also considered depending on the size and lipophilicity of compounds. The use of solubility in a biorelevant media and the consideration of formulation effects greatly improve the prediction of absorption from a PBPK model.[1] Thus, compound

11.2 EXAMPLES OF DATA INTEGRATION WITH PBPK MODELING 265

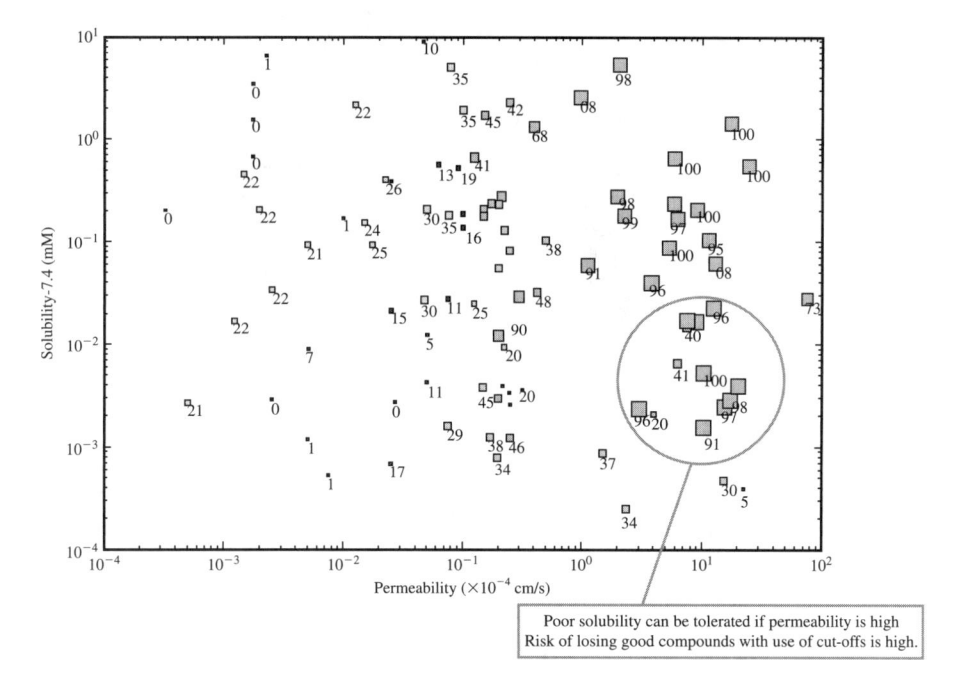

Figure 11.1. Dependence of absorption on measured buffer solubility and permeability. The numbers on the markers are represent the percentage of compound absorbed.

selection in the screening stage of drug discovery based on this integrated approach (see Fig. 11.2) is likely to be more reliable than conventional approaches based on solubility and permeability cutoffs. Similarly, the prediction of tissue distribution and renal clearance provide reliable results when appropriate data are integrated in a physiological context. However, this may not be true for prediction of hepatic metabolism, where *in vitro* measures of intrinsic clearance can be unreliable. The use of a sophisticated model such as PBPK should be avoided when the quality of the underlying data it uses is poor. For example, the value of an integrated whole-body PBPK model that combines all the ADME models to predict human pharmacokinetics can be limited by the lack of a good measure of intrinsic clearance. However, there are many examples in the literature where PBPK modeling has been used for human PK predictions.[2,3] Neither the use of a sophisticated model nor the incorporation of variability can compensate for the lack of *in-vivo*-relevant data.

As a project progresses from discovery to development, and more data becomes available, PBPK models could be iteratively refined, incorporating quantitative information of drug disposition from *in vivo* studies. Constant updating of the model with clinical data and integration with the pharmacodynamic (PD) models provide improved estimates of dose for the next clinical study. This **PBPK/PD** integration will be detailed in Chapter 14. Inclusion of

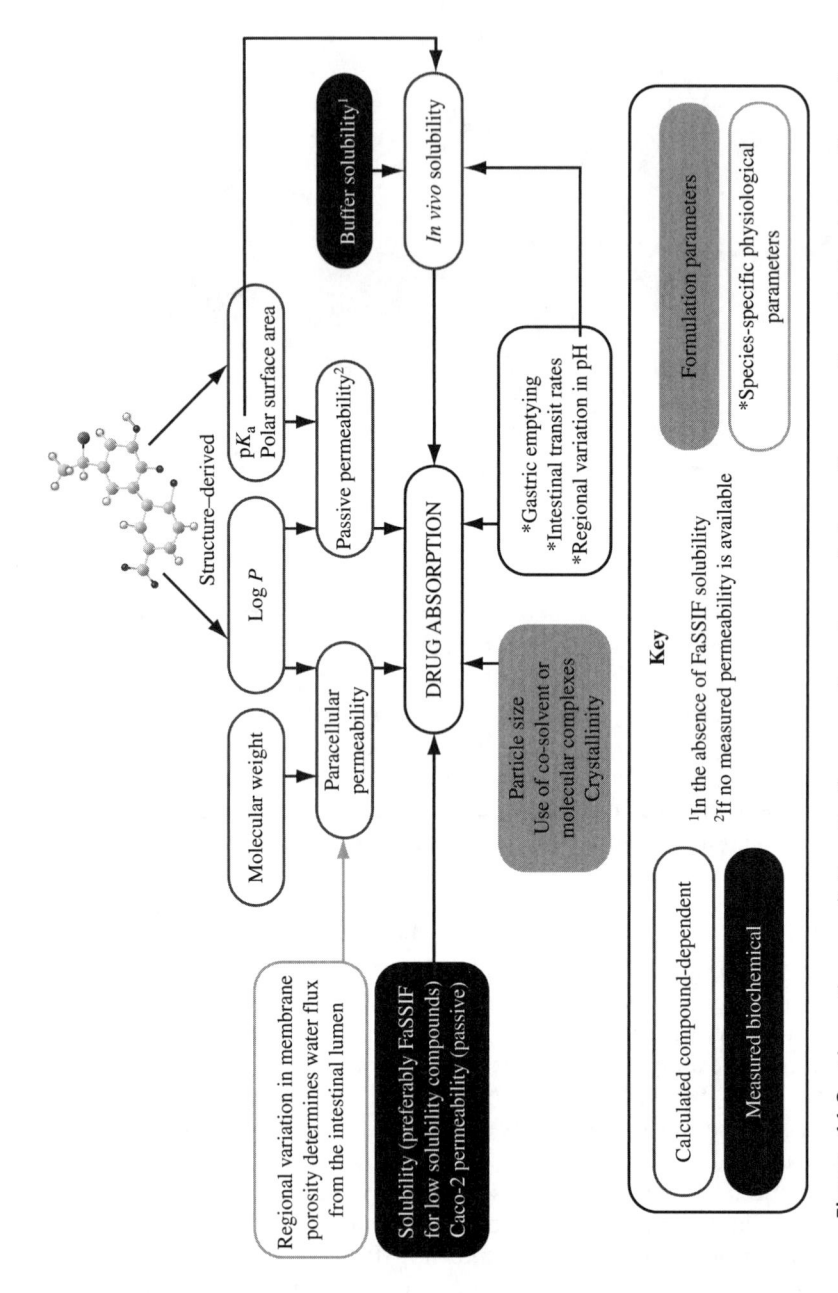

Figure 11.2. Absorption prediction in the screening stage of drug discovery through the integration of measured and calculated biochemical properties with formulation parameters in a physiological context.

all known changes in physiological parameters and enzymology due to the disease that the drug aims to treat will allow for the prediction of PK profiles for the target population. Similarly, incorporation of variability for the various physiological parameters and polymorphisms of relevant enzymes and transporters in the model will provide a means of identifying extremes in the population who may be at risk of not benefiting from the treatment or who may experience adverse reactions to treatment. This kind of analysis is critical if the drug is expected to have a narrow therapeutic window. Figure 11.3 summarizes the utility of the PBPK model as an integrative tool along the value chain. Integrated knowledge management facilitates the effective use of all available information while enhancing the reliability of conclusions drawn, thus paving the way for better go/no go decisions at transition points along the value chain.

11.3 EXAMPLES OF SENSITIVITY ANALYSIS WITH PBPK MODELING

Sensitivity analysis helps in the identification of the most influential model parameters that in turn can guide compound design. It can aid an understanding of the extent to which uncertainties in measured parameters or variability in physiological parameters will affect the concentration–time profile, as well as the PK parameters derived from it. Chapter 8 laid out the definitions and methodology of sensitivity analysis. This section provides examples of sensitivity analysis from the literature.

Sensitivity analyses allowed for a quantitative assessment of the impact of input parameters on the tissue concentrations of manganese (Mn) at different inhaled concentrations[4] from PBPK simulations. Normalized sensitivity coefficients calculated using the central difference method showed that Mn tissue concentrations in the models had dose dependencies in (1) biliary excretion of free Mn from the liver, (2) saturable tissue binding in all tissues, and (3) differential influx/efflux rates for tissues that preferentially accumulate Mn. Many such studies in environmental toxicology have been reported.[5,6]

A sensitivity analysis in which the sensitivity of time-dependent parameters such as tissue concentrations are examined as a function of time can provide vital information for experimental design. Figure 11.4 shows the time variation of the sensitivity of a time-dependent parameter with respect to input parameters. In this figure, the early time points are associated with a higher sensitivity of the time-dependent model output. If this model output is used for parameter estimation, then more measurements of this variable at earlier time points should provide better confidence in the estimated parameter. Clewell et al.[7] reported a time-dependent sensitivity analysis in which the parameters in the PBPK model of methylene chloride were systematically varied, and the resulting variation in a number of model outputs was determined as a function of time for mice and humans at several exposure concentrations. The authors found that the relative importance of the model parameters was a function

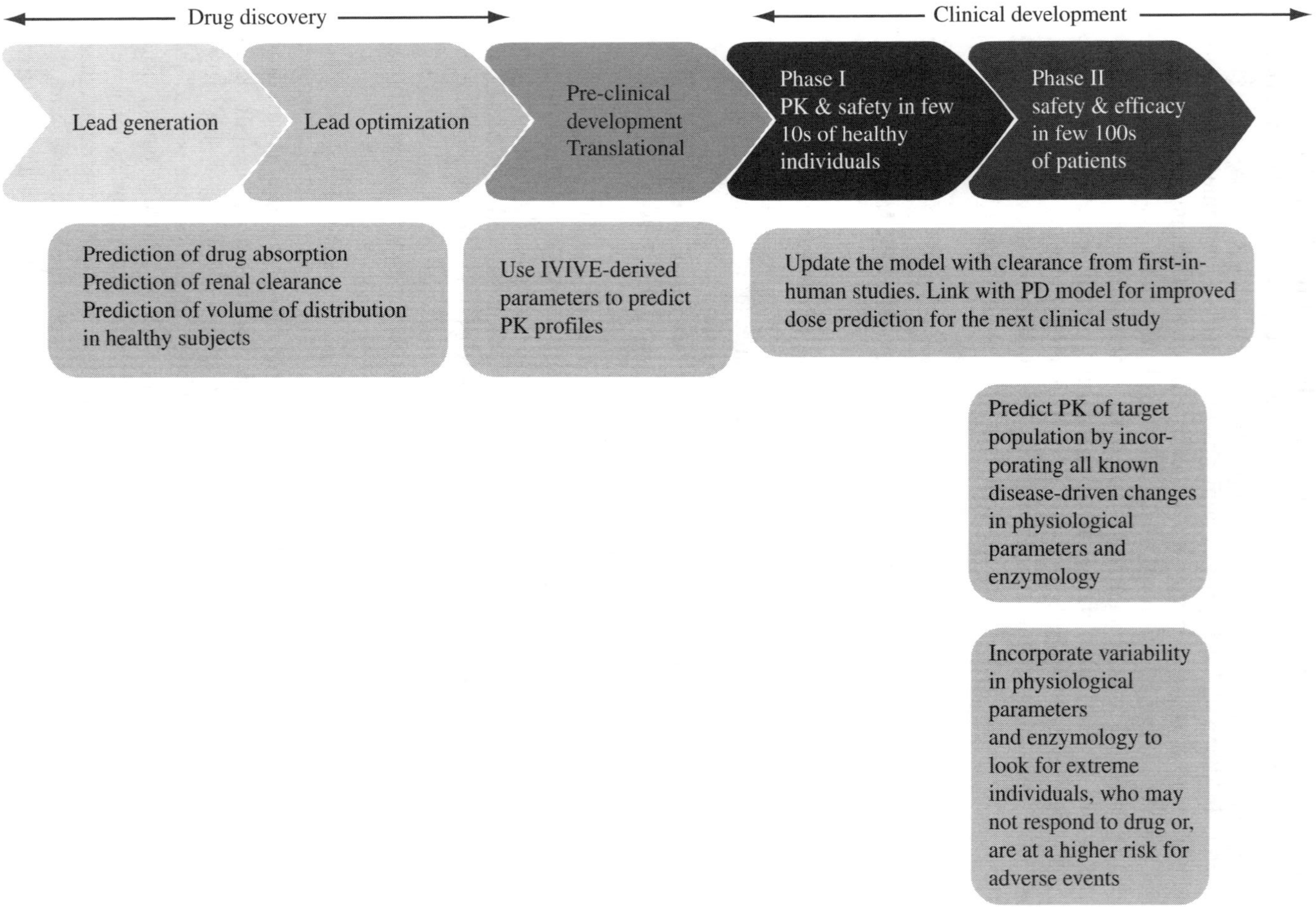

Figure 11.3. Applications of data integration in PBPK modeling along the value chain.

11.3 EXAMPLES OF SENSITIVITY ANALYSIS WITH PBPK MODELING

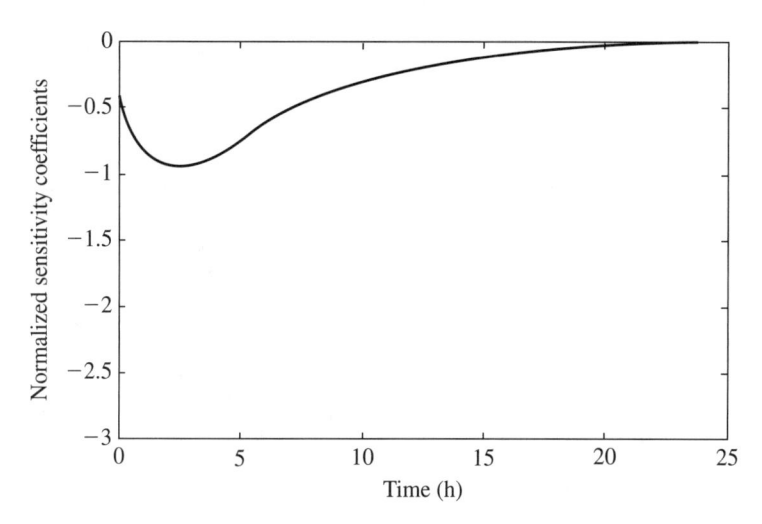

Figure 11.4. Normalized sensitivity coefficient of a time-dependent model output as a function of time.

of the conditions (concentration, species) for which the simulation was performed, the model output being considered as well as the model structure. All of the normalized sensitivity coefficients calculated ranged between −1.12 and 1, which meant that errors in the individual input parameters were not greatly amplified in the outputs. Time-dependent sensitivity analysis showed that the sensitivity of amount in the tissues at early time points is much higher compared to later time points. Thus, sensitivity analysis contributed to designing experimental conditions and sampling points to maximize parameter identifiability, so that the parameters estimated from the data were more reliable. Similarly, Evans et al.[8] applied sensitivity analysis to prioritize the impact of key model parameters on model predictions and structures. A sensitivity analysis confirmed that the permeability/diffusion-limited model structure improves the time course description over a blood flow-limited structure, particularly when early experimental time points are available. The sensitivity analysis for both structural cases studied illustrated the importance of collecting data at early experimental time points.

Sensitivity analysis identified the rate of transcapillary transport of a monoclonal antibody (mAb) to be a critical determinant of antibody penetration and localization in the tumor.[9] A sensitivity analysis on a respiratory PBPK model is described by Morris and Hubbs.[10] Yokley and Evans[11] used sensitivity analysis to selectively discriminate between two alternative PBPK model structures that differed in the compartmentalization of the metabolites of trichloroethylene (TCE), an environmental carcinogen. In one of the models, compartments existed for all chemicals, while in the other model only a generalized body compartment existed for each of the metabolites, along with multiple compartments for the parent (TCE). Sensitivities to cardiac output

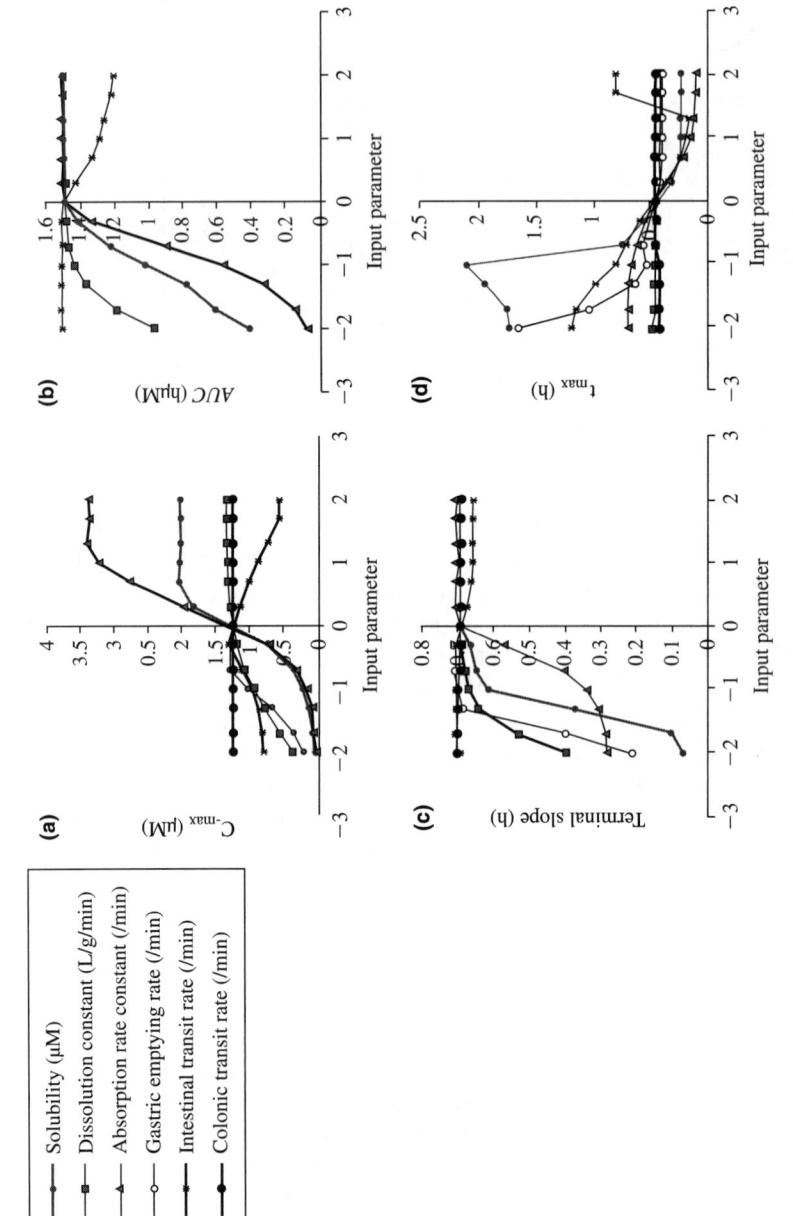

Figure 11.5. Parameter sensitivity analysis. X axis represents the logarithm of the factor by which the input parameter is multiplied in order to vary the parameter over a wide range of values. [Reprinted from "Early identification of drug-induced impairment of gastric emptying through pharmacokinetic (PBPK) simulation of plasma concentration-time profiles in rat." *J Pharmacokinet Pharmacodyn.* 2008;35(1):1–30. Copyright (2008). With permission from Springer.]

identified the former model as a better mathematical description of TCE metabolism. This study demonstrated the use of sensitivity analysis in model development, when considering physiological parameters that vary across the population.

A physiologically-based pharmacokinetic and pharmacodynamic (PBPK/PD) model[12] was recently developed to study the effect of diisopropylfluorophosphate (DFP) on the activity of acetylcholinesterase (AChE) in mice and rats. In order to identify parameters that contributed most to the variability of AChE dynamics for model optimization against data, the influence of the variability of the rate constants for synthesis and degradation of AChE, and regeneration and aging of inhibited AChE on the variability of AChE activity in mouse and rat venous blood and brain, was calculated by a global sensitivity analysis. The authors showed that the rate constants for synthesis and degradation of AChE were the most influential factors of AChE activity at shorter and longer durations, respectively, after DFP challenge.

A formal parameter sensitivity analysis (Fig. 11.5) was used to show that the modulation of gastric emptying rate alone and not any other absorption-affecting parameters could bring about the agreement of a predicted oral profile with the observed.[13] Such an approach was useful in the hypothesis testing with PBPK and to establish confidence in the conclusions drawn (see Chapter 12).

REFERENCES

1. Gao Y, et al. A pH-dilution method for estimation of biorelevant drug solubility along the gastrointestinal tract: Application to physiologically-based pharmacokinetic modeling. *Mol Pharm.* 2010;7(5):1516–1526.
2. Jamei M, Dickinson GL, Rostami-Hodjegan A. A framework for assessing interindividual variability in pharmacokinetics using virtual human populations and integrating general knowledge of physical chemistry, biology, anatomy, physiology and genetics: A tale of "bottom-up" vs "top-down" recognition of covariates. *Drug Metab Pharmacokinet.* 2009;24(1):53–75.
3. Jones HM, Parrott N, Jorga K, Lave T. A novel strategy for physiologically-based predictions of human pharmacokinetics. *Clin Pharmacokinet.* 2006;45(5):511–542.
4. Nong A, Taylor MD, Clewell HJ, 3rd, Dorman DC, Andersen ME. Manganese tissue dosimetry in rats and monkeys: Accounting for dietary and inhaled Mn with physiologically-based pharmacokinetic modeling. *Toxicol Sci.* 2009;108(1):22–34.
5. Campbell A. Development of PBPK model of molinate and molinate sulfoxide in rats and humans. *Regul Toxicol Pharmacol.* 2009;53(3):195–204.
6. Simmons JE, et al. A physiologically-based pharmacokinetic model for trichloroethylene in the male Long-Evans rat. *Toxicol Sci.* 2002;69(1):3–15.
7. Clewell HJ, 3rd, Lee TS, Carpenter RL. Sensitivity of physiologically-based pharmacokinetic models to variation in model parameters: Methylene chloride. *Risk Anal.* 1994;14(4):521–531.

8. Evans MV, Dowd SM, Kenyon EM, Hughes MF, El-Masri HA. A physiologically-based pharmacokinetic model for intravenous and ingested dimethylarsinic acid in mice. *Toxicol Sci.* 2008;104(2):250–260.
9. Davda JP, Jain M, Batra SK, Gwilt PR, Robinson DH. A physiologically-based pharmacokinetic (PBPK) model to characterize and predict the disposition of monoclonal antibody CC49 and its single chain fv constructs. *Int Immunopharmacol.* 2008;8(3):401–413.
10. Morris JB, Hubbs AF. Inhalation dosimetry of diacetyl and butyric acid, two components of butter flavoring vapors. *Toxicol Sci.* 2009;108(1):173–183.
11. Yokley KA, Evans MV. An example of model structure differences using sensitivity analyses in physiologically-based pharmacokinetic models of trichloroethylene in humans. *Bull Math Biol.* 2007;69(8):2591–2625.
12. Chen K, Teo S, Seng KY. Sensitivity analysis on a physiologically-based pharmacokinetic and pharmacodynamic model for diisopropylfluorophosphate-induced toxicity in mice and rats. *Toxicol Mech Methods.* 2009;19(8):486–497.
13. Peters SA, Hultin L. Early identification of drug-induced impairment of gastric emptying through physiologically-based pharmacokinetic (PBPK) simulation of plasma concentration-time profiles in rat. *J Pharmacokinet Pharmacodyn.* 2008;35(1):1–30.

12

HYPOTHESIS GENERATION AND PHARMACOKINETIC PREDICTIONS

CONTENTS

12.1 Introduction... 274
12.2 PBPK Simulations of Pharmacokinetic Profiles for Hypothesis Generation and Testing... 274
    12.2.1 Methodology... 274
    12.2.2 *In Vivo* Solubility... 278
    12.2.3 Delayed Gastric Emptying... 280
    12.2.4 Regional Variation in Intestinal Loss: Gut Wall Metabolism, Intestinal Efflux, and Luminal Degradation... 282
    12.2.5 Enterohepatic Recirculation... 284
    12.2.6 Inhibition of Drug-Metabolizing Enzymes... 286
    12.2.7 Inhibition of Hepatic Uptake... 286
    12.2.8 Inhibition of Hepatobiliary Efflux... 290
12.3 Pharmacokinetic Predictions... 293
    12.3.1 Human Predictions from Preclinical Data... 293
    12.3.2 Pharmacokinetic Predictions in Clinical Development... 293
    References... 294

Physiologically-Based Pharmacokinetic (PBPK) Modeling and Simulations: Principles, Methods, and Applications in the Pharmaceutical Industry, First Edition. Sheila Annie Peters.
© 2012 John Wiley & Sons, Inc. Published 2012 by John Wiley & Sons, Inc.

12.1 INTRODUCTION

Physiologically-based PK approach can be used for the prediction or simulation of pharmacokinetic (concentration–time) profiles. As a prediction tool, its value is limited by the lack of reliable input parameters, especially for clearance, where the *in vitro* measurements for intrinsic clearance rarely match up to the *in vivo*. The difficulty in getting experimental measures of tissue partition coefficients is a further deterrent to using PBPK modeling for predictions. In a simulation, however, the focus is not on quantitative predictions. Instead, the emphasis is on gaining valuable insights into processes driving the pharmacokinetics of a compound.

Pharmacokinetic profiles carry subtle information on the mechanisms driving the PK. By using prior information in PBPK models to simulate observed PK profiles and explaining any discrepancies from the observed profile mechanistically, it is possible to discern information contained in PK profiles. This approach provides the basis for a sound understanding of the real PK issues behind poor exposure. It ensures maximal extraction of information from available data to generate hypothesis, to trigger the right experiments for supporting hypothesis, and to answer key questions in discovery and development projects. PBPK simulation of preclinical profiles can help drug discovery projects that are in the screening stage to design away from potential limitations to drug disposition. A project that is close to candidate selection can apply the information learned from PBPK simulations to improve the quality of human PK predictions, once relevance to humans has been experimentally confirmed in human-specific *in vitro* systems. During clinical development, PBPK simulations of human PK profiles can aid a mechanistic understanding of the key processes driving pharmacokinetics in humans. This can help formulation efforts to improve exposure of the compound, trigger an alternative clinical trial design, and enable better PK predictions for subsequent trials. PBPK models are uniquely suited for optimal use and maximal understanding of available data, thus improving the efficiency and saving valuable resources through reduction of experiments and animals. Examples of hypothesis-driven PBPK modeling[1,2] are rare in the literature, but it is hoped that this powerful application will soon become the norm in the industry, as more examples are uncovered and published. In this chapter, examples from projects in the lead optimization (LO) stage of drug discovery are used to illustrate the applications of PBPK for hypothesis generation and testing. The advantages and limitations of PBPK-based extrapolation to humans are also discussed.

12.2 PBPK SIMULATIONS OF PHARMACOKINETIC PROFILES FOR HYPOTHESIS GENERATION AND TESTING

12.2.1 Methodology

Absorption, distribution, metabolism, and excretion models within PBPK are based on sound mechanistic principles incorporating known PK mechanisms

12.2 PBPK SIMULATIONS OF PHARMACOKINETIC PROFILES

and reflecting current understanding in human physiology, anatomy, and enzymology (Chapters 4–7). Generic PBPK models are, therefore, expected to predict PK profiles provided reliable estimates of compound-dependent parameters from *in vitro* studies are used as model inputs. Therefore, major discrepancies of an observed PK profile of a compound from the predicted could reflect an underlying flaw in the data used as input or an inadequate representation of relevant PK mechanisms in the model. A principal limitation to hypothesis-driven mechanistic understanding is poor quality of data, especially related to drug metabolism where *in vitro* assays are often underpredictive of *in vivo* reality. Similarly, although drug distribution models based on physicochemical properties[3-9] provide reasonable estimates of tissue distribution for many drugs, certain compounds can prove to be more challenging than others. Involvement of transporters can complicate matters further as they can introduce nonlinearity and modulate metabolism and distribution. Even though *in vitro* models are available for evaluation of compounds as transporter substrates, current limitations in transporter-related knowledge make it difficult to make reliable predictions. In order to overcome these limitations, intravenous (IV) PK profiles can be used to extract metabolism and distribution parameters in order to interpret the more complex oral PK profiles.[10] To facilitate parameter extraction from IV profiles, a multiplicative factor (K_p factor) can be introduced that simply multiplies all the tissue partition coefficients. The unbound intrinsic clearance ($CL_{int,u}$) and K_p factor are modulated (see Fig. 12.1), keeping all other input parameters constant, until a best fit to the observed IV profile is obtained. Goodness of fit is judged by ensuring a good agreement of calculated AUC with the mean of the observed, as well as through calculation of reduced chi square statistic (equation 12.1):

$$\chi^2 = \frac{1}{N} \sum_{i=1}^{N} \left\langle \frac{\Delta_i^2}{\sigma_i^2} \right\rangle \tag{12.1}$$

where N is the number of observations and Δ is the difference between the observed and predicted values at the same time points; σ is the standard deviation at the corresponding time points. As the square of the deviations from the observed values at each time point is expected to agree with the observed variance (σ^2) at that time point, χ^2 should be close to 1 for a good fit. The best-fit parameters of $CL_{int,u}$ and K_p factor can then be used for oral PK profile simulation, as clearance and metabolism are expected to be invariant to the route of administration. It has been demonstrated that the use of PBPK-extracted distribution and metabolism parameters along with measures of solubility and permeability can accurately predict the oral PK profiles of compounds whose absorption is determined only by their solubility and permeability.[10] This implies that when a simulated plasma curve does not agree with the experimentally observed, information on the underlying PK processes can be obtained. Figure 12.2 shows a schematic representation of the procedure for parameter extraction and different interpretations when predicted

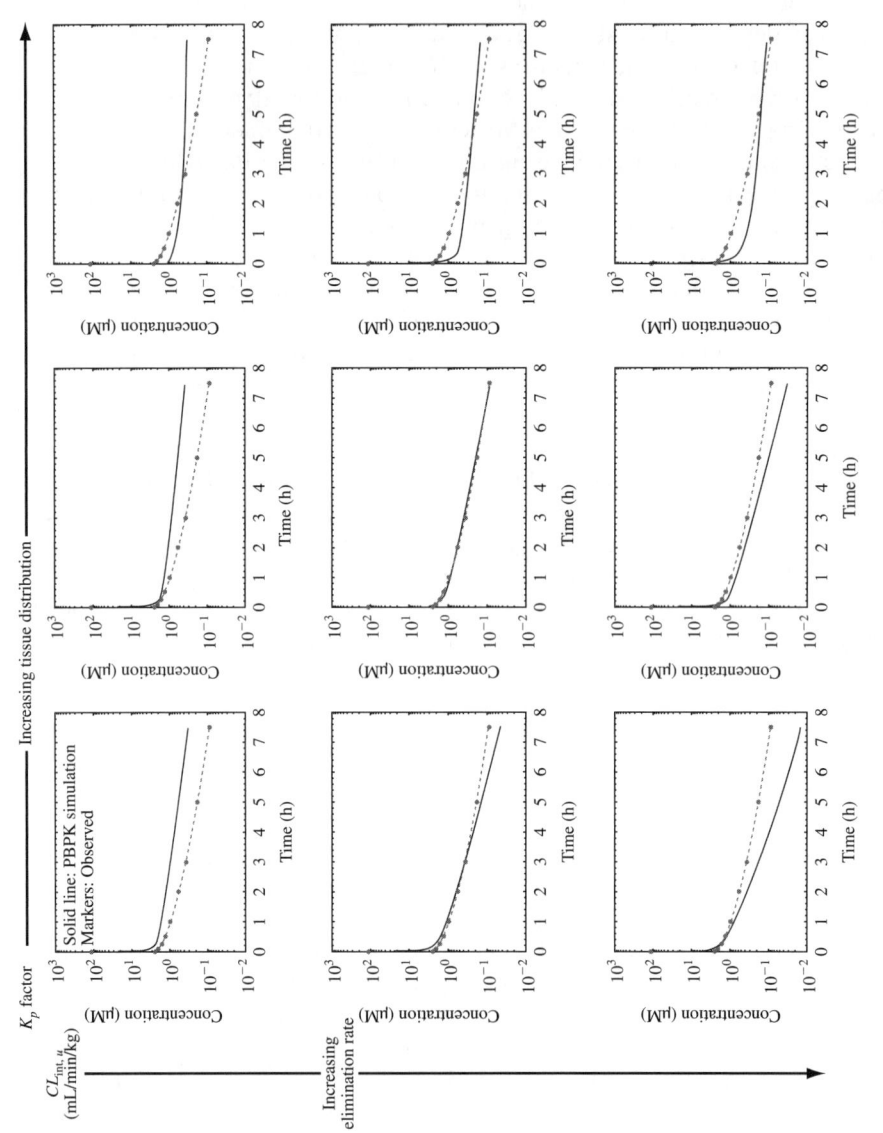

Figure 12.1. Effect of modulating $CL_{int,u}$ and K_p factor on an IV pharmacokinetic profile. As $CL_{int,u}$ is increased, the profile shifts down, with the shape remaining intact. The effect of increasing K_p factor is to change the shape.

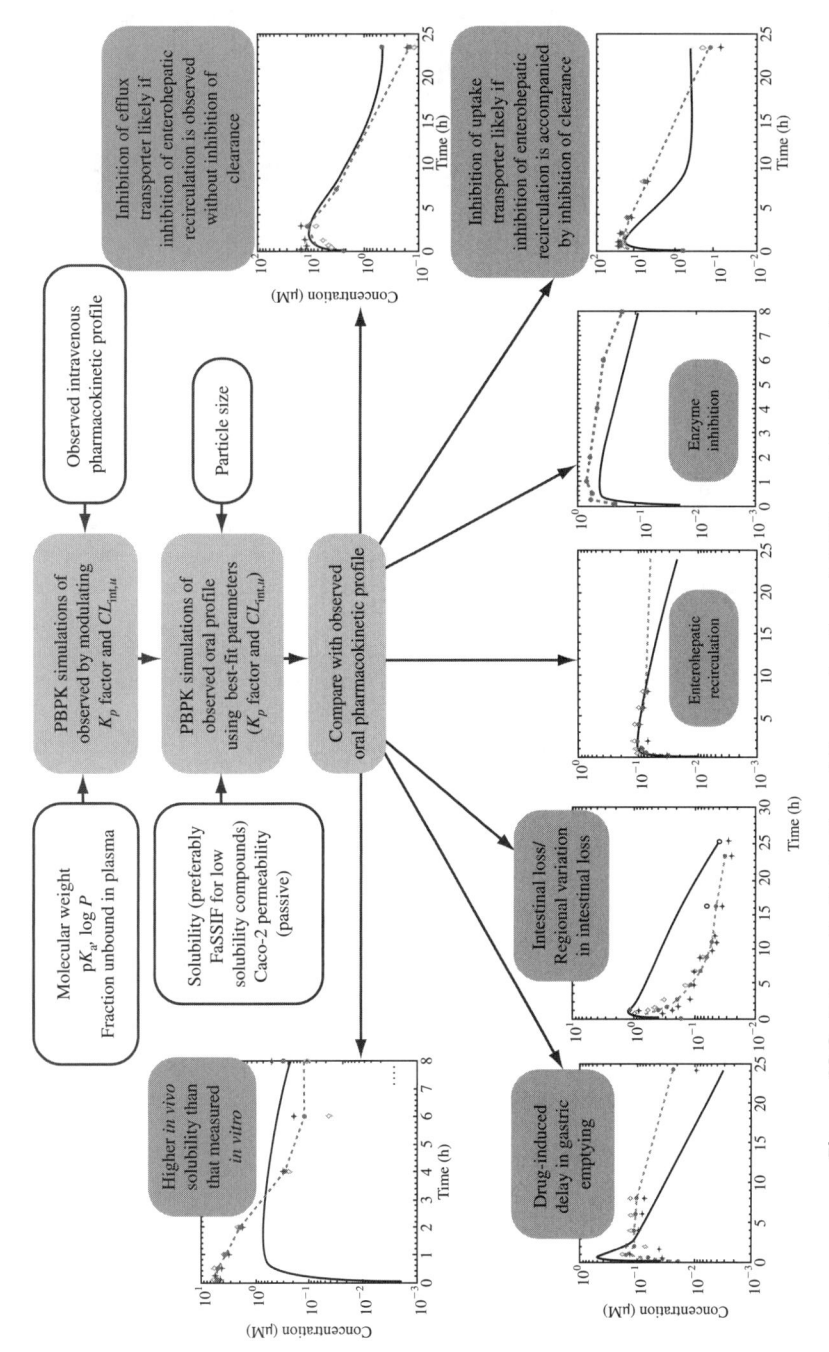

Figure 12.2. Schematic illustration of the application of PBPK modeling for oral line shape analysis. The signature discrepancies of predicted oral profile from the observed for the different factors that influence oral PK profile are also shown.

oral line shapes differ from the observed. Depending on the nature of discrepancy of a simulated profile from the observed, a hypothesis is generated. These could be related to solubility, drug-induced impairment of gastric emptying, regional variation in intestinal loss, enterohepatic recirculation, and enzyme or transporter inhibition. To test the hypotheses, appropriate parameters affecting oral disposition are modulated to get a good fit to the observed oral PK profile. By doing so, PBPK simulations of oral profiles can identify PK mechanisms that can affect disposition of an orally administered drug. Examples of each are provided below.

12.2.2 *In Vivo* Solubility

Only the dissolved fraction of a drug can be orally absorbed. *In vitro* measures of solubility are often poor models for *in vivo* intestinal solubility because of the very different media involved in the two situations. For poorly soluble drugs, even a small increase in *in vivo* solubility can mean a significant increase in drug absorption. The dissolution rates of a drug *in vivo* can vary widely depending on its formulation (see Chapter 4) and can be very high compared to *in vitro* dissolution rates. The need for *in vitro–in vivo* correlation (IVIVC) is, therefore, seen as vital for reliable predictions of *in vivo* performance of a drug.[11] Without an IVIVC, the risk of a compound or a series of compounds being discarded on the basis of poor solubility during LO is quite high. Poor solubility may also trigger a formulation development, leading to wastage of resources. Oral simulations employing the method described in Figure 12.2 can identify a higher *in vivo* solubility than a measured *in vitro* solubility in the preclinical species tested. If the measured *in vitro* solubility employed in the oral simulation is inadequate to explain the high initial plasma concentrations in the observed profile and no increase in permeability can explain it, a higher than measured solubility value that best fits the observed curve is employed in the model. Thus, an estimate of minimum *in vivo* solubility can be obtained. Since the intestinal fluid in humans is comparable to that in preclinical species, the estimated solubility is likely to apply to humans.

In a certain drug discovery project, buffer solubility in DMSO (dimethyl sulfoxide) was used in the screening stage of LO. A lead compound in the project had a measured buffer solubility of 0.9 mM. Figure 12.3a shows the best-fit PBPK simulation of the IV profile with modulated $CL_{int,u}$ and K_p factor and an enterohepatic recirculation (EHR) rate introduced. Fixing these parameters at the best-fit values, the measured solubility and the Caco-2 permeability of 70 cm/s are employed to simulate the oral profile (Fig. 12.3b). The high initial concentrations of the compound could not be simulated even with a hypothetically high permeability. Instead, a much higher solubility of 1 mM had to be used to simulate the observed profile (Fig. 12.3c), which is close to the FaSSIF solubility of the compound (0.4 mM). PBPK simulation suggests that the *in vivo* solubility should be at least 1 mM. The actual *in vivo* solubility could have been higher.

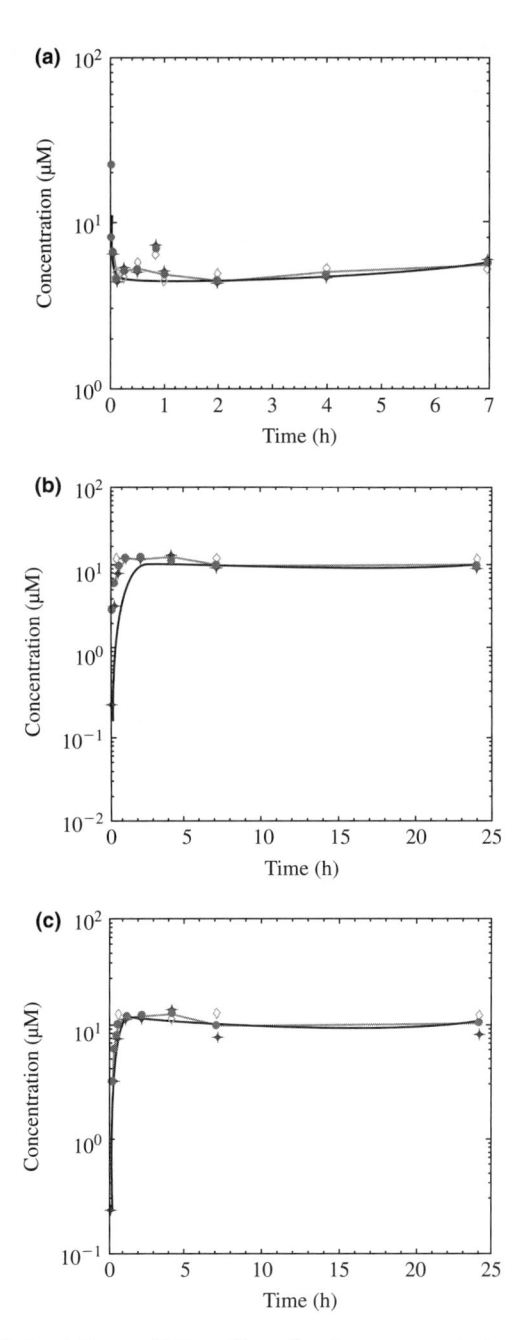

12.2 PBPK SIMULATIONS OF PHARMACOKINETIC PROFILES

Figure 12.3. PBPK simulations of PK profiles of a discovery project compound observed in dog. (a) IV simulation obtained by modulating $CL_{int,u}$, K_p factor, and enterohepatic recirculation (EHR) rate. (b) A hypothetically high permeability could not simulate the initial plasma concentrations, if buffer solubility is used. (c) A good fit to the oral profile could be obtained only with the use of a much higher value for solubility (1 mM).

This example demonstrates the unique value of PBPK simulations in bringing to light valuable information that is otherwise simply lost.

12.2.3 Delayed Gastric Emptying

Drug-induced impairment of gastric emptying leads to a longer drug absorption phase, thus increasing the duration of drug action. Although this may be a desirable feature in drug development, a drug-induced impairment of gastric emptying may not be desirable for other reasons. One reason is that it is difficult to reach high plasma concentrations with increasing dose as required for the toxicological studies. For drugs that may require fast action, the delayed onset of drug action is unfavorable. Drug-induced impairment of gastric emptying can also cause side effects such as stomach inflammation and ulceration as a result of prolonged retention of gastric contents. Delayed gastric emptying is also associated with a progressive dilatation of the proximal stomach, leading to a shortening of the lower esophageal sphincter. The shortened sphincter and the increased gastric contents following a defective emptying can lead to gastresophageal reflux disease (GERD). Several drugs in the market, such as adrenergic antagonists, anticholinergic drugs, and opioid agonists, can cause impairment of gastric emptying.[12-19] An early identification of drug-induced impairment can aid drug design and provide an understanding of why exposure is low at high doses. Relevance to humans is difficult to establish through experimentation. However, if PBPK identifies drug-induced gastric emptying impairment in two or more preclinical species, then the impairment is likely to occur in humans. The identification of drug-induced impairment of gastric emptying through PBPK simulations of oral PK profiles[1] is illustrated in Figure 12.4 with verapamil as an example. Verapamil exhibits gut wall metabolism[20] and delays gastric emptying in humans. The intravenous PK profile is first simulated by adjusting the $CL_{int,u}$ and the K_p factor in the PBPK model (Fig. 12.4a). The other model inputs (log $P = 3.95$, molecular weight $= 454.6$, base $pK_a = 8.7$, and fraction unbound in plasma $= 0.12$) are retained as such. The best-fit parameters ($CL_{int,u}$ and K_p factor) along with buffer solubility (10 mM) and Caco-2 permeability (10×10^{-6} cm/s) were employed in the model for simulation of the oral profile. Figure 12.4b shows that the predicted profile is very different from the observed. It is clear that the $CL_{int,u}$ used in IV simulation is inadequate to explain the observed AUC. This leads to the hypothesis that there could be additional metabolism in the gut. The unique stamp of delayed gastric emptying on the PK line shape is seen in Figure 12.2. The observed profile has a much lower initial and higher terminal plasma concentrations compared to what is expected. This is because the rate of absorption *in vivo* is limited by gastric emptying rate rather than by permeability. Gastric emptying also leads to a prolonged absorption phase resulting in the so-called flip-flop kinetics. PK profile of verapamil does show much lower initial concentration and elevated terminal concentrations. The hypotheses to be tested in verapamil are, therefore,

Figure 12.4. (a) Simulation of the intravenous PK profile by adjusting $CL_{int,u}$ and K_p factor in the PBPK model. (b) PBPK simulation of oral profile employing the best-fit parameters ($CL_{int,u}$ and K_p factor) along with buffer solubility (10 mM) and Caco-2 permeability (10×10^{-6} cm/s). (c) Intestinal loss rate constants are introduced and gastric emptying rate constant is reduced to get a best fit to the observed. (d) Effect of further refinement of permeability. Effects of adjusting either the intestinal loss (e) or gastric emptying parameters (f).

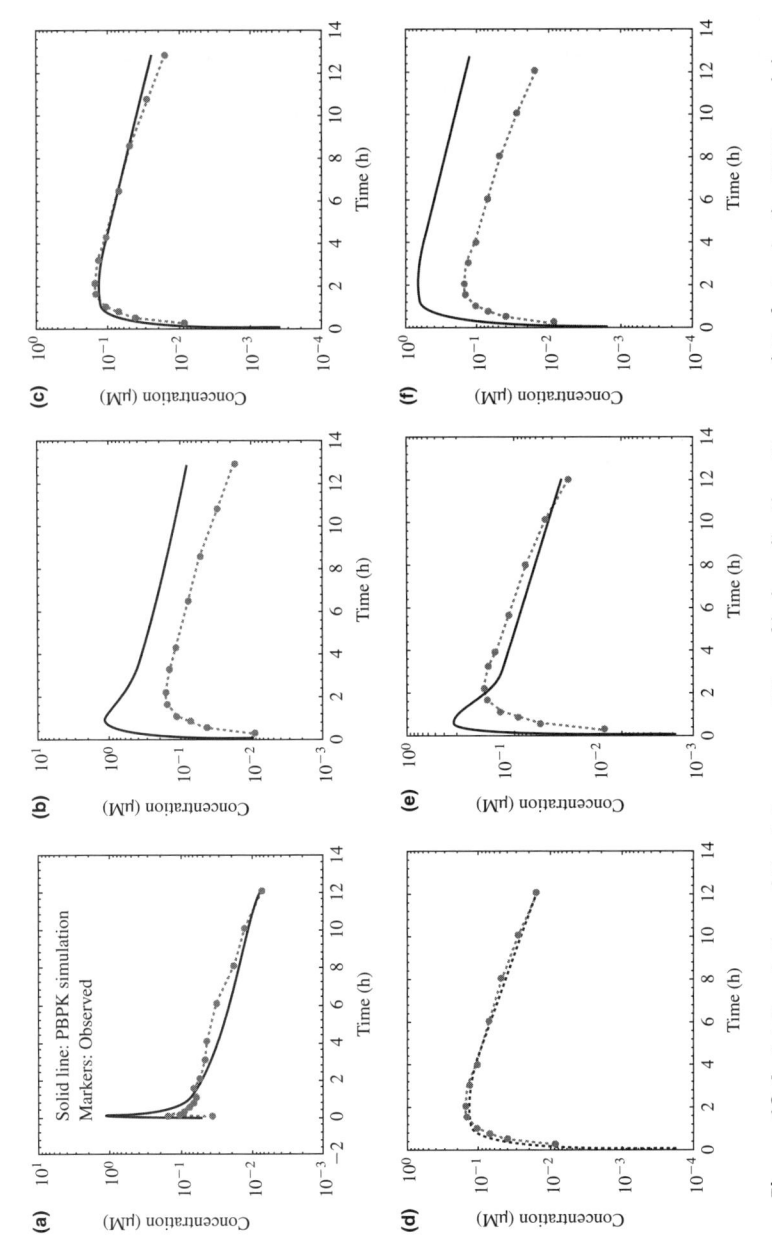

281

gut wall metabolism and delayed gastric emptying. This is done by introducing intestinal loss rate constants (Chapter 4) and by lowering the gastric emptying rate constant until a best fit to the observed is obtained (Fig. 12.4c). The Effect of further refinement of permeability is shown in Figure 12.4d. The effects of adjusting either the intestinal loss or gastric emptying parameters (Figs. 12.4e and 12.4f) are shown. These effects on the lineshape are independent and unique, providing confidence in the conclusions. However, a sensitivity analysis can be done to demonstrate that none of the other oral PK-profile-influencing parameters can lead to a good fit to the observed oral line shape.[1]

Lower observed initial concentrations compared to the expected could be misinterpreted as due to poor solubility. However, poor solubility cannot explain the higher terminal concentrations seen in the case of delayed gastric emptying. Line shape analysis with PBPK can thus distinguish between the two, as the effects of these on the line shapes are unique.

12.2.4 Regional Variation in Intestinal Loss: Gut Wall Metabolism, Intestinal Efflux, and Luminal Degradation

Intestinal loss of an orally administered drug could be attributed to gut wall metabolism, intestinal efflux, or luminal degradation. Although P-gp is highly expressed in intestinal epithelial cells, the impact of P-gp on drug absorption is not as significant as it is generally believed.[21] Luminal degradation is very compound structure specific and can easily be tested *in vitro*. Gut metabolism, therefore, is the most common reason for intestinal loss. It can lead to low and variable bioavailability with the possibility of harm to the GI tract from toxic metabolites. In the lead optimization stage, it is important to screen away from this problem. Substrates of CYP3A and UGTs are the most susceptible to gut metabolism. As seen in Section 12.2.3 with the example of verapamil, gut metabolism can be identified with PBPK simulations of oral PK profiles. Traditionally, the fraction of drug extracted in the gut can be determined using hepatic extraction and oral bioavailability (see equation 3.6) assuming that $f_{abs} = 1$. Such an assumption is not required in PBPK, as f_{abs} is determined from solubility and permeability. Another advantage with PBPK is that it can identify regional variation in intestinal loss. The pattern of regional loss can provide clues on the root cause for the intestinal loss. CYP3A-mediated gut loss is largest in the proximal region of the small intestine,[22] while P-gp efflux is likely to increase from proximal to distal small intestine.[23,24] In the colon neither of these is likely to predominate in humans. However, intestinal loss mediated by glucuronidation can be high right through the GI tract in humans.[25,26] Figure 12.5 demonstrates the use of PBPK simulations to identify regional variation in metabolism.[27] Figure 12.5a shows the IV simulation of a representative compound in a lead series, obtained by modulating $CL_{int,u}$, K_p factor, and the EHR rate. Figure 12.5b shows the PBPK prediction of oral PK using the best-fit parameters ($CL_{int,u}$, K_p factor, and EHR rate) from IV

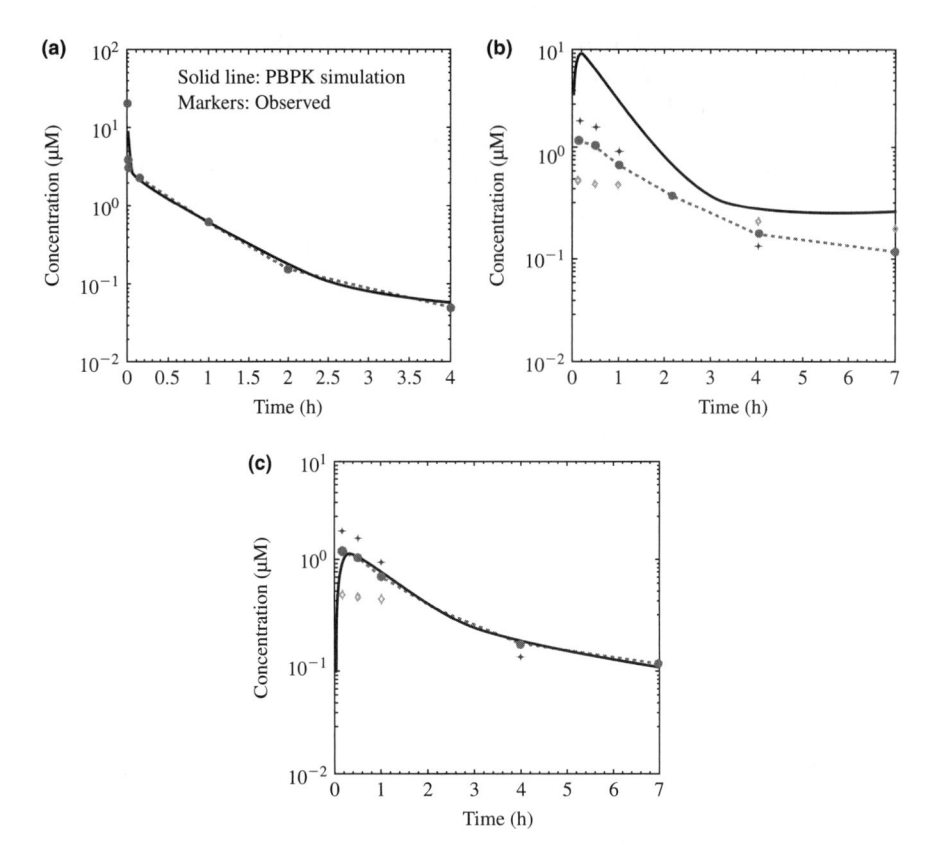

12.2 PBPK SIMULATIONS OF PHARMACOKINETIC PROFILES

Figure 12.5. PBPK simulations of PK profiles of a representative project compound observed in rat. (a) IV simulation obtained by modulating $CL_{int,u}$, K_p factor, and enterohepatic recirculation (EHR) rate. (b) PBPK prediction of oral PK using the best-fit parameters ($CL_{int,u}$, K_p factor, and EHR rate) from IV simulation and Caco-2 permeability (10×10^{-6} cm/s). (c) A good fit to the observed oral profile could be obtained by using multiplicative factors for intestinal permeability (0.09, 0.008, 0.008, 0.008, 0.008, 0.008, 0.008, 0.008, 0.004) in the place of default values of 1 in all small intestinal compartments and 0.1 in the colon.

simulation along with Caco-2 permeability (10×10^{-6} cm/s^{-2}), using the default permeability multipliers of 1 in all small intestinal compartments and 0.1 in the colon. The compound has very high solubility. Figure 12.5c shows a good fit to the observed oral profile by using multiplicative factors for intestinal permeability (0.09, 0.008, 0.008, 0.008, 0.008, 0.008, 0.008, 0.008, and 0.004) in the place of default values. This implies that the compound is extensively lost in both the small intestine and colon. Since the compound contains a hydroxyl group, a likely site for glucuronidation, the extensive intestinal loss seen in this compound could be attributed to gut metabolism. Since the compound had a very high solubility and had reasonably high permeability, it has a narrow window of absorption in the duodenum (where expression of drug-metabolizing enzymes (DME) is generally considered low), resulting in peaks in plasma concentration profiles. It was, therefore, of interest in the project to explore the possibility for an extended release formulation. To facilitate this optimization, PBPK was used as a screening tool to identify compounds that had the least gut metabolism in the colon.[27] The rat is generally a good model for CYP3A-mediated gut metabolism. Thus, compounds which are metabolized in the rat gut are likely to be metabolized in the human gut too. However, glucuronidation can be very extensive in the rat gut compared to human and vice versa. It is therefore important to confirm PBPK findings in the rat or other preclinical species through experimentation. In order to check the relevance of gut metabolism in humans, permeability in small sections of human intestinal jejunal and colonic tissues were determined in Ussing chamber experiments (see Chapter 4) for compounds that progressed to the profiling stage in lead optimization. Very often low bioavailability due to gut metabolism can be misinterpreted as due to poor solubility or precipitation of the drug *in vivo*. PBPK modeling can help distinguish between the two, aiding a focused approach to lead optimization.

12.2.5 Enterohepatic Recirculation

Many large, lipophilic, and ionizable compounds are capable of emptying into the bile and then reabsorbed in the GI tract. Apart from parent recirculation, some phase II metabolites are emptied into bile and then into intestinal lumen, converted to parent by gut microflora and reabsorbed as parent. Enterohepatic recirculation of either parent or metabolite leads to an apparent increase in terminal concentrations in the IV PK profile that is not accountable by modulating $CL_{\text{int},u}$ and K_p factor (see Fig. 12.6a). By introducing an EHR rate, a better fit of the IV profile is obtained (Fig. 12.6b). The EHR rate is expected to be the same for oral simulation. A higher EHR in oral simulation compared to IV is indicative of higher metabolite availability in the oral simulation coming from gut metabolism. In these cases, the transport into bile is not the rate-limiting factor for EHR. A lower EHR in oral simulations compared to IV is not desirable as it indicates inhibition of hepatic uptake transporter or bile efflux transporter as will be discussed in the following sections.

Both EHR and delayed gastric emptying can lead to higher terminal concentrations in the oral profile, resulting in longer apparent oral half-life. However, unlike drug-induced delay in gastric emptying, EHR is not accompanied by reduced initial concentrations. EHR is seen in IV profiles as well. PBPK simulations can thus answer the following question: Is the observed long apparent oral half-life due to enterohepatic recirculation or gastric emptying delay? The subtle changes in PK line shapes due to EHR and drug-induced delay in gastric emptying can easily be distinguished in PBPK simulations.

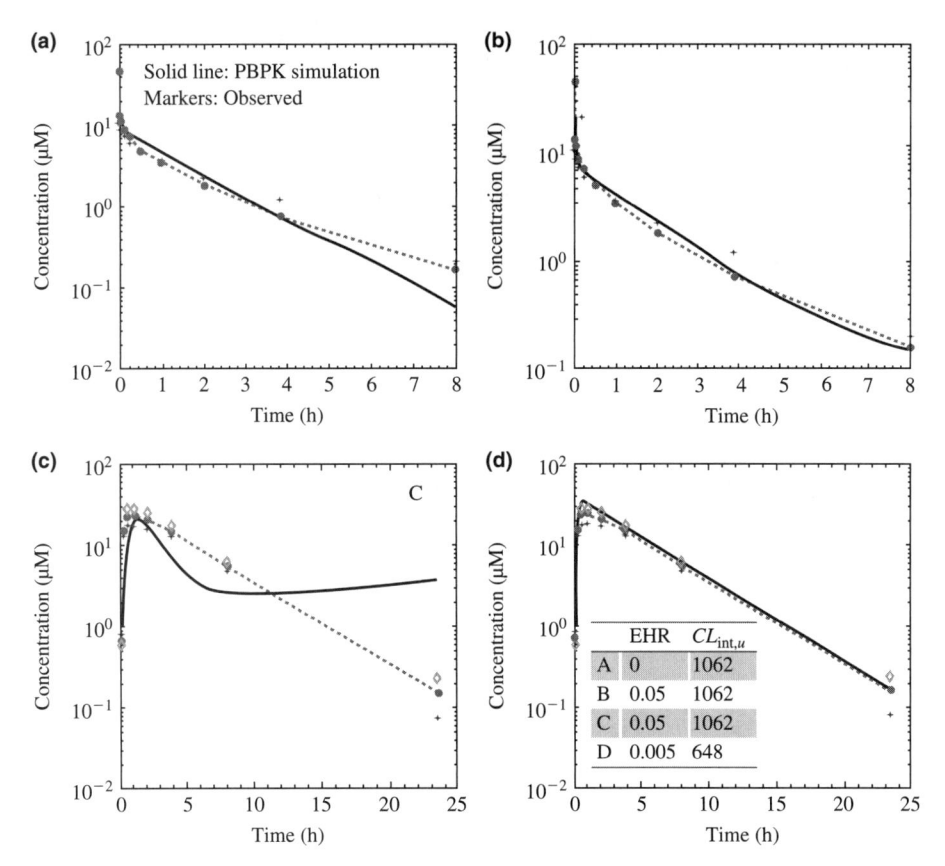

12.2 PBPK SIMULATIONS OF PHARMACOKINETIC PROFILES

Figure 12.6. PBPK simulations of PK profiles of a discovery project compound observed in Sprague–Dawley rat at 2 and 10 μmol/kg IV and oral doses. (a) Simulation of IV profile with modulated K_p factor and $CL_{int,u}$ (1062 mL/min/kg) to get the best fit to the observed profile. Enterohepatic recirculation (EHR) rate = 0. (b) A good fit to the observed terminal concentrations of the IV profile achieved by including an EHR rate = 0.05. (c) Prediction of oral PK profile retaining the K_p factor, $CL_{int,u}$, and EHR rate used in (b). (d) An improved fit to the observed oral PK profile achieved with a reduced $CL_{int,u}$ (648 mL/min/kg) and EHR rate = 0.

EHR is much more common in rats compared to humans and, therefore, it is important to consider the relevance of the EHR of a compound in humans. If EHR is not likely in humans, there is a risk of overestimating bioavailability in humans with only rat PK. If PBPK simulation identifies EHR in a second preclinical species, then it could be significant in humans as well. Thus, a quantitative estimation of EHR in rat would allow us to arrive at a more realistic predicted human bioavailability. Biliary efflux transport is generally associated with high interindividual variability due to genetic abnormalities, disease states, age, transporter expression, and activity. These factors should be considered when extrapolations from average healthy to subpopulations are carried out.

12.2.6 Inhibition of Drug-Metabolizing Enzymes

In preclinical species, oral PK studies are done at higher doses compared to IV in order to achieve similar plasma concentrations in the two routes, expecting the oral bioavailability to be <100. A higher hepatic concentration is, therefore, likely during oral drug administration compared to IV. Even if the IV and oral doses are similar, hepatic drug concentrations will be higher in the oral route due to first pass through the liver. Thus, if a compound inhibits DME at therapeutic concentrations, a higher extent of inhibition might be expected in the oral compared to the IV route. Inhibition of clearance in the oral route relative to IV can be identified with PBPK simulations. The predicted oral profile, using the best fit parameters derived from IV PBPK simulation has a lower AUC compared to the observed (see Fig. 12.2). A higher $CL_{int,u}$ to that obtained through IV fit is then employed to reproduce the observed profile. Once nonlinear kinetics is identified in preclinical species, it can be confirmed in appropriate *in vitro* experiments and its relevance to humans can be tested in human-specific *in vitro* systems.

12.2.7 Inhibition of Hepatic Uptake

An inhibition of DME accompanying inhibition of EHR could be interpreted in one of the ways shown in Table 12.1. Either the parent compound or one

TABLE 12.1. Plausible Explanations for Reduced $CL_{int,u}$ and EHR Rate in Oral Compared to IV Simulations

Observation	Plausibile Explanations
Clearance inhibition	Inhibition by parent or metabolite of • UGT enzyme • Other metabolizing enzyme
Reduced rate of EHR of parent, metabolite, or both at high doses	Inhibition by parent or metabolite of • UGT enzyme • Biliary efflux transporter
An apparent inhibition of clearance and efflux	Inhibition by parent or metabolite of • Hepatic uptake transporter

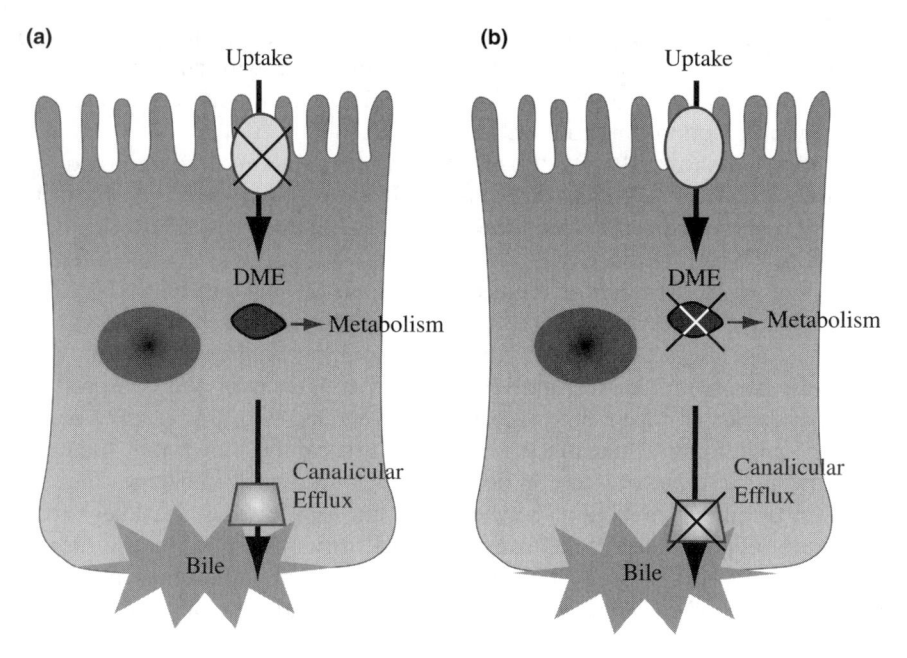

Figure 12.7. Two possible scenarios that can explain the reduction in EHR and inhibition of clearance seen in oral PBPK simulations. (a) An uptake transporter is inhibited. (b) A drug-metabolizing enzyme and efflux transporter are inhibited by the parent or one of its metabolites.

of its metabolites inhibit an uptake transporter leading to an apparent clearance inhibition and EHR inhibition, or the DME and an efflux transporter were inhibited by the parent or one of its metabolite. These possible scenarios are illustrated in Figure 12.7. If the compound is a carboxylic acid, it is likely that organic anion transporting polypeptides (OATP) hepatic uptake transporters are involved. An OATP inhibition assay needs to be done. In addition, the major isoform of metabolizing enzyme and the efflux transporter needs to be identified and inhibition assays carried out.

Inhibition of enzymes entails the potential risk for drug–drug interactions (DDI). Inhibition of hepatic uptake or efflux transporters may have consequences for efficacy and toxicity of drugs that are mainly eliminated by the hepatobiliary system.[28] Uptake of drugs into hepatocytes is primarily mediated by OATPs 1B1, 1B3, and 2B1 and organic cation transporters (OCTs), which are expressed on the sinusoidal membrane in the liver. The OATPs[29–31] mediate the hepatocellular uptake of many drugs and endogenous compounds such as bile salts, conjugates of steroids, hormones, and other large amphiphilic organic anions. The substrate specificities of various OATPs overlap considerably, although unique features of individual transporters have been demonstrated.[30] Cyclosporin A and rifampicin are potent inhibitors of OATPs.[32-34] An increasing number of DDIs observed with statins have been attributed to inhibition of OATPs.[35-37] Several *in vitro* models are available to assess the interaction potential arising from OATPs.[38-40]

Figure 12.6 shows the PBPK simulations of an example compound that exhibits both decreased clearance and EHR rates.[41] Figure 12.6c shows the simulation of oral profile using the $CL_{int,u}$, K_p factor, and EHR rate from the best-fit IV simulation.

The best fit to the observed values is obtained through reducing the $CL_{int,u}$ and EHR rate, implying an inhibition of a DME and an efflux transporter. However, as suggested earlier, this combination could also result from inhibition of an uptake transporter. If the metabolism of the compound is uptake-transporter-dependent (e.g., a carboxylic acid with low passive permeability or high intrinsic clearance or both) or if it can be shown that the major metabolizing enzyme of the compound is a high capacity enzyme, then the observed apparent reductions in clearance and EHR rates are likely to be due to inhibition of the uptake transporter. Identifying an inhibition of hepatic uptake transporter using PBPK simulations relies on differences in hepatic concentrations of the inhibitor between IV and oral administrations of the compound. In the example in Figure 12.6, the oral dose of the compound is much higher compared to its IV dose. Consequently, the higher hepatic concentrations during oral administrations lead to a higher level of inhibition of the uptake transporter. The reverse situation in which the hepatic drug concentrations are higher during IV rather than oral administration of the drug can also happen. Although oral doses are generally kept higher compared to IV doses in order to achieve similar systemic drug concentrations in both routes, the concentrations reaching the liver of an orally administered drug can be greatly limited by gut metabolism, resulting in a lower concentration of the drug from an oral dose. Atorvastatin is known to inhibit Oatp/OATP in both rats[28,42] and humans.[35] The compound is a CYP3A substrate and undergoes extensive intestinal metabolism in rats.[43] PBPK simulations of atorvastatin are shown in Figure 12.8. The signature discrepancy of the predicted profile from the observed (Fig. 12.8b) is different from the previous example (Fig. 12.7). Thus, there are two possible signature or fingerprint discrepancies associated with hepatic uptake transporter inhibition. It has to be pointed out that the absorption phase of the observed oral profile is a net effect of both metabolism and absorption rates, and it is not possible to quantitatively distinguish between gut metabolism and hepatic uptake transporter inhibition. As both gut metabolism and hepatic uptake transporter inhibition influence the *AUC* of the oral PK profile, any quantitative estimation of gut extraction using PBPK simulations is likely to be flawed. It would still be possible to identify these because no one change (neither a decrease in permeability multipliers nor an increase in $CL_{int,u}$) can simulate the oral profile.

Identification of autoinhibition of hepatic uptake or DME and biliary efflux by PBPK simulation of PK profiles in preclinical species should trigger a

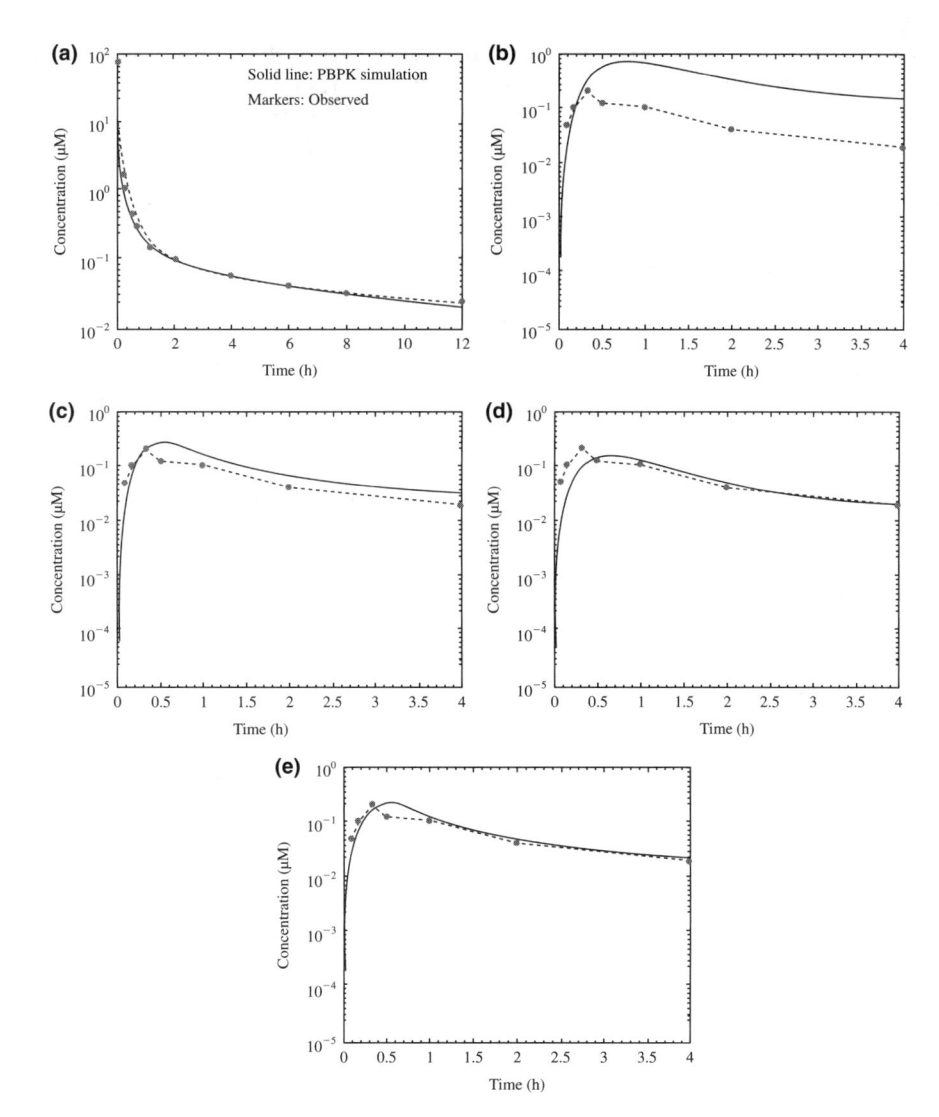

12.2 PBPK SIMULATIONS OF PHARMACOKINETIC PROFILES

Figure 12.8. PBPK simulations of PK profiles of atorvastatin observed in rat at 3.58 and 17.9 μmol/kg iv and oral doses respectively. (a) Simulation of IV profile with modulated K_p factor and $CL_{int,u}$ (2998 mL/min/kg) to get the best fit to the observed profile. (b) Prediction of oral PK profile retaining the K_p factor and $CL_{int,u}$ used in (a). (c) Gut permeability multipliers in the small intestinal and colonic were altered to 1.2 in the first small intestinal compartment and zeros in all of the others. (d) Oral simulation with an increased $CL_{int,u}$ (20,000 mL/min/kg) provides good fit to the later part of the curve but not the initial absorption phase. (e) Oral simulations with altered permeability multipliers as in (c) and an increased $CL_{int,u}$ (4140 mL/min/kg) gives the best fit to the observed.

variety of confirmatory experiments in human *in vitro* systems. Both the parent compound and its major metabolites should be tested in inhibition assays for OATP, major DME and efflux transporters. For the example shown in Figure 12.6, the parent compound was shown to inhibit the uptake transporter, OATP1B1.

12.2.8 Inhibition of Hepatobiliary Efflux

If the best fit to observed oral profile requires the use of a lower EHR compared to that obtained from IV (see Fig. 12.2), it implies that one or more efflux transporters are inhibited by the parent compound or one of its metabolites. Again, relevance to humans should be tested in human-specific *in vitro* systems.[44]

Inhibition of efflux transporters may not contribute to an altered clearance of its substrates, unless biliary elimination plays a major role in the total clearance of a compound. Consequently, efflux inhibition cannot mediate DDIs. However, for some compounds, the increased intracellular concentrations following efflux inhibition can mean an additional role for low-affinity, high-capacity enzymes in their metabolism. Efflux inhibition can cause cholestatic drug-induced liver injury (DILI).[45] In addition, accumulation of conjugated metabolites of endogenous and exogenous compounds that rely on the inhibited efflux transporter for biliary elimination can lead to cell toxicity. Different mechanisms in drug-induced hepatotoxicity has been reviewed recently.[46] By enabling an early identification of the inhibitions, PBPK simulations can trigger the right experiments to support the hypotheses. Efflux inhibitions may be caused by parent drugs or by metabolites like acyl glucuronides.[47–49] *In vitro* methods to evaluate the extent of efflux inhibition rely on the measurement of accumulation of appropriate fluorescent-labeled transporter substrates at different inhibitor concentrations. However, these studies are employed only in the later stages of lead optimization. An early identification of potential PK issues in a compound series can be valuable to screen away from the problem.

The examples in this section have shown that PBPK simulations are uniquely suited to deconvolute information from PK profiles. PBPK is a powerful tool for understanding why we see the profiles we see. Since different PK mechanisms uniquely impact the PK profiles (Fig. 12.2), the information from PBPK is highly reliable even if multiple factors simultaneously impact the line shape. PBPK simulations provide an opportunity for optimum use of available data with an excellent potential to reduce animal testing and focus resources on confirmatory rather than exploratory tests during all phases of drug optimization and development. Table 12.2 summarizes the various PK mechanisms that can be identified with PBPK simulations and the benefits it brings to drug discovery projects.

TABLE 12.2. Summary of PK Mechanisms Identified through Hypothesis Testing with PBPK Simulations and their Benefits to Drug Discovery Projects

PK Mechanism Identified by PBPK	Benefits to Drug Discovery Projects
Compound-induced gastric emptying delay	• Delayed gastric emptying causes delayed onset of drug action and GI inflammation. If identified in more than one species, likely to be relevant in man. If high uptake rate is critical in a project, then important to de-risk through design changes in LO. • No need for imaging experiments in toxicology.
In vivo solubility which is often much higher compared to the *in vitro* measured value	• Lead compounds need to be lipophilic in some projects. *In vitro* solubility then becomes an important screening parameter. For potential leads, PBPK can identify, if solubility is really an issue *in vivo* and help de-risk a compound. • *In vivo* solubility from PBPK can be used for predicting human fraction absorbed, as *in vivo* solubility tend to be similar across species. • Focus resources on right issues. Often poor bioavailability due to other reasons is blamed on solubility. With PBPK identification of *in vivo* solubility, efforts on costly formulations can be avoided. • FASSIF experiments need not be carried out.
Gut metabolism	• Alert for the possibility for low, variable bioavailability in man and high uncertainty in human dose predictions. Depending on the enzymes involved, relevance to human is determined. Experiments are initiated in human specific systems, which can lead to cost avoidance or de-risking • PBPK can identify regions in the gut where metabolism is high. This can enable possible protection through formulation.

(*Continued*)

TABLE 12.2. (Continued)

PK Mechanism Identified by PBPK	Benefits to Drug Discovery Projects
Autoinhibition of drug metabolizing enzymes	• If a compound inhibits its own metabolism, it could be a perpetrator of drug-drug interactions. UGT1A1 inhibition can lead to hepatotoxicity due to interference with bilirubin conjugation. In LO, efforts to keep effective dose low should be a high priority for a lead series with this problem.
Autoinhibition of uptake transporters	• Inhibition of uptake transporters can also lead to DDI when co-administered with drugs that depend on the inhibited uptake transporter for their clearance (example, statins). During the screening stage of LO, experiments to identify uptake transporter inhibition should be initiated.
Autoinhibition of efflux transporters	• Efflux inhibition can cause hepatotoxicity due to accumulation of toxic conjugates in hepatocytes. Quantification of *in vivo* efflux inhibition has not been possible so far. For the first time, this is now possible with PBPK. This allows for *in vitro in vivo* correlations as well as species correlations to be made. Better understanding of these correlations would lead to better confidence in human prediction with fewer experiments.

12.3 PHARMACOKINETIC PREDICTIONS

12.3.1 Human Predictions from Preclinical Data

Traditionally, translation to humans is based on allometry,[50–53] which rely on size as the principal differentiator between species. Allometry can be a useful tool for scaling drugs that are predominantly cleared renally. However, the large species differences in enzymes and transporters (expression levels, types of enzymes, and therefore affinity of drug and metabolic pathways) are not related to size. For the majority of drugs for which the predominant elimination route is hepatic metabolism, allometry does do well.[51] It tends to seriously fail for low clearance compounds, for which differences in enzymes and transporters play a critical role, rather than size-dependent hepatic blood flow. Absorption is also not predicted well by allometry.

Physiologically-based PK scaling using human *in vitro* data has been considered as an alternative to allometry. It has been applied both in the pharmaceutical sector[54–56] and in human health risk assessment.[57–59] Its main advantage is that it can provide the temporal profile of PK. Another advantage is that PBPK models facilitate parametric and nonparametric extrapolations across routes of administration (e.g., inhalation exposures to ingestion), species, and different populations. Any dose can be simulated, provided the kinetics is linear in the dose range of interest. However, the PBPK approach can be compromised by lack of good *in vitro* data for metabolism and for protein binding (especially for highly bound compounds).

The oral route is the preferred route of drug administration. Predicting the oral bioavailability is, therefore, very important, as concentrations at the target site can be significantly limited by oral bioavailability. Knowledge learned from preclinical PK should, therefore, be applied to predict human oral bioavailability, if found relevant to humans (Fig. 12.9).

12.3.2 Pharmacokinetic Predictions in Clinical Development

Section 12.2 dealt with the use of PBPK for data interpretation within a species. These principles can be applied to humans, once the first-in-human PK profiles are available. However, it is important to have IV data in addition to the oral PK data. The identification of different PK mechanisms in humans can be useful in the development of drug formulation. Apart from aiding clinical trial design, an understanding of factors that influence disposition and their modulation by age, disease, gender, and genetics at a mechanistic level can be very useful for extrapolation to different subpopulations (see Fig 12.9).

This chapter has shown how to maximize the utility of preclinical data in understanding the PK issues that could be relevant in humans and how PK dose predictions to humans can be improved through a maximal exploitation of available data. Mechanisms underlying PK profiles observed in humans can be

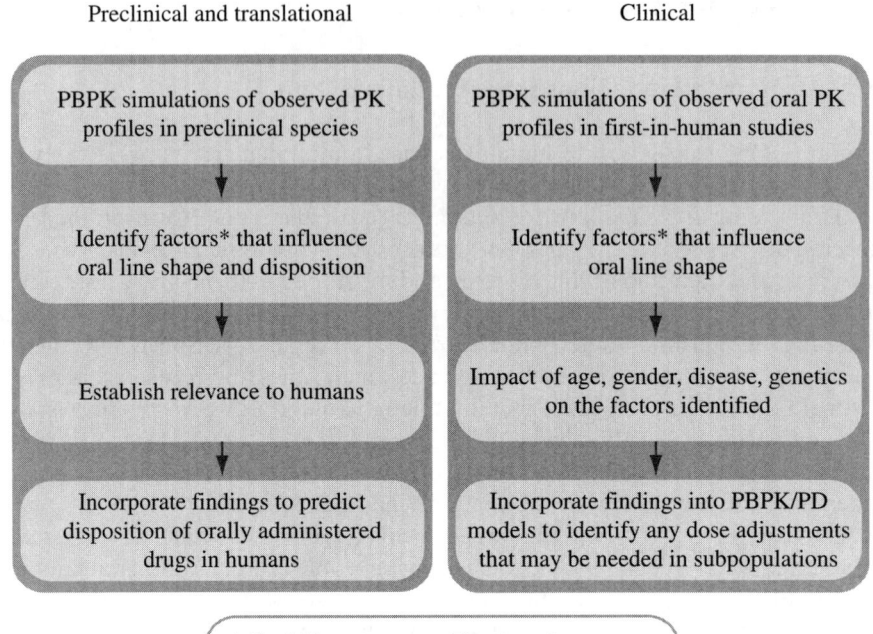

```
                    Preclinical and translational              Clinical

              ┌──────────────────────────────┐   ┌──────────────────────────────┐
              │ PBPK simulations of observed PK │   │ PBPK simulations of observed oral PK │
              │ profiles in preclinical species │   │ profiles in first-in-human studies   │
              └──────────────┬───────────────┘   └──────────────┬───────────────┘
                             ▼                                    ▼
              ┌──────────────────────────────┐   ┌──────────────────────────────┐
              │ Identify factors* that influence │   │ Identify factors* that influence │
              │ oral line shape and disposition  │   │ oral line shape                  │
              └──────────────┬───────────────┘   └──────────────┬───────────────┘
                             ▼                                    ▼
              ┌──────────────────────────────┐   ┌──────────────────────────────┐
              │ Establish relevance to humans │   │ Impact of age, gender, disease, genetics │
              │                               │   │ on the factors identified                │
              └──────────────┬───────────────┘   └──────────────┬───────────────┘
                             ▼                                    ▼
              ┌──────────────────────────────┐   ┌──────────────────────────────┐
              │ Incorporate findings to predict │ │ Incorporate findings into PBPK/PD │
              │ disposition of orally administered │ │ models to identify any dose adjustments │
              │ drugs in humans                 │ │ that may be needed in subpopulations    │
              └──────────────────────────────┘   └──────────────────────────────┘
```

* Could be one or combination of
1. Higher *in vivo* solubility
2. Drug-induced delay in gastric emptying
3. Gut metabolism
4. Enterohepatic recirculation
5. Inhibition of drug metabolizing enzyme
6. Inhibition of uptake or efflux transporters

Figure 12.9. Hypotheses-driven translational and clinical PK predictions.

unraveled with hypotheses-driven PBPK simulations, reserving experimentation for confirmation only. The enormous resource-saving potential of this approach in drug discovery and development is obvious.

REFERENCES

1. Peters SA, Hultin L. Early identification of drug-induced impairment of gastric emptying through physiologically-based pharmacokinetic (PBPK) simulation of plasma concentration-time profiles in rat. *J Pharmacokinet Pharmacodyn*. 2008;35(1): 1–30.
2. Peters SA. Identification of intestinal loss of a drug through physiologically-based pharmacokinetic simulation of plasma concentration-time profiles. *Clin Pharmacokinet*. 2008;47(4):245–259.

REFERENCES

3. Poulin P, Theil FP. Prediction of pharmacokinetics prior to in vivo studies. 1. mechanism-based prediction of volume of distribution. *J Pharm Sci.* 2002;91(1): 129–156.
4. Poulin P, Theil FP. Prediction of pharmacokinetics prior to in vivo studies. II. Generic physiologically-based pharmacokinetic models of drug disposition. *J Pharm Sci.* 2002;91(5):1358–1370.
5. Rodgers T, Leahy D, Rowland M. Physiologically-based pharmacokinetic modeling. 1: Predicting the tissue distribution of moderate-to-strong bases. *J Pharm Sci.* 2005;94(6):1259–1276.
6. Rodgers T, Rowland M. Physiologically-based pharmacokinetic modeling. 2: Predicting the tissue distribution of acids, very weak bases, neutrals and zwitterions. *J Pharm Sci.* 2006;95(6):1238–1257.
7. Bjorkman S. Prediction of drug disposition in infants and children by means of physiologically-based pharmacokinetic (PBPK) modelling: Theophylline and midazolam as model drugs. *Br J Clin Pharmacol.* 2005;59(6):691–704.
8. Poulin P, Ekins S, Theil FP. A hybrid approach to advancing quantitative prediction of tissue distribution of basic drugs in human. *Toxicol Appl Pharmacol.* 2011;250(2): 194–212.
9. Peyret T, Poulin P, Krishnan K. A unified algorithm for predicting partition coefficients for PBPK modeling of drugs and environmental chemicals. *Toxicol Appl Pharmacol.* 2010;249(3):197–297.
10. Peters SA. Evaluation of a generic physiologically-based pharmacokinetic model for lineshape analysis. *Clin Pharmacokinet.* 2008;47(4):261–275.
11. Lu Y, Kim S, Park K. In vitro-in vivo correlation: Perspectives on model development. *Int J Pharm.* 2011;418(1):142–148.
12. Wallden J, Thorn SE, Wattwil M. The delay of gastric emptying induced by remifentanil is not influenced by posture. *Anesth Analg.* 2004;99(2):429–34.
13. Calatayud S, et al. Downregulation of nNOS and synthesis of PGs associated with endotoxin-induced delay in gastric emptying. *Am J Physiol Gastrointest Liver Physiol.* 2002;283(6):G1360–1367.
14. Cho SH, Park H, Kim JH, Ryu YH, Lee SI, Conklin JL. Effect of sildenafil on gastric emptying in healthy adults. *J Gastroenterol Hepatol.* 2006;21(1 Pt 2):222–226.
15. Djaldetti R, Ziv I, Melamed E. Impaired absorption of oral levodopa: A major cause for response fluctuations in Parkinson's disease. *Isr J Med Sci.* 1996;32(12): 1224–1227.
16. Harasawa S, Kikuchi K, Senoue I, Nomiyama T, Miwa T. Gastric emptying in patients with gastric ulcers—effects of oral and intramuscular administration of anticholinergic drug. *Tokai J Exp Clin Med.* 1982;7(5):551–559.
17. Bozkurt A, Deniz M, Yegen BC. Cefaclor, a cephalosporin antibiotic, delays gastric emptying rate by a CCK-A receptor-mediated mechanism in the rat. *Br J Pharmacol.* 2000;131(3):399–404.
18. Murphy DB, Sutton A, Prescott LF, Murphy MB. A comparison of the effects of tramadol and morphine on gastric emptying in man. *Anaesthesia.* 1997;52(12): 1224–1229.
19. Murphy DB, Sutton JA, Prescott LF, Murphy MB. Opioid-induced delay in gastric emptying: A peripheral mechanism in humans. *Anesthesiology.* 1997;87(4):765–770.

20. Vogelgesang B, Echizen H, Schmidt E, Eichelbaum M. Stereoselective first-pass metabolism of highly cleared drugs: Studies of the bioavailability of L- and D-verapamil examined with a stable isotope technique. *Br J Clin Pharmacol.* 1984; 18(5):733–740.
21. Lin JH, Yamazaki M. Clinical relevance of P-glycoprotein in drug therapy. *Drug Metab Rev.* 2003;35(4):417–454.
22. Thelen K, Dressman JB. Cytochrome P450-mediated metabolism in the human gut wall. *J Pharm Pharmacol.* 2009;61(5):541–558.
23. Bruyere A, et al. Effect of variations in the amounts of P-glycoprotein (ABCB1), BCRP (ABCG2) and CYP3A4 along the human small intestine on PBPK models for predicting intestinal first-pass. *Mol Pharm.* 2010;7(5):1596–1607.
24. Mouly S, Paine MF. P-glycoprotein increases from proximal to distal regions of human small intestine. *Pharm Res.* 2003;20(10):1595–1599.
25. Ritter JK. Intestinal UGTs as potential modifiers of pharmacokinetics and biological responses to drugs and xenobiotics. *Expert Opin Drug Metab Toxicol.* 2007; 3(1):93–107.
26. Fisher MB, Paine MF, Strelevitz TJ, Wrighton SA. The role of hepatic and extrahepatic UDP-glucuronosyltransferases in human drug metabolism. *Drug Metab Rev.* 2001;33(3–4):273–297.
27. Peters SA, Leifeng C, Åsa S, Ungell A. Identification of regional variation in intestinal glucuronidation through physiologically-based pharmacokinetic (PBPK) simulation of plasma concentration-time profiles in rat. To be published.
28. Lau YY, Wu CY, Okochi H, Benet LZ. Ex situ inhibition of hepatic uptake and efflux significantly changes metabolism: Hepatic enzyme-transporter interplay. *J Pharmacol Exp Ther.* 2004;308(3):1040–1045.
29. Kalliokoski A, Niemi M. Impact of OATP transporters on pharmacokinetics. *Br J Pharmacol.* 2009;158(3):693–705.
30. Hagenbuch B, Meier PJ. Organic anion transporting polypeptides of the OATP/SLC21 family: Phylogenetic classification as OATP/SLCO superfamily, new nomenclature and molecular/functional properties. *Pflugers Arch.* 2004;447(5):653–665.
31. Hagenbuch B, Gui C. Xenobiotic transporters of the human organic anion transporting polypeptides (OATP) family. *Xenobiotica.* 2008;38(7–8):778–801.
32. Treiber A, Schneiter R, Hausler S, Stieger B. Bosentan is a substrate of human OATP1B1 and OATP1B3: Inhibition of hepatic uptake as the common mechanism of its interactions with cyclosporin A, rifampicin, and sildenafil. *Drug Metab Dispos.* 2007;35(8):1400–1407.
33. Parasrampuria R, Mehvar R. Dose-dependent inhibition of transporter-mediated hepatic uptake and biliary excretion of methotrexate by cyclosporine A in an isolated perfused rat liver model. *J Pharm Sci.* 2010;99(12):5060–5069.
34. Shitara Y, Nagamatsu Y, Wada S, Sugiyama Y, Horie T. Long-lasting inhibition of the transporter-mediated hepatic uptake of sulfobromophthalein by cyclosporin a in rats. *Drug Metab Dispos.* 2009;37(6):1172–1178.
35. Lau YY, Huang Y, Frassetto L, Benet LZ. Effect of OATP1B transporter inhibition on the pharmacokinetics of atorvastatin in healthy volunteers. *Clin Pharmacol Ther.* 2007;81(2):194–204.

REFERENCES

36. Nakagomi-Hagihara R, Nakai D, Tokui T, Abe T, Ikeda T. Gemfibrozil and its glucuronide inhibit the hepatic uptake of pravastatin mediated by OATP1B1. *Xenobiotica.* 2007;37(5):474–486.
37. Shitara Y, Itoh T, Sato H, Li AP, Sugiyama Y. Inhibition of transporter-mediated hepatic uptake as a mechanism for drug-drug interaction between cerivastatin and cyclosporin A. *J Pharmacol Exp Ther.* 2003;304(2):610–616.
38. Soars MG, Webborn PJ, Riley RJ. Impact of hepatic uptake transporters on pharmacokinetics and drug-drug interactions: Use of assays and models for decision making in the pharmaceutical industry. *Mol Pharm.* 2009;6(6):1662–1677.
39. Backman JT, Luurila H, Neuvonen M, Neuvonen PJ. Rifampin markedly decreases and gemfibrozil increases the plasma concentrations of atorvastatin and its metabolites. *Clin Pharmacol Ther.* 2005;78(2):154–167.
40. Poirier A, Funk C, Lave T, Noe J. New strategies to address drug-drug interactions involving OATPs. *Curr Opin Drug Discov Devel.* 2007;10(1):74–83.
41. Peters SA, Lutz M, Ungell A. Identification of hepatic uptake inhibition through physiologically-based pharmacokinetic (PBPK) simulation of plasma concentration-time profiles in rat. To be published.
42. Lau YY, Okochi H, Huang Y, Benet LZ. Pharmacokinetics of atorvastatin and its hydroxy metabolites in rats and the effects of concomitant rifampicin single doses: Relevance of first-pass effect from hepatic uptake transporters, and intestinal and hepatic metabolism. *Drug Metab Dispos.* 2006;34(7):1175–1181.
43. Lau YY, Okochi H, Huang Y, Benet LZ. Multiple transporters affect the disposition of atorvastatin and its two active hydroxy metabolites: Application of in vitro and ex situ systems. *J Pharmacol Exp Ther.* 2006;316(2):762–771.
44. Peters SA, Stahl S, Latham L, Larsson L, Ungell A. Identification of efflux inhibition and a higher *in vivo* solubility of a drug through physiologically-based pharmacokinetic (PBPK) simulations of plasma concentration-time profiles in pre-clinical species. To be published.
45. Pauli-Magnus C, Stieger B, Meier Y, Kullak-Ublick GA, Meier PJ. Enterohepatic transport of bile salts and genetics of cholestasis. *J Hepatol.* 2005;43(2):342–357.
46. Russmann S, Kullak-Ublick GA, Grattagliano I. Current concepts of mechanisms in drug-induced hepatotoxicity. *Curr Med Chem.* 2009;16(23):3041–3053.
47. Boelsterli UA. Mechanisms of NSAID-induced hepatotoxicity: Focus on nimesulide. *Drug Saf.* 2002;25(9):633–648.
48. Boelsterli UA. Xenobiotic acyl glucuronides and acyl CoA thioesters as protein-reactive metabolites with the potential to cause idiosyncratic drug reactions. *Curr Drug Metab.* 2002;3(4):439–450.
49. Spahn-Langguth H, Benet LZ. Acyl glucuronides revisited: Is the glucuronidation process a toxification as well as a detoxification mechanism? *Drug Metab Rev.* 1992;24(1):5–47.
50. Tang H, Mayersohn M. A global examination of allometric scaling for predicting human drug clearance and the prediction of large vertical allometry. *J Pharm Sci.* 2006;95(8):1783–1799.
51. Boxenbaum H. Interspecies scaling, allometry, physiological time, and the ground plan of pharmacokinetics. *J Pharmacokinet Biopharm.* 1982;10(2):201–227.

52. Mahmood I. Allometric issues in drug development. *J Pharm Sci.* 1999;88(11): 1101–1106.
53. Shim HJ, et al. Interspecies pharmacokinetic scaling of DA-8159, a new erectogenic, in mice, rats, rabbits and dogs, and prediction of human pharmacokinetics. *Biopharm Drug Dispos.* 2005;26(7):269–277.
54. Kawai R, Lemaire M, Steimer JL, Bruelisauer A, Niederberger W, Rowland M. Physiologically-based pharmacokinetic study on a cyclosporin derivative, SDZ IMM 125. *J Pharmacokinet Biopharm.* 1994;22(5):327–365.
55. Jones HM, Parrott N, Jorga K, Lave T. A novel strategy for physiologically-based predictions of human pharmacokinetics. *Clin Pharmacokinet.* 2006;45(5):511–542.
56. Luttringer O, Theil FP, Poulin P, Schmitt-Hoffmann AH, Guentert TW, Lave T. Physiologically-based pharmacokinetic (PBPK) modeling of disposition of epiroprim in humans. *J Pharm Sci.* 2003;92(10):1990–2007.
57. Bruckner JV, Keys DA, Fisher JW. The acute exposure guideline level (AEGL) program: Applications of physiologically-based pharmacokinetic modeling. *J Toxicol Environ Health A.* 2004;67(8–10):621–634.
58. Kirman CR, Sweeney LM, Meek ME, Gargas ML. Assessing the dose-dependency of allometric scaling performance using physiologically-based pharmacokinetic modeling. *Regul Toxicol Pharmacol.* 2003;38(3):345–367.
59. Young JF, Wosilait WD, Luecke RH. Analysis of methylmercury disposition in humans utilizing a PBPK model and animal pharmacokinetic data. *J Toxicol Environ Health A.* 2001;63(1):19–52.

13

INTEGRATION OF PBPK AND PHARMACODYNAMICS

CONTENTS

13.1 Introduction.................................... 300
13.2 Pharmacodynamic Principles........................ 300
    13.2.1 Pharmacological Targets and Drug Action........... 300
    13.2.2 Functional Adaptation Processes: Tolerance, Sensitization, and Rebound (Fig. 13.2)............. 301
13.3 Pharmacodynamic Modeling......................... 307
    13.3.1 Concentration–Effect, Dose–Response Curves, and Sigmoid E_{max} Models....................... 307
    13.3.2 Mechanism-Based PD Modeling................... 315
    13.3.3 Simple Direct Effects........................ 315
    13.3.4 Models Accommodating Delayed Pharmacological Response................................ 321
    13.3.5 Models Accommodating Nonlinearity in Pharmacological Response with Respect to Time................... 332
13.4 Pharmacokinetic Modeling: Compartmental PK and PBPK...... 335
13.5 Integration of PK or PBPK with PD Modeling.............. 335
13.6 Reasons for Poor PK/PD Correlation.................... 339

Physiologically-Based Pharmacokinetic (PBPK) Modeling and Simulations: Principles, Methods, and Applications in the Pharmaceutical Industry, First Edition. Sheila Annie Peters.
© 2012 John Wiley & Sons, Inc. Published 2012 by John Wiley & Sons, Inc.

13.7 Applications of PK or PBPK/PD Modeling in
Drug Discovery and Development.................. 340
  13.7.1 Need for a Mechanistic PBPK/PD Integration......... 341
  13.7.2 Applications of PK or PBPK/PD in Drug Discovery.... 342
  13.7.3 Applications of PK or PBPK/PD in Drug Development.. 360
13.8 Regulatory Perspective................................. 370
13.9 Conclusions... 371
  Keywords... 372
  References... 376

13.1 INTRODUCTION

Pharmacokinetics (PK) provides an understanding of factors that affect absorption, distribution, metabolism, and excretion of an administered drug, all of which determine its exposure or concentration at the target organ (effect site). Relation of this exposure to the onset, intensity, and duration of drug action is determined by pharmacodynamics (PD). As Leslie Benet stated succinctly, "pharmacokinetics may be simply defined as what the body does to the drug, as opposed to pharmacodynamics which may be defined as what the drug does to the body."[1] A well-defined, quantitative relationship between drug concentrations in biological fluids and pharmacodynamic effect is fundamental to predicting a dosing regimen that best fits the requirements of effect. Linking physiologically-based pharmacokinetic (PBPK) modeling with mechanistic PD enables an understanding of the time course of drug effects under a variety of physiological and pathological conditions and makes it possible to extrapolate across different animal species and dosing routes. PBPK also enables the linking of target tissue concentrations to PD. PK/PD modeling techniques have made rapid strides since their establishment in the late 1970s. Improved definitions of PD effects, better availability of modeling techniques, and an improved understanding of mechanisms underlying drug action have all contributed to the advance of PK/PD modeling. In this chapter, a brief overview of PD principles and PD modeling will be followed by a discussion on linking PK or PBPK with PD, with examples of applications in different stages of drug discovery and development.

13.2 PHARMACODYNAMIC PRINCIPLES

13.2.1 Pharmacological Targets and Drug Action

Pharmacological effect often involves the modulation of an intrinsic physiological process by a drug that binds to a target protein. Pharmacological targets can be receptors and proteins involved in regulatory pathways, enzymes, structural proteins, nuclear receptors, transporter proteins, or ion channels (Table 13.1). The binding of a drug to a target protein can initiate several responses depending on the type of target protein, its location, and mechanism of action (Fig. 13.1). For example, a drug binding to a G-protein-coupled receptor (GPCR) activates a **second messenger**, which, in turn, initiates a series of biochemical reactions. Drugs act by either stimulating the activity of the receptor to which they bind (**agonist**) or by blocking an endogenous agonist from stimulating the receptors, thus reducing the rate of activity of the receptors or enzymes on which they act (**partial agonist, competitive antagonist, noncompetitive (unsurmountable) antagonist, reversible or irreversible inhibitors**). **Inverse agonists** appear to act in an opposite manner to agonists at a receptor site. For example, studies[2] suggest that there are inverse agonists at the benzodiazepine–GABA binding site. Unlike an agonist, which produces an increase in receptor activity, or an antagonist, which blocks this activity, an inverse agonist produces an effect that reverses receptor activity by binding to constitutively active receptors that are coupled to second messenger pathways even in the absence of an agonist. Activated or blocked receptors trigger a response, either directly on the body or through the release of hormones and/or other endogenous ligands to stimulate a particular response or effect. No drug produces a single effect. Small-molecule drugs are selective rather than specific. Even where a drug acts on a single receptor, this receptor may be ubiquitously expressed and may, therefore, be present in organs where the drug is not intended to act, or the targeted receptor may exist in other isoforms that when affected, can produce adverse effects. Additionally, drugs may act on several receptor classes. A primary effect is the desired therapeutic effect. Secondary effects include all other effects besides the desired effect and may be either beneficial (which is rare) or harmful. With appropriately designed PK and PD studies, one can potentially interpret and predict the outcome of primary and secondary effects. PD involves a study of the relationships between plasma drug concentrations, receptor occupancy, receptor activation, and the pharmacological effect, aided by the biochemical and physiological mechanisms of drug action. For drugs that have low membrane permeability or that are substrates of drug transporters, target tissue concentrations rather than plasma concentrations should correlate well with the observed pharmacological effect.

13.2.2 Functional Adaptation Processes: Tolerance, Sensitization, and Rebound (Fig. 13.2)

One of the perils of drug development is the occurrence of time-dependent modulations in pharmacological activity upon repeat administration of certain drugs, arising from functional **adaptation** processes such as **tolerance** and **sensitization**.

TABLE 13.1. Classes of Pharmacological Targets

Pharmacological Target Class	Examples of Targets	Location	Endogenous Ligands	Type of Drug
G-protein-coupled receptors (GPCRs)	Adrenoreceptor, metabotropic neurotransmitter receptors (metabotropic glutamate receptors or mGluRs, muscarinic acetylcholine receptors, GABA-B receptor, 5HT1, 5HT2, CCR, opioid, NK	Transmembrane, cell surface or intracellular	Neurotransmitters[a] and hormones[b]	Competitive, noncompetitive, and irreversible antagonist, agonists, partial agonist, or inverse agonist.
Receptor tyrosine kinase	EGF, VEGF, PDGF, FGF, HGF, TIE, Trk, MuSK, RYK, and insulin receptor families	Cell surface	Growth factors,[c] insulin	Competitive, noncompetitive, and irreversible antagonist, agonists, partial agonist, or inverse agonist.
Guanylyl cyclase receptors	GC-A, GC-B, GC-C	Cell surface	Guanosine-5′-triphosphate (GTP)	Competitive, noncompetitive, and irreversible antagonist, agonists, partial agonist, or inverse agonist.
Nuclear receptors	Thyroid hormone-like, estrogen-like, glucocorticoid receptors, PPAR, FXR, LXR, and other orphan receptors	Cytosol or translocated to nucleus	Retinoic acid, vitamin D_3, hormones	Competitive, noncompetitive, and irreversible antagonist, agonists, partial agonist, or inverse agonist.
Ligand-gated ion channels or ionotropic receptors	Glutamate cationic receptor, nicotinoid receptor (5-HT3, nicotinic acetylcholine receptor, GABA-A)	Synapses	Neurotransmitters	Competitive, noncompetitive, and irreversible antagonist, agonists, partial agonist, or inverse agonist.

Voltage-gated ion channels	NaV, KV, and CaV	Transmembrane ion channels found along the axon and at the synapse in neurons and in other cells	Respond to changes in voltage	Blocker, inhibitor
Enzymes and factors	Transferases such as RNA directed DNA-polymerase; oxidoreductases such as HMG-CoA reductase and cyclooxygenases Peptidases such as ACE, trypsin, thrombin, clotting factors, renin, HIV protease, cathepsin, and caspase	Cell surface, intracellular, or membrane spanning	Large variety of endogenous and exogenous substrates	Inhibitor activator
Structural proteins	$\alpha, \beta, \gamma, \delta,$ and ε tubulin	Extracellular or intracellular	GTP	Inhibitor

[a]Neurotransmitters are amino acids such as glutamate, aspartate, serine, GABA, and glycine; monoamines such as dopamine, 5-HT, adrenaline, noradrenaline, and melatonin; neuropeptides; acetylcholine, adenosine, histamine, anadamide, etc.

[b]Hormones could be derivatives of the amino acids tyrosine and tryptophan. Examples are catecholamines and thyroxine. Peptide hormones consist of chains of amino acids. Examples are TRH and vasopressin. Examples of protein hormones include insulin and growth hormone. More complex protein hormones bear carbohydrate side chains and are called glycoprotein hormones. Luteinizing hormone, follicle-stimulating hormone, and thyroid-stimulating hormone are examples of glycoprotein hormones. Lipid and phospholipid-derived hormones are derivatives of lipids such as linoleic acid and arachidonic acid and phospholipids. Steroid hormones are derived from cholesterol and eicosanoids. Examples of steroid hormones are testosterone and cortisol. Steroid hormones such as calcitriol are a homologous system. The adrenal cortex and the gonads are primary sources of steroid hormones. Examples of eicosanoids are the widely studied prostaglandins.

[c]Growth factors: Fibroblast growth factors comprise the largest family of growth factor ligands at 23 members. The natural alternate splicing of four fibroblast growth factor receptor (FGFR) genes results in the production of over 48 different isoforms of FGFR. Vascular endothelial growth factor (VEGF) is one of the main inducers of endothelial cell proliferation and permeability of blood vessels; The platelet-derived growth factors PDGF-A and -B have been recognized as important factors regulating cell proliferation, cellular differentiation, cell growth, development, and many diseases including cancer. Neurotrophins are critical to the functioning of the nervous system. Angiopoietins are protein growth factors required for the formation of blood vessels (angiogenesis). Others include epidermal growth factor (EGF) and hepatocyte growth factor (HGF).

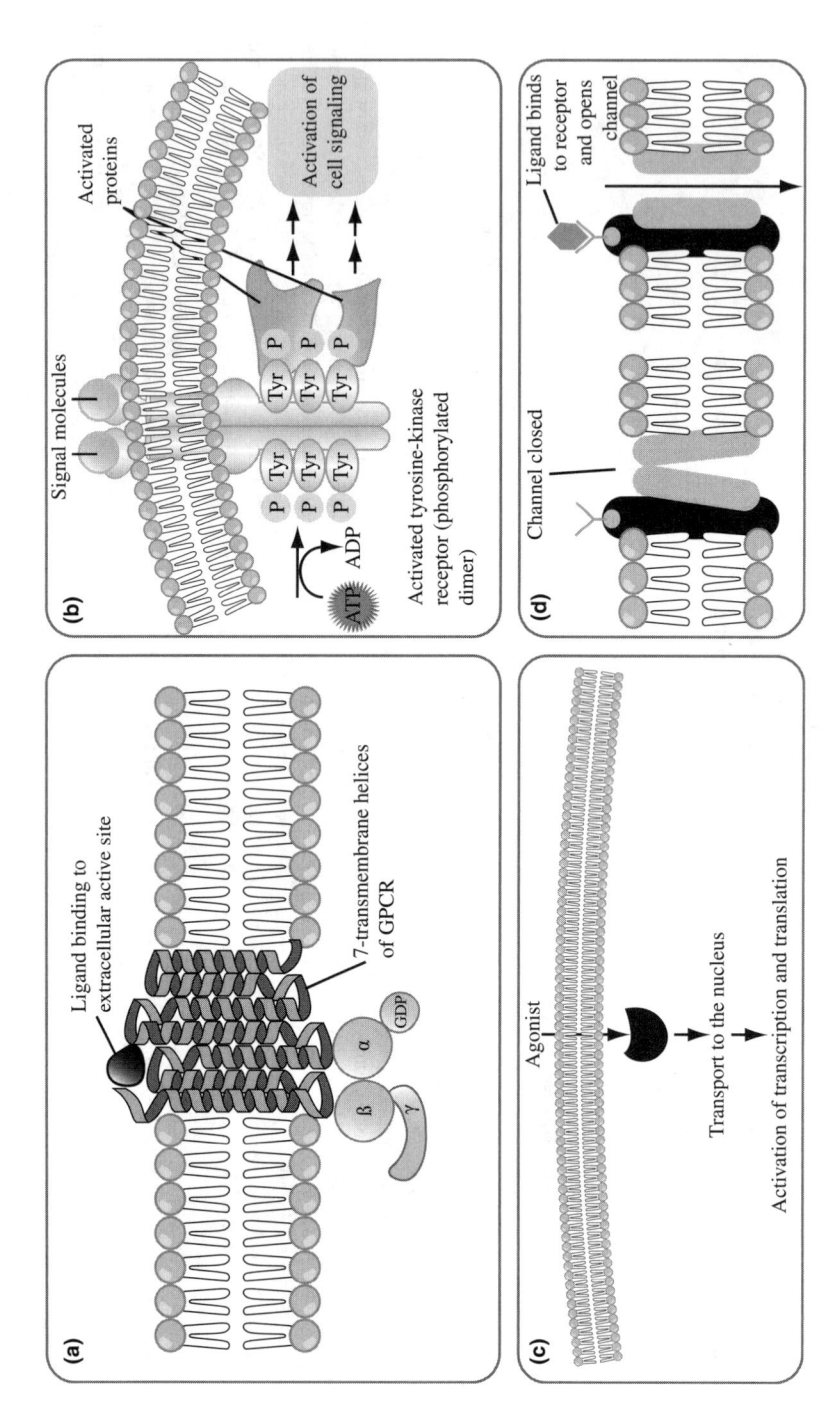

Figure 13.1. Target classes. (a) G-protein-coupled receptor (GPCR), (b) tyrosine kinase, (c) nuclear receptor, and (d) ion channel.

13.2 PHARMACODYNAMIC PRINCIPLES

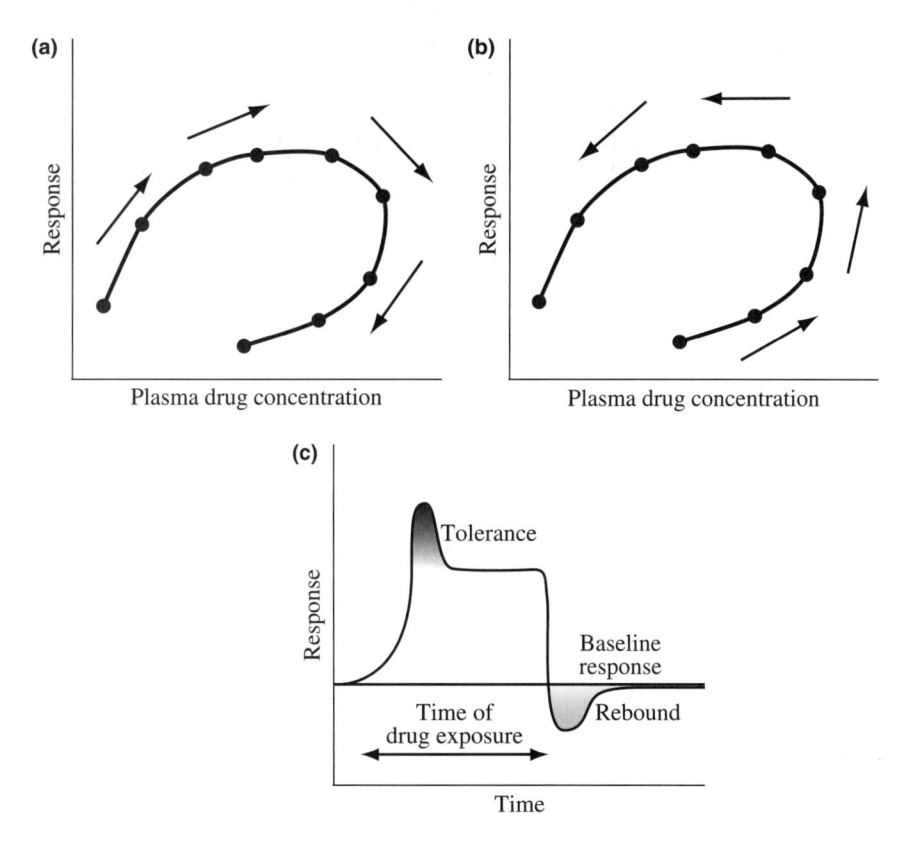

Figure 13.2. Functional adaptation processes for a stimulatory drug. (a) Loss of receptor-activated function on persistent exposure to agonist. Tolerance is characterized by a clockwise hysteresis. (b) Sensitization, an increase in drug effect despite a constant concentration at the target, is characterized by a counterclockwise hysteresis for a stimulatory drug. (c) Time course of response showing tolerance and rebound of a stimulatory drug. Arrows in (a) and (b) indicate direction of increasing time.

13.2.2.1 Desensitization, Tolerance, and Tachyphylaxis. Receptor-mediated responses to drugs often desensitize with time. This reduced response to an agonist, which is usually reversible upon cessation of treatment, is called **desensitization**. A second exposure to the agonist after a lapse usually restores response. Agonists that desensitize receptors due to conditioning mechanisms can trigger behavioral tolerance and addiction. Some examples of medicines that are considered to be addictive are tranquilizers and sedatives (sleeping pills). Many of these belong to a group of similar substances called benzodiazepines, which allosterically potentiate GABA type A receptors. Some pain killers and other medications can cause addiction after long-term use. Desensitization is often used to suggest a reduced response by any mechanism—receptor phosphorylation, uncoupling, antagonistic metabolites, or negative physiological feedback—that occurs over a relatively short period of time, while tolerance is reserved for reduced responsiveness developing over longer periods, usually because of receptor down-regulation. **Tachyphylaxis** is a rapidly developing desensitization or tolerance after just one or two administrations of the drug. Often this is used to describe decreased response to a drug that occurs soon after administration when active drug levels might imply that sufficient drug is available to induce a full response. Most receptors can undergo agonist-induced regulation. When an agonist occupies a receptor, it can result in a series of events that can lead to **internalization** of the receptor. The signals for internalization include phosphorylation of the receptor, which sets the stage for adapter molecules to bind to the receptor. The agonist is then removed and the receptor is targeted for cycling back to the cell membrane or is trafficked into a degradation pathway. Cycling of the receptor is associated with resensitization of the receptor, while degradation results in a loss of total receptor density. When receptor density is decreased, it is usually associated with a shift to the right in the dose–effect curve for agonists. An example of desensitization is the high **affinity**, desensitized, closed state of the nicotinic acetylcholine receptors (nAChRs), induced by chronic exposure to acetylcholine (ACh) or nicotinic drugs, leading to a gradual decrease in the rate of opening of K^+, Na^+, and sometimes Ca^{2+} cationic channels (milliseconds to minutes).

Sensitization has the opposite effect of tolerance, where an increase in drug effect is observed after repeated administration of certain drugs. When the body tries to return to homeostasis following a sudden discontinuation of a drug, the receptors, which are deprived of their agonists/blockers, become hypersensitive to an agent that targets it, causing a further exacerbation of the symptoms/conditions that triggered the use of the drug in the first place. This **rebound** effect can be minimized by a gradual rather than a sudden discontinuation of the drug. Several anxiolytics and hypnotics have a rebound effect. For example, benzodiazepine withdrawal can cause severe anxiety and insomnia, worse than the original insomnia or anxiety disorder. Other examples causing rebound effects include sedatives such as Lunesta and Ambien, the short-acting hypnotic, triazolam (due to its high potency and ultra-short half-life), stimulants such as methylphenidate or dextroamphetamine, antidepressants such as serotonin-selective reuptake inhibitors (SSRIs), and alpha-2 adrenergic agents such as clonidine and guanfacine. Rebound on drug withdrawal can be a factor in the chronic use of medications and drug dependence, with patients taking the medications only to ward off withdrawal or rebound withdrawal effects.

Time-dependent changes in drug action arising from tolerance, sensitization, and rebound cause a loss of consistency in the concentration–effect relationship with increasing number of doses, leading to the serious consequences of reduced pharmacological effect or safety with time, which can spell disaster for drugs with narrow therapeutic windows.

13.3 PHARMACODYNAMIC MODELING

13.3.1 Concentration–Effect, Dose–Response Curves, and Sigmoid E_{max} Models

The pharmacological effects of a drug in a species could be direct or indirect, delayed or rapid, and may or may not induce functional adaptations (Fig. 13.3). The dose levels and route of administration of the drug play an important role in the initiation and sustenance of these effects. The relation between drug dose and the clinically observed response is complicated by the time-dependent variation in disposition of the drug in plasma and in the target tissue. However, in carefully controlled *in vitro* systems, the relationship between drug concentration and its effect is often simple. The concentration–response (or effect) curves for

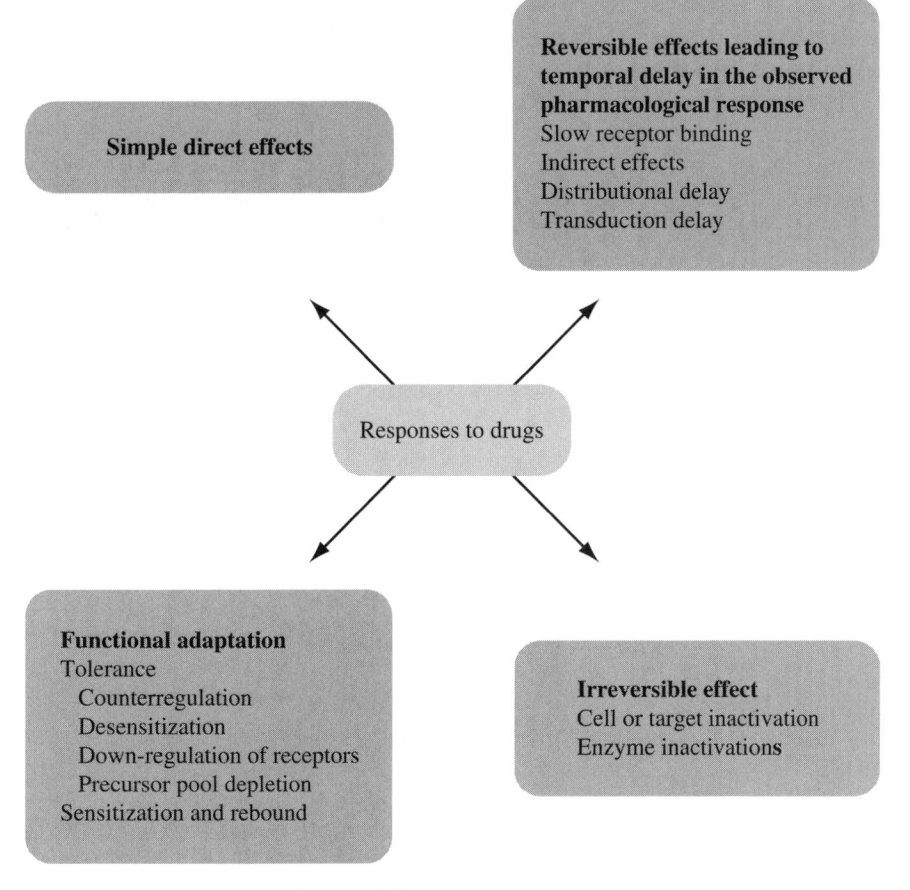

Figure 13.3. Responses to drugs.

an agonist, competitive antagonist, noncompetitive antagonist, and partial agonist are shown in Figures 13.4–13.7.

The pharmacology of drug–receptor/enzyme interaction is a saturable process. The efficacious or toxic **drug effect**, E, of a drug is described by the sigmoid E_{max} model, also called the Hill equation[3,4]:

Inhibitory drug:

$$\text{Effect}(E) = E_0 - \frac{E_{max} \times C^\gamma}{EC_{50}^\gamma + C^\gamma}$$

Stimulatory drug:

$$\text{Effect}(E) = \frac{E_{max} \times C^\gamma}{EC_{50}^\gamma + C^\gamma} \tag{13.1}$$

where E_{max} is the observed maximum drug effect (capacity), also called **efficacy** or **intrinsic activity**. E_{max} can be masked by a physiological maximum and, therefore, may not be observable *in vivo*. C is the unbound concentration of a drug (only the unbound concentration drives pharmacological response) at steady-state. The steady-state relation between drug concentration and its pharmacological effect may not always apply in patients. In that case, the effect and concentration are replaced by area under the effect vs. time curve (AUE) and area under C-time curve (AUC) respectively. EC_{50} is the unbound drug concentration at which the pharmacological effect is half the maximal effect (sensitivity). It is also referred to as **potency** and is a measure of how tightly the drug binds to the target and how sensitive the target is to the drug. E_0 is the **baseline value** of effect in the absence of the drug, and its inclusion is necessary for inhibitory but not for stimulatory effect. γ is a unitless parameter called the Hill factor, which determines the slope of the sigmoid concentration–effect curve. When γ is 1 and E_0 is 0, equation 13.1 is called the E_{max} model, and it is identical to the Michaelis–Menten equation for enzyme kinetics. A steep concentration–effect curve could result from cooperative interactions of several different actions of a drug or due to effector systems that

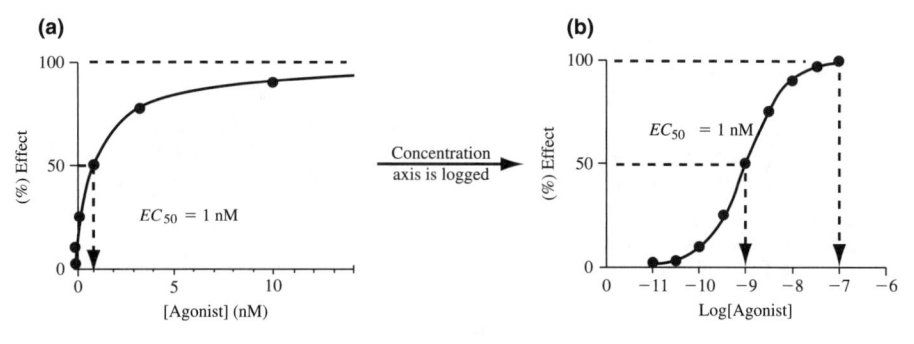

Figure 13.4. Concentration–response curve of an agonist in (a) linear and (b) log scale.

$$RC \underset{k_{on}}{\overset{k_{off}}{\rightleftharpoons}} C+R;\ R+A \underset{k_2}{\overset{k_1}{\rightleftharpoons}} RA$$

(a) Agonist A plus increasing concentrations of competitive antagonist C

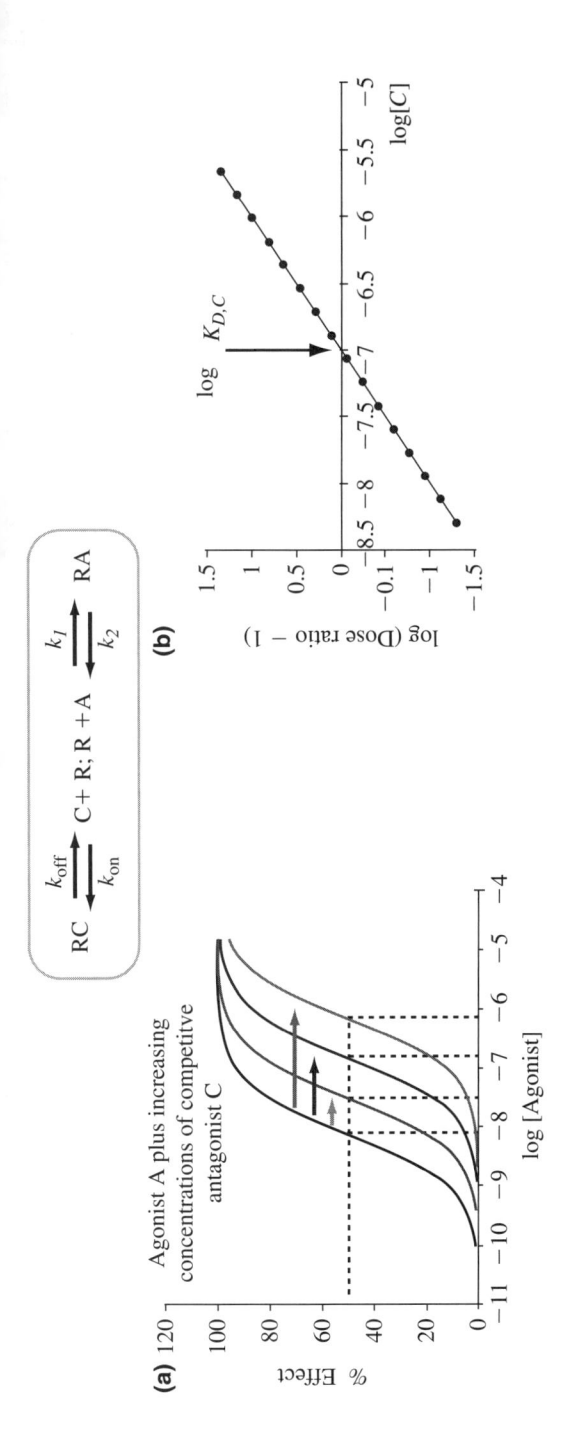

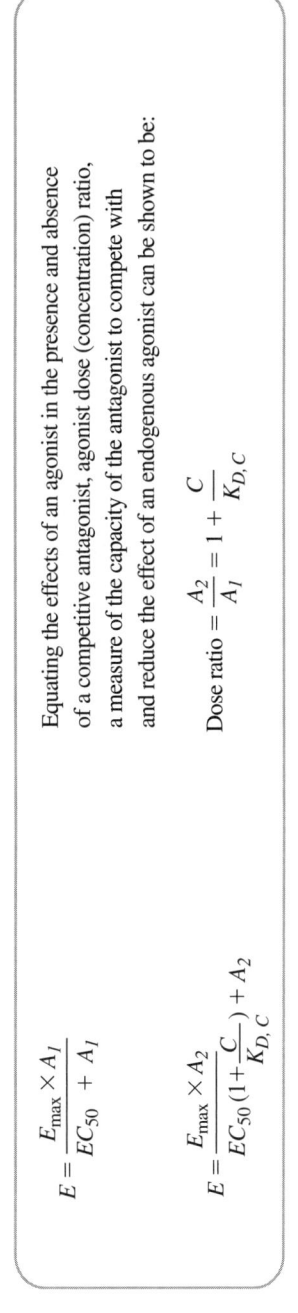

$$E = \frac{E_{max} \times A_1}{EC_{50} + A_1}$$

$$E = \frac{E_{max} \times A_2}{EC_{50}(1+\frac{C}{K_{D,C}}) + A_2}$$

(b)

Equating the effects of an agonist in the presence and absence of a competitive antagonist, agonist dose (concentration) ratio, a measure of the capacity of the antagonist to compete with and reduce the effect of an endogenous agonist can be shown to be:

$$\text{Dose ratio} = \frac{A_2}{A_1} = 1 + \frac{C}{K_{D,C}}$$

Figure 13.5. Dissociation constant of competitive antagonist $K_{D,C} = k_{off}/k_{on}$; A_1: Agonist concentration required to produce an effect E in the absence of the competitive antagonist; A_2: Agonist concentration required to produce the same effect E in the presence of the competitive antagonist. C: concentration of the competitive antagonist. EC_{50}: potency of agonist. (a) Shift in the concentration–response curve of an agonist with increasing concentrations of competitive antagonist. (b) Schild plot aids in extracting $K_{D,C}$. It is useful in the development of improved antagonists. The equation that forms the basis for the Schild plot is derived.

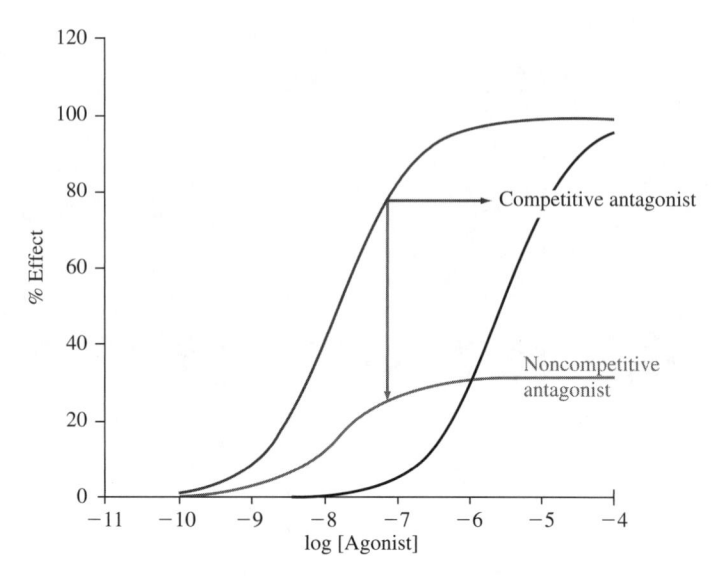

Figure 13.6. Concentration–response curve of an agonist and its shift by noncompetitive (unsurmountable) and competitive antagonists.

require a high occupancy before the response is observed. Due to nonlinearity (with respect to drug concentration) of most biosensor processes, a wide range of concentrations are needed to extract EC_{50} and E_{max}. However, experimental limitations often do not permit the characterization of the entire curve *in vivo* or *in vitro*. Therefore, simplified submodels (Fig. 13.8) to the E_{max} model are usually employed to characterize concentration–response. There are several. One is the linear model

$$E = E_0 + \text{slope} \times C \qquad (13.2)$$

which describes the concentration–effect curve at very low concentrations ($C \ll EC_{50}$). The slope approaches E_{max}/EC_{50}. As concentrations approach EC_{50}, the model needs to be modified to

$$E = E_0 + \text{slope} \times C^n \qquad (13.3)$$

in order to explain the much steeper effect profile with increasing concentrations. Log transformation of the right-hand side yields

$$E = n \times \ln(C + C_0) \qquad (13.4)$$

This log-linear model explains the concentration–effect curve between 20 and 80% of maximum effect, but the full sigmoidal profile is described by the E_{max} model or the Hill equation.

13.3 PHARMACODYNAMIC MODELING

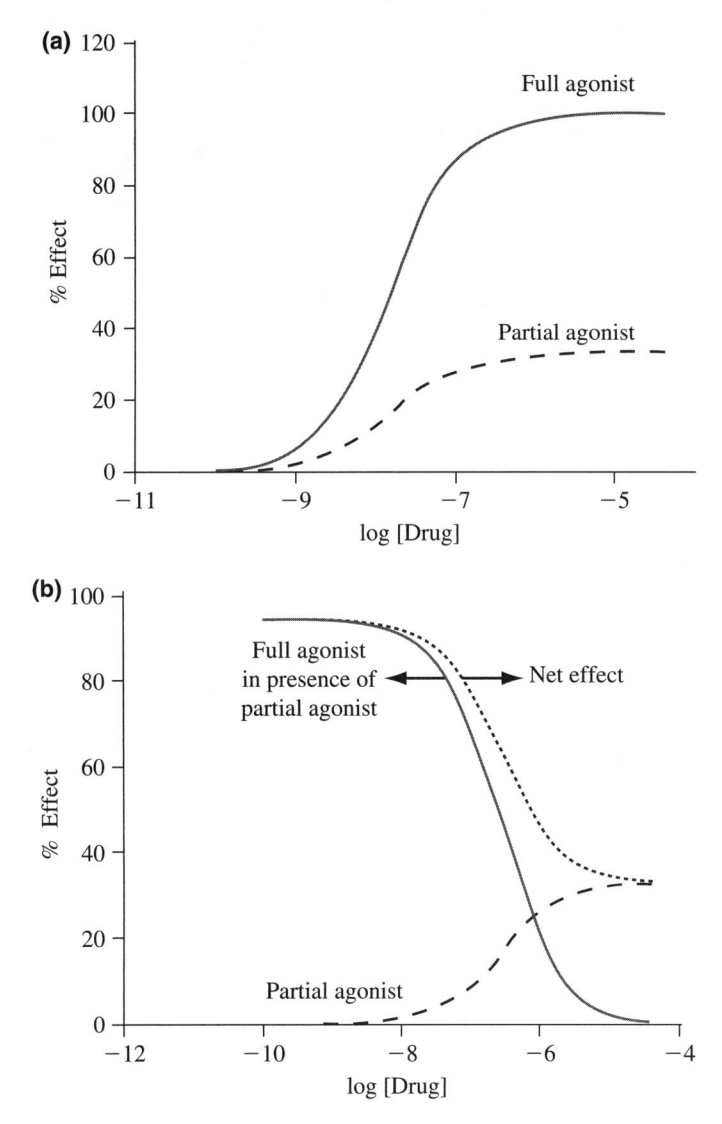

Figure 13.7. (a) Concentration–response curves of a partial agonist and a full agonist both of which have a similar affinity toward the receptor. (b) Partial agonists may actually act as functional antagonists when in competition with higher efficacy agonists.

The pharmacological effect of a drug or the response that it evokes in a species could be quantified using a pharmacodynamic **biomarker** (a PD endpoint or a pharmacological outcome), a characteristic that can objectively be measured and evaluated. The nature of biomarkers is dependent on whether they are meant for early screening assays such as binding or cell-based assays,

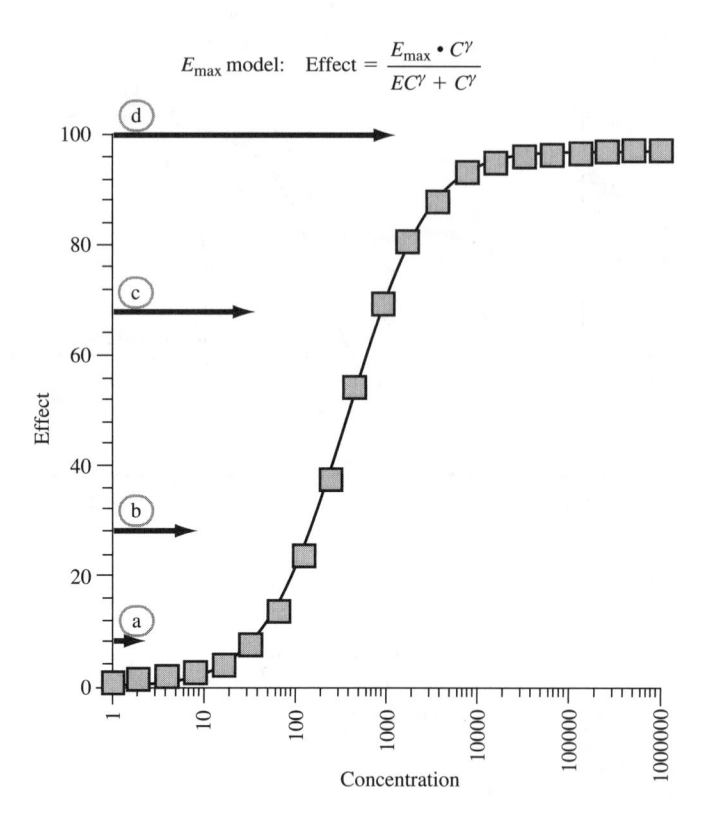

E_{max} model: Effect $= \dfrac{E_{max} \cdot C^\gamma}{EC^\gamma + C^\gamma}$

Figure 13.8. Submodels to the E_{max} model: (a) Linear, (b) power, (c) log-linear model, and (d) full E_{max} models.

or whether they are used later in the value chain in *in vivo/ex vivo* preclinical or clinical development. They can provide great predictive value if they reflect the mechanism of drug action (target site–drug exposure, drug–target interaction, target activation, signal transduction, and homeostatic feedback mechanisms in normal and disease populations) and if the biomarker levels needed to reach the desired pharmacological effect are known. A classification based on mechanism of drug action and drug–disease interaction has been proposed.[5] Type 0 biomarkers (or pharmacogenomic biomarkers) are measurable DNA/RNA characteristics that identify a PK- or PD-related genotype/phenotype of an individual, which determines drug response in that individual. For example, mutation of the epidermal growth factor receptor (EGFR) gene is reported to be associated with clinical responsiveness to gefitinib, a EGFR kinase inhibitor. Phosphorylated-EGFR for gefitinib assessed by immunohistochemistry assays is a good example of type 0 biomarker. Other examples include phosphorylated-CRKL for Gleevec and mRNA gene expression based biomarkers for anticancer drugs in cell-based assays. Type 1 to type 4 constitute PD biomarkers at the different levels of PD effect/response. Type 1 biomarker is a measure of drug

13.3 PHARMACODYNAMIC MODELING

exposure (plasma or more PD-relevant target tissue concentrations of a drug). Type 2 biomarkers reflect receptor occupancy and can be useful in cases where target occupancy correlates well with therapeutic response.[6] Type 3 biomarkers quantify target site activation, which is determined by the intrinsic efficacy of the drug and receptor density. Intrinsic efficacy of the drug determines the extent of occupancy needed for target activation, while receptor density determines the system maximum, which, if different between sites, will determine the selectivity of drug action. An example of type 3 biomarkers is the quantitative electroencephalogram (EEG) parameters reflecting the OP_3 opioid receptor activation for synthetic opioids.[7] Type 2 and/or type 3 biomarkers are an indicator of **target engagement**, target function, or pharmacological response. Type 4 biomarkers refer to physiological measures in the integral biological system. Disease biomarkers serving as functional endpoints of disease progression (such as tumor size in cancer) at a physiological level constitute the type 5 biomarkers. biomarkers need not be directly related to the clinical outcome. Those that are predictive of clinical outcome are considered as **surrogate biomarkers**. However, even generally accepted surrogate endpoints are unlikely to capture all of the therapeutic benefits and the potential adverse effects that a drug will have in a diverse patient population.[8] Accordingly, combinations of biomarkers will probably be needed in order to provide a more complete characterization of the spectrum of pharmacologic response. A **clinical endpoint** (type 6 biomarker) is the ultimate measure of efficacy that quantifies the direct benefit to a patient. However, the long periods of time needed to achieve it make it an impractical measure during the short-term clinical trials. All six types of biomarkers need not be available. Depending on the drug, some are more readily available.[9] Table 13.2 distinguishes biomarkers from surrogate biomarkers and clinical endpoints.

Data from genomics and proteomics, differentiating healthy from disease states, lead to biomarker discovery.[10] Most biomarkers are endogenous macromolecules that are measured in biological fluids such as whole blood plasma, serum, urine, saliva, buccal mucosa samples, sweat or tissues, tumor and the like. It is crucial to develop and evaluate (doing a retrospective analysis of how well a biomarker predicted the outcome) robust, sensitive, quantitative, and reproducible biomarkers.[11] Biomarkers provide the basis for lead candidate selection, contribute to mechanistic understanding, and identify the subtype of the disease for which an intervention is most suited.

Equation 13.1 suggests that high potency (low EC_{50}) and high efficacy (large E_{max}) are the most desired pharmacodynamic characteristics for a drug. However, a third desired characteristic for a drug, not evident from the equation, is a slow turnover of effect, which is possible for irreversible inhibitors and for drugs targeting receptors with slow transduction mechanisms. The Hill equation does not provide insight into the factors that determine the shape and location of the concentration–effect curve as it is an empirical rather than a mechanistic model. The limitations of the Hill equation are the assumption of equilibrium conditions and the inability to explicitly distinguish

TABLE 13.2. Biomarker, Surrogate Marker, and Clinical Endpoint

	Biomarker (Biological Marker)	Surrogate Biomarker	Clinical Endpoint
Definition	A characteristic that is objectively measured and evaluated as an indicator of normal biologic process or pharmacologic response.	A biomarker intended to substitute for a clinical endpoint and expected to predict the effect of a therapy. Selection of surrogate is based on epidemiologic, therapeutic, pathophysiologic, or other scientific evidence for predicting clinical endpoint.	A characteristic or variable that reflects how a patient feels or functions or how long a patient survives.
Value	Mainly in early efficacy and safety evaluation in in vitro studies, in vivo animal models, and early clinical trials to establish proof of concept. Need not be directly related to clinical outcome.	Biomarkers that are readily observed and easily quantified. Predicts clinical outcome. Clinical relevance of the surrogate is generally well validated.	Assess benefit (cure or reduced morbidity) of therapeutic intervention to the patient. Ultimate measure of efficacy but difficult to quantify.
Examples	Blood cholesterol concentrations for assessing risk of heart disease.	Pupil dilation for narcotics, biochemical tumor markers for anticancer drugs, exercise tolerance tests in chronic stable angina, for myocardial infarction. QT interval as a surrogate for torsades de Pointes. HbA1c for diabetes. Blood pressure, body weight for obesity; viral load of HIV, hepatitis C or B virus for assessing level of infection. CD4 cell counts.	Chest pain for a medication aimed at prevention of heart attack. Survival. Recurrence of cancer. Stroke. Occurrence of infections in HIV.

Biochemical and Clinical Biomarkers

Disease	Biochemical Biomarker	Clinical Biomarker
Asthma and chronic obstructive pulmonary disease	Leukotrienes, chemokines, and cytokines	Pulmonary function tests, exacerbations
Type I or II diabetes; diabetic retinopathy/nephropathy	Glucose, fructosamine, glycosylated albumin glycated hemoglobin (HbA1c) and cytokines	Retinal evaluation, nephropathy measures, and peripheral neuropathy assessments
Hypertension	Angiotensin I, angiotensin II, plasma renin, aldosterone, and ACE activity	Blood pressure and heart rate measures

between drug-specific receptor affinity and activation from system-specific parameters such as receptor density, transducer function, and the like. These limitations may restrict the ability of the equation for use in extrapolation and prediction. For example, the Hill equation parameters derived for a receptor in a particular tissue cannot be extended to the same receptor in another tissue with different receptor density or another stimulus–response relationship. A mechanism-based PD model resolving the drug and system-specific parameters is better suited for wider extrapolations. The study of mechanisms underlying receptor-mediated signaling also shed light on why some drug actions persist for different durations, how cytosolic signal transduction pathways explain **spare receptors**, why chemically similar drugs can exhibit remarkable selectivity, and pave the way for identification of new targets for drug development.

13.3.2 Mechanism-Based PD Modeling

Whenever the mechanism of a pharmacological response is known (even partially), a mechanism-oriented approach should be adopted to better understand the mechanism underlying pharmacological action and to enable extrapolation across species, routes of administration, dosing regimens, and different clinical situations. A wide variety of mechanistic PD models, both explanatory and predictive, are reported in the literature.[12] These aid the understanding of concentration–response relationship under steady-state and non-steady-state conditions. Mechanism-based models for each of the drug effects (Fig. 13.3) are examined below.

13.3.3 Simple Direct Effects

An observed response is expected to be proportional to the concentration of the drug in the target tissue (biophase). Under steady-state conditions, there is a rapid equilibrium between the plasma and tissue drug concentrations and their ratio remains constant. Thus, plasma concentrations can be a surrogate for target tissue concentration. When the observed pharmacological effect is determined by the target site concentration and a minimal delay is involved in establishing equilibrium between plasma and tissue drug concentrations, the time course of response follows that of the systemic concentration (Fig. 13.9a).[13] In this case, the classical receptor theory is applicable. More commonly, however, the response profile is delayed with respect to the concentrations, leading to an anticlockwise hysteresis in the concentration–response curve for a stimulatory drug (Fig. 13.9b), requiring models that accommodate the time delay.

13.3.3.1 Classical Receptor Theory for Simple Direct Effects. The classical receptor theory has been a powerful source of pharmaceutical innovation,

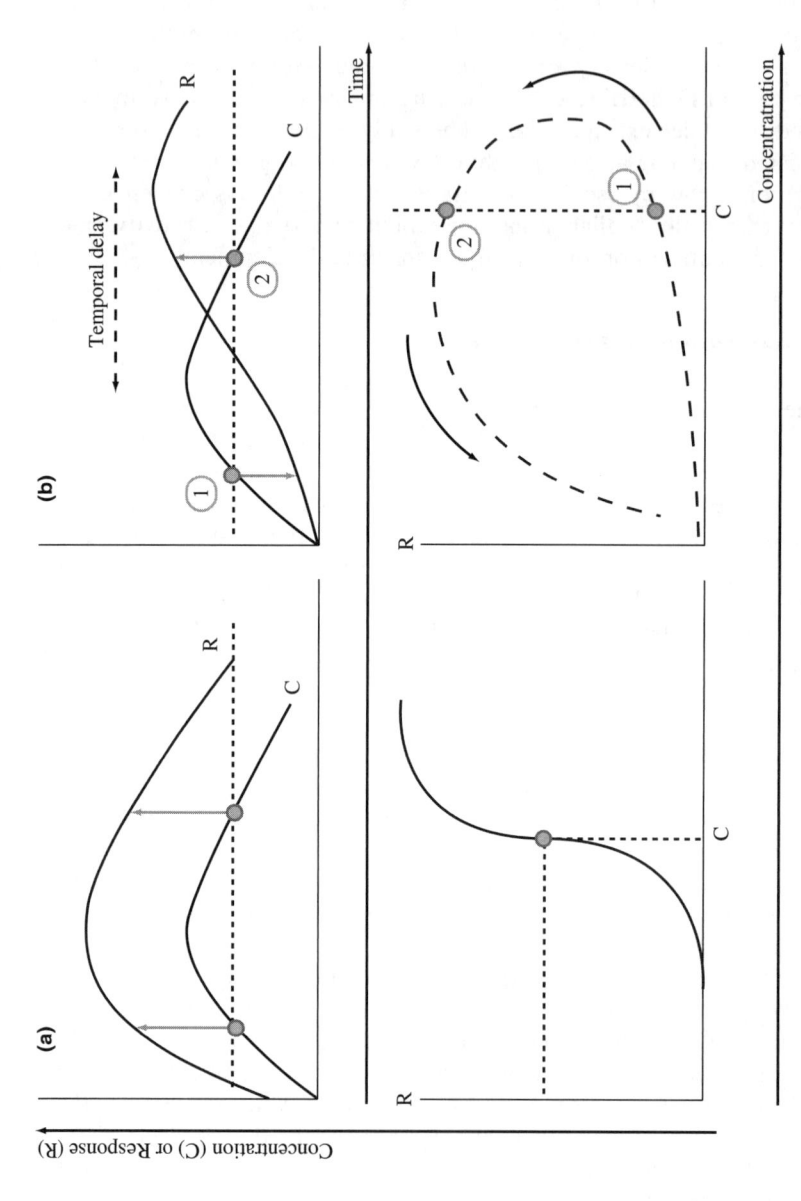

Figure 13.9. (a) Direct drug effects in which the time course of response and concentration follow each other and (b) indirect drug effects leading to temporal delay in the observed pharmacological response with respect to drug concentration. Indirect effect could be due to slow receptor binding, distributional delay, or transduction delay and is characterized by an anticlockwise hysteresis of the concentration effect curve for a stimulatory drug. [Reprinted from "Early integration of pharmacokinetic and dynamic reasoning is essential for optimal development of lead compounds: Strategic considerations." *Drug Discov Today* **14**:358–372. Copyright (2009). With permission from Elsevier.]

13.3 PHARMACODYNAMIC MODELING

and it aims to describe the pharmacological response to a drug in terms of drug-dependent properties (receptor affinity and intrinsic efficacy) and biological system-specific parameters (receptor density and a transducer function that relates the receptor acivation to pharmacological response). The receptor theory[14] was first proposed by John Newton Langley in 1878 as a by-product of his research on the physiology of the autonomic nervous system. An elaborate theory around the receptor idea[15] was further developed by Paul Ehrlich, a Nobel Prize winner in medicine or physiology for his work on immunity. Based on the receptor principles, the receptor occupancy model[16] was first proposed by Alfred Joseph Clark in 1933 to explain the activity of drugs at receptors and to quantify the relationship between drug concentration and the observed effect. It is based on mass action kinetics and attempts to link the action of a drug and the proportion of receptors occupied by that drug at equilibrium. A detailed historical perspective[17] is available in the literature.

The receptor theory combines the two independent drug-specific and system-specific components to describe drug action. The binding of a ligand, L (endogenous or drug), to a receptor (R) to form a ligand–receptor complex (LR) depends solely on the number of receptors, the concentration of ligand, C, and is given by

$$L + R \underset{k_{off}}{\overset{k_{on}}{\rightleftharpoons}} LR$$

where k_{on} is the number of binding events per unit time (second-order rate of association) and k_{off} is the number of dissociation events per unit time (first-order rate of dissociation).

Then at steady-state conditions, according to the law of mass action (which assumes equal accessibility of all receptors to ligand, reversible and complete binding of ligand to a receptor, and an unaltered ligand upon binding):

$$\frac{LR}{C \times R} = \frac{k_{on}}{k_{off}} = K_A = \frac{1}{K_D} \qquad (13.5)$$

where C, R, and LR are the effect site concentrations of ligand, receptor, and ligand–receptor complex, respectively. K_A is the affinity constant and is the inverse of the dissociation constant, K_D. Two drugs with the same affinity or K_D can still have very different response–time profiles because of differences in k_{off}. This is illustrated in Figure 13.10.[13] If k_{off}, the rate at which the drug dissociates from the receptor is very slow, it implies that the target residence time and therefore the duration of effect is high. In other words, the turnover of effect is low and the pharmacodynamic response half-life ($t_{1/2}$ response) is high. Thus, the *in vitro* measurement of dissociative half-life[18] is vital to the understanding of the system-specific response behavior of the target. The long pharmacodynamic half-lives of irreversible inhibitors (Fig. 13.11) are only limited by the degradation of the [LR] complex. Phenoxybenzamine is an irreversible,

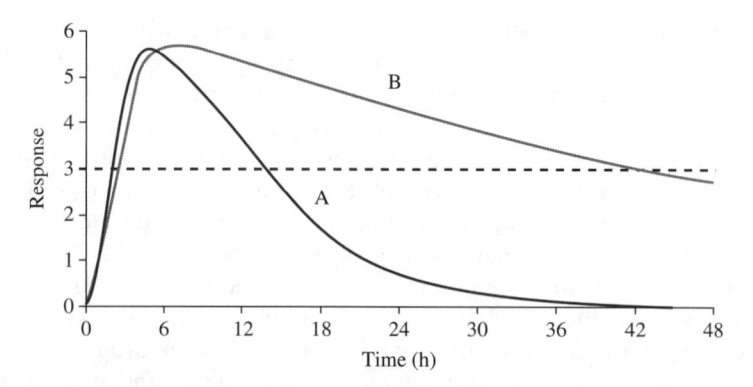

Figure 13.10. Simulated time course of response when K_D is the same, but curve A has 10-fold higher k_{off} and consequently shorter duration of effect compared to curve B. [Reprinted from "Early integration of pharmacokinetic and dynamic reasoning is essential for optimal development of lead compounds: Strategic considerations." *Drug Discov Today* **14**:358–372. Copyright (2009). With permission from Elsevier.]

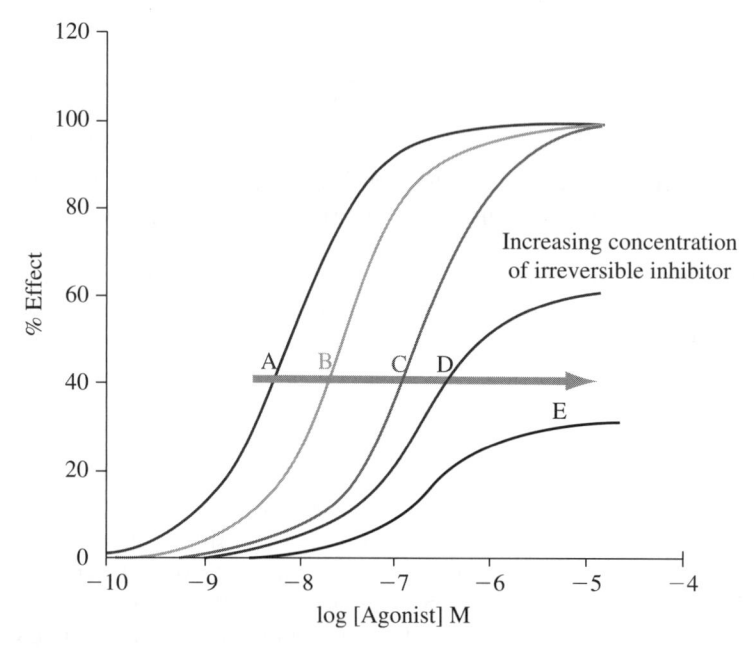

Figure 13.11. Concentration–response curve of an agonist in the presence of increasing concentrations of an irreversible inhibitor.

α-adrenoreceptor antagonist used in the control of hypertension caused by catecholamine release from tumors of the adrenal medulla (pheochromocytome). Control by phenoxybenzamine is maintained even during large bursts of catecholamine release, although in the event of an overdose phenoxybenzamine

13.3 PHARMACODYNAMIC MODELING

cannot be displaced from the receptor. The effects must be antagonized physiologically by using vasopressin, which targets the angiotensin II mechanism.

Covalent acetylation of platelet membrane cyclooxygenase by acetyl salicylic acid (ASA or aspirin) causing irreversible effects on platelet aggregation and gastric secretion that lasts for the lifetime of the platelet (7–10 days) is another example. Similarly, omeprazole produces an irreversible effect on the proton pump in pareital cells, causing an inhibition of gastric acid secretion that lasts for around 24 h although the PK half-life is less than 1 h. The advantage of a drug with a long response $t_{1/2}$ is that drug action can persist long after the compound has been eliminated from the system. Half-life response is relatively independent of the drug elimination rate described by the pharmacokinetic $t_{1/2}$.

Receptor binding, receptor activation, and target cell responses are usually proportional to the degree of receptor occupancy or fractional occupancy, f_{LR}, defined as the ratio of bound receptors to the total receptor density (bound + unbound), R_0. Then, substituting from equation 13.5 for LR,

$$f_{LR} = \frac{LR}{R + LR} = \frac{LR}{R_0} = \frac{C}{K_D + C} \qquad (13.6)$$

where R_0 is the total receptor density. Although the target cell response curve and target binding curve are expected to be similar, the ligand concentration at which half of the receptors are occupied by ligand (K_D) is often lower than the concentration required for eliciting a half-maximal biological response (EC_{50}). This is because the response may not be elicited until occupancy is very high (Fig. 13.12). Alternatively, the transduction process between receptor

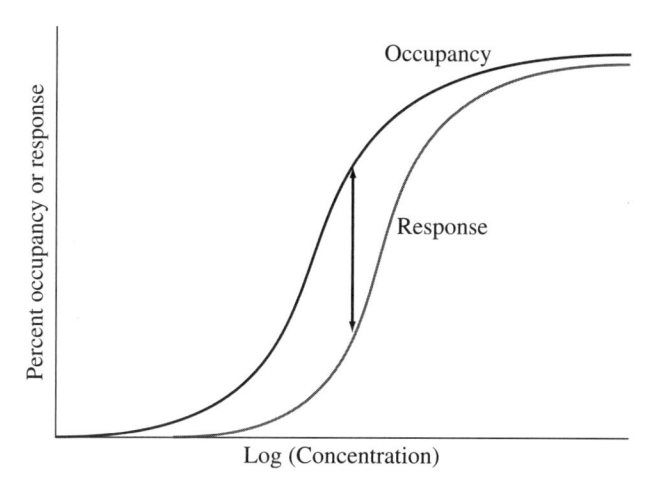

Figure 13.12. Receptor occupancy and drug response as a function of drug concentration. A high percentage of receptor occupancy need not result in a high percentage of response.

occupancy and drug response may not be efficiently coupled. Efficient coupling can result from a **receptor reserve** or existence of spare receptors. Spare receptors are said to exist when maximal response is elicited by an agonist at a concentration that does not produce full occupancy of the available receptors.

13.3.3.2 Semiparametric Receptor Model.
To explain the discrepancy between K_D and EC_{50}, Stephenson in 1956 introduced the concept of stimulus, defined as

$$S = \frac{eC}{K_D + C} \quad (13.7)$$

where e is a dimensionless proportionality factor denoting the power of a drug to produce a response in a tissue; e is related to the intrinsic efficacy of the drug ε and the receptor density R_0 by

$$\varepsilon = \frac{e}{R_0} \quad (13.8)$$

The response (the ratio of observed drug effect, E to the maximum observed effect, E_{max}) to the stimulus is then assumed to be a function of the stimulus, $f(S)$ and expressed in terms of either drug-specific (ε and drug affinity) (equation 13.9) or system-specific (receptor density) parameters:

$$\frac{E}{E_{max}} = f(S) = f\left(\frac{\varepsilon \times R_0 \times C}{K_D + C}\right) \quad (13.9)$$

Knowledge of how a stimulus varies with receptor occupancy and how the response is related to the stimulus are obtained by a simultaneous analysis of *in vivo* concentration–effect relationships of several ligands to a receptor. This approach is considered as semiparametric PD modeling, as it uses a parametric function to describe the receptor activation and a nonparametric function to describe transduction. The semiparametric approach has been successfully applied in the PK/PD analysis of GABA$_A$ receptor agonists.[19]

13.3.3.3 Full Parametric Receptor Model.
The full parametric approach defines a parametric expression for the transducer function that relates the receptor activation and effect. A good example is Black and Leff's operational model of agonism (OMA). The model allows for receptor reserve in biological systems in the hyperbolic parametric transducer function given by

$$\frac{E}{E_m} = \frac{LR^n}{K_E^n + LR^n} \quad (13.10)$$

13.3 PHARMACODYNAMIC MODELING

where E_m is system maximum and K_E is the stimulus that can bring about half-maximal response, also defined as the value midpoint location of the transducer function. Substituting for LR from equation 13.6, and defining τ as the transducer ratio R_0/K_E, which is a measure of coupling (of receptor activation into pharmacological effect) efficiency, the above equation becomes

$$\frac{E}{E_m} = \frac{\tau^n \times C^n}{(K_D + C)^n + \tau^n \times C^n} \tag{13.11}$$

Now comparing equation 13.11 with equation 13.1 for a stimulatory drug,

$$E_{\max} = \frac{E_m \times \tau^n}{\tau^n + 1} \tag{13.12}$$

and

$$EC_{50} = \frac{K_D}{(2 + \tau^n)^{1/n} - 1} \tag{13.13}$$

Thus the parametric approach provides a theoretical rationale for the mixed PD parameters $E_{\max}$ and EC_{50} in terms of drug-specific and system-specific parameters in equations 13.12 and 13.13. Successful applications of the OMA include PK/PD correlations of A1 adenosine receptor agonists,[7,20] μ opioid receptor agonists,[21,22] 5HT1A receptor agonists,[23] and QT interval prolongation.[24] Although nonparametric and semiparametric models have fewer assumptions, full parametric models are well established and can be of great value.

Receptor theory works when there is a rapid equilibrium between plasma and target drug concentrations, there is an instantaneous binding to the receptor, and the pharmacological response is linear with tespect to time and follows receptor occupation without delay. Consequently, peak effects occur simultaneously to peak drug concentrations in plasma. More realistic models should capture delays and nonlinearity with respect to time in the observed pharmacological responses.

13.3.4 Models Accommodating Delayed Pharmacological Response

An apparent time lag between plasma drug concentration and pharmacological response could distort the relationship between the two (Fig. 13.9). The delayed antipsychotic effect of neuroleptics with a $t_{1/2}$ response of a few weeks is an example. The concentration response curve is then an anticlockwise hysteresis (Fig. 13.9). Several mechanisms can explain a delayed pharmacological response.

13.3.4.1 Distributional Delay. Plasma concentrations are generally used as surrogate for target tissue exposure because drug disposition is assumed to be tissue nonspecific. However, nonparallel blood and tissue PK profiles are possible under non-steady-state conditions, when the equilibrium between the plasma and target concentrations needs constant readjustment. Depending on the distribution process, the tissue–drug concentrations can be lower than that in blood/plasma, leading to a pharmacological response that lags behind plasma–drug concentrations. Unique organ perfusions, differential tissue and plasma compositions, complex patterns of tissue binding, local metabolism at target site, and intracellular sequestration are some reasons why the surrogacy of plasma concentrations is not valid. Whether plasma concentrations can be used as a direct indicator of receptor occupancy and a predictor of PD response depends on the nature of the drug.

The apparent dissociation between drug concentration and effect was rationalized by Giorgio Segre[25] who in 1968 proposed the use of a hypothetical effect compartment to account for the time lag. Later, Sheiner et al.[26] further developed the effect compartment concept in their biophase distribution model and applied it to describe the PD of tubocurarine. The rate of change of effect site concentration following an IV bolus, for a one-compartment PK model can be defined as

$$\frac{dC_e}{dt} = k_{1e} \times C_p - k_{e0} \times C_e \qquad (13.14)$$

where C_e, the effect site concentration is linked to a PD model instead of plasma concentration (Fig. 13.13); k_{1e} is the first-order distribution rate constant; k_{e0}, is the first-order rate constant describing the disappearance of the drug from the effect compartment. In practice, however, since k_{1e} is nonidentifiable, it is simply considererd as equal to k_{e0}, which ensures a steady-state concentration of the drug in the tissue. Also k_{e0} is obtained through an iterative procedure that involves the collapsing of the concentration–response hystereris loop. Although such an approach could account for the delayed pharmacological response, it does not discriminate between different mechanistic steps that could lead to the delayed pharmacological response. When the delay is of the order of hours rather than minutes, the mechanism of delay is probably not tissue distribution. Other reasons for delay include slow receptor binding, indirect pharmacological response, and signal transduction mechanisms.

13.3.4.2 Slow Receptor Binding of the Drug. Drug binding to a target protein is generally assumed to lead to a rapid and reversible equilibrium. However, association and dissociation to some ion channels have been shown to be slow. If k_{on} is low (slow association), the receptor binding is said to be slow. If k_{off} is low (slow dissociation) enough for receptor binding to be irreversible, then response half-lives can be much higher compared to PK half-lives.

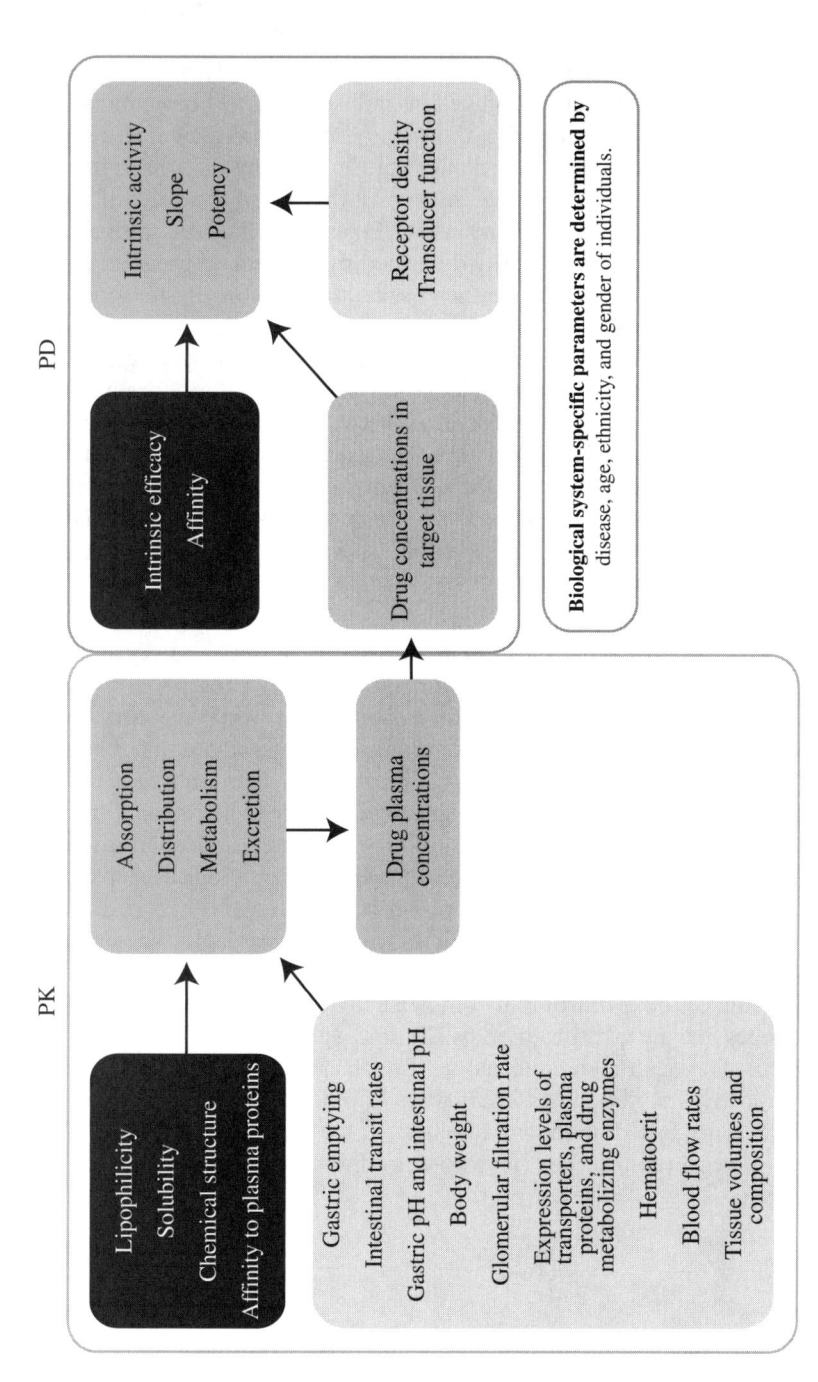

Figure 13.13. Drug-specific and biological-system-specific parameters that determine PK and PD.

In fact, the slow kinetics can be in any one of the multistep cascade of events between receptor binding and eliciting a pharmacological response (as in signal transduction). Cellular transduction mechanisms are generally nonlinear, with operating rate constants ranging from a few milliseconds to hours and sometimes even days. A time-independent transducer function is sufficient to describe a transduction mechanism that does not cause a delay in observing the pharmacological effect. However, when rate constants are of the order of hours, transduction mechanisms can lead to a higher PD response half-life compared to the PK half-life and can be empirically described by indirect response models.

In considering delays caused by either tissue distribution or slow receptor binding, the pharmacological response is still assumed to be directly proportional to target concentrations or receptor occupation respectively. Any intermediate steps between receptor occupation and the response are assumed to be fast enough to give a good correlation between effect site concentration (or receptor occupation) and response. However, if there is no reason to believe that the temporal dissociation between concentration and effect is caused by direct effects, mechanism-based indirect response models should be applied in which the observed pharmacological effect is secondary to one or more intermediary response steps.

13.3.4.3 Mechanism-Based Indirect Response Models. A delay in pharmacological response could result when a drug indirectly causes the response. Drugs might induce their effects not by direct interaction with receptors but by determining the fate of endogenous compounds that mediate the observed response. For example, many drugs act by stimulating or inhibiting a physiological process such as the synthesis or degradation (turnover) of an endogenous substance, which precedes the observation of response. Mechanism-based indirect response models aim to incorporate distinct physiological steps in an attempt to understand the link between plasma concentrations and pharmacological response. Nagashima et al.[27] first developed this approach to describe the anticoagulant effect of warfarin, where drug concentration directly impacted the synthesis rate of the prothrombin complex activity, which in turn caused the desired changes in the prothrombin with time (pharmacological response). Following this, Dayneka et al.[28] developed four basic indirect response models that systematically describe the different types of indirect response. The indirect response models are based on the premise that, in the absence of a drug, the rate of change of a measured response (R) whose production and loss are determined by an endogenous ligand is given by:

$$\frac{dR}{dt} = k_{in} - k_{out} \times R \quad (13.15)$$

where k_{in} represents the zero-order rate constant for the production of response and k_{out}, is the first-order rate constant for loss of the response. It is assumed that k_{in} and k_{out} fully account for the production and loss of the response.

13.3 PHARMACODYNAMIC MODELING

The pharmacological response R is distinguished from the drug effect (E) on the synthesis or degradation of endogenous ligands. By recognizing that the rate of change of response is zero at time zero, it is clear from equation 13.15 that the ratio k_{in}/k_{out} is equal to the initial or baseline value of response, R_0, which is usually known. A drug may inhibit or stimulate one or several physiological processes (Figs. 13.14–13.17) that ultimately lead to the modulation of k_{in} or k_{out} resulting in the following variations of equation 13.15:

$$\frac{dR}{dt} = k_{in} \times I(t) - k_{out} \times R \tag{13.16}$$

$$= k_{in} - k_{out} \times I(t) \times R \tag{13.17}$$

$$= k_{in} \times S(t) - k_{out} \times R \tag{13.18}$$

$$= k_{in} - k_{out} \times S(t) \times R \tag{13.19}$$

where $I(t)$ and $S(t)$ are the inhibitory and stimulatory modulations brought about by drug action and are given by

$$I(t) = 1 - \frac{C_p(t)}{C_p(t) + IC_{50}} \tag{13.20}$$

$$S(t) = 1 + \frac{E_{max} \times C_p(t)}{C_p(t) + EC_{50}} \tag{13.21}$$

Where $C_p(t)$ is the time-dependent plasma concentrations of the drug. Initial estimates of IC_{50} (or EC_{50}), k_{in}, and k_{out} are obtained from observed response-time and concentration-time data, by applying limiting conditions to the model equations 13.16–13.19 as shown in Table 13.3. Solution to a differential equation provides estimates of IC_{50} (or EC_{50}), k_{in}, and k_{out}, which give the best fit to the observed effect–time data. These estimates, apart from providing an excellent insight into the underlying PD mechanisms, are also useful for extrapolation to humans. Examples of applications of indirect response models are presented in Figures 13.14–13.17. Since many physiological functions are regulated by homeostatic mechanisms, inclusion of lower (in equations 13.16 and 13.19) and upper limits (in equations 13.17 and 13.18) to the responses provides a more realistic characterization of drug responses for turnover systems, which are maintained within a certain range.[29] Thus, equation 13.15 becomes

$$\frac{dR}{dt} = k_{in} - k_{out} \times R \left(1 - \frac{R_l}{R}\right) \tag{13.22}$$

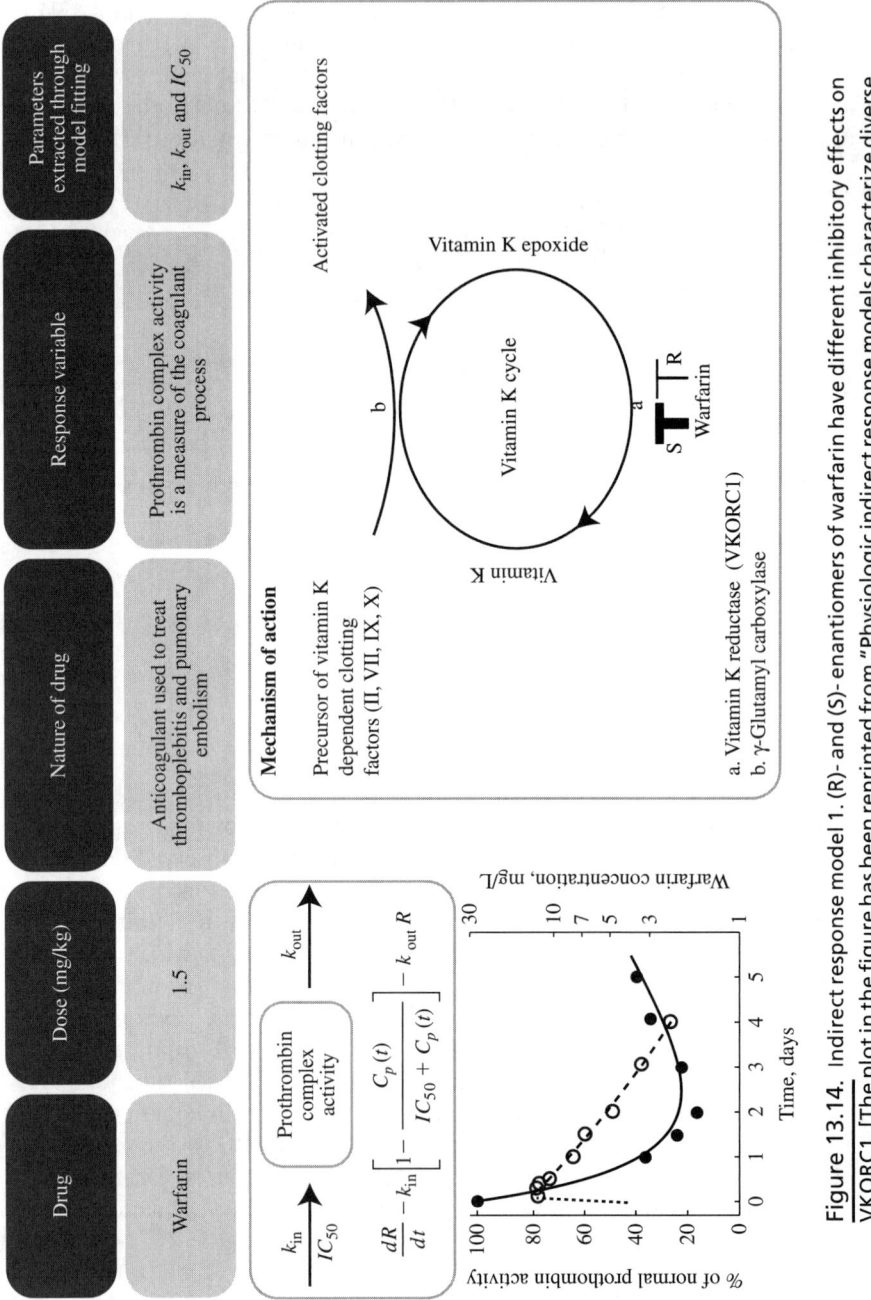

Figure 13.14. Indirect response model 1. (R)- and (S)- enantiomers of warfarin have different inhibitory effects on VKORC1. [The plot in the figure has been reprinted from "Physiologic indirect response models characterize diverse types of pharmacodynamic effects." *Clin Pharmacol Ther* **56**:406–419. Copyright (1994). With permission from Nature Publishing Group.]

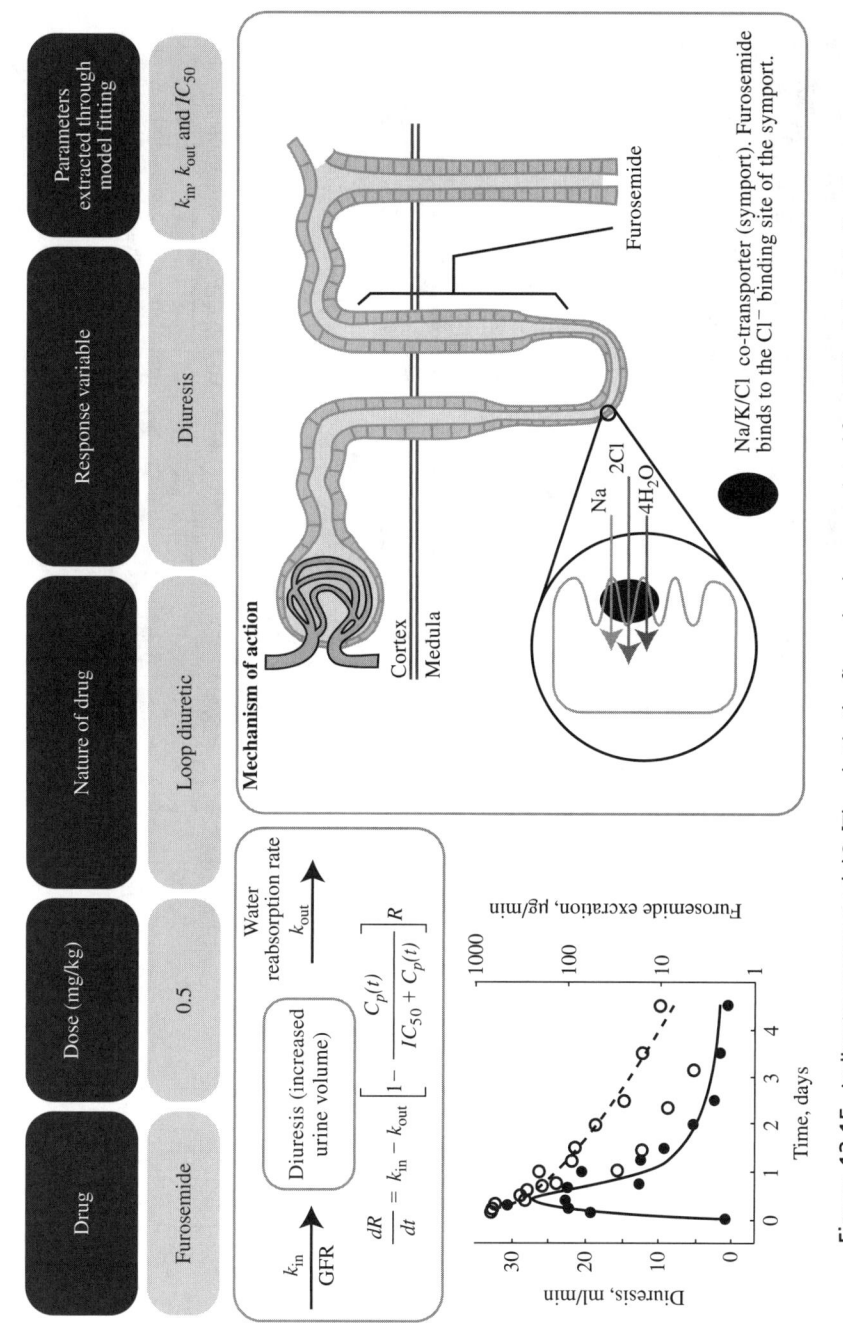

Figure 13.15. Indirect response model 2. [The plot in the figure has been reprinted from "Physiologic indirect response models characterize diverse types of pharmacodynamic effects." *Clin Pharmacol Ther* **56**:406–419. Copyright (1994). With permission from Nature Publishing Group.]

327

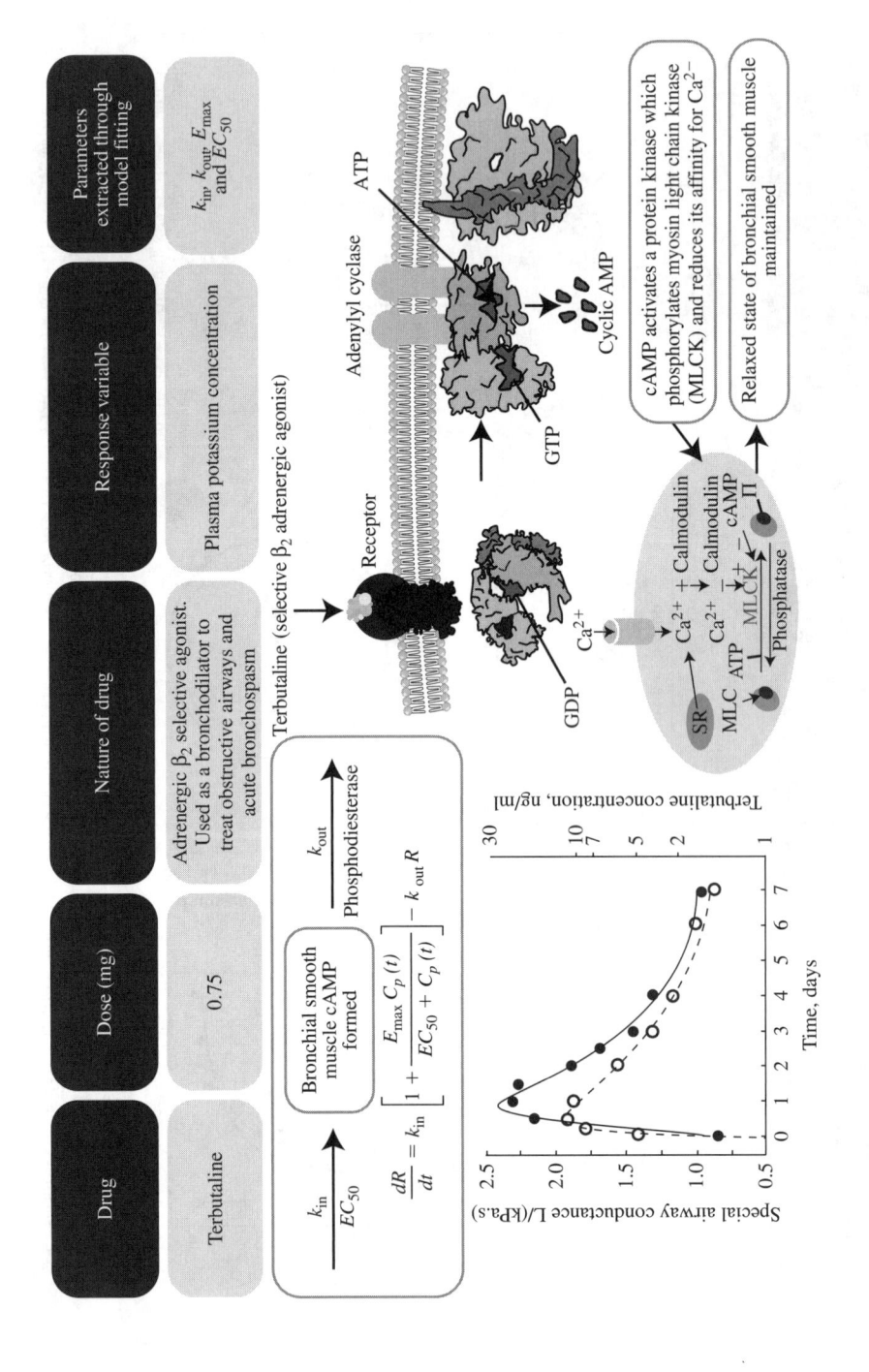

Drug	Dose (mg)	Nature of drug	Response variable	Parameters extracted through model fitting
Terbutaline	0.75	Adrenergic β₂ selective agonist. Used as a bronchodilator to treat obstructive airways and acute bronchospasm	Plasma potassium concentration	k_{in}, k_{out}, E_{max} and EC_{50}

$$\frac{dR}{dt} = k_{in}\left[1 + \frac{E_{max} C_p(t)}{EC_{50} + C_p(t)}\right] - k_{out} R$$

Figure 13.16. Indirect response model 3. The activation of the β_2-adrenergic receptor by terbutaline causes the G-protein-linked receptor to bind to the intracellular G protein, which results in the G protein expelling its guanine diphosphate (GDP) molecule and replacing it with guanine triphosphate (GTP). The α subunit, with its bound GTP, then moves along the membrane until it finds the enzyme adenylyl cyclase and activates it. The activated adenylyl cyclase then produces the second messenger, cyclic adenine monophosphate (cAMP) from ATP (adenine triphosphate), converting the GTP back to the resting state of GDP. SR is the signal recognition particle receptor. One major advantage of this approach is that it allows the signal to be amplified. In the signaling chain shown here, a single molecule of adrenaline can stimulate the production of many molecules of cAMP. By incorporating an enzyme (like adenylyl cyclase) into the chain, a weak signal from outside the cell can be translated into a strong signal throughout the inside of the cell. [The plot in the figure has been reprinted from "Physiologic indirect response models characterize diverse types of pharmacodynamic effects." *Clin Pharmacol Ther* **56**:406–419. Copyright (1994). With permission from Nature Publishing Group.]

◄────────────────

where R_l is the lowest possible response or

$$\frac{dR}{dt} = k_{\text{in}} \times \left(1 - \frac{R}{R_h}\right) - k_{\text{out}} \times R \qquad (13.23)$$

where R_h is the highest response allowed, depending on whether the physiological limit applies for the loss or production of response. Accordingly, when the rate of change of response (left-hand side of equations 13.22 and 13.23) is zero (when time = 0), the baseline response R_0 is given by either

$$R_0 = \frac{k_{\text{in}}}{k_{\text{out}}} + R_l \qquad (13.24)$$

or

$$R_0 = \frac{k_{\text{in}}}{k_{\text{out}} + \frac{k_{\text{in}}}{R_h}} \qquad (13.25)$$

The rationale for equations 13.22 and 13.23 can be understood as follows. Both an inhibition of production of response (equation 13.16) and a stimulation of loss of response (equation 13.19) by a drug lead to a decrease in response with time. When the response decreases to the lowest level R_l, there is no further reduction in response.

Mathematically, this condition is realized by multiplying the second terms on the right-hand side of equations 13.16 and 13.19 by $1 - R_l/R$. Similarly, both

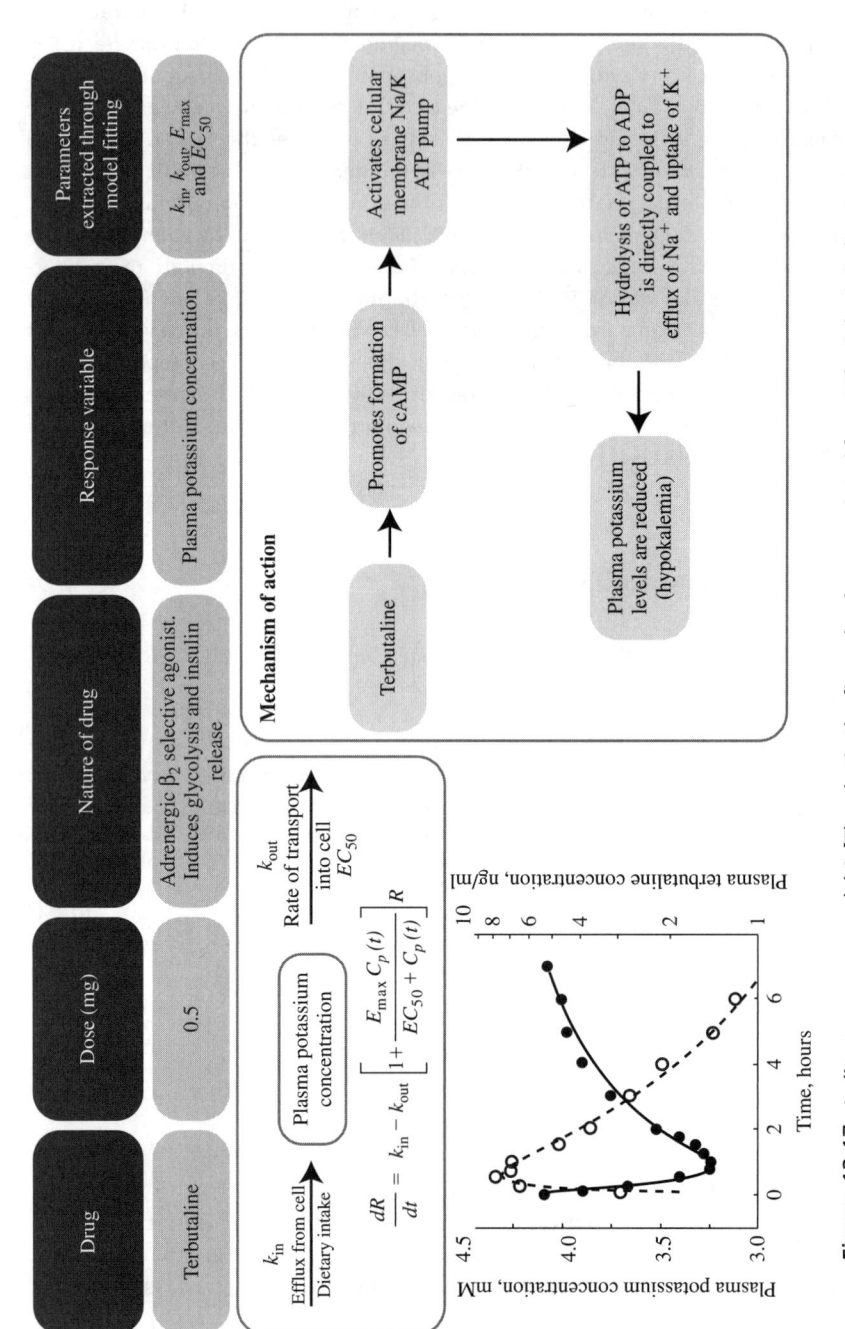

Figure 13.17. Indirect response model 4. [The plot in the figure has been reprinted from "Physiologic indirect response models characterize diverse types of pharmacodynamic effects." *Clin Pharmacol Ther* **56**:406–419. Copyright (1994). With permission from Nature Publishing Group.]

TABLE 13.3. Derivation of Initial Estimates of IC_{50} (or EC_{50}), k_{in}, and k_{out} for the Four Indirect Response Models

Limiting Condition	Model 1	Model 2	Model 3	Model 4
(1) $C_p \gg IC_{50}$ (or EC_{50}) (soon after IV bolus)	$\dfrac{dR}{dt} = k_{in}\left[1 - \dfrac{C_p(t)}{IC_{50} + C_p(t)}\right] - k_{out} R$ $\dfrac{dR}{dt} = -k_{out} R_0$ $\dfrac{d\ln R}{dt} = -k_{out}$ k_{out} is obtained from initial slope in $\ln R$ vs. time plot	$\dfrac{dR}{dt} = k_{in} - k_{out}\left[1 - \dfrac{C_p(t)}{IC_{50} + C_p(t)}\right] R$ $\dfrac{dR}{dt} = k_{in}$ k_{in} is obtained from initial > slope in R vs. time plot	$\dfrac{dR}{dt} = k_{in}\left[1 + \dfrac{E_{max} C_p(t)}{EC_{50} + C_p(t)}\right] - k_{out} R_0$ $\dfrac{dR}{dt} = k_{in}(1 + E_{max}) - k_{out} R_0$ $\dfrac{dR}{dt} = k_{in} E_{max}$ k_{in} is obtained from initial slope in R vs. time plot	$\dfrac{dR}{dt} = k_{in} - k_{out}\left[1 + \dfrac{E_{max} C_p(t)}{EC_{50} + C_p(t)}\right] R$ $\dfrac{dR}{dt} = k_{in} - k_{out}(1 + E_{max}) R_0$ $\dfrac{dR}{dt} = -k_{out} R_0 E_{max}$ k_{out} is obtained from initial slope in R vs. time plot
(2) When $R = R_{max}$, $t = t_{max}$ and $dR/dt = 0$	$k_{in}\left[1 - \dfrac{C_p(t_{max})}{IC_{50} + C_p(t_{max})}\right] = k_{out} R_{max}$ which rearranges to $\dfrac{IC_{50}}{IC_{50} + C_p(t_{max})} = \dfrac{R_{max}}{R_0}$ after substituting for k_{in}	$k_{in} = \left[1 - \dfrac{C_p(t_{max})}{IC_{50} + C_p(t_{max})}\right] k_{out} R_{max}$ which rearranges to $\dfrac{IC_{50}}{IC_{50} + C_p(t_{max})} = \dfrac{R_0}{R_{max}}$ after substituting for k_{in}	$k_{in}\left[1 + \dfrac{E_{max} C_p(t_{max})}{EC_{50} + C_p(t_{max})}\right] = k_{out} R_{max}$ which rearranges to $1 + \dfrac{E_{max} C_p(t_{max})}{EC_{50} + C_p(t_{max})} = \dfrac{R_{max}}{R_0}$ after substituting for k_{in}	$k_{in} = \left[1 + \dfrac{E_{max} C_p(t_{max})}{EC_{50} + C_p(t_{max})}\right] k_{out} R_{max}$ which rearranges to $1 + \dfrac{E_{max} C_p(t_{max})}{EC_{50} + C_p(t_{max})} = \dfrac{R_0}{R_{max}}$ after substituting for k_{in}

a stimulation of production of response (equation 13.18) and an inhibition of loss of response (equation 13.17) by a drug result in an increased response compared to the baseline value, and, when the response increases to R_h, homeostasis prevents any further increase. This condition is taken into consideration by multiplying the first terms of equations 13.17 and 13.18 by $1 - R/R_h$. Further modifications to the basic models (equations 13.16–13.19) to accommodate more complex mechanisms of drug action such as time-dependent transduction leading to delays in observed response,[30] partial inhibition/stimulation or joint effects should also be considered, depending on the need.[31] Although a simple biophase distribution model could explain any temporal dissociation of plasma concentration and effect, a blind application of that approach should be avoided[32] in recognition of the fact that several mechanisms apart from distributional delay can lead to the temporal dissociation. A delayed response leading to an anticlockwise hysteresis in the response versus agonist concentration curve (Fig. 13.9) can be caused by a single or multiple concurrent mechanisms, detailed above and summarized in Figure 13.18. The inclusion of all known mechanisms in PD modeling whenever possible allows for wider applicability and enhanced functionality of a model. The use of functionally appropriate mechanistic models permits extrapolation across dose, species, and individuals. The effects of disease, gender, age, and ethnicity on the physiological mechanisms leading to the observation of the response can also be taken into account.

13.3.5 Models Accommodating Nonlinearity in Pharmacological Response with Respect to Time

Functional adaptation processes such as tolerance, sensitization, and rebound lead to time-modulated pharmacological responses in which an increase or decrease in the sensitivity to a drug-initiated stimulus is observed on continuous drug exposure, even though the effect site drug concentrations remain unchanged. These time-dependent changes can be in EC_{50} (drug affinity) and/or in E_{max} (down or up-regulation of target expression levels). Drug tolerance is a frequently observed phenomenon. Although current understanding of its mechanisms is poor, mechanistic models of tolerance[31] have been proposed based on counterregulation, desensitization, up- or down regulation of target expression levels and precursor pool depletion.

13.3.5.1 Counterregulation. A counterregulation mechanism relies on a body response (M) that counteracts the drug-induced pharmacological response (R), such that the rate of change in M increases as R increases but decreases as M itself increases (negative feedback). This is represented by the following differential equation:

$$\frac{dM}{dt} = k_1 \times (R - k_2 \times M) \tag{13.26}$$

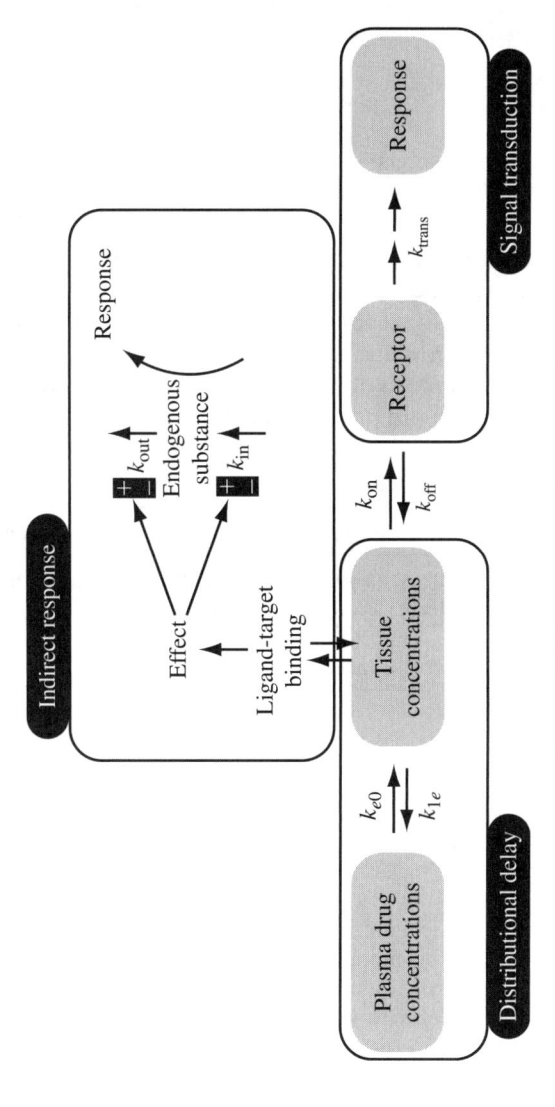

Figure 13.18. Summary of PK/PD models.

where k_1 and k_2 are the first-order rate constants for the production and loss of the counterresponse, M. The net response will reflect the difference $(R - M)$.

13.3.5.2 Desensitization. Receptors can get desensitized due to internalization leading to an apparent decrease in drug affiity with time. One of the ways in which this can be considered is to allow for a temporary loss of drug receptors into an inactive pool (R_i):

$$\frac{dR_i}{dt} = k_d \times (R_t - R_i) \qquad (13.27)$$

where k_d is a first-order rate constant and R_t is the receptor pool at any time t. Altered receptor density (up- or down-regulation of receptors) can be handled by an indirect response model.[33] Indirect response models can also be applied for precursor-depletion-driven tolerance. If a drug action involves the stimulation of response through the liberation of an endogenous precursor that is required for the production of response, then a continuous exposure to the drug can lead to a kind of tolerance due to depletion of the precursor pool. Further examples of time-variant models for tolerance driven by modulation of drug effect or drug concentration are available in the literature (Table 13.4).

TABLE 13.4. Examples of Tolerance Models

Tolerance Model	Tolerance Mechanism	Reference
Hypokalemia resulting from the β$_2$-agonist terbutaline	Increase in EC_{50} over time due to decreased receptor numbers	Jonkers et al.[34]
Supression of endogenous cortisol by the corticosteroid triamcinolone acetonide during prolonged therapy	Decrease in E_{max} over time	Meibohm et al.[35]
Progressive attenuation of the chronotropic effect of cocaine	Modulation of effect system	Ambre[36] and Chow et al.[37]
Cardioaccelerating effect of nicotine	Effect modulated through a hypothetical noncompetitive antagonist whose formation is driven by the concentration of the agonist (nicotine).	Porchet et al.[38]
Hemodynamic tolerance toward nitroglycerin	Driven by the effect of the drug, where the counterregulatory vasoconstrictive effect is determined by the extent of nitroglycerin-induced vasodilation	Bauer and Fung[39]
Diuretic and natriuretic effect of furosemide	Counterregulation	Wakelkamp et al.[40]

13.4 PHARMACOKINETIC MODELING: COMPARTMENTAL PK AND PBPK

The time course of drug concentrations in plasma is traditionally described by empirical compartmental models (see Chapter 2 for a description), which characterize the drug transfer between interconnected hypothetical compartments attempting to mimic the absorption, distribution, and elimination processes of a drug. Empirical models are case specific, and, although extremely useful for descriptive purposes, the potential for credible extrapolations across species, pathophysiological conditions within a species or route of administration is limited. On the other hand, in mechanistic models such as PBPK, the compartments have a physiological meaning. PBPK models are prespecified to capture the underlying mechanisms of the system they represent to the extent known, with parameters that correspond to some physical entities of the system. These models can, therefore, be used to predict the next set of data and are ideal for extrapolations across species, routes of administration, and different subpopulations.

13.5 INTEGRATION OF PK OR PBPK WITH PD MODELING

The PD models described in Section 13.3 relate the observed pharmacological effect to effect site drug concentration (exposure). If static measures of concentration such as C_{max} or area under the plasma concentration–time curve (AUC) are used as exposure variable in PD models, it amounts to discarding information about the time course of concentration, which is described by PK models. In isolation, PK or PD modeling adds little value. PK/PD modeling aims to integrate the changes in concentration over time following administration of a drug, as assessed by PK to the relationship between the concentration at the effect site and the intensity of the observed response as quantified by pharmacodynamics (Fig. 13.19),[41] resulting in a time course of response. A drug can be considered a perturbation of the normal homeostasis of the body and a PK/PD model should reveal the pharmacological properties of the drug and the rate-limiting steps such as turnover, transduction, or tolerance in the biological system.

Both the PK and PD of a drug are determined by a number of drug-specific parameters that are often identical between species and biological system-specific parameters that vary between species, individuals (with demographic and pathophysiologic variabilities), and experimental conditions (Fig. 13.13). While *in vivo* experiments provide the best estimates of system-specific parameters, *in vitro* assays are sufficient to determine drug-dependent properties. In the traditional PK/PD modeling, empirical models such as the compartmental PK model and the Hill equation are used to describe the pharmacokinetics and the typical hyperbolic shape of the *in vivo* drug concentration–effect

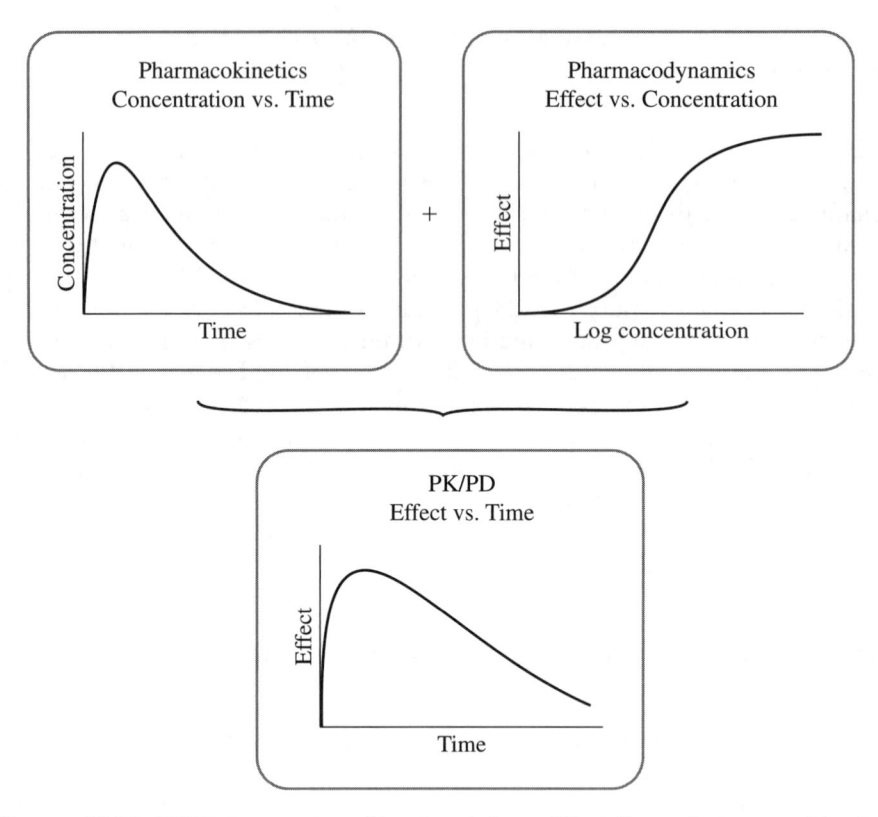

Figure 13.19. PK/PD integration. [Reprinted from "Modeling of pharmacokinetic/pharmacodynamic (PK/PD) relationships: Concepts and perspectives." *Pharm Res* **16**:176–185. Copyright (1999). With permission from Springer.]

relationships. These models, as stated earlier, have limited potential for extrapolative predictions as they do not distinguish between drug- and system-specific parameters. Mechanistic PD models and PBPK models on the other hand intrinsically differentiate system-specific and drug-specific parameters and have the potential to integrate drug mass transfer, physiological processes, and mechanisms of drug action. A combination of mechanism-based PD modeling and PBPK modeling (PBPK/PD) offers great flexibility, maximizing the scope for extrapolation and prediction as well as allowing for a mechanistic understanding of PK and PD through hypothesis testing. PBPK/PD integration is achievable in two ways. The differential equations of a whole-body PBPK model (Chapters 4–7) are solved simultaneously along with the appropriate differential equations in Section 13.3, depending on the PD mechanism, to get a response versus time, which is then integrated with the concentration–time profile of the target organ to get the response versus concentration. A simultaneous solving of the differential equations of PBPK and PD models as opposed to sequential PK/PD modeling is essential when there is a bidirectional dependence of PK and PD as, for example, in target-mediated drug disposition explained in Chapter 10. Another approach to PBPK/PD integration is combining the time course of tissue concentrations obtained from PBPK with a PD model.

PBPK/PD integration allows for the use of tissue concentrations rather than plasma concentrations. Plasma concentrations are not relevant for targets located in poorly vasculatured tumors. Also, pharmacological targets of drug action are often located within tissues rather than exposed directly to plasma, thus making target tissue concentrations much more relevant for PK/PD integration. Although the pharmacologically relevant unbound drug concentrations are expected to be the same in plasma and tissues for small molecules, this is not true for targets behind transporter barriers in central nervous system (CNS), liver, and kidney. The increase or decrease in target site concentrations with respect to plasma concentrations in these organs are determined by the expression and functionality of specific transporters. PBPK models are thus uniquely suited to predict target exposure of CNS drugs such as serotonin reuptake inhibitors and semisynthetic opioids.[42–45] For drugs that exert pharmacological action on receptors by allosteric modification of receptor configuration from within cells or act on targets within the cells in a tissue, unbound intracellular concentrations are the most relevant. Statins are HMG-CoA reductase inhibitors for which the pharmacologically relevant concentration is the unbound hepatic intracellular drug concentrations. Many statins, such as atorvastatin, rely on uptake transporters to reach the target enzyme and tend to get accumulated within the hepatocytes, leading to a higher than expected effect. Tissues are not homogenous and the concentrations in the pharmacologically relevant compartments within a tissue can be different from the unbound tissue concentration either because of transporters or because of differences in pH. Physiological models are also ideally suited for pre-systemic targets located in the lung, gut etc, for which plasma concentrations are not relevant. The tissue concentration of large biological drug molecules such as proteins and antibodies are also expected to be very different from their plasma concentrations, being limited by membrane permeability. The maternal and fetal plasma concentrations of tretinoin (active ingredient of a teratogenic topical antiwrinkle cream) and its metabolites estimated by a PBPK model[46] formed the basis for FDA approval of the cream. The target organ could be the site of pharmacological action or a site where toxic effects are expected. For example, the heart for QT prolongation. By specifying different organs as individual compartments and solving the differential equations for each of these organs in a PBPK model, the time profile of tissue concentrations can be obtained. Provided the estimates of tissue partition coefficients are derived from reliable sources, target tissue concentrations can be combined with PD data, eliminating the use of distributional delay models, unless it is mechanistically appropriate. Hypothesis testing with PBPK can also identify any distributional delay, in which case a permeability-limited tissue distribution model can be used in the place of the generally default perfusion-limited model.

To summarize, the advantages of combining PBPK with PD include (i) mechanistically derived, quantitative variations in target concentrations that drive pharmacodynamic effect and provide better correlations of PK with PD, (ii) better extrapolation capability, (iii) use of permeability-limited distribution, if hypothesis testing identifies a distributional delay, (iv) hypothesis testing for understanding mechanisms underlying an observed response, and (v) examination of the sensitivity of a system to parameter variation. These ideas are summarized in Figure 13.20. Fingolimod (FTY 720),[47] a spingosine-1-phosphate receptor agonist that can be of potential use in the treatment of autoimmune diseases such as multiple sclerosis, acts by preventing lymphocytes from exiting lymph nodes, thus exerting an effect on the immune system that is thought to be beneficial in multiple sclerosis. It may also act on blood vessels and directly on CNS tissue. A human-scaled PBPK model of fingolimod has been combined with an indirect response PD model that characterized the time course of lymphocytes in the blood. The simulated effect versus concentration compared well with that observed, after single-dose administration in humans. Examples of PBPK/PD in the literature[48] are still rare. PBPK or PK and PD integration of toxicokinetic data with toxicological response[49] should take into consideration the possibility of nonlinear PK as well as that of PD because of the large doses involved in such studies.

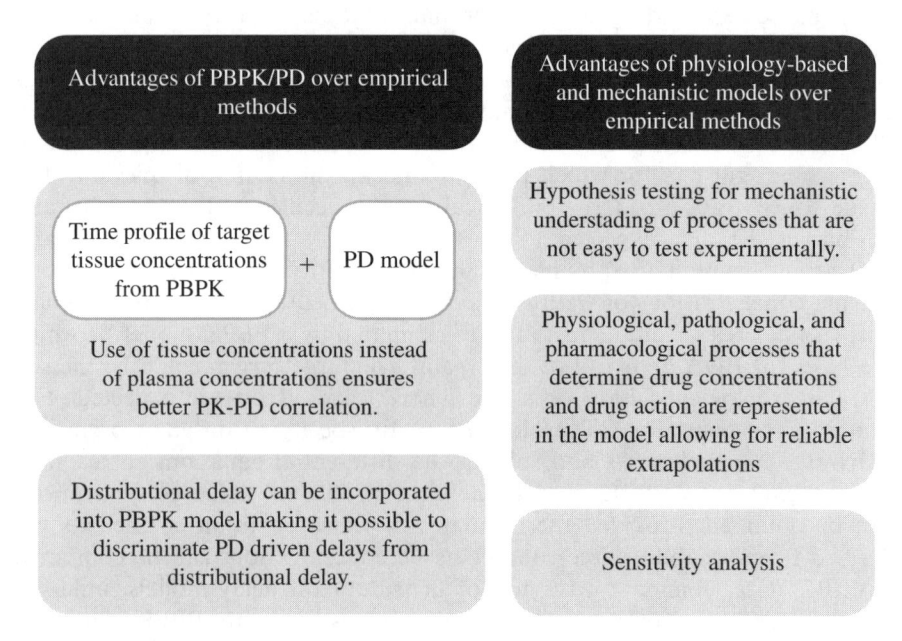

Figure 13.20. Advantages of mechanistic models over empirical models.

13.6 REASONS FOR POOR PK/PD CORRELATION

A good PK/PD correlation between the plasma drug concentrations and the time course of drug action is a prerequisite to performing reliable extrapolations and predictions. A good correlation can result if the pharmacological response is influenced only by the dosage, route of administration, formulation, and the PK of a drug. It is not always easy to establish a PK/PD relationship as there are often several other complicating factors (Table 13.5) that can modulate the pharmacological response, which need to be factored in, before a correlation can become obvious. For example, investigators have had difficulty in linking tumor growth inhibition to blood concentrations of cancer drugs because the intensity of effect lags behind the concentrations and time course of cancer drugs. In addition, the effects of lysergic acid diethylamide (LSD) on mental performance and the inotropic effects of digoxin do not correlate with plasma concentrations. In these cases, mechanistic models involving either a

TABLE 13.5. Mechanisms Leading to Poor PK/PD Correlations

Complexities that Cause an Apparent Distortion in Plasma Drug Concentration-effect Relationship
Slow and/or irreversible receptor binding leading to large response half-lives
Slow signal transduction leading to large response half-lives
Drug action involving the inhibition or stimulation of a time-consuming physiological process leading to clinical response
Target concentrations that are different from those in plasma due to:
• Local metabolism at target site
• Complex pattern of tissue binding
• Intracellular sequestration of drug
• Unique blood perfusion, tissue composition, and transporter distribution on different organs
• Active metabolites
Modulation of pharmacological response with respect to time.
• Active and interactive metabolites
• Functional adaptation processes such as sensitization, tolerance, and rebound
• Time-dependent signal transduction mechanisms
Target-mediated drug disposition (common for biologicals)
• Significant and high-affinity binding of a drug to a capacity-limited target leading to a nonlinear PK and thus PD
Study limitations
• Poorly defined biomarkers of measured PD/clinical endpoints
• Insensitive and/or nonspecific analytical methods
• Poor study designs

distributional delay or a physiological turnover delay[50] are capable of describing delayed effects and have been widely used in many areas other than cardiovascular drugs to explain slow onset of observed drug response. These mechanistic approaches have been outlined in Section 13.3. The use of inappropriate models and inadequate experimental designs for observing the time course of tumor response can also result in a poor PK/PD relationship.

Usually, pharmacological response to a drug correlates with the unbound plasma drug concentrations. However, this need not always be true. For example, when local metabolism occurs at the target organ leading to an active metabolite, drug response correlates better with unbound target organ concentrations of metabolite and drug rather than unbound plasma drug concentrations. As an example, L-DOPA easily penetrates the blood–brain barrier compared to its active metabolite dopamine.[51] So, L-DOPA is used as a prodrug and it gets decarboxylated to the active metabolite in the brain. Similarly, oral administration of heroin leads to formation of the active metabolite morphine via extensive first-pass metabolism, but morphine shows poor brain penetration.[52] However, an intravenous administration of the more lipophilic heroin ensures rapid brain penetrations. In the brain, it is hydrolyzed to morphine and to 6-acetyl-morphine, both of which are pharmacologically active. Thus, it is the sum of the active metabolite concentrations and any parent concentration at the target site that can be expected to correlate with response rather than the unbound plasma heroin concentrations. Active metabolites may have similar or different pharmacological profiles to that of parent. A mechanism-based PBPK/PD that considers the metabolite PK, target-specific PD mechanism, and the physiological components of drug concentrations and drug action, which in turn are affected by disease, gender, age (Fig. 13.21), species differences, and interaction with other drugs, would provide meaningful correlations and extrapolations.

Lack of correlation in the early stages of drug discovery is no reason to give up PK/PD analysis altogether. A poor relationship should prompt the right questions to be asked and should help to shape studies that can answer them. The use of mechanism-based PBPK/PD approaches permit a continuous improvement in the PK/PD relationship, as more mechanistic knowledge is gained along the value chain. As PBPK modeling predicts drug concentrations at the site of pharmacological or toxicological action (exposure at target organ or organ where toxic effects are seen), combining this approach with PD holds the promise of obtaining a better correlation.

13.7 APPLICATIONS OF PK OR PBPK/PD MODELING IN DRUG DISCOVERY AND DEVELOPMENT

Less than 10% of drugs in phase I clinical trials make it to the approval phase. Two key reasons why drugs fail at late stages are a lack of understanding of the relationship between dose–concentration and response or unanticipated safety events. It is critical, therefore, to have enabling tools that help predict how a

13.7 APPLICATIONS OF PK OR PBPK/PD MODELING IN DRUG DISCOVERY

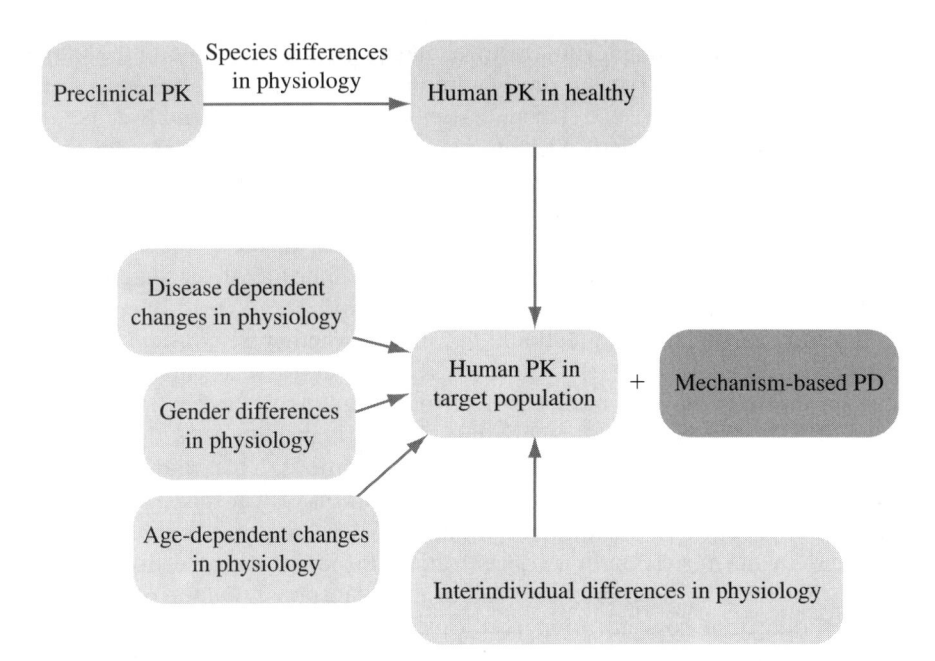

Figure 13.21. Mechanism-based PBPK/PD integration. PK impacting physiological parameters such as pH, transit time, gastric emptying rates, dimensions of organs, cardiac output, tissue blood flow rates, enzyme and transporter expression levels, and turnover rates are influenced by species, age, gender, interindividual differences arising from genetic variations and disease states. Mechanistic considerations and incorporation of these differences in physiology in PBPK models result in reliable extrapolations of PK in the target population, which can then be combined with mechanistic PD models.

drug will perform *in vivo* and assist in the success of a drug candidate. Drug discovery and development are becoming increasingly complex, time consuming, and cost intensive, and pharmaceutical companies are looking to make better use of PK and PD data to eliminate flawed candidates at the beginning of the process and, equally, to identify those with the best chance of clinical success. PK/PD integration is thus typically performed at every stage of the drug discovery and development process. Integration of PBPK with PD adds further value, by providing a means for improved PK/PD correlations through consideration of target concentration profile, and for enhanced predictive and extrapolative capacity through incorporation of mechanistic PK. Examples of PBPK/PD abound in toxicology[53,54] but not in drug discovery.

13.7.1 Need for a Mechanistic PBPK/PD Integration

Traditionally, pharmacological response is related to the administered dose (dose–response) rather than to the PK-driven exposure of drug at the target organ. This amounts to ignoring the mechanisms that are underlying the shape of the dose–response curve. An orally administered dose, for example, needs to be absorbed and metabolized during first pass through the gut and liver before it can reach the target organ. Depending on the mechanism of drug action, a series of other events lead to the observed response. To illustrate this, a drug targeting a GPCR needs to have sufficient receptor occupancy before it is activated for signal transduction to ultimately result in the desired response. The shape of the dose–response curve is thus a composite of the shapes of all the intermediate relationships as seen in Figure 13.22. If these intermediate relationships are linear, then it should not matter whether we considered dose–response or concentration–response curves. However, these are not necessarily linear; a doubling in dose need not lead to a doubling of concentration.[51] In Figure 13.22a for example, it is obvious that at higher doses, a doubling of dose will lead to more than doubling of concentrations. This kind of a dose–concentration relationship can arise from saturation of transporters or drug-metabolizing enzymes. In this situation, the dose–response curve appears steeper than the corresponding concentration–response curve, leading one to believe that the drug dose causes an "all or none" response. On the other hand, a hyperbolic dose–concentration relationship is also possible, due to auto-induction, saturated plasma protein binding, or due to solubility limitation, and, in this case, the dose–response curve is shallower than the corresponding concentration–response curve. The shape of a dose–response curve could also be misunderstood as arising from drug tolerance. In general, the steepness or shallowness of the slope of dose–response curves can originate from PK or PD mechanisms. To distinguish between these, it is necessary to link the unbound drug concentrations from PK to the response. A mechanistic breakdown of dose–response is crucial for extrapolation and is afforded by PBPK and mechanistic PD models (as detailed in Section 13.3).

The mechanism of duration of drug action can be driven by pharmacokinetics or pharmacodynamics or both. For example, off-rate kinetics (k_{off}) as seen in the preceding sections can cause a prolonged duration of effect. PBPK/PD will enable appropriate translation to humans as it can identify mechanisms such as drug-induced impairment of gastric emptying rate, which can also affect duration of action.[55] Receptor off-rate kinetics and extent of impairment of gastric emptying can be very different across species and the use of species-specific parameters is necessary for a good translation to humans.

13.7.2 Applications of PK or PBPK/PD in Drug Discovery

Pharmacokinetic and PD integration is vital at every stage of drug discovery and development for informed decisions to be made. The desirable PK and PD characteristics of a drug are summarized in Table 13.6. Traditionally, optimization of a lead series to arrive at these PK and PD characteristics has been performed independent of one other. The problem with this approach is that it

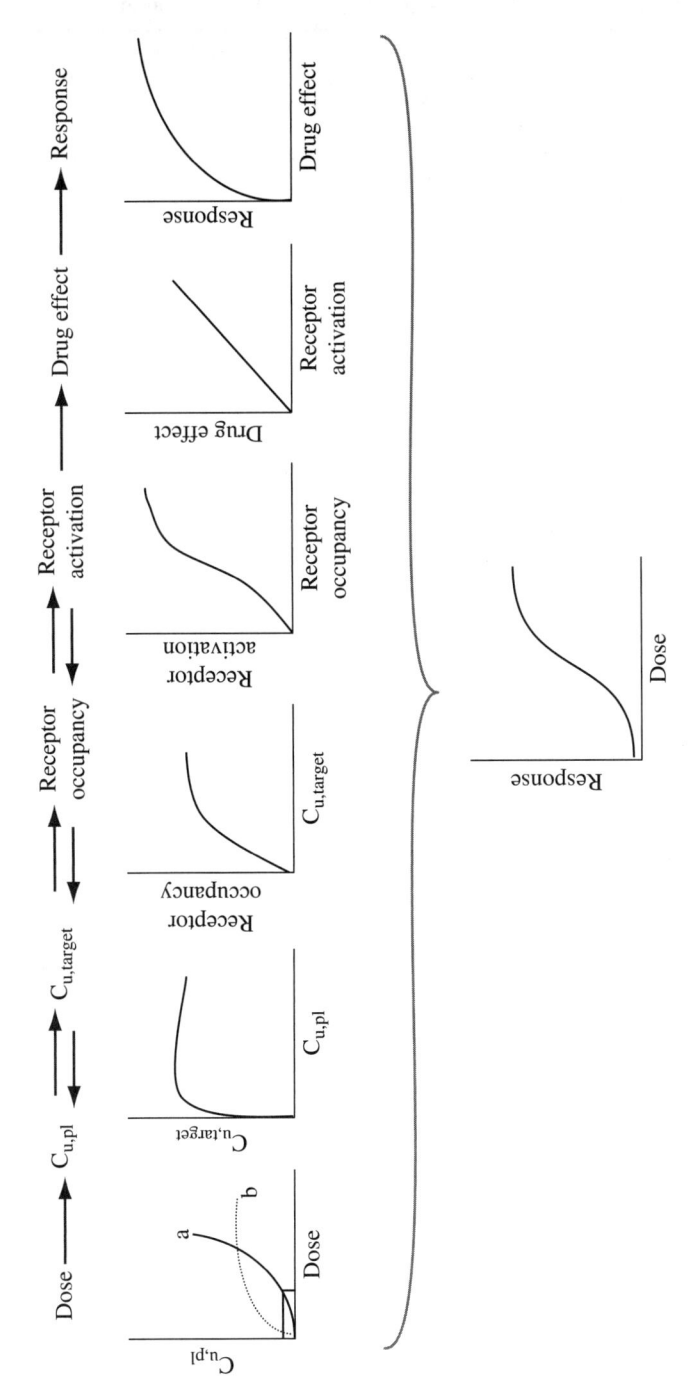

Figure 13.22. Dose–response. $C_{u,pl}$: unbound drug concentration in plasma; $C_{u,target}$: unbound drug concentration in target organ. (a) doubling of dose leads to more than doubling of concentration. (b) Doubling of dose leads to less than doubling of concentration.

TABLE 13.6. Desirable PK and PD Characteristics for a Lead Compound

Pharmacokinetics		Pharmacodynamics	
Desirable Characteristics	Example	Desirable Characteristics	Example
Adequate $t_{1/2}$ (low clearance and high volume of distribution)	Amlodipine ($t_{1/2}$ = 40 h) Methadone, etc.	High efficacy	Sulfadoxine-pyrimethamine, Amodiaquine
Good bioavailability	Gemfibrozil, glipizide, fluconazole, etc.	High potency	Felodipine (EC_u = <10 nM in humans) Ergotamine
High rate of absorption if rapid action is needed	NSAIDs	Slow turnover of response (as in depression and psychosis)	Quetiapine (Seroquel) ($t_{1/2}$ response = 1–2 weeks)
Active metabolites	Many CNS drugs	Irreversible binding to target protein for drugs with poor PK properties.	Omeprazole (Prilosec) ($t_{1/2}$ response = 15–20 h)

ignores the interdependence of PK and PD leading to an inadvertent loss of potentially good drug candidates. A good PK and safety profile allows a certain level of relaxation on potency, while a highly potent drug or a drug with slow turnover of response does not need excellent PK. For example, omeprazole has a pharmacokinetic $t_{1/2}$ of <1 h in humans but, because of irreversible binding to the target, the response $t_{1/2}$ is very long. A process-driven generic approach where PK is optimized independent of PD would have resulted in an erroneous decision of discarding compounds like omeprazole during lead optimization (LO). A mechanistic characterization and integration of all available *in vitro* and *in vivo* PK and PD data is, therefore, essential at different stages of drug discovery (Fig. 13.23). The following paragraphs explain how the above principles can be applied along the value chain, depending on the available data at each stage.

13.7.2.1 Candidate Drug–Target Profile Definition. At the start of a lead generation program, a desired profile for a potential candidate drug is defined to guide the screening of compounds through various PK and PD assays. This definition should ideally be tailor-suited to the target and aided by **PBPK/PD** integration of forerunner compounds, through which some idea of the onset, intensity, and duration of drug effect for the class of compounds is known. Forerunner information, if available, is highly valuable in setting the PK and PD criteria for a candidate drug. For certain targets, the mechanism of action

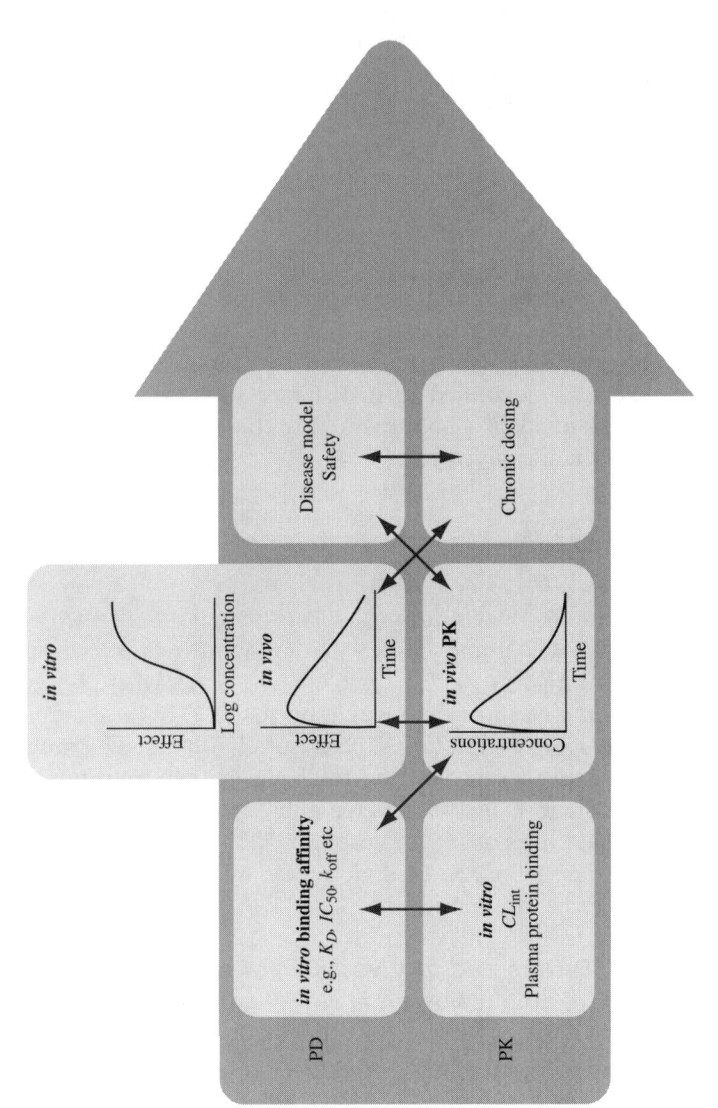

Figure 13.23. Preclinical PK/PD along the value chain.

might necessitate long PK or PD half-life in order to avoid unwanted effects such as tolerance and rebound. In others, PBPK/PD integration for a forerunner compound might have identified an irreversible target binding or a PD mechanism that leads to a PD half-life that is much longer than PK half-life. In such cases, it is immaterial if the compound has a short PK half life (arising from low volume of distribution and high clearance). This implies that once the target is occupied, the effect persists long after the drug is removed from the body, so while defining the target profile for a potential candidate drug, criteria on PK can be relaxed. Additionally, since drug plasma concentrations need not be high, the risk for DDI and reactive metabolites is rather low. The *in vivo* to *in vitro* potency ratio for the target, if known, should be factored into the definition of target candidate drug profile.

13.7.2.2 In Vitro PK and PD. In the lead generation phase of drug discovery (objectives of this phase are defined in Table 13.9), *in vitro* measures of potency (IC_{50}) and metabolic stability (CL_{int}) are measured for a series of compounds structurally related to hits identified against a target of interest. Plasma protein binding of these compounds is also measured routinely. Optimization is done on the dosing rate, which is given by

$$\text{Dosing rate} = \frac{\text{Dose}}{\text{Dosing interval}} = CL_u \times C_u$$

where C_u is the unbound concentration that must equal an expected effective concentration (eg., the IC_{50}), and CL_u is the unbound clearance expected in humans, estimated from $CL_{int,u}$. $CL_{int,u}$ is $CL_{int}/f_{u,\text{incubation}}$ where $f_{u,\text{incubation}}$ is the fraction of drug unbound in the incubation media. This assumes that we have a good *in vitro–in vivo* correlation. The use of C_u (obtained by multiplying the fraction unbound in plasma and the total plasma concentration) can be appreciated by realizing that, *in vivo*, it is the unbound drug concentration that must equal IC_{50} (or multiples of IC_{50}) to produce 50% effect. Similarly, it is the unbound clearance, CL_u (obtained by dividing CL by the fraction unbound in plasma), that needs to be minimized.

13.7.2.3 In Vitro PD Data and in Vivo PK Data. In the lead generaration and early LO phases of drug discovery, *in vivo* PK data are measured for the few compounds that have good *in vitro* potency, at least in one preclinical species. Based on allometry, predicted human PK parameters and profile are obtained. At this stage, PK and PD integration can provide useful information about the onset, intensity, and duration of effect (see Fig. 13.24a), all of which have an impact on the dosing regimen. PK/PD integration can aid rank ordering of compounds in a series in the screening stage of LO by optimizing on IC_{50}/C_u, or during candidate selection, based on the estimation of human dose using equation 13.28 (see Fig. 13.24b):

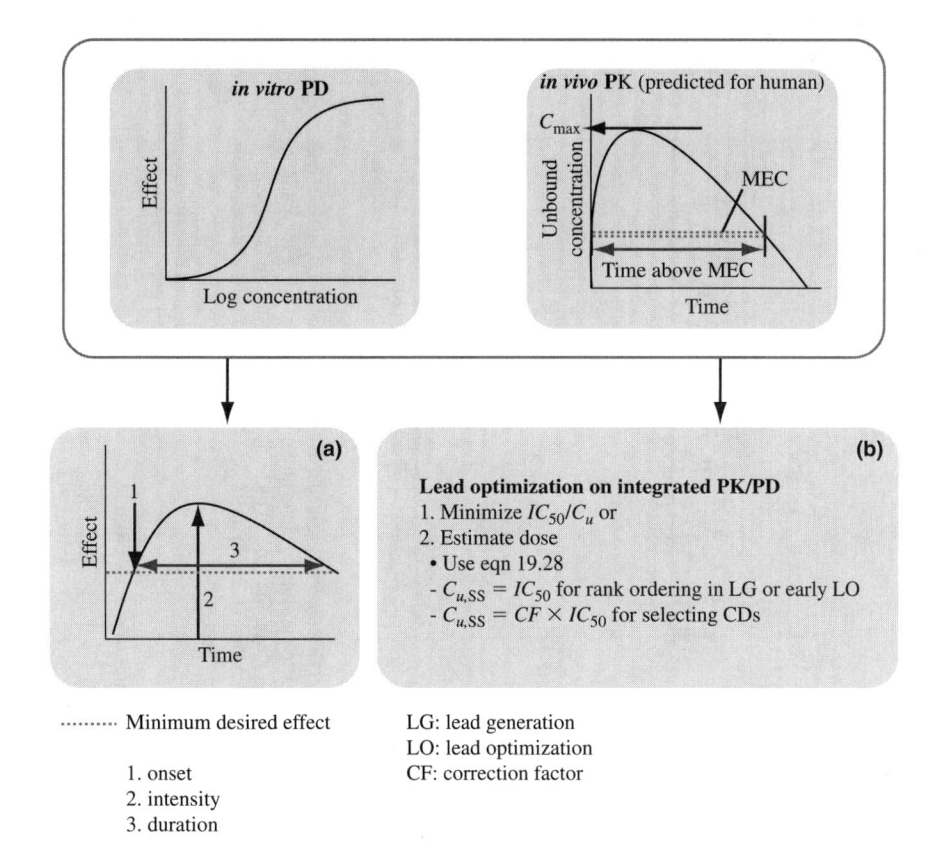

13.7 APPLICATIONS OF PK OR PBPK/PD MODELING IN DRUG DISCOVERY

Figure 13.24. PK/PD when *in vivo* PK and *in vitro* PD are available.

$$\text{Therapeutic dose} = \frac{C_{u,\text{SS}} \times CL \times \text{dose interval}}{\text{Bioavailability}} \quad (13.28)$$

where $C_{u,\text{SS}}$ is the unbound concentration at steady state. For candidate selection, the *in vitro* measure of potency in a human system is treated as the unbound steady state concentration required for equation 13.28. If any correction factor (CF) for the *in vitro* potency is available from knowledge of effective unbound concentrations in human for tool compounds, it is factored into the calculation. An estimated human CL either from allometry or from *in vitro–in vivo* extrapolation (IVIVE) is used. Dose estimation using a human PBPK model gives the time course of the predicted human concentrations (Fig. 13.25). The optimization of lead compounds on integrated PK and PD, through human dose estimation rather than on single properties such as clearance, PK half-life, or IC_{50}, is absolutely vital to minimize the risk of losing good compounds that may still have a good pharmacological effect overall,

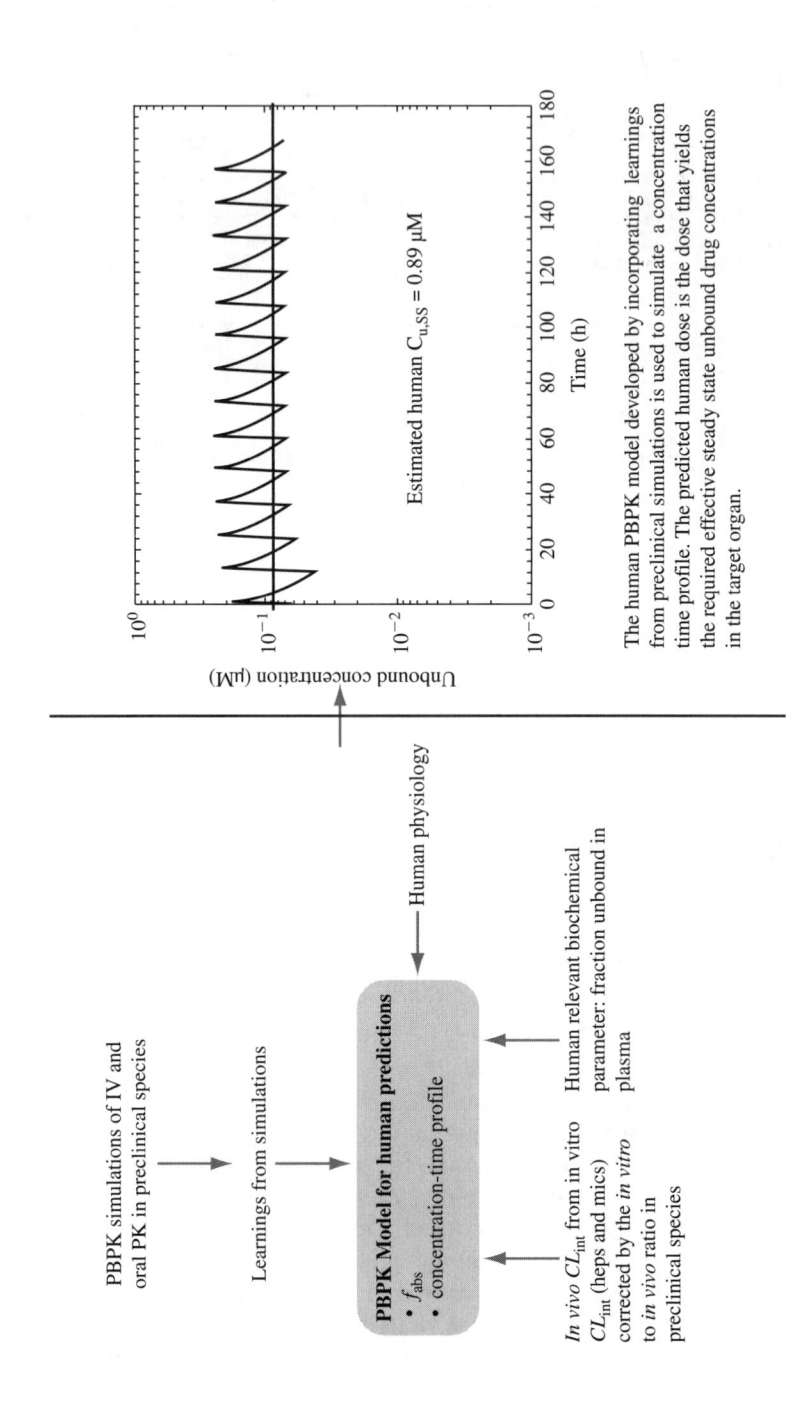

Figure 13.25. Human PK predictions with PBPK.

although they may have poor PK (free plasma exposure is low). This approach guarantees that the estimated dose frequency is the minimum necessary to maintain disease control.

The design of a dosing regimen for antibiotic therapy for preventing the emergence of resistance can illustrate the importance of PK/PD integration. The primary measure of antibiotic activity is the minimum inhibitory concentration (MIC), which is the lowest concentration of an antibiotic that completely inhibits the growth of a microorganism *in vitro*. While the MIC is a good indicator of the potency of an antibiotic, it indicates nothing about the time course of antimicrobial activity. For that, MIC needs to be linked to the PK of the antibiotic. Integration of PK parameters with the MIC yields three PK/PD parameters, which quantify the activity of an antibiotic: C_{max}/MIC ratio, the time above MIC, and the 24-h *AUC*/MIC ratio, which is determined by dividing the 24-h *AUC* by the MIC.

For type I antibiotics (aminoglycosides, fluoroquinolones, daptomycin, and the ketolides), the higher the concentration the more extensive and the faster is the degree of killing. Therefore, the 24-h *AUC*/MIC ratio, and the C_{max}/MIC ratio are important predictors of antibiotic efficacy. For aminoglycosides, it is best to have a C_{max}//MIC ratio of at least 8–10 to prevent resistance.

Type II antibiotics (β-lactams, clindamycin, erythromcyin, and linezolid) demonstrate contrasting properties. The ideal dosing regimen for these antibiotics is one that maximizes the duration of exposure. The time over MIC is the parameter that best correlates with efficacy. For β-lactams and erythromycin, maximum killing is seen when the time above MIC is at least 70% of the dosing interval.

Type III antibiotics (vancomycin, tetracyclines, azithromycin, and the dalfopristin–quinupristin combination) have mixed properties; they have time-dependent killing and moderate persistent effects. The ideal dosing regimen for these antibiotics maximizes the amount of drug received. Therefore, the 24-h *AUC*/MIC ratio is the parameter that correlates with efficacy. For vancomycin, a 24-h *AUC*/MIC ratio of at least 125 is necessary (some researchers recommend a ratio of 400 or more for problem bugs). Dose predictions for pharmacological and toxicological animal studies should also be guided by an integration of all available PK and PD information.

13.7.2.4 In Vivo PK and PD Data. Ideally, plasma drug concentration measurements should accompany the PD response measurements at the different time points. This ensures that the PK/PD correlation remains unaffected by differences in drug formulation and study conditions and allows investigations of potential time delays between exposure and response. If this is not possible, it is necessary to ensure that the PK and PD data are collected under similar conditions. If a disease model rat is used for PD, the same model should be employed for obtaining *in vivo* PK, if there is a possibility that the disease could affect the clearance or volume of distribution of the drug. Once the PK and PD data are available, overlaying the plots of PD response versus time and unbound concentration versus time will show how the PD half-life compares to the PK half-life. A higher PD half-life compared to PK half-life should trigger the measurement of k_{off} to check the target residence time. If the measured k_{off} is very small, it implies that there is an irreversible binding to the target. A slow k_{off} or turnover systems with long PD half-lives are beneficial, as they reduce the demands on systemic exposure. A plot of PD response versus time will also identify the onset, duration, and intensity of the effect (Fig. 13.24a). The overlay plot will also identity any delay in response (Fig. 13.26). If a delay in response is observed, it has implications on the design of any subsequent PK or PD study in guiding the dosing, duration, and sampling times. As discussed in Section 13.3.3, a PD delay is one of the reasons for the observation of a hysteresis in the effect versus concentration plot. This means that it becomes difficult to extract a single value of EC_{50} required for the estimation of dose. It is, therefore, necessary to explore the reasons for the delay in pharmacological response and to use an appropriate model (see Section 13.3) to extract the correct PD parameters (EC_{50}, E_{max}, and other model-specific parameters) for the estimation of dose (equation 13.28). For example, hypothesis testing in PBPK can identify a potential delay in distribution of the drug to the target organ. In such cases, PBPK-derived target tissue concentrations should be plotted against pharmacological response to get the PD parameters. However, in order to do this, reliable estimates of tissue partition coefficients and transmembrane permeability rates are necessary. EC_{50} obtained using an appropriate model from effect versus concentration (unbound) plot is assumed to be the same across different species after correcting for differences in protein binding and used in equation 13.28. As before, CL is the estimated human CL either from allometry or IVIVE. Another use of the EC_{50} obtained through *in vivo* studies is correlation with an *in vitro* value for a compound series. Such comparisons will enable an assessment of the predictivity of the *in vitro* measure of potency and give an idea of what *in vitro* potency to aim for (while setting the target profile for a candidate drug) in subsequent LO screening.[56] The *in vivo* to *in vitro* potency ratio varies widely depending on the target. According to the fundamental free ligand hypothesis, the average free effective concentration at steady state *in vivo* should correlate with the intrinsic unbound potency determined from an *in vitro* assay. However, an *in vitro*–*in vivo* correlation is often confounded by nonphysiological conditions and nonspecific binding in the *in vitro* assay as well as the use of inappropriate PK/PD model for the *in vivo* data. Generally, a potent enzyme inhibitor or a receptor antagonist *in vitro* works only at a much higher inhibitor concentration *in vivo*, if it works at all. On the other hand, receptor agonists generally require a lower level of target engagement for efficacy. Unless this discrepancy is corrected through a well-established *in vitro* —*in vivo* correlation, dose estimations using the *in vitro* measures of effective concentrations (instead of $C_{u,SS}$) in equation 13.28 are highly unreliable.

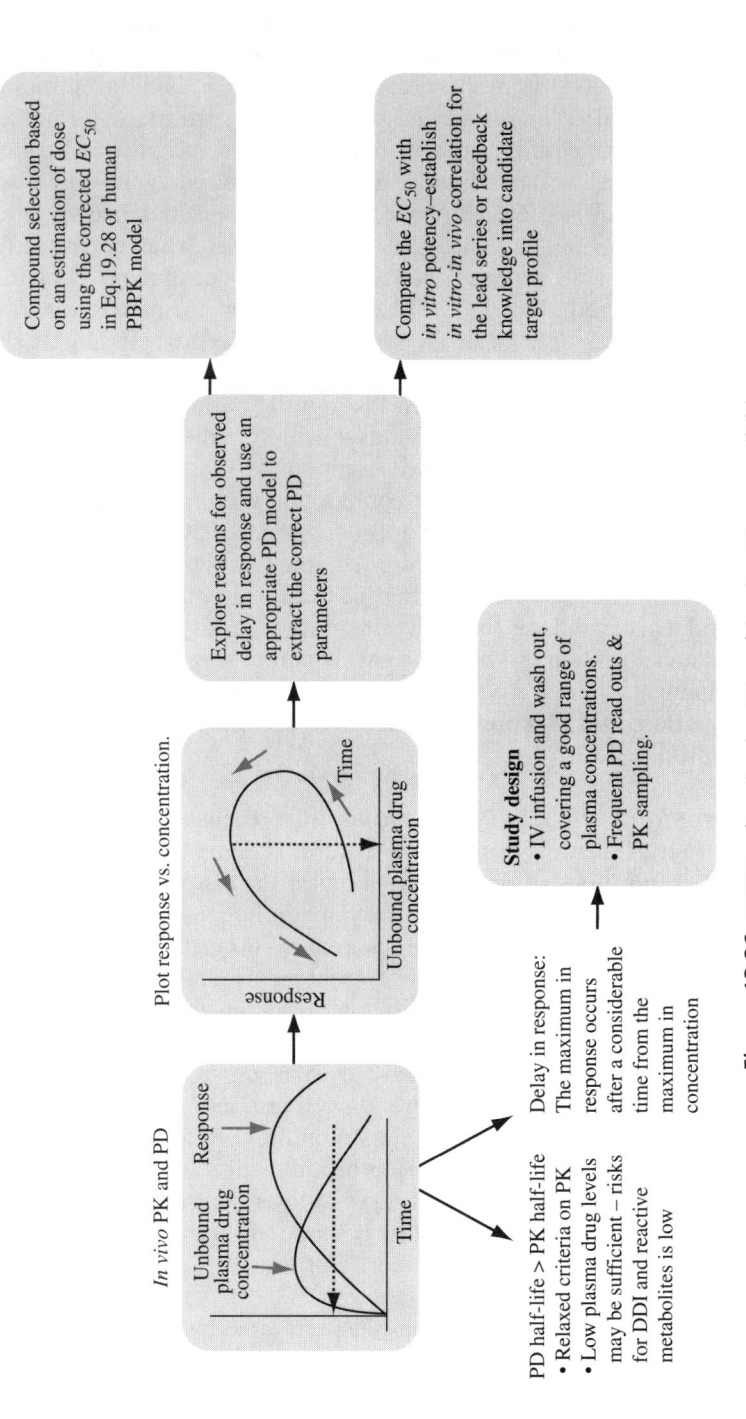

Figure 13.26. PK/PD when *in vivo* PK and *in vivo* PD are available.

Sometimes, an excellent correlation can be established between *in vitro* and *in vivo* potency. In a drug discovery program for obesity that targeted a protein located behind the blood–brain barrier, plasma and brain concentrations measured in rats 1 h after administering doses ranging from 1–20 µmol/kg, were found to be in rapid equilibrium with each other. The receptor occupancy, measured *ex vivo*, served as the PD read out and was plotted against unbound plasma concentrations. The EC_{50} extracted from these plots matched very well with *in vitro* measures of potency for three compounds and when corrected for differences in their *in vitro* potency (plotting receptor occupancy versus C_u/EC_{50}), the plots collapsed. This kind of excellent agreement between *in vitro* and *in vivo* shows that it is free plasma levels that drive receptor binding. It also provides better confidence in the prediction of target binding and optimization (by maximizing C_u/EC_{50}) based on screen PK and PD parameters. The *ex vivo* receptor occupancy for different compounds when plotted versus the times at which they were observed, distinguished compounds that had a prolonged binding from those that did not. The compounds that exhibited high and prolonged receptor occupancy showed a good body weight reduction in rats compared to those with poor receptor occupancy, which were also good substrates for P-gp. This demonstrates that CNS exposure is clearly necessary for efficacy, and it also validates the target. In another obesity project where the target was a channel protein, no body weight reduction was seen in *ob/ob* mice on chronic administration of a compound that had free exposure several-fold above the IC_{50}. Based on this observation, the target was devalidated and the project was closed.

13.7.2.5 Chronic in Vivo PKPD[51] Studies.
In preclinical species, chronic *in vivo* PKPD studies extending over a few weeks are necessary to establish steady-state plasma concentrations of the drug, to monitor the disease-curing effect of candidate drugs in animals, and to take into account the long-term changes in drug action. Steady-state concentrations can be achieved by intravenous infusion or by the use of osmotic minipumps. Chronic (repeated) dosing aims to measure pharmacological response that is valid for long-term administration, by providing sufficient time for target engagement and for slow turnover of pharmacological response and allowing for long-term adaptive mechanisms such as tolerance, and for temporal differences between concentration and response.

For example, repeated dosing of a cannabinoid-1 (CB1) inverse agonist is essential for establishing a good drug–receptor interaction in order to observe the expected response of fat reduction. It takes 3–4 weeks of dosing to establish a pharmacodynamic steady state of a test compound, even if the PK steady state is reached much earlier due to its short half-life. This is an example of a slow turnover of pharmacological response.

Rapid plasma kinetics, which is responsible for a rapid loss of PD effect, can cause tolerance on repeated dosing leading to a time-dependent loss of response and a rebound effect on sudden discontinuation of the drug. This is an example of PK-driven changes in PD effects with time.

13.7 APPLICATIONS OF PK OR PBPK/PD MODELING IN DRUG DISCOVERY

The therapeutic mechanism of drug action is probably dictated by the long-term changes brought about by the drug, and therefore PK/PD integration with chronic dosing is essential to understand these mechanisms. Examples include antidepressant action of tricyclic antidepressants, SSRIs and monoamine oxidase inhibitors. The basis for long-term changes in drug action for these compounds is a time delay in the antidepressant effect and their neuroadaptive effects. Another example of temporal difference between plasma concentration and pharmacological response is a compound where inhibition of the synthesis of γ-secretase in the brain leads to reduced formation of neurofibrillary tangles. The C_{max} for the compound is 0.5 h but the inhibitory effect was seen at 2 h—a delay of 1.5 h. Thus, sampling the pharmacological response at the C_{max} would have resulted in a biased estimate of potency. This example highlights the importance of a PK/PD approach to design the study.

13.7.2.6 Translational PBPK/PD. The quantitative prediction of PK and PD (pharmacological effect and safety) properties of drugs in humans, using prior information from preclinical *in vitro* and *in vivo* studies, requires the recognition of species differences (Table 13.7) in the pathophysiological mechanisms underlying PK and PD and correction for differences in the plasma protein binding. PBPK and mechanistic PD models developed in a preclinical setting can easily incorporate these species differences to make reliable human predictions with respect to both safety and pharmacological effects. It is important, however, to have translatable biomarkers. Applications of translational science for the selection of compounds with best safety margins and for the estimation of first-in-human dose are discussed below.

Zhang et al.[57] characterized acetylcholinesterase (AChE) inhibition of carbofuran by developing a PBPK/PD model in rats. A PBPK/PD model for ansamycin benzoquinone and its metabolite in tumor-bearing mice was built to describe their combined action on the oncoproteins Raf-1 and p185.[58] In addition, a murine PBPK model for docetaxel was extrapolated to humans to predict the human plasma concentration.[59] The resulting concentrations were then coupled to a low-order neutrophil model from the literature[60] and individual patient absolute neutrophil count (ANC) predictions were compared with actual ANC data. Plasma docetaxel concentration versus time data were obtained from 75 patients who were given 1-h infusions and sampled out to 48 h. Clearance rate from the mouse PBPK model were used unchanged as parameters in the scaled model. The PBPK model was used to drive the ANC model, which had been developed with data from patients treated on a q3wk (once in 3 weeks) schedule.

Selecting Compounds with the Best Safety Margins. Safety margins estimated for humans can be derived from PBPK/PD-integrated effect–concentration curves, incorporating species differences in physiology. A workshop convened by the Safety Pharmacology Society in February 2007 emphasized the need to incorporate mechanistic PK/PD modeling into safety pharmacology to improve the predictability of preclinical investigations for human outcomes.

TABLE 13.7. Species Differences in the Physiological Factors Impacting PK and PD

PK/PD Property	Physiological Differences Between Species
Absorption	Gastric and intestinal pH and transit times Intestinal intercellular porosity Blood flow rate to GI Uptake and efflux transporters in enterocytes
Metabolism	Abundance and differences in isoforms of intestinal and hepatic drug-metabolizing enzymes (DMEs) and transporters contribute to differences in substrate specificity, metabolizing pathways, and active or nonactive metabolites. Drugs whose metabolism is primarily flow-limited scale well.

Species	Glucuronidation	Acetylation	Sulphation
Cat	Poor		
Rat		Good	
Dogs		Poor	
Pigs			Poor
Human	Good	Mediocre	

Drug–drug interaction potential	Large species differences in drug activation of constitutive androstane receptor (CAR) and pregnane X receptor (PXR), which regulate the expression of DMEs and transporters
Biliary elimination	Biliary flow and efflux transporters in bile canaliculi. Humans, primates, guinea pigs, and rabbits are poor biliary excretors. Mice, rats and dogs are good biliary excretors, while cats are intermediate
Renal elimination	Glomerular filtration rate and number of nephrons, generally scale well allometrically. Thus, hydrophilic drugs such as penicillins and antibiotics for which renal elimination is the most important clearance pathway, scale well
Target cells and systems	Expression levels, tissue distribution and regulation of receptors, first and second messengers. Differences in target turnover rates. Differences in receptor affinity between species can come from differences in one or more amino acids in their receptors. For example, the sensitivity of the dog, sheep, and human Na^+/K^+ ATPase transporters to cardiac glycosides is ∼1000 times greater than in mice and rats. Although PK is generally more species specific than PD, even few mutations in target proteins can cause a profound change in response

Some of the case studies presented at the meeting have been reported.[61] In one example, a predicted QTc prolongation in humans was derived for dofetilide from a PK/PD integration of *in vitro* (hERG channel inhibition) and preclinical *in vivo* data (QTc prolongation in dogs). An *in vitro* to *in vivo* (dog) correlation identified a transducer function for dofetilide, which in combination with the

13.7 APPLICATIONS OF PK OR PBPK/PD MODELING IN DRUG DISCOVERY

unbound concentrations in humans indicated a five-fold higher potency in humans compared to the *in vitro* result. This supports the clinical finding that a QT_{10} rather than a QT_{50} explains the observed 20-ms prolongation in humans. In another example, it was demonstrated that steady-state plasma concentration studies for a fast distributing, slow eliminating test compound, using a loading dose IV bolus injection, followed by maintenance infusions in dogs, produced exactly the same EC_{50} as that from an intratracheal administration. The loading dose route does not permit frequent sampling that is required for the development of a good PK/PD model, but with the intratracheal route this was possible. An indirect response model was fitted to the data to get an estimate for EC_{50}, which is more reliable compared to the loading dose route. This model could then be confidently employed for exatrapolation to humans. Safety margin is the difference between effective and safe concentrations. This is obtained by plotting response versus concentration for both pharmacological effect and safety (Fig. 13.27). Often measures of potency such as EC_{30}, EC_{50}, EC_{90}, and so forth are derived from one preclinical species and safety measures such as QT_{10} are derived from others. In such cases, the safety margins need to take into account differences in protein binding between the two species and so response versus free concentrations (C_u) provide better estimates of safety margins.[13] The use of *ex vivo* measurements of protein binding in individuals has been recommended over standard *in vitro* measurements in pooled plasma samples for more reliable predictions. Again, the effects of temporal differences between plasma concentration and response (delayed response) should be considered for both the primary pharmacological effect and the secondary safety effect to ensure better estimates of safety margins.[13]

Prediction of Dose for First-in-Human Studies. A pharmacologically active dose is obtained by using equation 13.28 as described in Section 13.7.2 or by using PBPK (Fig. 13.25) taking the following[13] into consideration:

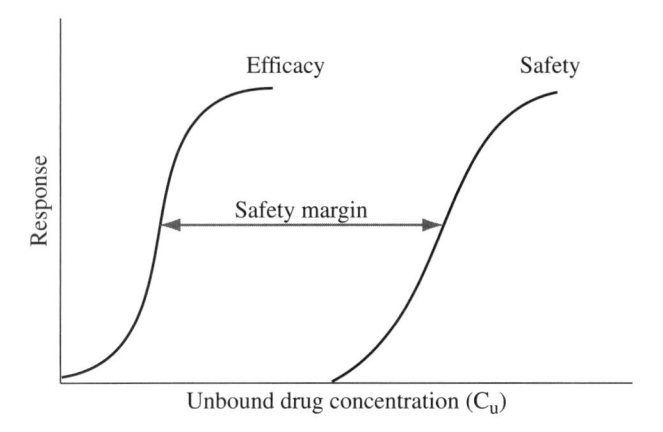

Figure 13.27. Safety margins.

- Novelty of the drug and mechanism of action—a more conservative dose estimation should be used for a novel mechanism as forerunner information is not available.
- All available knowledge of preclinical and human-specific PD (measured or extrapolated).
- *In vivo* potency in preclinical species
- Receptor occupancy in humans
- Knowledge of species specificity of receptors and PK (Table 13.7)
- *In vitro* concentration—response curves of biological effects in humans and animals
- Quantitative methods of PK/PD reasoning.
- Safety margins based on unbound and total concentrations.
- Physiological variability and uncertainty in the predicted measures of human exposure and response.

Physiologically-based PK models combined with mechanistic PD models are uniquely suited to incorporate species differences in deriving a first-in-human dose. PBPK offers a natural way of considering variability in the physiological parameters in the predictions of dose to humans. Conventionally, the impact of physiological variability and uncertainty in the predictions of human CL and effective concentrations are addressed by constructing a dose nomogram (Table 13.8).

The estimated dose in the example shown in Table 13.8 is 11.5 mg but can range anywhere from 4 to 22.5 mg depending on the physiological variability and uncertainties in the predictions of effective concentration and clearance. In PBPK/PD, physiological variability with respect to enzyme abundance, receptor densities, blood flow rates, and tissue composition in any population (healthy, disease, pediatric, etc.) is derived from a database of known measurements, while uncertainties in the predictions of effective concentrations and CL can be incorporated through standard deviations in the measurement of input parameters.

TABLE 13.8. Dose Nomogram for Hypothetical Drug Whose Predicted Human CL Is 400 mL/min and Predicted Effective Plasma Concentration Is 0.05 μM

Predicted Human Doses (mg) Assuming a 30% Variability in the Predicted Human CL and 50% in the Effective Plasma Concentration in Human

Predicted Human CL (mL/min)	Effective Concentration Range (μM)		
	0.015	0.05	0.1
250	4	8	12
400	6	11.5	17
500	7.5	15	22.5

13.7.2.7 Demonstrate Market Differentiation. Through PK/PD integration it is possible to establish that a candidate selected for development is the best among those short-listed or has a competitive edge over drugs already available on the market for the treatment of a disease. Based on the concentration–effect relationship, a hypertension drug was shown to be superior to a marketed comparator because of its higher potency and efficacy at high concentrations.[62] This was later proved in clinical studies.

13.7.2.8 Decreasing the Risk of Adverse Effects and Maximizing Pharmacological Response of Medications through PBPK/PD Simulations of Hypothetical Scenarios. PBPK/PD modeling is ideal for simulating a desired concentration–response profile using hypothetical PK and PD characteristics. A good example is provided in the literature.[13] Nicotinic acid has been recommended as a lipid lowering agent. It acts by suppressing the lipolysis (fat mobilization) of triglycerides (TG) in the adipose tissue and inhibiting the transfer of cholesteryl ester from high-density lipoproteins (HDL) to very low density lipoproteins (VLDL) through its high-affinity interaction with nicotinic acid receptor, a GPCR with tissue distribution in spleen, adipose, and macrophages (see Fig. 13.28). Although effective in the short term, in the long term it is ineffective due to the development of tolerance. There is also a rebound effect, once dosing is discontinued. It causes a secondary adverse effect of flushing, which lasts 30–60 minutes in humans. So nicotinic acid is not widely used. The underlying mechanisms for the undesirable effects remain elusive. However, the rapid rise in the plasma concentration following the administration of nicotinic acid has been implicated in flushing. This can be overcome with an extended release (ER) formulation, with a slower absorption rate, which is, however, associated with the risk of shifting the metabolic pathway from a relatively safe glycine conjugation route (a low-affinity, high-capacity, and saturable pathway important at high concentrations) to the nicotinamide route (a high-affinity, low-capacity pathway that takes over at low concentrations). Tolerance and rebound have been attributed to the rapid plasma kinetics of nicotinic acid, which is responsible for the rapid loss of the PD effect. With these PK and PD characteristics, it is unlikely that nicotinic acid can ever become a useful drug. However, novel agonists for the nicotinic acid receptors without the undesirable effects of tolerance, rebound, and the flushing would be an attractive option. PK–PD integration can be useful in simulating different input scenarios to see what kind of PK profile for the novel agonist can provide the best pharmacological response with the least risk for the unwanted effects. The fall of non-esterified fatty acid (NEFA) levels is a direct consequence of the suppression of lipolysis. Its time course following a multiple constant rate infusion of nicotinic acid to four rats showed a rebound effect in the wash-out phase. A PK/PD model that predicted the observed response was used to simulate the profile using an exponential decay of the infusion rate beginning at 110 min, instead of a sudden termination of infusion. The simulated profile shows a reduced rebound effect. Thus, a compound with a long PK or response half-life that mimics the slow withdrawal of the drug is likely

Mechanism of action

Adipocyte
Blood vessel
Macrophage

TG in adipose

Lipolysis ⊕ Hormone sensitive lipase (HSL)
● Nicotinic acid

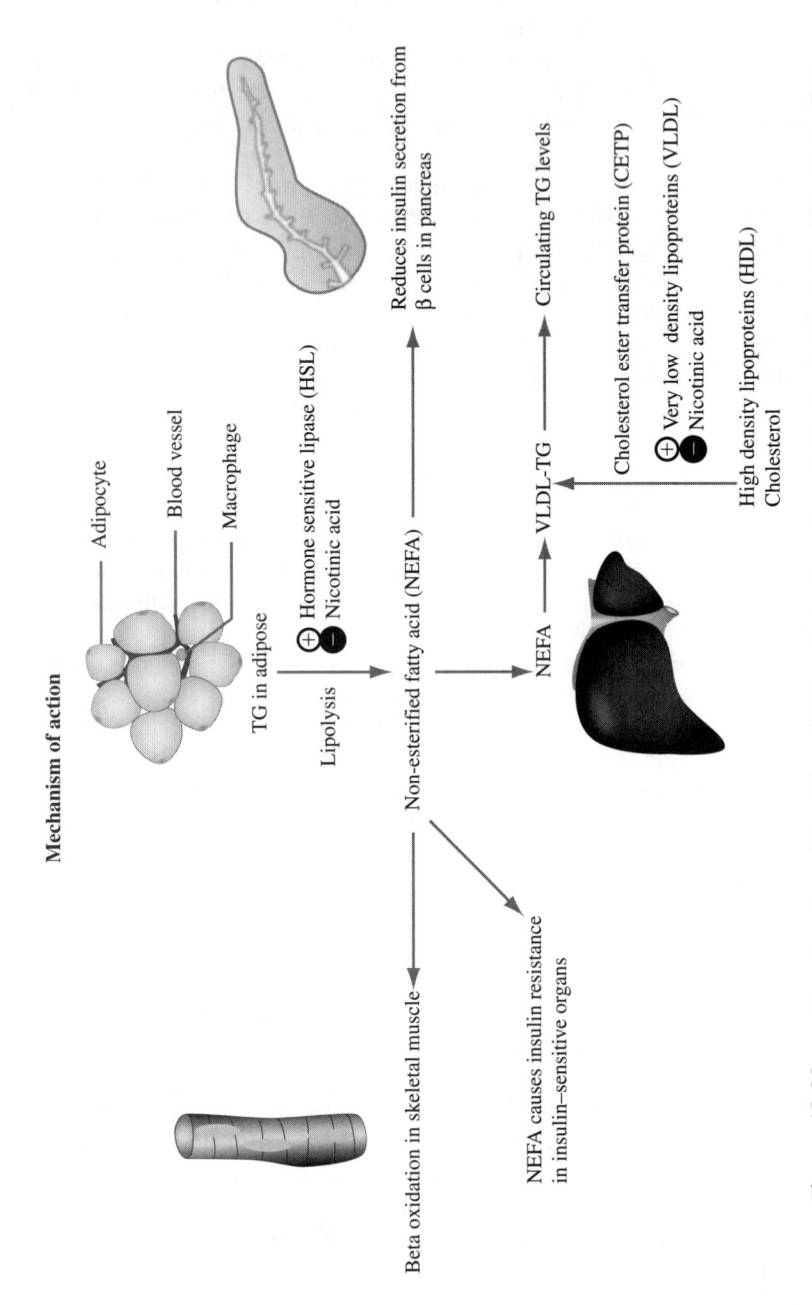

Non-esterified fatty acid (NEFA) → Reduces insulin secretion from β cells in pancreas

NEFA → VLDL-TG → Circulating TG levels

Cholesterol ester transfer protein (CETP)
⊕ Very low density lipoproteins (VLDL)
● Nicotinic acid

High density lipoproteins (HDL)
Cholesterol

Beta oxidation in skeletal muscle

NEFA causes insulin resistance in insulin-sensitive organs

Figure 13.28. Antilypotic action of nicotinic acid. PD endpoint: lowering of NEFA levels. Nicotinic acid is a high-affinity agonist to nicotinic acid receptor, a GPCR with tissue distribution at spleen, adipose, and macrophages. It is associated with adverse effect of flushing due to rapid increase in plasma levels.

13.7 APPLICATIONS OF PK OR PBPK/PD MODELING IN DRUG DISCOVERY

to be associated with a reduced rebound effect and, therefore, better suited as a drug compared to nicotinic acid. PK/PD simulation has thus been successful in demonstrating the benefits of a gradual rather than a rapid loss of PD effect for reducing the undesired effects of nicotinic acid. This information could be valuable in a new drug hunting project targeting the nicotinic acid receptor. Testing hypothetical dose or route of administration or other PK-relevant hypothetical scenarios are even more feasible with mechanistic models such as PBPK, as discussed in Chapter 12. Table 13.9 summarizes the applications of PBPK/PD

TABLE 13.9. Summary of Applications of PBPK/PD Integration in Drug Discovery

Drug Discovery Phase	Objectives	Applications
Target validation	Evaluate target to enable informed decision for commencing drug discovery program.	PBPK/PD-aided target validation. If exposure at target, relative to *in vitro* potency correlates with observed efficacy in the *in vivo* model, target engagement is confirmed. In the absence of a good correlation (no effect observed in the *in vivo* model even when exposure is several-fold of *in vitro* potency), target is devalidated.
Lead generation	Synthesize lead compounds with improved potency, physiochemical, metabolic properties, and reduced off-target activities.	Develop a rational target profile for a candidate lead, with a good understanding of target engagement through PBPK/PD integration (Section 13.7.2.1). Assess lead series by combining *in vitro* PK and PD data (Section 13.7.2.2).
Lead optimization (LO screening stage)	Optimize the safety margins based on PK and PD (efficacy and safety).	Rank order compounds by combining *in vivo* PK and *in vitro* or *in vivo* safety and efficacy PD data, rather than on single properties like CL, $t_{1/2}$, EC_{50}, etc. (Section 13.7.2.3). Dose predictions for animal studies (Section 13.7.2.3). Use PBPK/PD simulations to understand what modulation of the compound characteristics is necessary to achieve a desired profile, with the least adverse effect (Section 13.7.2.8). Apply PK/PD principles to predict safety margins for lead compound and compare it to reference compound in human.

(*Continued*)

TABLE 13.9. (*Continued*)

Drug Discovery Phase	Objectives	Applications
Lead optmization (LO candidate selection)	Select for development, a high-quality candidate drug that satisfies a preset target profile.	Dose predictions for animal studies (Section 13.7.2.3). Candidate drug selection based on human dose predictions (Sections 13.6.2.3–5) using PBPK and superiority over comparator (Section 13.7.2.7). Translation of safety preclinical information into humans (Section 13.7.2.6). Rational assessment of therapeutic window from efficacy and risk profiles (Section 13.7.2.6).

integration in drug discovery, discussed in this section. As is evident from this Table, PBPK/PD-driven drug discovery promises to be efficient, focused, and leaner, while supporting well-informed, rational decisions at the various drug discovery milestones and delivering high-quality drug candidates for further development.

13.7.3 Applications of PK or PBPK/PD in Drug Development

A PK/PD-guided drug development[47,63,64] can streamline the development process by enhancing the effective use of resources and lead to a lower number of postmarketing drug dosage changes. PBPK/PD integration in early clinical studies is used to confirm information learned from discovery, to refine preclinical PBPK/PD models, and to update dose predictions for subsequent clinical trials. Lack of efficacy seen at phase II trials is often the reason for failure of new chemical entities (NCEs). Success at this stage relies on having the right patients in the study (benefit-to-risk ratio may not be uniform across different populations due to pharmacogenomic variability), good target engagement of the compound as indicated by a valid biomarker, validity of the target mechanism (proof of mechanism), and efficacy of the drug. A quantitative PK/PD analysis of an NCE can identify the extent of target engagement. In discovery, depending on the stage, various biomarkers of target engagement can be useful—ratio of free plasma concentration to IC_{50}, physiological response, or other downstream pharmacological events. A PBPK/PD model that has been validated in preclinical species, translated to humans, and updated after phase I studies has the potential to identify a lack of target engagement early enough to avert a costly failure at late stages of development. A short-listed candidate drug should ideally have the highest level of evidence from a direct estimate of target engagement afforded by imaging techniques such as ligand-displacement positron emission tomography (PET) or single-photon emission-computed tomography (SPECT). These methods, however, are expensive.

13.7.3.1 Optimizing Effective and Safe Dosage Regimen in Subsequent Clinical Trials. PBPK and mechanistic models aid in arriving at the best predictions of concentration–time and response–time curves (therefore, concentration–response curves). As the development of the drug advances, mechanistic models provide means of incorporating all available knowledge at that stage, making it possible to arrive at the best knowledge-integrated concentration–response curve for predictions of subsequent trials. The shape of the concentration–response relationship is important in determining the optimal dosage regimen. A few years ago, determining dose regimens for clinical trials was often empirical or semiempirical, and there were many failures in phase III trials due to suboptimal dosing, which ultimately resulted in poor efficacy or safety. Hence, several drug development companies started using predictive models earlier in the development process so as to have better outcomes in phase II and phase III trials.

13.7.3.2 Population PBPK/PD Models. Population modeling is primarily based on nonlinear mixed effects regression models introduced by Sheiner and co-workers that characterizes dose concentration–effect relationships in populations rather than in individuals.[65,66] It has been successfully applied to describe the PK/PD of many drug.[67–70] but is limited in its ability to extrapolate. Contrary to this top-down approach, PBPK and mechanism-based PD constitute a bottom-up[71] approach in which known variability in demographics (gender, age, body size, ethnicity, etc.), societal factors (environmental effects, dietary habits, etc.), concurrent medication, ethnic differences in genotype frequencies and levels of CYP and UGT enzymes, physiological differences (e.g., liver size and hepatic blood flow), biological components within relevant disease populations can be incorporated in PK and PD models in order to elucidate and differentiate sources of interindividual variability in response. Such an approach to population PK is incorporated in SimCyp, a population-based, commercial, PK simulation software. Understanding the mechanisms by which covariates influence PK and thereby PD, provides an opportunity to anticipate events in groups of patients who cannot be studied in a clinical setting (children, elderly, etc.). Any clinical data that becomes available for these groups are then confirmatory rather than informative and can be obtained from a much smaller number of patients. This, however, requires extensive libraries on demographics, developmental physiology and the ontogeny of receptor densities, drug elimination pathways, enzymes and transporters, and the like. Use of population-based PK/PD analyses can help explain differences in response among individuals receiving the same dose (e.g., covariate analyses). It can help identify subpopulations at risk, define risk–benefit ratios, and/or risk management strategies and guide the use of pharmacogenomics to individualize patient drug therapy. Population PK can obviate the need for selected clinical trials (age, gender, renal impairment). For example, captopril, a well-tolerated drug causes dose and concentration-related agranulocytosis. At daily doses of 75–150 mg, a substantial accumulation in the renally impaired can cause agranulocytosis. This kind of toxicity could easily be identified through PBPK modeling, in which the concentration profiles can be predicted for the renally impaired population, using a reduced glomerulation filtration rate, with implications on appropriate dose adjustments to this special population. A schematic illustrating the application of mechanistic PBPK/PD for dose adjustments in subpopulations is shown in Figure 13.29.

Preclinical studies, along with phase I, II, and III clinical trials demonstrate the pharmacokinetics, pharmacodynamics, safety, and efficacy of a new drug under well-controlled circumstances in relatively homogeneous populations. However, these types of studies generally do not answer important questions about variability in specific factors that predict pharmacokinetic and pharmacodynamic (PK/PD) activity, in turn affecting safety and efficacy. Extrapolation of clinical results for population from one region to another region with major cultural and racial differences requires bridging studies to provide the additional pharmacologic data. PBPK/PD predictions of dose–response curve account for differences in the physiological and genetic characteristics of a virtual target population, thereby reducing the number of recruits in bridging studies for confirmatory purposes. Capecitabine is an orally administered prodrug of 5-fluorouracil (5-FU), designed to exploit tissue-specific differences in metabolic enzyme activities in order to enhance efficacy and safety. It undergoes extensive metabolism in multiple physiologic compartments, and presents particular challenges for predicting PK and PD activity in humans. Physiologically-based pharmacokinetic and pharmacodynamic models were developed to characterize the activity of capecitabine and its metabolites and the clinical consequences under varying physiological conditions such as creatinine clearance or the activity of key metabolic enzymes.[72] The results of the modeling investigations were consistent with capecitabine's rational design as a triple prodrug of 5-FU. PK/PD and PBPK/PD modeling approaches used in capecitabine development provide a more thorough understanding of the key predictors of its activity and how variability in these predictors may affect its PK/PD and, thereby, clinical outcomes.

For drugs already approved and in the market (phase IV), PBPK/PD can help monitor medications with a narrow therapeutic index for drug–drug interactions, drug disease interactions, or other covariates that interfere with the effect or safety of the drug. Postmarketing surveillance of moclobemide therapy identified that the adverse event frequency was concentration dependent and that it was 1.4 times more frequent in females than in males.[73] Here again, PBPK models can add value by identifying the physiological risk factors that can make a group more vulnerable to an adverse event.

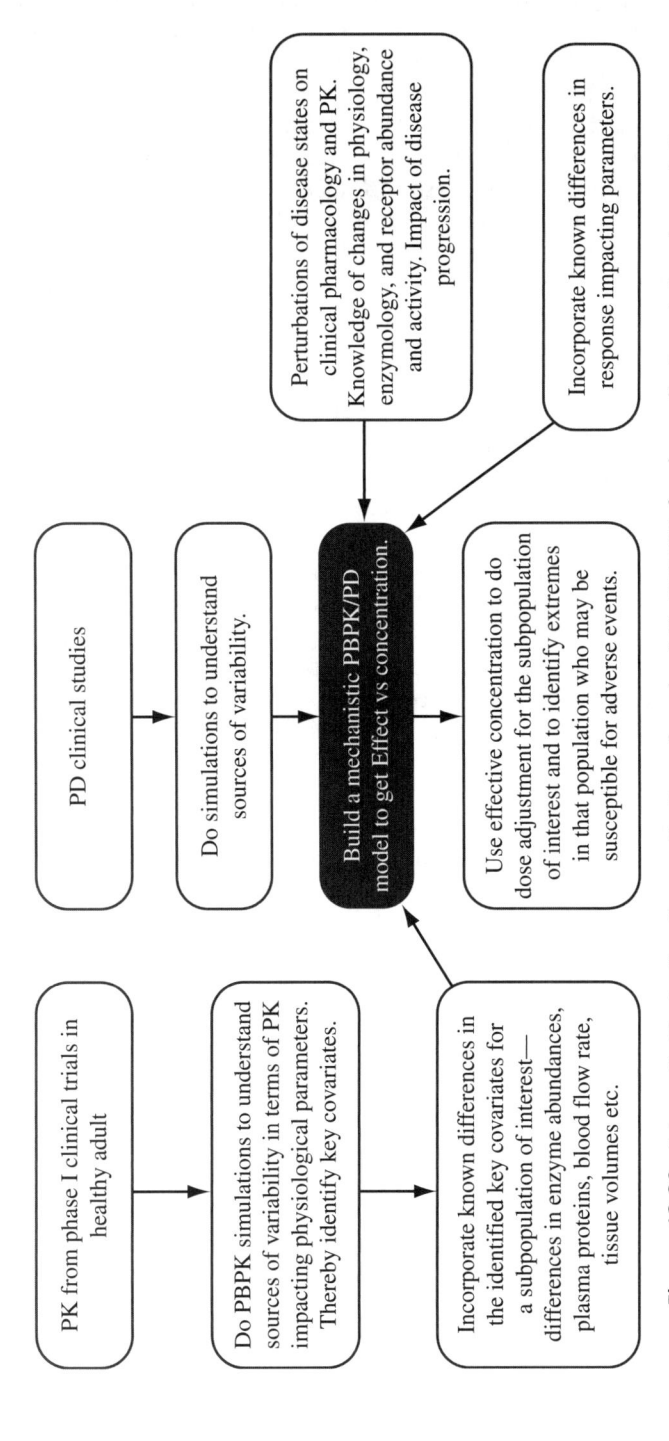

Figure 13.29. Schematic illustrating the application of mechanistic PBPK/PD for dose adjustments in subpopulations.

Physiologically-based PK/PD modeling can be valuable in the identification and validation of biomarkers. Glycyrrhizic acid, a widely applied sweetener in food and tobacco, is metabolized to glycyrrhetic acid (GA) *in vivo*. Habitual consumption of the sweetener causes the metabolite to selectively inhibit 11-β-hydroxysteroid dehydrogenase2 (11-β-HSD2), an enzyme involved in the conversion of cortisol to cortisone (Fig. 13.30). Inhibition of this enzyme causes cortisol levels to increase, producing the adverse effects of increased blood pressure and electrolyte imbalances. The effect of 130 mg GA/day for 5 days (to mimic the continuous use) on the inhibition of 11-β-HSD2 was studied in 12 healthy male subjects, using the urinary cortisol (F) cortisone (E) ratio as a biomarker for assessing the GA-induced impairment of 11-β-HSD2 activity *in vivo*. The observed variability could be attributed to the interindividual differences in GI transit time of GA's metabolites. A 2-fold increase in this ratio was observed and it took 4 days for baseline levels to be restored post-cessation. A population PBPK/PD (indirect response) model that described the effect of GA on the urinary F/E ratio after consumption of GA showed that it was possible to forecast the change in the F/E ratio, under variable, literature-derived exposure scenarios (dose and length), thereby making it suitable as a noninvasive biomarker for identifying individuals at risk for adverse effects associated with glycyrrhizic acid overconsumption.[74]

13.7.3.3 Dosage Regimens for Subpopulations. Dosing in children and in the elderly needs to take into account changes in CYP enzyme systems, liver blood flow, liver size, and plasma protein levels with age. Some CYPs are expressed in the fetal liver and become down-regulated during development. Glucuronidation is reduced in the neonate and reaches adult levels by the age of 3. In young patients, there is rapid maturation of organs that are involved in drug absorption, distribution, and elimination necessitating a different dosage regimen. The activity of major CYPs do not change throughout adulthood, but the liver blood flow and liver size decline with age in the elderly. Serum albumin decreases slightly with age. The elderly are more susceptible to pharmacological effects of agents such as proarrhythmic anticholinergic dopaminergic drugs. Comorbidities and polypharmacy also need to be specifically addressed in the elderly.

Pharmacokinetic and PD behavior can be modeled in infants, neonates, children, pregnant women, and the elderly to provide valuable information relevant to first-time dosing decisions and the design of clinical studies. Efficacy and safety data from adults can be extrapolated to obtain PK information for pediatric patients. When PK studies are to be carried out in a pediatric population, population PK studies with sparse sampling methodology is adopted. PBPK/PD gives the option for adequate and efficient interpretation of the data.

13.7.3.4 Drug Disease Model. A drug disease model is occasionally developed during phase II, describing the time course of disease progression, in order to understand the perturbations of disease states on the clinical pharmacology

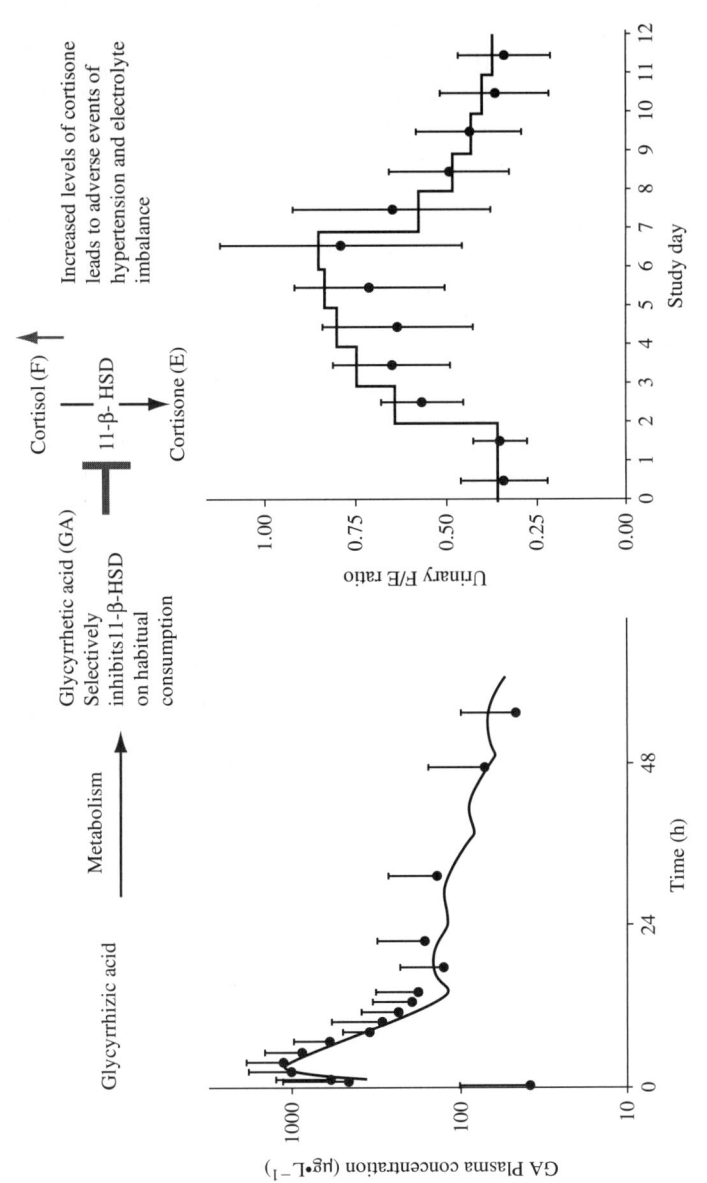

Figure 13.30. Example of population PBPK/PD. [Reprinted from "A population physiologically-based pharmacokinetic/pharmacodynamic model for the inhibition of 11-beta-hydroxysteroid dehydrogenase activity by glycyrrhetic acid." *Toxicol Appl Pharmacol* **170**:46–55. Copyright (2001). With permission from Elsevier.]

of the drug.[75] These trials are used to make clinical trial design recommendations and to view benefit versus risk for special patient subcategories such as genotype, gender, and ethnicity. Interpatient variability in response can sometimes be attributed to the initial severity of the disease. A mechanism-based model[76] describing glucose and insulin disposition and regulation was developed, based on several types of glucose provocation experiments in healthy volunteers and in type II diabetes patients. In these experiments, the homeostasis between glucose and insulin is disturbed by administering glucose orally or intravenously in order to study the system under non-steady-state conditions, making interpretations of regulatory mechanisms possible. Important differences between the healthy volunteers and the patients were identified and quantified based on established physiological knowledge.[77] The use of this model for the optimization of the intravenous glucose tolerance test in type II diabetes patients demonstrated the potential to reduce the number of patients, number of samples, and total sample time. Such a model can also be employed to quantify drug effects in an anti-diabetic treatment and to gain mechanistic understanding of drug action. Combining the control mechanisms of insulin and glucose with a PBPK model will also allow incorporation of physiological and pathological influences that can further modulate the concentrations of the drug and therefore its action. The use of the model can be extended to long term studies of disease progression models[78, 79] that can capture the changes in fasting plasma glucose (FPG), glycosylated hemoglobin (HbA1c), and in β-cell mass.

13.7.3.5 Optimize Clinical Study Design with Respect to Dose or Exposure Range and Sampling Schemes[80] through Clinical Trial Simulations.
Clinical trial simulation (CTS) allows virtual exploration of dosing regimens, for optimizing trial designs, and offers the advantages of fewer, focused studies, improved efficiency, and cost effectiveness. For drugs with a narrow therapeutic window, such as anti-cancer drugs, an optimal trial design assumes greater importance.[81] PBPK/PD models established during the preclinical and early clinical learning phases, and supplemented by population data analysis, can provide the backbone for these assessments. Physiological and pathophysiological processes, including disease development and variability could be factored in through PBPK (see Chapter 8) and mechanistic PD models to predict compliance and adverse events. Examples from the literature[82] are compiled in Table 13.10. Potential benefits of trial simulation include upfront comparison of alternative study designs, identification of appropriate dosing regimens to meet the study objectives, determination of the number of subjects needed for adequate power, and evaluation of drug interaction and disease effects. All of these can significantly reduce the number of phase II studies.

Physiologically-based PK/PD supports scheduling, enables the clinical development team to test hypothetical scenarios, and simulates trials to assess the likely range of trial outcomes. The effect of each variable, inclusion/exclusion criteria, and dropout rates can also be studied to design an optimal

13.7 APPLICATIONS OF PK OR PBPK/PD MODELING IN DRUG DISCOVERY

TABLE 13.10. Examples of Clinical Trial Simulations

Clinical Trial Simulations	Reference
Differentiate utility of effect-controlled and concentration-controlled clinical trials with sparse sampling.	Levy[32] and Ebling and Levy[83]
Age and lean body mass identified as significant covariates for the EEG effect of short acting opioid remifentanil. PK/PD simulation identified that bolus doses should be halved and infusion rates decreased to one-third in elderly compared with young patients.	Minto et al.[84,85]
Simulation of fibrinogen receptor occupancy including interindividual PK/PD variability to determine appropriate dosing regimens for achieving predefined receptor occupancy with antagonist fradafiban in phase IIa.	Brickl et al.[86]
PK/PD model including cytokine processes and disease progression that describes development of drug resistance toward reverse transcriptase inhibitors and aspartile protease inhibitors to explore ways of delaying resistance development to antiretroviral agents.	Jackson[87]
Optimization of study design for randomized concentration-controlled trial evaluation of immune suppressive agent mycophenolate mofetil.	Hale et al.[88]
PK/PD-based simulation of a clinical trial for an oral anticancer drug applied to select appropriate dosing regimen and evaluate consequences of dose adaptation rules on efficacy and adverse events.	Gieschke et al.[89]

trial. Trial simulations can be easily updated as new relevant clinical data become available. Added benefits are smaller, less costly trials, with more information gleaned from a given subject population and where subjects are faster to enroll. Some companies are taking trial simulation into earlier stages of development, running very high replicate simulations to investigate early-signals of drug effect. Though such large, early-stage simulations are computationally intensive and can consume resources up front, they can also result in huge savings in the long term as better informed "go" or "no-go" decisions can be made. Knowledge-based decision making increases confidence in those decisions, due to an enhanced understanding of the PD and safety margins of the drugs in question.

13.7.3.6 Cost Benefits. Pharmacokinetic or PBPK/PD integration can bring about tremendous cost savings to a drug development program through the following means:

1. Reduce sample sizes associated with exposure–response versus dose–response approaches. PBPK/PD modeling at the phase IIb stage can help eliminate the need for an additional dose group of patients, which

TABLE 13.11. Summary of Applications of PK or PBPK/PD Integration in Drug Development

Drug Development Phase	Test Population	Size	Purpose of Study	PBPK/PD Application
Phase 0 (preclinical)	Laboratory and animal studies		Demonstrate biological activity in disease animal models. Generate toxicology data to support initial dosing in human. Establish dosing range for first time in human Phase I studies.	• Predict clinical potency. • Predict human oral bioavailability, clearance, and potential for drug–drug interaction using PBPK (Sections 13.7.2.3–13.7.2.5) • Provide PBPK/PD driven guidance for dose range and optimal sampling based on the above (Sections 13.7.2.3–13.7.2.5). • Assess safety margins (Section 13.7.2.6).
Phase I (single and multiple ascending dose and tolerance studies)	Healthy	20–100	Determine safety and dosage.	• Determine MTD. • Refinement of preclinical PBPK/PD models based on the human PK and PD data. • Develop bottom-up population PBPK model. Predict expected variability in exposure (Section 13.7.3.2). • Arrive at an optimal dosing regimen and sampling design that will provide the desired target exposure for PoC study with PBPK/PD (Section 13.7.3.5).
Phase IIa (proof of concept, PoC)	Representative patient population	Small, 100–500	Demonstrate measurable benefit in targeted patients.	• Develop drug disease model (Section 13.7.3.4). • Arrive at an optimized clinical study design, dosing, and sampling schemes using PBPK/PD (Section 13.7.3.5).

Phase IIb	Representative patient population	Small	Optimal use in target population-Establish doses for Phase III. Maximize benefit-to-risk ratio.	• Quantitation of benefit-to-risk ratio (Section 13.7.3.2). • PBPK/PD-derived dose selection for Phase III (Section 13.7.3.1). • Assess efficacy/safety profile relative to comparators based on PBPK/PD (Section 13.7.2.7).
Phase III (2 or more long-term comparative, randomized, single or double blind, parallel, placebo-controlled trials, often multicenter and multinational)	Patient population	Large 1000–5000	Confirm acceptable benefit-to-risk ratio.	• Understand response-segmentation in different populations (demographics, disease, concomitant medication, etc.) (Section 13.7.3.2). • Validate population PBPK/PD model (Section 13.7.3.2). • Confirm dose–exposure response relationship in target subpopulation with PBPK/PD (Section 13.7.3.3). • Assess need for dose adjustment in special population using PBPK/PD (Section 13.7.3.2).
Phase IV (postmarketing surveillance)	Patient	Larger than Phase III	Ongoing technical support after approval and safety surveillance to monitor adverse events.	• Monitor medications with a narrow therapeutic index for drug–drug interactions, drug disease interactions, etc. that interfere with the effect or safety of the drug (Section 13.7.3.4).

translates to large time and cost savings. Selection of an optimal dose may lower sample size requirements for noninferiority trials.
2. Aid higher quality submissions that could:
   a. Facilitate regulatory review
   b. Enhance relationship with regulatory authorities
   c. Minimize postsubmission questions
   d. Facilitate transition to novel dosage forms based on PK studies only, if the PBPK/PD relationship is known

Table 13.11 summarizes the applications of PBPK/PD integration in drug development discussed in this section.

In summary, the benefits of PBPK/PD modeling extend to all phases of drug discovery and development and broadly fall into the following categories:

1. Data integration and exploratory data analysis to maximize knowledge gain.
2. Mechanistic understanding, hypothesis testing, and estimation of inaccessible system variables.
3. Data-driven responses to regulatory questions and/or "what-if" scenarios.
4. Integrated and informed decision making at every stage of drug discovery and development.
5. Design of PK and PD experiments (derive optimal sampling schemes for exposure and response) for future trials.
6. Extrapolation (making predictions of system responses under new conditions that were not originally used to build the model):
   a. *In vitro* to *in vivo*
   b. Experimental animals to humans
   c. Healthy volunteers to patients or special populations such as children or the elderly.
   d. One mode of administration to another
   e. One dose, dosing regimen, or formulation to another[90]
7. Demonstrate competitive edge of NCE over existing drugs and develop approval criteria by providing evidence of effectiveness and safety through PBPK/PD. Provide basis for data-driven market differentiation.

13.8 REGULATORY PERSPECTIVE

Regulatory decisions include approval of **investigational new drug (IND)**, protocol design at the end of phase IIa, **new drug application (NDA)** for small-molecule drugs and **biologics license application (BLA)** for biologics, pediatric exclusivity, formulation of dosing instructions for labeling purposes, public warnings and precautions for safety issues or black-box label warnings, and product withdrawal from market. Regulatory attention to the mechanistic PK/PD modeling approach is on the increase. The U.S. FDA, through its critical path initiative,[91] emphasizes the need for a quantitative data analysis of PK/ or PBPK/PD to understand efficacy, safety, and dosing regimen design through dynamic learn–apply cycles, and to aid regulatory decision making.[92–95] An exploratory analysis of PK, PD, and disease progression data is referred to as **pharmacometrics**. Section 115 of FDAMA (FDA Modernization Act of 1997) provides for "new drug approval based upon evidence from a single adequate and well controlled trial, supported by confirmatory scientific evidence from other studies (e.g., Phase II PK/PD studies) in the NDA." Ketorolac is an example of such an approval by the FDA.[9] Machado et al.[96] present examples of applications of PK/PD for the purposes of drug labeling, resolving safety concern, and for improving therapeutic monitoring of anaesthetic depth during surgery. Applications of PK/PD to a variety of activities in regulatory decision making[97] are discussed in the literature. Case studies in which PK/PD have been pivotal in regulatory decisions[98] and led to labeling instructions have been presented. These were about half of the 42 cases that had pharmacometric analysis among the 244 NDAs submitted between 2000 and 2004, according to an interesting FDA survey. The FDA recommends special end-of-phase IIa meetings with the sponsor to discuss the results of PK/PD. The FDA is also building an information library of empirical and mechanistic disease models based on information from the literature that captures the time course of disease progression for various diseases. Guidance from regulatory bodies such as the U.S. FDA and EMA (formerly EMEA) provide recommendations on study design, data analysis, and report writing for population PK or PK/PD studies.[99–101] Model-based analysis forms the basis of regulatory consultation or decisions regarding safety and dosing. PK/PD models can influence decisions on market access, labeling, trial, and pediatric exclusivity.

13.9 CONCLUSIONS

Pharmacokinetics and PD are not mutually exclusive disciplines. A complete characterization of drug interaction with its target is possible only through integration of these two complementary disciplines, each providing a piece of the dose–response picture. Empirical dose–response models and mechanism-based PBPK/PD models represent two different ways of performing PK/PD integration. The latter offers precious insight into the mechanisms underlying dose–exposure, exposure–target stimulation, and target stimulation–response relationships, with enhanced predictive and extrapolative values. Mechanistic PBPK/PD methods maximize benefits from experimental protocols and reduce animal studies through the optimization of protocols, enable hypothesis testing, and integrated knowledge management, guaranteeing optimum use of all available data and models. They offer a huge potential but still remain relatively unexplored and unexploited. Applications of PBPK/PD integration in drug discovery are in their infancy. However, there is a growing recognition that mechanism-based approaches with their explicit distinction between drug-specific and biological-system-specific parameters provide the best opportunity for extrapolations across species, route of administration and populations.

KEYWORDS

Adaptation: Cellular and molecular adaptations such as receptor down- and up-regulation are likely explanations for functional adaptations such as tolerance and rebound.

Affinity: Tenacity by which a drug binds to its receptor.

Agonist: Compound that binds to a receptor and functionally mimics or enhances the action of an endogenous ligand.

Baseline Level: Or baseline effect is the physiological effect in the absence of drug dosing. It can change as a function of time of day or food intake.

Biomarkers: Biomarkers are defined as biological characteristics that can be objectively measured and evaluated as indicators of normal/pathogenic biological processes or pharmacological responses to a therapeutic intervention.[102,103] Common biomarkers include readily detectable plasma or cellular molecules (typically, gene mutations, proteins, or RNAs) or imaging technologies. Depending on their use, biomarkers are called pharmacogenomic, pharmacodynamic (PD), disease biomarkers. Pharmacogenomic biomarkers are measurable DNA and/or RNA characteristics that identify genotype/phenotype variations affecting pharmacokinetic or PD (safety and efficacy), which in turn aids in the prediction of response to medications or susceptibility to adverse events but not necessarily health outcomes. Pharmacodynamic biomarkers are indicators of target engagement, pharmacological or safety response. Disease biomarkers in the diagnosis of a disease, in monitoring the progression or regression of a disease or in the assessment of treatment outcome. An ideal biomarker is one that is quantitative, reproducible, specific, sensitive, well-validated for the purpose and associated with low variability.

Clinical Endpoint: A clinical endpoint refers to an overall target outcome in a clinical trial that a protocol is designed to evaluate. Common endpoints include severe toxicity, disease progression or survival.

Competitive Antagonist: Acts competitively at the same binding site as an agonist.

Coupling: Transduction process between the receptor occupancy and drug response.

Desensitization: Agonist-induced loss of receptor responsiveness triggered by a persistent exposure of receptors to an agonist by any mechanism—receptor phosphorylation, uncoupling, antagonistic metabolites, or negative physiological feedback that occurs over a relatively short period, receptor down-regulation or receptor internalization or sequestration. For an agonist, it is characterized by a clockwise hysteresis loop in the concentration–effect plot. It is modeled by time-dependent EC_{50}. Homologous desensitization is induced by prior activation of that receptor system. Heterologous desensitization is a result of activation of another receptor system.

Drug Effect: Any drug-induced change in a physiological parameter in comparison with the predose baseline value.

Efficacy: Drug's ability to bind a receptor and elicit a functional response.

Homologous Receptor Down-Regulation: Changes in either receptor synthesis (i.e., reduction in the steady-state level of receptor mRNA) and/or receptor degradation. Receptor down-regulation is also cell specific and involves the complex interaction between multiple cellular events to influence receptor expression.

Internalization: Sequestration of agonist–receptor complexes into endosomal compartments within cells. Removal of receptors from the plasma membrane prevents further stimulation by intercellular signals.

Intrinsic Activity: Property of a drug that determines the amount of pharmacological effect produced per unit of drug–receptor complex formed. Two ligands combining with equivalent sets of receptors may not produce equal effects even if both are given in maximally effective doses as they differ in their intrinsic activities. Meperidine and morphine presumably combine with the same receptors to produce analgesia, but regardless of dose, the maximum degree of analgesia produced by morphine is greater than that produced by meperidine.

Inverse Agonists: Bind to constitutively active receptors (receptors that are coupled to second-messenger pathways in the absence of agonists—mostly GPCRs) and shift the equilibrium to the formation of the inactive conformer.

Investigational New Drug (IND) Application: After completing preclinical testing, a company files an IND with the U.S. FDA to test a new drug in clinical investigation. The IND becomes effective if the FDA does not disapprove it within 30 days. The IND shows results of previous experiments; how, where, and by whom the new studies will be conducted; the chemical structure of the compound; how it is thought to work in the body; any toxic effects found in the animal studies; and how the compound is manufactured.

Irreversible Inhibitor or Antagonist (as opposed to reversible): Covalently modifies and inactivates a target protein by forming an irreversible complex. Example: alkylating agents. Irreversible inhibitors are characterized by low turnover of effect or a long response half-life and are long acting by virtue of receptor occupancy rather than PK half-life. Duration of effect is determined only by turnover of target protein.

New Drug Application (NDA)/Biologic License Application (BLA): Following the completion of all three phases of clinical trials, a company analyzes all of the data and files an NDA or BLA with the FDA if the data successfully demonstrate both safety and effectiveness. The applications contain all of the scientific information that the company has gathered.

Noncompetitive (Unsurmountable) Antagonist: A drug that blocks or inhibits the production of a pharmacological response by the agonist, via its effects at a binding site different from the agonist. The binding could trigger a conformational change in the receptor protein and interrupt the linkage between receptor and effector. A noncompetitive antagonist can be reversible or irreversible. This will usually reduce the slope and the maximum of the log-concentration–response curve of the agonist. Both noncompetitive and irreversible inhibitors decrease the receptor count that is available for the endogenous agonist and thereby reduce its efficacy.

Partial Agonist: Agonist that does not elicit a full effect, even when bound to all receptors. Partial agonists may be full agonists for some effects and actually function as antagonists in some cases. Usually, as receptor density in a tissue increases, the efficacy of a partial agonist also increases. In some cases, a partial agonist can become a full agonist if receptor density increases sufficiently. The use of long-acting, partial opioid agonist buprenorphine for stabilization and medical withdrawal of heroin is based on this principle. Up-regulation of a receptor in a tissue increases the efficacy of a partial agonist, even to becoming a full agonist. Similarly, a full agonist may become a partial agonist with receptor down-regulation. Receptor alkylation can also easily convert a full agonist into a partial agonist. Theories that explain partial agonism:

1. The partial agonist may fit the receptor binding site well but coupling between receptor occupancy, receptor conformational change, and transduction may not be efficient.
2. The receptor may isomerize between an active and an inactive form. If the drug has a higher affinity to the inactive form than the active, then it may appear to be a partial agonist.

Pharmacodynamic Response Half-Life ($t_{1/2}$ response): Time required for 50% reduction in pharmacodynamic response.

Pharmacometrics: Science that deals with quantifying disease and pharmacology, and influencing drug development and regulatory decisions.

Potency: Ability of a drug to cause a measured functional change.

Rebound: Occurs when the body tries to return to homeostasis following a sudden discontinuation of a drug. Receptors that are suddenly deprived of their agonists/blockers become hypersensitive to an agent that targets it, causing a further exacerbation of the symptoms/conditions that triggered the use of the drug in the first place. This rebound effect can be minimized by a gradual rather than a sudden discontinuation of the drug.

Receptor Reserve: Exists when spare receptors are available.

Second Messenger: Relay molecules that target molecules in the cytosol and/or nucleus in response to receptors that are activated at the cell surface by hormones, growth factors etc. Second messengers can greatly amplify the strength of the signal by causing large changes in the biochemical activities within the cell. Major classes of second messengers include cyclic nucleotides such as cAMP and cGMP, inositol trisphosphate (IP3), diacylglycerol (DAG), and calcium ions Ca^{2+}.

Sensitization: Occurs in drug addiction and is defined as an increased effect of drug following repeated doses. Drug-induced sensitization is the opposite of tolerance. For an agonist, it is characterized by a counterclockwise hysteresis loop in the concentration–effect plot. Sensitization can result from accumulation of parent or active metabolites, positive feedback, up-regulation of receptors, or enhanced sensitivity of receptors to a ligand.

Spare Receptors: Receptors in excess of those required to produce a full effect. A high receptor reserve causes maximal responses to be elicited by an agonist at less than maximal receptor occupancy. Their existence may be demonstrated by the undiminished maximal response by a high-affinity agonist, even after a portion of the receptor pool is occupied by quasi-irreversible antagonist. Example, catecholamines exert maximal effect, when 90% of β-adrenoreceptors are occupied by a quasi-irreversible antagonist, implying that the myocardium has a large proportion of spare adrenoreceptors.

Surrogate Biomarker or Endpoint: A surrogate biomarker is a laboratory substitute for a clinical endpoint.

Tachyphylaxis: Rapidly developing desensitization or tolerance to a drug. It is often used to describe decreased response to a drug that occurs readily with regular exposure to short-acting $β_2$-adrenoceptor agonists.

Target Engagement: Describes how robustly a drug interacts with the mechanism of action. Exposure of a drug at the effect site determines the binding to the target. The translation of target binding/receptor occupancy to the expected target function and to its pharmacological response should be monitored by appropriate biomarkers to secure better confidence in the ability of a drug to modulate an intended therapeutic target or pathway. PET imaging provides a direct measure of target binding. Other indirect measures include downstream pharmacological events or the ratio of unbound concentration to IC_{50}.

Target-Mediated Drug Disposition: Significant and high-affinity binding of a drug to a capacity-limited target leading to nonlinear PK (V_{ss} and clearance) and thus nonlinear PD. This is more relevant for biologicals than for small molecules.

Target Validation: Evaluate a selected drug target (an enzyme, receptor, ion channel, etc.) to demonstrate its critical involvement in a disease process, such that any modulation of its activity will bring about a substantial therapeutic effect. Target validation enables informed decision for commencing a drug discovery program.

Tolerance: Loss of receptor-activated function following prolonged exposure to a drug, usually due to receptor down-regulation. For an agonist drug, tolerance is characterized by a clockwise hysteresis loop in the concentration–effect plot. Tolerance causes an empirical shift to the right in the dose–response curve with an associated increase in the ED_{50} or EC_{50} and may also be accompanied by a reduction in the maximal effect (E_{max}). Tolerance is reserved for reduced responsiveness developing over longer period while desensitization over shorter period. Nicotine increases the stimulation of nicotinic receptors. The excessive and chronic activation of these receptors is balanced by a down-regulation in the number of active receptors causing the psychotropic effect of nicotine. It is modeled by a time-dependent E_{max}.

Transduction: Cascade of often nonlinear processes that govern the time course of the pharmacological (biological or physiological) response *in vivo*, following drug-induced target activation.

REFERENCES

1. Benet LZ. Pharmacokinetics: Basic principles and its use as a tool in drug metabolism. In: Mitchell JR, Horning MG, eds. *Drug Metabolism and Drug Toxicity*. Raven Press: New York, 1984, p. 199–211.
2. Sieghart W. Pharmacology of benzodiazepine receptors: An update. *J Psychiatry Neurosci*. 1994;19(1):24–29.
3. Hill AV. The possible effects of the aggregation of the molecules of hemoglobin on its dissociation curves. *J.Physiol*. 1910;40:iv–vii.
4. Goutelle S, et al. The hill equation: A review of its capabilities in pharmacological modeling. *Fundam Clin Pharmacol*. 2008;22(6):633–648.
5. Danhof M, Alvan G, Dahl SG, Kuhlmann J, Paintaud G. Mechanism-based pharmacokinetic-pharmacodynamic modeling—a new classification of biomarkers. *Pharm Res*. 2005;22(9):1432–1437.
6. Nordstrom AL, et al. Central D_2-dopamine receptor occupancy in relation to antipsychotic drug effects: A double-blind PET study of schizophrenic patients. *Biol Psychiatry*. 1993;33(4):227–235.
7. Van der Graaf PH, Van Schaick EA, Math-ot RA, Ijzerman AP, Danhof M. Mechanism-based pharmacokinetic-pharmacodynamic modeling of the effects of N6-cyclopentyladenosine analogs on heart rate in rat: Estimation of in vivo

REFERENCES

operational affinity and efficacy at adenosine A1 receptors. *J Pharmacol Exp Ther.* 1997;283(2):809–816.

8. Lesko LJ, Atkinson AJ, Jr. Use of Biomarkers and surrogate endpoints in drug development and regulatory decision making: Criteria, validation, strategies. *Annu Rev Pharmacol Toxicol.* 2001;41:347–366.

9. Peck CC, Rubin DB, Sheiner LB. Hypothesis: A single clinical trial plus causal evidence of effectiveness is sufficient for drug approval. *Clin Pharmacol Ther.* 2003;73(6):481–490.

10. Colburn WA. Biomarkers in drug discovery and development: From target identification through drug marketing. *J Clin Pharmacol.* 2003;43(4):329–341.

11. Colburn WA, Lee JW. Biomarkers, validation and pharmacokinetic-pharmacodynamic modeling. *Clin Pharmacokinet.* 2003;42(12):997–1022.

12. Colburn WA. Simultaneous pharmacokinetic and pharmacodynamic modeling. *J Pharmacokinet Biopharm.* 1981;9(3):367–388.

13. Gabrielsson J, Dolgos H, Gillberg PG, Bredberg U, Benthem B, Duker G. Early integration of pharmacokinetic and dynamic reasoning is essential for optimal development of lead compounds: Strategic considerations. *Drug Discov Today.* 2009;14(7–8):358–372.

14. Langley JN. On the physiology of salivary secretion. *J. Physiol.* 1878;1:339–367.

15. Ehrlich P. Chemotherapeutic scientific principles, methods and results. *Lancet.* 1915:445–451.

16. Clark AJ. *The Mode of Action of Drugs on Cells.* London: Edward Arnold. 1933.

17. Maehle AH, Prull CR, Halliwell RF. The emergence of the drug receptor theory. *Nat Rev Drug Discov.* 2002;1(8):637–641.

18. Copeland RA, Pompliano DL, Meek TD. Drug-target residence time and its implications for lead optimization. *Nat Rev Drug Discov.* 2006;5(9):730–739.

19. Tuk B, van Oostenbruggen MF, Herben VM, Mandema JW, Danhof M. Characterization of the pharmacodynamic interaction between parent drug and active metabolite in vivo: Midazolam and alpha-OH-midazolam. *J Pharmacol Exp Ther.* 1999;289(2):1067–1074.

20. Van der Graaf PH, Van Schaick EA, Visser SA, De Greef HJ, Ijzerman AP, Danhof M. Mechanism-based pharmacokinetic-pharmacodynamic modeling of antilipolytic effects of adenosine A(1) receptor agonists in rats: Prediction of tissue-dependent efficacy in vivo. *J Pharmacol Exp Ther.* 1999;290(2):702–709.

21. Cox EH, Kerbusch T, Van der Graaf PH, Danhof M. Pharmacokinetic-pharmacodynamic modeling of the electroencephalogram effect of synthetic opioids in the rat: Correlation with the interaction at the mu-opioid receptor. *J Pharmacol Exp Ther.* 1998;284(3):1095–1103.

22. Garrido M, Gubbens-Stibbe J, Tukker E, et al. Pharmacokinetic-pharmacodynamic analysis of the EEG effect of alfentanil in rats following beta-funaltrexamine-induced mu-opioid receptor "knockdown" in vivo. *Pharm Res.* 2000;17(6): 653–659.

23. Zuideveld KP, et al. Mechanism-based pharmacokinetic-pharmacodynamic modeling of 5-HT1A receptor Agonists: Estimation of in vivo affinity and intrinsic efficacy on body temperature in rats. *J Pharmacol Exp Ther.* 2004;308(3): 1012–1020.

24. Jonker DM, Kenna LA, Leishman D, Wallis R, Milligan PA, Jonsson EN. A pharmacokinetic-pharmacodynamic model for the quantitative prediction of dofetilide clinical QT prolongation from human ether-a-go-go-related gene current inhibition data. *Clin Pharmacol Ther*. 2005;77(6):572–582.
25. Alfieri-Rolla G. The effect of carbon tetrachloride on the kinetics of bromosulphalein on the rabbit. *Eur. J. Pharmacol*. 1968;3(4):330–336.
26. Sheiner LB, Stanski DR, Vozeh S, Miller RD, Ham J. Simultaneous modeling of pharmacokinetics and pharmacodynamics: Application to d-tubocurarine. *Clin Pharmacol Ther*. 1979;25(3):358–371.
27. Nagashima R, O'Reilly RA, Levy G. Kinetics of pharmacologic effects in man: The anticoagulant action of warfarin. *Clin Pharmacol Ther*. 1969;10(1):22–35.
28. Dayneka NL, Garg V, Jusko WJ. Comparison of four basic models of indirect pharmacodynamic responses. *J Pharmacokinet Biopharm*. 1993;21(4):457–478.
29. Yao Z, Krzyzanski W, Jusko WJ. Assessment of basic indirect pharmacodynamic response models with physiological limits. *J Pharmacokinet Pharmacodyn*. 2006;33(2):167–193.
30. Mager DE, Jusko WJ. Pharmacodynamic modeling of time-dependent transduction systems. *Clin Pharmacol Ther*. 2001;70(3):210–216.
31. Mager DE, Wyska E, Jusko WJ. Diversity of mechanism-based pharmacodynamic models. *Drug Metab Dispos*. 2003;31(5):510–518.
32. Levy G. Mechanism-based pharmacodynamic modeling. *Clin Pharmacol Ther*. 1994;56(4):356–358.
33. Ramakrishnan R, DuBois DC, Almon RR, Pyszczynski NA, Jusko WJ. Pharmacodynamics and pharmacogenomics of methylprednisolone during 7-day infusions in rats. *J Pharmacol Exp Ther*. 2002;300(1):245–256.
34. Jonkers RE, Braat MC, Koopmans RP, van Boxtel CJ. Pharmacodynamic modeling of the drug-induced downregulation of a beta 2-adrenoceptor mediated response and lack of restoration of receptor function after a single high dose of prednisone. *Eur J Clin Pharmacol*. 1995;49(1–2):37–44.
35. Meibohm B, Hochhaus G, Derendorf H. Time dependency of the pharmacological response to glucocorticoids. *Clin. Pharmacol. Ther*. 1997;61:155.
36. Ambre JJ. Acute tolerance to pressor effects of cocaine in humans. *Ther Drug Monit*. 1993;15(6):537–540.
37. Chow MJ, Ambre JJ, Ruo TI, Atkinson AJ,, Jr, Bowsher DJ, Fischman MW. Kinetics of cocaine distribution, elimination, and chronotropic effects. *Clin Pharmacol Ther*. 1985;38(3):318–324.
38. Porchet HC, Benowitz NL, Sheiner LB. Pharmacodynamic model of tolerance: Application to nicotine. *J Pharmacol Exp Ther*. 1988;244(1):231–236.
39. Bauer JA, Fung HL. Pharmacodynamic models of nitroglycerin-induced hemodynamic tolerance in experimental heart failure. *Pharm Res*. 1994;11(6):816–823.
40. Wakelkamp M, Alvan G, Gabrielsson J, Paintaud G. Pharmacodynamic modeling of furosemide tolerance after multiple intravenous administration. *Clin Pharmacol Ther*. 1996;60(1):75–88.
41. Derendorf H, Meibohm B. Modeling of pharmacokinetic/pharmacodynamic (PK/PD) relationships: Concepts and perspectives. *Pharm Res*. 1999;16(2):176–185.

REFERENCES

42. Liefaard LC, et al. Population pharmacokinetic analysis for simultaneous determination of B (max) and K (D) in vivo by positron emission tomography. *Mol Imaging Biol.* 2005;7(6):411–421.
43. Geldof M, Freijer J, van Beijsterveldt L, Danhof M. Pharmacokinetic modeling of nonlinear brain distribution of fluvoxamine in the rat. *Pharm Res.* 2008;25(4): 792–804.
44. Groenendaal D, Freijer J, de Mik D, Bouw MR, Danhof M, de Lange EC. Population pharmacokinetic modeling of nonlinear brain distribution of morphine: Influence of active saturable influx and P-glycoprotein mediated efflux. *Br J Pharmacol.* 2007;151(5):701–712.
45. Groenendaal D, Freijer J, de Mik D, Bouw MR, Danhof M, de Lange EC. Influence of biophase distribution and P-glycoprotein interaction on pharmacokinetic-pharmacodynamic modeling of the effects of morphine on the EEG. *Br J Pharmacol.* 2007;151(5):713–720.
46. Clewell HJ,, 3rd, Andersen ME, Wills RJ, Latriano L. A physiologically-based pharmacokinetic model for retinoic acid and its metabolites. *J Am Acad Dermatol.* 1997;36(3 Pt 2):S77–85.
47. Dingemanse J, Appel-Dingemanse S. Integrated pharmacokinetics and pharmacodynamics in drug development. *Clin Pharmacokinet.* 2007;46(9):713–737.
48. Chen H, Wang X, Chen XZ. A novel PBPK/PD model with automatic nervous system in anesthesia. *Conf Proc IEEE Eng Med Biol Soc.* 2005;1:66–69.
49. Smith DA. Pharmacokinetics and pharmacodynamics in toxicology. *Xenobiotica.* 1997;27(5):513–525.
50. Simeoni M, et al. Predictive pharmacokinetic-pharmacodynamic modeling of tumor growth kinetics in xenograft models after administration of anticancer agents. *Cancer Res.* 2004;64(3):1094–1101.
51. Gabrielsson J, Green AR. Quantitative pharmacology or pharmacokinetic pharmacodynamic integration should be a vital component in integrative pharmacology. *J Pharmacol Exp Ther.* 2009;331(3):767–774.
52. Bouw MR, Xie R, Tunblad K, Hammarlund-Udenaes M. Blood-brain barrier transport and brain distribution of morphine-6-glucuronide in relation to the antinociceptive effect in rats—pharmacokinetic/pharmacodynamic modeling. *Br J Pharmacol.* 2001;134(8):1796–1804.
53. Ling MP, Liao CM. A human PBPK/PD model to assess arsenic exposure risk through farmed tilapia consumption. *Bull Environ Contam Toxicol.* 2009;83(1): 108–114.
54. Andersen ME. Development of physiologically-based pharmacokinetic and physiologically-based pharmacodynamic models for applications in toxicology and risk assessment. *Toxicol Lett.* 1995;79(1–3):35–44.
55. Peters SA, Hultin L. Early identification of drug-induced impairment of gastric emptying through physiologically-based pharmacokinetic (PBPK) simulation of plasma concentration-time profiles in rat. *J Pharmacokinet Pharmacodyn.* 2008; 35(1):1–30.
56. Yu LJ, et al. Establishment of correlation between in vitro enzyme binding potency and in vivo pharmacological activity: Application to liver glycogen phosphorylase a inhibitors. *J Pharmacol Exp Ther.* 2006;317(3):1230–1237.

57. Zhang X, et al. A physiologically-based pharmacokinetic/pharmacodynamic model for carbofuran in Sprague-Dawley rats using the exposure-related dose estimating model. *Toxicol Sci.* 2007;100(2):345–359.
58. Xu L, Eiseman JL, Egorin MJ, D'Argenio DZ. Physiologically-based pharmacokinetics and molecular pharmacodynamics of 17-(allylamino)-17-demethoxygeldanamycin and its active metabolite in tumor-bearing mice. *J Pharmacokinet Pharmacodyn.* 2003;30(3):185–219.
59. Florian JA, Jr, et al. A physiologically-based pharmacokinetic (PBPK) and pharmacodynamic (PD) model of docetaxel (doc) and neutropenia in humans. *J. Clin. Oncol.* 2007;25(18S):2567.
60. Kloft C, Wallin J, Henningsson A, Chatelut E, Karlsson MO. Population pharmacokinetic-pharmacodynamic model for neutropenia with patient subgroup identification: Comparison across anticancer drugs. *Clin Cancer Res.* 2006;12(18):5481–5490.
61. Cavero I. Using pharmacokinetic/pharmacodynamic modeling in safety pharmacology to better define safety margins: A regional workshop of the safety pharmacology society. *Expert Opin Drug Saf.* 2007;6(4):465–471.
62. Chien JY, Friedrich S, Heathman MA, de Alwis DP, Sinha V. Pharmacokinetics/pharmacodynamics and the stages of drug development: Role of modeling and simulation. *AAPS J.* 2005;7(3):E544–559.
63. Gallo JM. Pharmacokinetic/pharmacodynamic-driven drug development. *Mt Sinai J Med.* 2010;77(4):381–388.
64. Rajman I. PK/PD modeling and simulations: Utility in drug development. *Drug Discov Today.* 2008;13(7–8):341–346.
65. Hashimoto Y, Sheiner LB. Designs for population pharmacodynamics: Value of pharmacokinetic data and population analysis. *J Pharmacokinet Biopharm.* 1991;19(3):333–353.
66. Sheiner LB, Ludden TM. Population pharmacokinetics/dynamics. *Annu Rev Pharmacol Toxicol.* 1992;32:185–209.
67. Karlsson MO, Molnar V, Bergh J, Freijs A, Larsson R. A general model for time-dissociated pharmacokinetic-pharmacodynamic relationship exemplified by paclitaxel myelosuppression. *Clin Pharmacol Ther.* 1998;63(1):11–25.
68. Yuh L, et al. Population pharmacokinetic/pharmacodynamic methodology and applications: A bibliography. *Biometrics.* 1994;50(2):566–575.
69. Vozeh S, et al. The use of population pharmacokinetics in drug development. *Clin Pharmacokinet.* 1996;30(2):81–93.
70. Mentre F, et al. Population pharmacokinetic-pharmacodynamic analysis of fluindione in patients. *Clin Pharmacol Ther.* 1998;63(1):64–78.
71. Jamei M, Dickinson GL, Rostami-Hodjegan A. A framework for assessing interindividual variability in pharmacokinetics using virtual human populations and integrating general knowledge of physical chemistry, biology, anatomy, physiology and genetics: A tale of "bottom-up" vs "top-down" recognition of covariates. *Drug Metab Pharmacokinet.* 2009;24(1):53–75.
72. Blesch KS, Gieschke R, Tsukamoto Y, Reigner BG, Burger HU, Steimer JL. Clinical pharmacokinetic/pharmacodynamic and physiologically-based pharmacokinetic modeling in new drug development: The capecitabine experience. *Invest New Drugs.* 2003;21(2):195–223.

73. Guentert TW, Banken L, Hilton S, Holford NH. Moclobemide: Relationships between dose, drug concentration in plasma, and occurrence of adverse events. *J Clin Psychopharmacol.* 1995;15(4 Suppl 2):84S–94S.
74. Ploeger B, Mensinga T, Sips A, Deerenberg C, Meulenbelt J, DeJongh J. A population physiologically-based pharmacokinetic/pharmacodynamic model for the inhibition of 11-beta-hydroxysteroid dehydrogenase activity by glycyrrhetic acid. *Toxicol Appl Pharmacol.* 2001;170(1):46–55.
75. Lehne G, Nordal KP, Midtvedt K, Goggin T, Brosstad F. Increased potency and decreased elimination of lamifiban, a GPIIb-IIIa antagonist, in patients with severe renal dysfunction. *Thromb Haemost.* 1998;79(6):1119–1125.
76. Silber H. *Integrated Modeling of Glucose and Insulin Regulation Following Provocation Experiments.* Uppsala, Sweden: Uppsala University, 2009.
77. Jauslin-Stetina P. *Mechanism-Based Modeling of Glucose-Insulin Regulation During Clinical Provocation Experiments.* PhD thesis. Uppsala, Sweden: Uppsala University; 2008.
78. Ribbing J, Hamrén B, Svensson MK, Karlsson MO. Modelling the dynamics of glucose, insulin, insulin sensitivity and beta-cells in subjects with insulin resistance and patients with type 2 diabetes. Abstr 1257 www.page-meeting.org/?abstract=1257, 2008:17.
79. Topp B, Promislow K, deVries G, Miura RM, Finegood DT. A model of beta-cell mass, insulin, and glucose kinetics: Pathways to diabetes. *J Theor Biol.* 2000;206(4):605–619.
80. Lieberman R, McMichael J. Role of pharmacokinetic-pharmacodynamic principles in rational and cost-effective drug development. *Ther Drug Monit.* 1996;18(4):423–428.
81. van Kesteren C, Mathot RA, Beijnen JH, Schellens JH. Pharmacokinetic-pharmacodynamic guided trial design in oncology. *Invest New Drugs.* 2003;21(2):225–241.
82. Meibohm B, Derendorf H. Pharmacokinetic/pharmacodynamic studies in drug product development. *J Pharm Sci.* 2002;91(1):18–31.
83. Ebling WF, Levy G. Population pharmacodynamics: Strategies for concentration- and effect-controlled clinical trials. *Ann Pharmacother.* 1996;30(1):12–19.
84. Minto CF, et al. Influence of age and gender on the pharmacokinetics and pharmacodynamics of remifentanil. I. Model development. *Anesthesiology.* 1997;86(1):10–23.
85. Minto CF, Schnider TW, Shafer SL. Pharmacokinetics and pharmacodynamics of remifentanil. II. Model application. *Anesthesiology.* 1997;86(1):24–33.
86. Brickl R, Heinzel G, Weisenberger H, Schubert H, Rutsch W, Roth W. PK/PD simulations as a tool for rational design of clinical dosage regimens: An example with fradafiban. *Int J Clin Pharmacol Ther.* 1997;35(10):475–480.
87. Jackson RC. A pharmacokinetic-pharmacodynamic model of chemotherapy of human immunodeficiency virus infection that relates development of drug resistance to treatment intensity. *J Pharmacokinet Biopharm.* 1997;25(6):713–730.
88. Hale MD, et al. The pharmacokinetic-pharmacodynamic relationship for mycophenolate mofetil in renal transplantation. *Clin Pharmacol Ther.* 1998;64(6):672–683.
89. Gieschke R, Burger HU, Reigner B, Blesch KS, Steimer JL. Population pharmacokinetics and concentration-effect relationships of capecitabine metabolites in colorectal cancer patients. *Br J Clin Pharmacol.* 2003;55(3):252–263.

90. Lee H, Yim DS, Zhou H, Peck CC. Evidence of effectiveness: How much can we extrapolate from existing studies? *AAPS J.* 2005;7(2):E467–74.
91. U.S. FDA. Available at: http://www.fda.gov/ScienceResearch/SpecialTopics/CriticalPathInitiative/default.htm.
92. Powell JR, Gobburu JV. Pharmacometrics at FDA: Evolution and impact on decisions. *Clin Pharmacol Ther.* 2007;82(1):97–102.
93. Peck CC. Quantitative clinical pharmacology is transforming drug regulation. *J Pharmacokinet Pharmacodyn.* 2010;37(6):617–628.
94. Oo C, Chen YC. The need for multiple doses of 400 mg ketoconazole as a precipitant inhibitor of a CYP3A substrate in an in vivo drug-drug interaction study. *J Clin Pharmacol.* 2009;49(3):368–9; author reply 370.
95. Rowland M, Peck C, Tucker G. Physiologically-based pharmacokinetics in drug development and regulatory science. *Annu Rev Pharmacol Toxicol.* 2010;51:45–73.
96. Machado SG, Miller R, Hu C. A regulatory perspective on pharmacokinetic/pharmacodynamic modeling. *Stat Methods Med Res.* 1999;8(3):217–245.
97. Gobburu JV, Marroum PJ. Utilisation of pharmacokinetic-pharmacodynamic modeling and simulation in regulatory decision-making. *Clin Pharmacokinet.* 2001;40(12):883–892.
98. Bhattaram VA, et al. Impact of pharmacometrics on drug approval and labeling decisions: A survey of 42 new drug applications. *AAPS J.* 2005;7(3):E503–12.
99. EMA. Guideline on the role of pharmacokinetics in the development of medicinal products in the paediatric population. Available at: www.emea.europa.eu/pdfs/human/ewp/14701304en.pdf2006.
100. U.S. Food and Drug Administration. Pharmacokinetics in pregnancy—study design, data analysis, and impact on dosing and labelling 2004—draft guidance. Available at: http://www.fda.gov/ForConsumers/ByAudience/ForWomen/WomensHealthTopics/ucm117976.htm.
101. European Medicines Agency. EMA guideline on reporting the results of population pharmacokinetic analysis. Available at: http://www.emea.europa.eu/pdfs/human/ewp/18599006enfin.pdf.
102. Naylor S. Biomarkers: Current perspectives and future prospects. *Expert Rev Mol Design.* 2003;3:525.
103. Biomarkers Definitions Working Group. Biomarkers and surrogate endpoints: Preferred definitions and conceptual framework. *Clin Pharmacol Ther.* 2001;69(3):89–95.
104. Stephenson RP. A modification of receptor theory. *Br J Pharmac Chemother.* 1956;11:379–392.

14

PHYSIOLOGICALLY-BASED PHARMACOKINETIC MODELING OF POPULATIONS

CONTENTS

14.1 Introduction...383
14.2 Population Modeling with PBPK 384
14.3 Healthy to Target Patient Population:
      Impact of Disease on Pharmacokinetics..................386
14.4 Modeling Subpopulations: Impact of Age, Gender,
      Co-morbidities, and Genetics on Pharmacokinetics............389
14.5 Personalized Medicine with PBPK/PD....................392
      Keyword...395
      References...396

14.1 INTRODUCTION

The power of the physiologically-based pharmacokinetic (PBPK) method lies in simultaneous consideration of several impacting factors and in the mechanistic understanding that it affords. This chapter shows how to exploit this

---

Physiologically-Based Pharmacokinetic (PBPK) Modeling and Simulations: Principles, Methods, and Applications in the Pharmaceutical Industry, First Edition. Sheila Annie Peters.
© 2012 John Wiley & Sons, Inc. Published 2012 by John Wiley & Sons, Inc.

power to understand sources of population variability and to enable extrapolations across different populations.

Physiologically-based PK models use Monte Carlo methods for a priori modeling of inter- and intra-individual variability of biological parameters in a virtual population. Prospective predictions of population variability can guide study design through improved inclusion/exclusion criteria, dose selection, and sampling for PK and pharmacodynamics (PD) studies. Once clinical data becomes available, it is possible to begin to understand sources of variability and to identify covariates influencing safety and efficacy. Such a mechanistic understanding of the sources of variability would enable the PK prediction of a drug in a wide range of subpopulations, without the need for recruiting large numbers in clinical trials, saving valuable time and money.

Potential modulators of sensitivity to drug response such as disease states, age, gender, genetics, and interactions with food, chemicals, and drugs shift either the therapeutic window or the pharmacokinetic profile (relative to the therapeutic window) (Fig. 14.1). Thus, pharmacokinetics accounts for some of the variability in the dose–response. For drugs with narrow therapeutic windows, the risk of losing drug efficacy or introducing adverse events is even higher. PBPK along with systems biology methods allows for simultaneous consideration of the impact of various factors on target tissue exposure and drug response, enabling a tailor-suited dosing regimen for different individuals. Incorporation of relevant patient characteristics in these mechanistic models allows for individualized therapy.[1]

14.2 POPULATION MODELING WITH PBPK

Population pharmacokinetics[2,3] seeks to identify and quantify the pathophysiological factors that cause changes in the dose–concentration relationship, so that any resulting clinically significant shifts in the therapeutic index can be addressed through appropriate dose adjustments. Population PK analysis requires a relatively large number of subjects that many clinical trials cannot afford. PBPK models can complement the population PK approach, leading to more efficient data collection. In PBPK models, a virtual population can be generated using Monte Carlo (MC) simulations by incorporating known demographic, environmental, and genetic variations in a population (see Chapter 8). The PBPK-estimated interindividual variability in concentration–time profiles of the virtual population can then be compared with that observed clinically. When the model predicts the observed data, it implies that the sources of variability considered in the model were sufficient. When the prediction is poor, then it suggests that unidentified sources are more important. Observed variability in intravenous profiles in a first-in-human trial is simulated by incorporating polymorphism- and gender-driven differences in enzyme and transporter abundances in the model. Through a series of PBPK

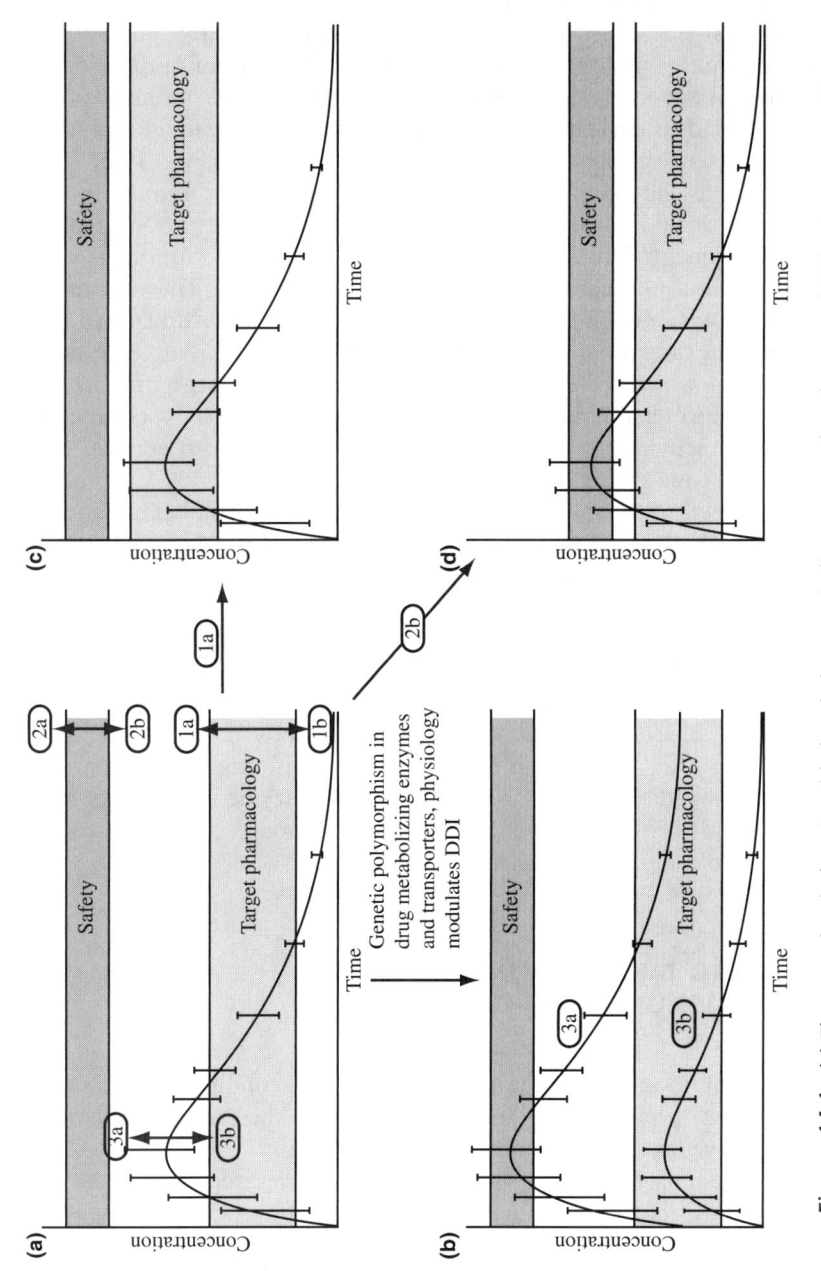

Figure 14.1. (a) Therapeutic window. Double-headed arrows indicate upward or downward shifts that are possible for safety, target pharmacology, or pharmacokinetics due to pharmacogenomics, physiology modulators, or drug–drug interactions. Effects of these factors in shifting (b) the efficacy, (c) PK profile, and (d) safety and its consequences are also illustrated.

simulations of the observed profiles, the influences of other covariates can be identified. A similar approach to identify sources of variability in oral PK profiles can be more challenging considering the much larger number of variables affecting oral profiles. Covariates identified from a healthy population are then appropriately employed for patient population, and an optimal drug dosing for the patient study is established. Such an a priori refinement of drug dosing can lead to maximal efficacy, safety, and overall patient and trial success. Continuous cycles of predict–confirm/learn–refine (Fig. 14.2) will provide a systematic understanding of the sources of variability and aid in the planning of larger scale trials as the drug progresses through the various phases of clinical development. PBPK simulations of physiological variability can identify extremes in a population at risk of adverse reactions. These extreme individuals can be excluded in the next clinical trial. PBPK/PD simulations aid dose selection. The advantage of this bottom-up approach over empirical approaches is that while the empirical method relies on a large number of subjects in the trial to reliably identify covariates, the PBPK-based mechanistic method needs a much smaller number of subjects in the trial to confirm the simulated profiles, thus saving valuable resources.

The PK of docetaxel, simulated in an MC-generated virtual population[4] compared well the clinically observed. Other examples of a priori assessment of interindividual variation using PBPK modeling[5] based on known or estimated variability in physiological parameters for clinical development and toxicological risk assessment[6–12] are available in the literature. In order to generate a virtual population, PBPK modeling tools should have access to large built-in databases of patient-specific anatomical and physiological characteristics such as height, weight, blood flow rate, organ volumes, body composition, and processes affecting drug elimination for populations differing in age, gender, and race. Population databases with information on enzyme and transporter polymorphisms and their expression levels in different organs are also needed. Several resources of data are available.[13–16]

14.3 HEALTHY TO TARGET PATIENT POPULATION: IMPACT OF DISEASE ON PHARMACOKINETICS

If the target patient population for a drug are those with conditions that might affect the PK, then the physiology and enzymology in the target population needs to be employed to predict the PK in this population.

The obese subpopulation,[17] with a large body surface area is expected to have an increased cardiac output, altered tissue compositions, elevated levels of CYP2E1, and alpha acid glycoprotein (AAG) with important consequences to the pharmacokinetic parameters.[18,19] PBPK extrapolations from healthy to obese[7,20] were done by increasing the volume of fat according to a weight-to-height ratio. Similarly, disposition of drugs are likely to be affected by liver and

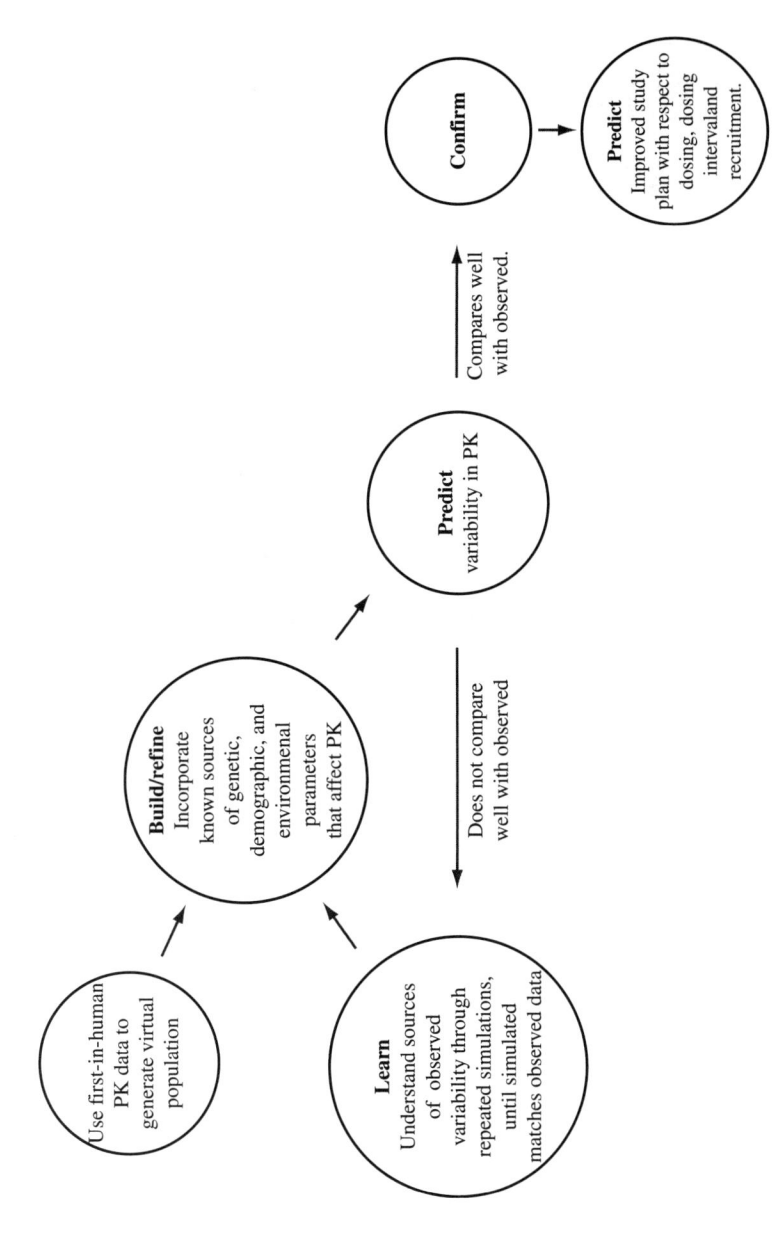

Figure 14.2. Predict–learn/confirm–refine cycles in gaining an understanding of the sources of variability.

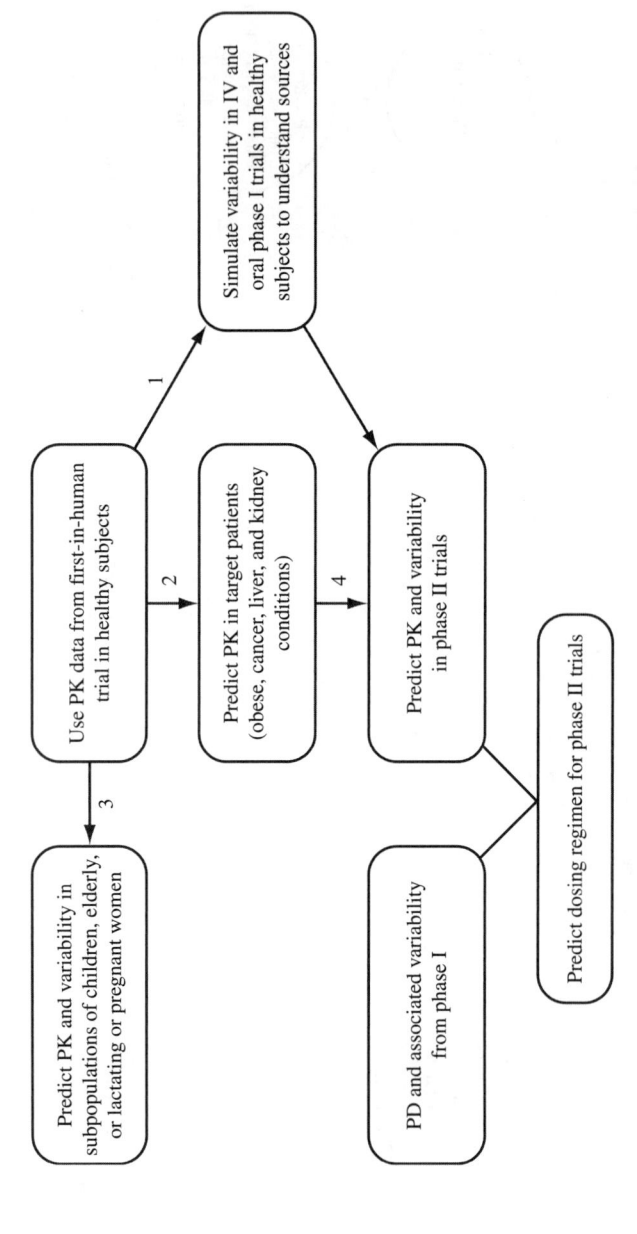

```
┌─────────────────────────┐                    ┌─────────────────────────┐
│ Predict PK and          │                    │ Simulate variability    │
│ variability in          │        1           │ in IV and oral phase I  │
│ subpopulations of       │ ─────────────────► │ trials in healthy       │
│ children, elderly,      │                    │ subjects to understand  │
│ or lactating or         │                    │ sources                 │
│ pregnant women          │                    └─────────────────────────┘
└─────────────────────────┘                                 ▲
            ▲                                               │
          3 │                                               │
            │                                               │
┌─────────────────────────┐                                 │
│ Use PK data from        │                                 │
│ first-in-human trial    │ ────────────────────────────────┘
│ in healthy subjects     │          1
└─────────────────────────┘
            │
          2 │
            ▼
┌─────────────────────────┐
│ Predict PK in target    │
│ patients (obese,        │        4           ┌─────────────────────────┐
│ cancer, liver, and      │ ─────────────────► │ Predict PK and          │
│ kidney conditions)      │                    │ variability in          │
└─────────────────────────┘                    │ phase II trials         │
                                                └─────────────────────────┘
                                                            │
┌─────────────────────────┐                                 │
│ PD and associated       │                                 │
│ variability from        │ ────────────────────────────────┤
│ phase I                 │                                 │
└─────────────────────────┘                                 ▼
                                                ┌─────────────────────────┐
                                                │ Predict dosing regimen  │
                                                │ for phase II trials     │
                                                └─────────────────────────┘
```

1. Reflects variability arising from enzyme and transporter genetic polymorphism and gender
2. PBPK model for target population: physiology and enzymology in target population
3. Subpopulation modeling
4. Reflects varibility in patient's comorbidities, age, obesity

Figure 14.3. Modeling population variability and subpopulations with PBPK.

kidney disorders. The variability associated with the altered physiology, if different from that of the healthy would impact the overall variability in the target population. Any learnings from the simulations of the phase I trial can also be incorporated to predict the dosing regimen and associated variability for the phase II trial in patients (Fig. 14.3) and to guide recruitment numbers. The cycle of predict–confirm/learn–refine will give valuable understanding of sources of variability associated with drug response, which at this stage reflects the differences in comorbidities, age, and obesity in patients.

14.4 MODELING SUBPOPULATIONS: IMPACT OF AGE, GENDER, CO-MORBIDITIES, AND GENETICS ON PHARMACOKINETICS

Subpopulations with a specific drug response behavior differ from the average healthy, caucasian adult due to the impact of age, ethnicity, co-morbidities and genetics on physiology and enzymology, anatomical differences, and enzyme and transporter polymorphisms. Changes in anatomy, physiology, enzymology, and transport proteins are readily incorporated in existing PBPK models for simulating the PK and associated variability that can be expected for subpopulations in clinical trials. Ethical need to minimize the number of clinical studies on sensitive subpopulations such as children, lactating or pregnant women, and elderly has meant that extrapolations of PK from the average are necessary for optimal design of clinical investigations. With PBPK clinical trials can be conducted to confirm rather than to explore. The strong age dependence of clearance makes children and elderly major subpopulations with a difficult-to-predict drug response. Children are the most well-established subpopulation in whom the use of modeling and simulation has been encouraged by regulatory authorities.[21] Both allometric predictions and PBPK extrapolations have been carried out in children. Allometric predictions are based on age-dependent liver volume[22,23] and body weight.[24] In addition to body weight and liver volume, PBPK extrapolations can account for differences in plasma protein and the ontogeny of drug-metabolizing enzymes and transporters that are important especially for children less than 2 years of age. *In vitro–in vivo* (IVIVE) extrapolations for children and neonates have been based on the physiological and enzymatic changes with age and adult *in vitro* data on metabolism. These approaches can be extended by PBPK to give a PK profile.[22,25–27] Children overall show higher exposure compared to healthy adults[28–30] not only for hepatically cleared compounds, but also for drugs that are primarily cleared in the kidney, as the glomerular filtration rate, tubular secretion, and renal blood flow are all deficient in neonates and children below 2 years of age with varying maturation rates.[31] Prediction of oral disposition of drugs in children is made difficult by the varying rates of maturation of organ size, composition, blood flow, and length of gastrointestinal tract.[31a,31b] A bottom-up assessment of differential age-related dosing for children has been described. The potential application of PBPK models to save time and effort in pediatric clinical drug development has been reviewed. Extrapolation to the geriatric population needs to consider decreases in renal and hepatic clearance with increasing age due to the reduction in liver mass,[22,23] microsomal protein, and hepatocellularity.[32] Using a PBPK model, it was shown that the reduced clearance of doxorubicin in the elderly[33] correlated with altered blood flows. PBPK modeling was used for extrapolation of midazolam PK from normal to elderly patients who suffered blood loss and fluid replacement during surgery and therefore changes in albumin.[28,34]

Midazolam is a highly bound drug (97% bound), whose PK is expected to be affected by the loss of albumin. A modest increase in clearance and a marked increase in volume of distribution resulting from the higher unbound fraction of the drug were correctly predicted. Interindividual variation in the disposition of midazolam could thus, in part, be related to the physiological characteristics of the patients and the unbound fraction of the drug in their plasma.

Use of PBPK to predict exposure differences between races and gender has been described in human health risk assessment.[8,35]

The loss of functional hepatocytes, altered blood flow due to the development of portacaval shunts, and the lower plasma protein levels in liver cirrhosis patients contribute to varying degrees of impaired systemic clearance and first-pass metabolism of drugs,[36] depending on the interplay between these parameters. Quantitative measures of the pathophysiological changes associated with Child–Pugh class A, B, and C liver cirrhosis, in organ blood flows, cardiac index, plasma binding protein concentrations, hematocrit, functional liver volume, hepatic enzymatic activity, and glomerular filtration rate were incorporated in a whole-body PBPK model to correctly predict the increased elimination half-life of theophylline.[37] Renal disease differentially affects uptake and efflux transporters and metabolic enzymes in the liver and GI tract. In patients with renal disease, this can cause exposure-related undesirable effects even for drugs that are not renally cleared.[38] The impact of inflammation on variability in the PK and PD of drugs can be accounted for by considering changes in the metabolism and transport of drugs, as well as the expression of receptors accompanying inflammation.[39]

Lack of efficacy seen at phase II trials is often the reason for failure of NCEs. Success at this stage relies on having good target engagement of the compound as indicated by a valid biomarker, validity of the target mechanism (proof of mechanism), and the translation of a valid target mechanism into high efficacy of the drug as indicated by a significant clinical impact. However, the benefit-to-risk ratio may not be uniform across different populations due to pharmacogenomic variability. The advent of human genomics and epigenetics triggered a large ongoing global effort to catalog genetic variations or polymorphisms that are dispersed throughout the human genome. In pharmacokinetics, genetic polymorphisms in enzymes[40–47] and transporters[48–50] can vary the dose by a factor of 25 or more. There are 29 polymorphisms and mutations in the CYP2D6[46,47] gene, resulting in poor, intermediate, extensive, or ultrarapid metabolizers. CYP2D6 ultrarapid metabolizers have 3–13 copies of the allele, so

14.4 MODELING SUBPOPULATIONS 391

they make much more enzyme, and metabolize the drug much more rapidly. The 2 polymorphisms in the CYP2C19 gene result in poor or extensive metabolizers. There are about 100 drugs on the market that rely on CYP2D6 especially for psychosis, depression, arrhythmia, blood pressure, and cough. Debrisoquine is metabolized by CYP2D to 4-hydroxy (OH)-debrisoquine, which is eliminated in urine. The urinary ratio of debrisoquine/4-OH-debrisoquine, which is a measure of the CYP2D activity, increases as the number of alleles decreases. Tamoxifen is a prodrug that must undergo biotransformation by CYP2D6 to the potent antiestrogen endoxifen. In a retrospective study of medical records of postmenopausal women enrolled in a trial (designed to test tamoxifen as an adjuvant for the treatment of early breast cancer), poor metabolizers of CYP2D6 (presence of the variant allele CYP2D6*4) showed earlier cancer relapse, compared with the absence of the PM metabolizer variant allele of this isoenzyme.[44] Thus, genotyping the patients could help identify those who are likely to have a better benefit-to-risk ratio.

The disposition of irinotecan is quite complex and involves numerous metabolic enzymes and transporter proteins. SN-38, the active metabolite of

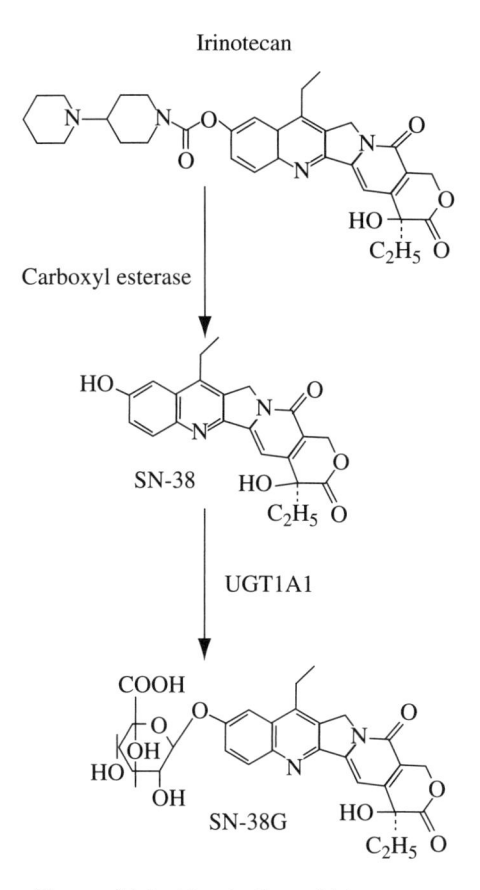

Figure 14.4. Metabolism of irinotecan.

irinotecan, is primarily eliminated via UGT1A1-mediated metabolism to SN-38G (Fig. 14.4), a biologically inactive glucuronide conjugate, which is then cleared via biliary excretion. The natural function of UGT1A1 is the catalysis of bilirubin glucuronidation. A genetic polymorphism in the UGT1A1 promoter (UGT1A1*28) results in enzyme underexpression, causing an impairment of bilirubin metabolism (reduced glucuronidation), clinically recognized as Gilbert's syndrome (UGT1A17/7 genotype). Case reports describing severe diarrhea and neutropenia in Gilbert's patients receiving standard starting doses of irinotecan suggested that SN-38 to SN-38G metabolism was reduced in patients with UGT1A1 polymorphism and was responsible for the observed irinotecan toxicity.[51]

As the gene regulation, expression, and activity of specific enzymes in the liver are reasonably well established, PBPK models can be used to determine the impact of differences in key metabolic enzymes, whether due to multiple genotypic expression, such as polymorphisms, or just due to normal variation in enzyme activities within the general population. Identification and characterization of a large number of genetic polymorphisms (pharmacogenomic biomarkers) in drug-metabolizing enzymes and drug transporters in an ethnically diverse group of individuals may provide substantial knowledge about the mechanisms of interindividual differences in drug response.

14.5 PERSONALIZED MEDICINE WITH PBPK/PD

Physiologically-based PK models that are linked to mechanistic PD models and incorporating an individual's physiology, enzymology, receptor (or target) expression levels, and polymorphism can predict an appropriate dose for that individual. Chapter 13 outlined how PBPK/PD modeling can aid decision making with regard to dose selection, optimized sampling strategy, through simultaneous consideration of expression levels, and turnover rates of drug target, enzymes, and transporters. PBPK models integrated with system biology can provide an excellent framework for the simulation of simultaneous effects of PK- and PD-driven factors on drug response. By prior identification of responders, clinical trials can focus on a smaller segment of patients, which translates to a tremendous saving of resources. They can also provide tailor-suited dosing regimen for different individuals, according to their specific intrinsic and extrinsic determinants of drug response. Differential response of individuals to the same drug has been known for a long time. In fact, some drugs are effective in less than half of the treated patients.[52] Poor efficacy rates of drugs have been attributed to the one-size-fits-all approach that has resulted in nonselective, nonspecific drugs on the market. A patient's response to a drug depends on the patient's age, gender, physical condition, phenotype, or genetic makeup (Fig. 14.5). Similarly, adverse drug reaction could be due to multiple factors such as disease determinants and environmental and genetic factors. Personalized

14.5 PERSONALIZED MEDICINE WITH PBPK/PD

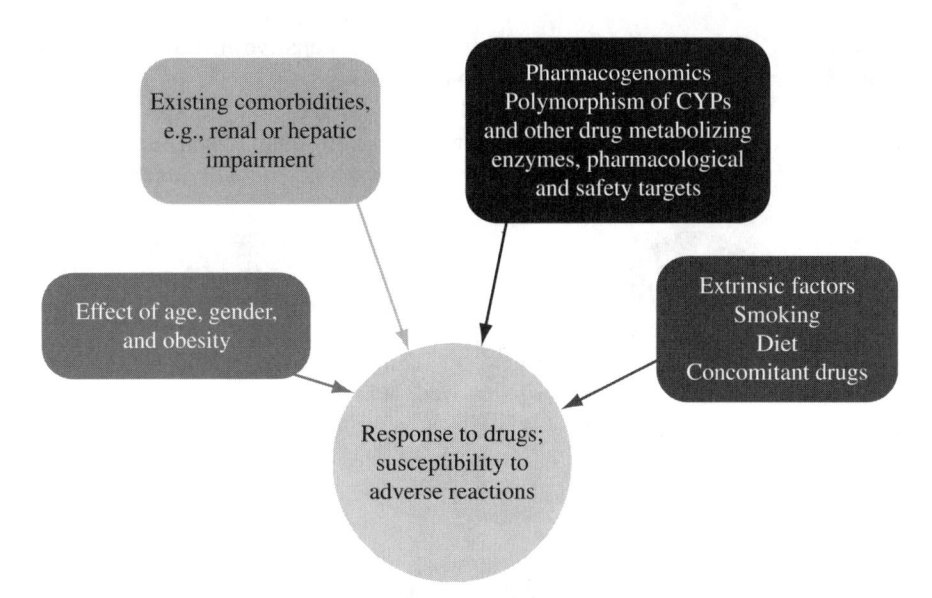

Figure 14.5. Intrinsic and extrinsic factors impacting response to drugs and susceptibility to adverse events.

medicine[53] through a systematic consideration of these determinants and using the accumulating knowledge of human genomic variation aims to decrease the number of adverse drug reactions and increase the efficacy of drug treatment.

In Section 14.4, the role of genetic diversity in determining the pharmacokinetic variability was outlined. Pharmacogenetic variation in pharmacological or safety targets could lead to loss of efficacy or toxic side effects as illustrated in Figure 14.1. Selecting the best drug for an individual patient relies on having a record of patient characteristics and useful predictive **Pharmacogenomic biomarkers** to determine the genetic makeup of the individual. The potential complexity of polygenic traits and demographic diversity challenges the development of these biomarkers. Implementing rigorous scientific and logistical strategies to address these will be crucial in order to achieve meaningful success. The Critical Path initiative[54] launched by the U.S. FDA supports the discovery and qualification of Pharmacogenomic biomarkers that will reliably guide patient dosing and predict an individual patient's responses to drug treatment or susceptibility to serious adverse drug reactions. Pharmacogenomic biomarkers currently in use are listed by FDA.[55]

Genentech's trastuzumab (Herceptin) is a monoclonal antibody against HER/neu-2 receptor, approved in 1998 for the treatment of metastatic breast cancer. It is an example of a drug that exploited knowledge of a patient's genotype to identify those patients who were most likely to respond to therapy. Of breast cancer patients 25–30% carried a genotype that led to the overexpression of HER/neu-2. These patients test positive in Herceptest, a diagnostic test that measures overexpression of the protein and can be treated with the drug.[56] Other breast cancer patients who are negative in the test do not respond to trastuzumab and need alternative therapy.

Development in cancer genomics and molecular targeted therapy leads to a paradigm shift in management of advanced-stage non-small-cell lung cancer. Multiple randomized studies have shown that patients with activated mutation

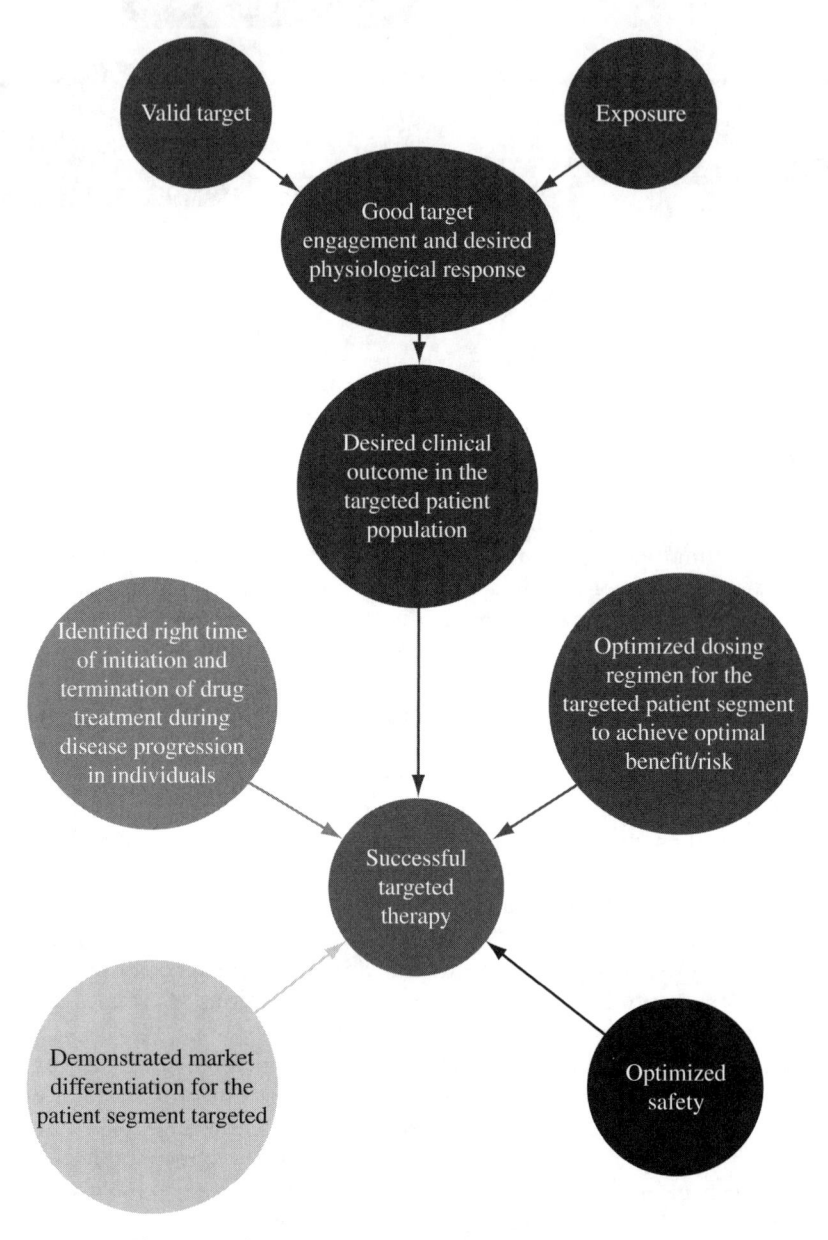

Figure 14.6. Elements of a successful targeted therapy.

of epidermal growth factor receptor (EGFR) responded dramatically to EGFR tyrosine kinase inhibitors such as gefitinib.[57,58] Personalized medicine should thus aim to recognize patient segmentation by providing a companion diagnostic to the targeted therapy.

While genetic profiling is a promising approach in oncology, there are other potential differential response indicators. For example, in developing a new compound to treat Alzheimer's, Lilly researchers hypothesized that patients without evident "plaque" (a waxy, translucent substance, composed primarily of protein fibers) were likely to respond very differently to drugs designed to treat the condition than patients with a significant plaque presence. Imaging techniques were used to characterize the plaque status of patients in clinical trials. Roughly 15% of patients with dementia, indistinguishable from Alzheimer's, did not demonstrate an amyloid plaque burden.

Apart from genomics and disease progression that distinguish responders from nonresponders, physiological perturbations related to age, gender, and comorbidities determine response to a drug. Extrinsic factors such as diet, smoking, and concomitant drugs also play a role. Drug–drug interactions from concomitant drugs can have a large effect in determining efficacy and safety response to a drug (see Chapter 9). Potential for drugs to interfere in each other's metabolism in patients with renal or hepatic impairment should guide DDI trial design.[59–62] Even small physiological perturbations from different sources have the potential to build up to a much larger net effect on drug response. A successful targeted therapy (Fig. 14.6) should aim to simultaneously consider all of these individual differences to provide optimal medication, tailor-suited to the individual with respect to efficacy and safety. In addition, it should also identify the right time of therapy intervention in an individual and demonstrate market differentiation for the patient segment targeted. Personalized medicine with PBPK/PD is expected to become a reality as our understanding of the factors distinguishing responders from nonresponders continues to grow. Biochemical, genomic, or imaging biomarkers enable capturing more responders, but knowledge of the root cause of disease, the existence of multiple targets, and the like remain largely unknown, emphasizing the need for integration with systems biology.

KEYWORD

Pharmacogenomic Biomarkers:[63] A genomic biomarker is defined as a measurable DNA and/or RNA characteristic that is an indicator of normal biologic processes, pathogenic processes, and/or response to therapeutic or other interventions. For example, a genomic biomarker could be a measurement of the expression, function, or regulation of a gene. DNA characteristics include, but are not limited to, single nucleotide polymorphisms (SNPs), variability of short sequence repeats, haplotypes, DNA modifications [e.g., methylation, deletions, or insertions of single nucleotide(s)], copy number variations, and cytogenetic rearrangements (e.g., translocations, duplications, deletions, or inversions). RNA characteristics include, but are not limited to, RNA sequences, RNA expression levels, RNA processing (e.g., splicing and editing), or micro-RNA levels. The definition of a genomic biomarker is not limited to human samples but includes samples from viruses and infectious agents as well as animal samples.

REFERENCES

1. Aarons L. Population pharmacokinetics: Theory and practice. *Br J Clin Pharmacol.* 1991;32(6):669–670.
2. Sheiner LB, Rosenberg B, Marathe VV. Estimation of population characteristics of pharmacokinetic parameters from routine clinical data. *J Pharmacokinet Biopharm.* 1977;5(5):445–479.
3. Sheiner LB, Beal S, Rosenberg B, Marathe VV. Forecasting individual pharmacokinetics. *Clin Pharmacol Ther.* 1979;26(3):294–305.
4. Hudachek SF, Gustafson DL. Customized in silico population mimics actual population in docetaxel population pharmacokinetic analysis. *J Pharm Sci.* 2011;100(3):1156–1166.
5. Bois FY, Jamei M, Clewell HJ. PBPK modeling of inter-individual variability in the pharmacokinetics of environmental chemicals. *Toxicology.* 2010;278(3):256–267.
6. Price PS, et al. Modeling interindividual variation in physiological factors used in PBPK models of humans. *Crit Rev Toxicol.* 2003;33(5):469–503.
7. Levitt DG, Schnider TW. Human physiologically-based pharmacokinetic model for propofol. *BMC Anesthesiol.* 2005;5(1):4.
8. Clewell HJ, Gentry PR, Covington TR, Sarangapani R, Teeguarden JG. Evaluation of the potential impact of age- and gender-specific pharmacokinetic differences on tissue dosimetry. *Toxicol Sci.* 2004;79(2):381–393.
9. Clewell HJ, 3rd, Andersen ME. Use of physiologically-based pharmacokinetic modeling to investigate individual versus population risk. *Toxicology.* 1996;111(1–3):315–329.
10. Willmann S, et al. Development of a physiology-based whole-body population model for assessing the influence of individual variability on the pharmacokinetics of drugs. *J Pharmacokinet Pharmacodyn.* 2007;34(3):401–431.
11. Yokley K, et al. Physiologically-based pharmacokinetic modeling of benzene in humans: A Bayesian approach. *Risk Anal.* 2006;26(4):925–943.
12. Krishnan K, Johanson G. Physiologically-based pharmacokinetic and toxicokinetic models in cancer risk assessment. *J Environ Sci Health C Environ Carcinog Ecotoxicol Rev.* 2005;23(1):31–53.
13. Thompson CM, et al. Database for physiologically-based pharmacokinetic (PBPK) modeling: Physiological data for healthy and health-impaired elderly. *J Toxicol Environ Health B Crit Rev.* 2009;12(1):1–24.
14. ILSI. Physiological information database. http://cfpub.epa.gov/ncea/cfm/recordisplay.cfm?deid=202847.

REFERENCES

15. International Commission on Radiological Protection (ICRP). *Basic Anatomical and Physiological Data for Use in Radiological Protection: Reference Values.* Amsterdam: ICRP publication 89, Elsevier Science, 2002.
16. Center for Disease Control and Prevention. Third national health and nutrition examination survey (NHANES III). Available at: http://www.cdc.gov/nchs/nhanes.htm.
17. Cheymol G. Effects of obesity on pharmacokinetics implications for drug therapy. *Clin Pharmacokinet.* 2000;39(3):215–231.
18. Hanley MJ, Abernethy DR, Greenblatt DJ. Effect of obesity on the pharmacokinetics of drugs in humans. *Clin Pharmacokinet.* 2010;49(2):71–87.
19. Blouin RA, Warren GW. Pharmacokinetic considerations in obesity. *J Pharm Sci.* 1999;88(1):1–7.
20. Edginton AN, Schmitt W, Willmann S. Application of physiology-based pharmacokinetic and pharmacodynamic modeling to individualized target-controlled propofol infusions. *Adv Ther.* 2006;23(1):143–158.
21. Manolis E, Pons G. Proposals for model-based paediatric medicinal development within the current European Union regulatory framework. *Br J Clin Pharmacol.* 2009;68(4):493–501.
22. Edginton AN, Schmitt W, Willmann S. Development and evaluation of a generic physiologically-based pharmacokinetic model for children. *Clin Pharmacokinet.* 2006;45(10):1013–1034.
23. Johnson TN, Tucker GT, Tanner MS, Rostami-Hodjegan A. Changes in liver volume from birth to adulthood: A meta-analysis. *Liver Transpl.* 2005;11(12):1481–1493.
24. Holford N. Dosing in children. *Clin Pharmacol Ther.* 2010;87(3):367–370.
25. Yang F, Tong X, McCarver DG, Hines RN, Beard DA. Population-based analysis of methadone distribution and metabolism using an age-dependent physiologically-based pharmacokinetic model. *J Pharmacokinet Pharmacodyn.* 2006;33(4):485–518.
26. Price K, Haddad S, Krishnan K. Physiological modeling of age-specific changes in the pharmacokinetics of organic chemicals in children. *J Toxicol Environ Health A.* 2003;66(5):417–433.
27. Ginsberg G, Hattis D, Russ A, Sonawane B. Physiologically-based pharmacokinetic (PBPK) modeling of caffeine and theophylline in neonates and adults: Implications for assessing children's risks from environmental agents. *J Toxicol Environ Health A.* 2004;67(4):297–329.
28. Bjorkman S. Prediction of drug disposition in infants and children by means of physiologically-based pharmacokinetic (PBPK) modeling: Theophylline and midazolam as model drugs. *Br J Clin Pharmacol.* 2005;59(6):691–704.
29. Johnson TN, Rostami-Hodjegan A, Tucker GT. Prediction of the clearance of eleven drugs and associated variability in neonates, infants and children. *Clin Pharmacokinet.* 2006;45(9):931–956.
30. Edginton AN, Schmitt W, Voith B, Willmann S. A mechanistic approach for the scaling of clearance in children. *Clin Pharmacokinet.* 2006;45(7):683–704.
31. DeWoskin RS, Thompson CM. Renal clearance parameters for PBPK model analysis of early lifestage differences in the disposition of environmental toxicants. *Regul Toxicol Pharmacol.* 2008;51(1):66–86.
31a. Edginton AN. Knowledge-driven approaches for the guidance of first-in-children dosing. *Paediatr Anaesth.* 2011;21(3):206–213.

31b. Khalil F, Läer S. Physiologically-based pharmacokinetic modeling: methodology, applications, and limitations with a focus on its role in **pediatric** drug development. *J Biomed Biotechnol.* 2011 Article ID 907461.
32. Barter ZE, et al. Scaling factors for the extrapolation of in vivo metabolic drug clearance from in vitro data: Reaching a consensus on values of human microsomal protein and hepatocellularity per gram of liver. *Curr Drug Metab.* 2007;8(1): 33–45.
33. Li J, Gwilt PR. The effect of age on the early disposition of doxorubicin. *Cancer Chemother Pharmacol.* 2003;51(5):395–402.
34. Bjorkman S, Wada DR, Berling BM, Benoni G. Prediction of the disposition of midazolam in surgical patients by a physiologically-based pharmacokinetic model. *J Pharm Sci.* 2001;90(9):1226–1241.
35. Brown EA, Shelley ML, Fisher JW. A pharmacokinetic study of occupational and environmental benzene exposure with regard to gender. *Risk Anal.* 1998;18(2): 205–213.
36. Johnson TN, Boussery K, Rowland-Yeo K, Tucker GT, Rostami-Hodjegan A. A semimechanistic model to predict the effects of liver cirrhosis on drug clearance. *Clin Pharmacokinet.* 2010;49(3):189–206.
37. Edginton AN, Willmann S. Physiology-based simulations of a pathological condition: Prediction of pharmacokinetics in patients with liver cirrhosis. *Clin Pharmacokinet.* 2008;47(11):743–752.
38. Nolin TD, Naud J, Leblond FA, Pichette V. Emerging evidence of the impact of kidney disease on drug metabolism and transport. *Clin Pharmacol Ther.* 2008; 83(6):898–903.
39. Schmith VD, Foss JF. Inflammation: Planning for a source of pharmacokinetic/pharmacodynamic variability in translational studies. *Clin Pharmacol Ther.* 2010; 87(4):488–491.
40. Ginsberg G, Smolenski S, Hattis D, Guyton KZ, Johns DO, Sonawane B. Genetic polymorphism in glutathione transferases (GST): Population distribution of GSTM1, T1, and P1 conjugating activity. *J Toxicol Environ Health B Crit Rev.* 2009;12(5–6):389–439.
41. Kusama M, Maeda K, Chiba K, Aoyama A, Sugiyama Y. Prediction of the effects of genetic polymorphism on the pharmacokinetics of CYP2C9 substrates from in vitro data. *Pharm Res.* 2009;26(4):822–835.
42. Neafsey P, Ginsberg G, Hattis D, Johns DO, Guyton KZ, Sonawane B. Genetic polymorphism in CYP2E1: Population distribution of CYP2E1 activity. *J Toxicol Environ Health B Crit Rev.* 2009;12(5–6):362–388.
43. Neafsey P, Ginsberg G, Hattis D, Sonawane B. Genetic polymorphism in cytochrome P450 2D6 (CYP2D6): Population distribution of CYP2D6 activity. *J Toxicol Environ Health B Crit Rev.* 2009;12(5–6):334–361.
44. Schroth W, et al. Association between CYP2D6 polymorphisms and outcomes among women with early stage breast cancer treated with tamoxifen. *JAMA.* 2009;302(13):1429–1436.
45. Walker K, Ginsberg G, Hattis D, Johns DO, Guyton KZ, Sonawane B. Genetic polymorphism in N-acetyltransferase (NAT): Population distribution of NAT1 and NAT2 activity. *J Toxicol Environ Health B Crit Rev.* 2009;12(5–6):440–472.

REFERENCES

46. Zhou SF. Polymorphism of human cytochrome P450 2D6 and its clinical significance: Part II. *Clin Pharmacokinet.* 2009;48(12):761–804.
47. Zhou SF. Polymorphism of human cytochrome P450 2D6 and its clinical significance: Part I. *Clin Pharmacokinet.* 2009;48(11):689–723.
48. Iwai M, Suzuki H, Ieiri I, Otsubo K, Sugiyama Y. Functional analysis of single nucleotide polymorphisms of hepatic organic anion transporter OATP1B1 (OATP-C). *Pharmacogenetics.* 2004;14(11):749–757.
49. Nishizato Y, et al. Polymorphisms of OATP-C (SLC21A6) and OAT3 (SLC22A8) genes: Consequences for pravastatin pharmacokinetics. *Clin Pharmacol Ther.* 2003;73(6):554–565.
50. Oswald S, et al. Disposition of ezetimibe is influenced by polymorphisms of the hepatic uptake carrier OATP1B1. *Pharmacogenet Genomics.* 2008;18(7):559–568.
51. Deeken JF, Slack R, Marshall JL. Irinotecan and uridine diphosphate glucuronosyltransferase 1A1 pharmacogenetics: To test or not to test, that is the question. *Cancer.* 2008;113(7):1502–1510.
52. Spear BB, Heath-Chiozzi M, Huff J. Clinical application of pharmacogenetics. *Trends Mol Med.* 2001;7(5):201–204.
53. Shastry BS. Pharmacogenetics and the concept of individualized medicine. *Pharmacogenomics J.* 2006;6(1):16–21.
54. U.S. FDA. Available at: http://www.fda.gov/ScienceResearch/SpecialTopics/CriticalPathInitiative/default.htm.
55. U.S. Food and Drug Administration. Table of pharmacogenomic biomarkers in drug labels. http://www.fda.gov/drugs/scienceresearch/researchareas/pharmacogenetics/ucm083378.htm.
56. Hertz DL, McLeod HL, Hoskins JM. Pharmacogenetics of breast cancer therapies. *Breast.* 2009;18(Suppl 3):S59–63.
57. Lam KC, Mok TS. Targeted therapy: An evolving world of lung cancer. *Respirology.* 2011;16(1):13–21.
58. Mok TS, Zhou Q, Leung L, Loong HH. Personalized medicine for non-small-cell lung cancer. *Expert Rev Anticancer Ther.* 2010;10(10):1601–1611.
59. Huang SM, et al. Therapeutic protein-drug interactions and implications for drug development. *Clin Pharmacol Ther.* 2010;87(4):497–503.
60. Huang SM, et al. New era in drug interaction evaluation: US food and drug administration update on CYP enzymes, transporters, and the guidance process. *J Clin Pharmacol.* 2008;48(6):662–670.
61. Zhang L, Zhang YD, Zhao P, Huang SM. Predicting drug-drug interactions: An FDA perspective. *AAPS J.* 2009;11(2):300–306.
62. Duan JZ, Jackson AJ, Zhao P. Bioavailability considerations in evaluating drug-drug interactions using the population pharmacokinetic approach. *J Clin Pharmacol.* 2010;51(7):1087–1100.
63. ICH. Guidance for industry: E15 definitions for genomic biomarkers, pharmacogenomics, pharmacogenetics, genomic data and sample coding categories. http://www.fda.gov/downloads/RegulatoryInformation/Guidances/ucm129296.pdf.

15

PBPK MODELS ALONG THE DRUG DISCOVERY AND DEVELOPMENT VALUE CHAIN

CONTENTS

15.1 Summary of Applications of PBPK Models along Value Chain... 401
15.2 Obstacles and Future Directions for PBPK Modeling.......... 403
     Keyword.. 405
     References... 405

15.1 SUMMARY OF APPLICATIONS OF PBPK MODELS ALONG VALUE CHAIN

Physiologically-based pharmacokinetic models predict the concentration–time profile of a drug, given the relevant drug-dependent properties (log P, pK_a, polar surface area, plasma protein binding, metabolic intrinsic clearance, permeability and solubility, etc.) and the physiology of the species (blood and urine flow rates, bile and gastric juice secretion, tissue volumes and composition, abundance and distribution of drug-metabolizing enzymes, membrane transporters and receptors, etc.). The parameter-intensive PBPK models demand a systematic consideration of the uncertainties associated with the large array of

Physiologically-Based Pharmacokinetic (PBPK) Modeling and Simulations: Principles, Methods, and Applications in the Pharmaceutical Industry, First Edition. Sheila Annie Peters.
© 2012 John Wiley & Sons, Inc. Published 2012 by John Wiley & Sons, Inc.

measured or calculated input data. Monte Carlo simulations are, therefore, built into PBPK models to estimate the overall uncertainty in the predicted PK profiles. Monte Carlo methods also allow for the simulation of physiological variability in PK profiles arising from the inherent variations in the physiological parameters. The extent of complexity of PBPK models can vary depending on the availability of data as well as on the purpose of the model, both of which depend on the stage of drug discovery or development. In early discovery (**target validation** and **hit identfication**), when very little pharmacokinetic data is available, PBPK models are of little value. During lead generation and lead optimization, however, PBPK models can provide an integrated view of all available data. It can also aid in compound design, based on an understanding of the sensitivity of a desired pharmacokinetic outcome on the different characteristics of the drug. With a progressively increasing model complexity, it is possible to consider the interplay of enzymes and transporters, provided reliable *in vitro* data and scaling factors are available. In the discovery stage, the lack of reliable data for intrinsic clearance and knowledge of clearance pathways can limit the value of PBPK models for prediction. However, PBPK is unique as a simulation tool, where it can help generate and test hypothesis for mechanistic understanding. During lead generation and lead optimization, with the availability of *in vivo* PK profiles in preclinical species for representative compounds in a series and for lead compounds, it is possible to gain mechanistic insights of the observed data and to apply the derived understanding for human predictions, depending on their relevance to humans. An important advantage of PBPK models over traditional methods is its prediction of drug concentrations as a function of time. A prediction of PK profile is possible at every stage of drug discovery and development, starting from lead optimization. A cycle of predict–learn/confirm–refine can continuously provide the best predictions of human PK profiles through the simultaneous consideration of multiple mechanisms on PK profiles and the integration of all available compound and system information at that stage. PBPK models are uniquely suited for addressing the complex interplay of physiology, compound properties, enzymes, transporters, and drug concentrations in impacting the PK and drug response. Integration of PBPK with PD enables the best estimates of dosing regimens for the first-in-human clinical studies. Preclinical PBPK/PD tumor models can be valuable in translating preclinical data to humans.[1]

During clinical development of an NCE, the availability of first-in-human PK data allows for updating the preclinical PBPK and PD models. From this point, a cycle of predict–confirm/learn–refine at the end of each trial aids in the selection of optimal dosing regimen and in the simulation of PK variability for the next trial. This bottom-up approach has the advantage of providing a mechanistic understanding of the sources of variability and in the identification of covariates.

Physiologically-based PK models can also simulate the effects of physiology and enzymology perturbations related to age, gender, disease conditions, genetics, and to environmental factors such as food, smoking, and concomitant drugs. This provides the possibility of using the PK data from a healthy, average population for extrapolation to other subpopulations. However, limitations in the current understanding of how age, disease, or organ dysfunction quantitatively affects the physiology and enzymology of a drug can hamper the application of these tools.

PBPK/PD models allow the combination of a variety of drug and patient characteristics to predict response outcome or to identify potential outliers in the population, who may be vulnerable to adverse drug reactions, making it a valuable tool in personalized medicine.

The success of PBPK approaches is sensitive to the validity of its assumptions. The assumption of well-stirred tissue compartments could fail, if within the tissue cells there are extra barriers/compartments that serve to keep out or accumulate the drug. The assumption that only unbound drug is available for metabolism, distribution, and excretion could also fail. For example, drugs characterized by a high rate of metabolism can have high clearance despite being highly bound to plasma proteins. Perfusion-limited models cannot explain the PK of large, hydrophilic drugs. In certain applications, use of a sophisticated tool may offer no real advantage to traditional methods because of poor quality of input data. The lack of good estimates for intrinsic clearance in the discovery stage is a clear limitation to the prediction of PK with PBPK models. Lack of information on the abundances and scalars for UGTs and transporters has been a deterrent too. When predictions of mean values are not reliable, then estimates of variability cannot be reliable too. PBPK models should, therefore, be validated for their intended use, to ensure that the expected value in the tool can actually be realized in practice.

In summary, PBPK models are useful to integrate data and models, understand underlying mechanisms, predict and simulate PK profiles and associated variability, extrapolate across species, route of administration, and across populations, and has applications in compound design, sensitivity analysis, hypothesis generation and testing, optimal dosing regimen selection, clinical trial design, and targeted therapies.[2-5] Figure 15.1 summarizes the applications of PBPK modeling and simulations along the drug discovery and development value chain. These applications can be valuable to reduce cost and to reduce animal studies. Ideally, PBPK models should be seen as a repository of knowledge and data, generated at various stages of drug discovery development. Strategic application and exploitation of such integrative tools promote informed decision making and facilitate efficient and cost-effective drug development.

Regulatory authorities are increasingly recommending the use of PBPK-based approaches. The FDA has started to employ PBPK modeling for new drug review.[6-9] PBPK approaches have been recommended in the FDA lactation guidance. Its use has been encouraged by both EMA[10] and FDA for application in pediatric medicines.

15.2 OBSTACLES AND FUTURE DIRECTIONS FOR PBPK MODELING

Application of PBPK has traditionally been confined to environmental toxicology and human health risk assessment, where clinical studies are not practical.

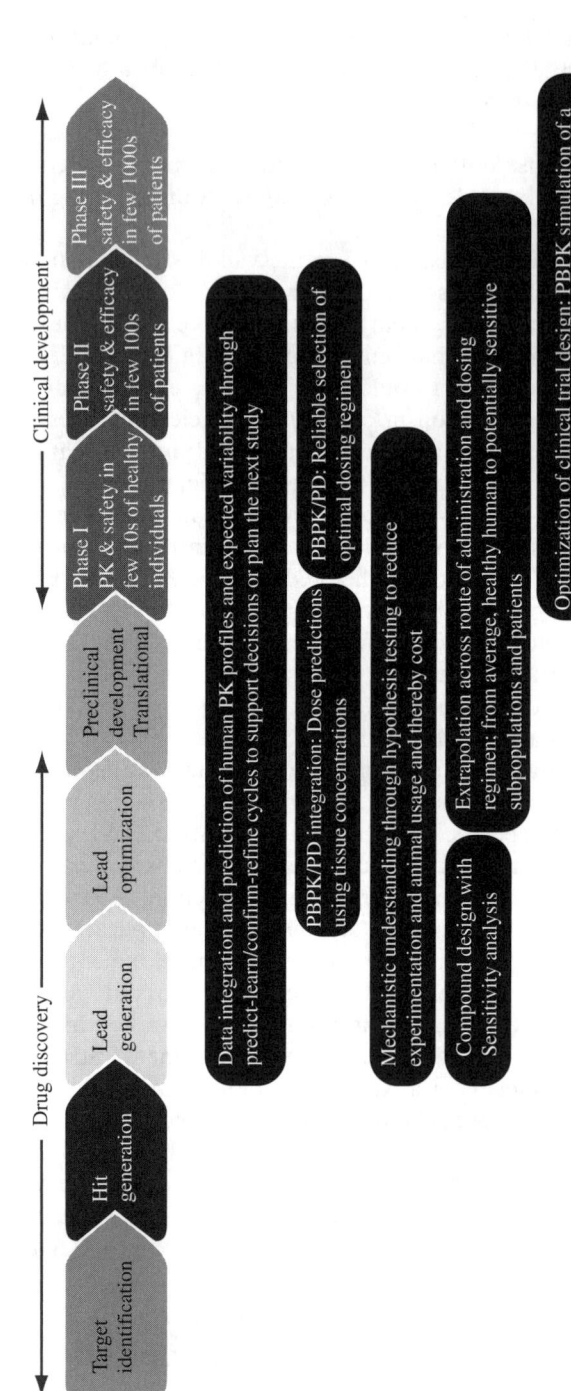

Figure 15.1. PBPK modeling applications along the drug discovery and development value chain.

Although the pharmaceutical industry has been slow to exploit the power of PBPK, rapid strides have been made in the last few years. Managerial and cultural barriers[3–5,11] that resisted a wider application of PBPK in drug discovery and development have largely been overcome in the last decade. Availability of user-friendly, commercial software has broadened the usage of PBPK models among the PK community. However, the risk of inappropriate interpretations of modeling results arising from a poor understanding of the principles, assumptions, and complexity of the models can be high, if adequate training is not provided to the users of ready-to-use tools. The use of commercial software also limits the flexibility to incorporate alternative PK mechanisms or different types of input data.

The preceding chapters have exposed the reader to the possibilities and limitations of PBPK modeling. The promise of better predictions based on a mechanistic understanding of PK processes, more rationale extrapolations, enhanced data integration, and improved efficiency achieved with a reduced use of animals and reduced cost has encouraged a wide range of applications across the value chain in the pharmaceutical industry. However, a number of obstacles need to be overcome before the full potential of the PBPK modeling is unleashed in drug discovery and development. The development of more stable and viable *in vitro* systems,[12] development of competitive binding assay for highly bound compounds,[13] use of human microdosing[14] for PK, use of imaging techniques for tissue partitioning, and an improved understanding of enzymes and transporters (*in vitro—in vivo* correlations) will enhance the quality of PBPK predictions in the future. Expanding the benefits of PBPK predictions to inhaled drugs requires the physiological modeling of lung and a knowledge of how transporters in the lung can affect disposition.[15] A combination of systems biology[16] (target expression) with PBPK and PD modeling[17] will further expand the scope of PBPK modeling especially in personalized medicine, which aims to improve individual patient outcomes to deliver quality, innovative medicines tailor-suited to a patient segment, at the right dose, and at the right time.

KEYWORD

Hit Identification: In early drug discovery, active compounds that were identified through high-throughput screening have to be subjected to a structure–activity relationship (SAR) evaluation in order to select the most promising compounds to progress to lead generation.

REFERENCES

1. Zhou Q, Gallo JM. The pharmacokinetic/pharmacodynamic pipeline: Translating anticancer drug pharmacology to the clinic. *AAPS J.* 2011;13(1):111–120.

2. Lupfert C, Reichel A. Development and application of physiologically-based pharmacokinetic-modeling tools to support drug discovery. *Chem Biodivers*. 2005; 2(11):1462–1486.
3. Rowland M, Balant L, Peck C. Physiologically-based pharmacokinetics in drug development and regulatory science: A workshop report (Georgetown University, Washington, DC, May 29–30, 2002). *AAPS J*. 2004;6(1):56–67.
4. Rowland M, Peck C, Tucker G. Physiologically-based pharmacokinetics in drug development and regulatory science. *Annu Rev Pharmacol Toxicol*. 2011; 51:45–73.
5. Edginton AN, Theil FP, Schmitt W, Willmann S. Whole body physiologically-based pharmacokinetic models: Their use in clinical drug development. *Expert Opin Drug Metab Toxicol*. 2008;4(9):1143–1152.
6. Zhao P, et al. Applications of physiologically-based pharmacokinetic (PBPK) modeling and simulation during regulatory review. *Clin Pharmacol Ther*. 2011; 89(2):259–267.
7. Oo C, Chen YC. The need for multiple doses of 400 mg ketoconazole as a precipitant inhibitor of a CYP3A substrate in an in vivo drug-drug interaction study. *J Clin Pharmacol*. 2009;49(3):368–369; author reply 370.
8. Zhao P, et al. Quantitative evaluation of pharmacokinetic inhibition of CYP3A substrates by ketoconazole: A simulation study. *J Clin Pharmacol*. 2009;49(3): 351–359.
9. Peck CC. Quantitative clinical pharmacology is transforming drug regulation. *J Pharmacokinet Pharmacodyn*. 2010;37(6):617–628.
10. Manolis E, Pons G. Proposals for model-based paediatric medicinal development within the current European Union regulatory framework. *Br J Clin Pharmacol*. 2009;68(4):493–501.
11. Nestorov I. Whole-body physiologically-based pharmacokinetic models. *Expert Opin Drug Metab Toxicol*. 2007;3(2):235–249.
12. Maguire TJ, et al. Design and application of microfluidic systems for in vitro pharmacokinetic evaluation of drug candidates. *Curr Drug Metab*. 2009;10(10): 1192–1199.
13. Schuhmacher J, Kohlsdorfer C, Buhner K, Brandenburger T, Kruk R. High-throughput determination of the free fraction of drugs strongly bound to plasma proteins. *J Pharm Sci*. 2004;93(4):816–830.
14. Lappin G, Garner RC. The utility of microdosing over the past 5 years. *Expert Opin Drug Metab Toxicol*. 2008;4(12):1499–1506.
15. Bosquillon C. Drug transporters in the lung—do they play a role in the biopharmaceutics of inhaled drugs? *J Pharm Sci*. 2010;99(5):2240–2255.
16. Ideker T, Galitski T, Hood L. A new approach to decoding life: Systems biology. *Annu Rev Genomics Hum Genet*. 2001;2:343–372.
17. Claudino WM, Quattrone A, Biganzoli L, Pestrin M, Bertini I, Di Leo A. Metabolomics: Available results, current research projects in breast cancer, and future applications. *J Clin Oncol*. 2007;25(19):2840–2846.

APPENDICES

APPENDIX A PHYSIOLOGICAL PARAMETERS IN PRECLINICAL SPECIES

Sprague–Dawley Rat

Radius of Small Intestine (cm)	Length of Jejunum (cm)	Flux of Water From Mucosal to Serosal Side (J_{ms}) (mL/min/cm)
0.18	90.00	0.00093

Intestinal Compartment	Intestinal Transit Rate Constant (k_t) (/min) Fed	Fasted	Volume of Intestinal Compartments (mL) Fed	Fasted	Intestinal pH Fed	Fasted	Permeability Multipliers
Stomach	0.024	0.37	3	3	3.5	3	0.0001
1	0.0410	0.0853	0.6	0.6	6.5	7.1	1
2	0.0410	0.0853	0.66	0.66	6.9	7.3	1
3	0.0410	0.0853	0.66	0.66	6.9	7.5	1
4	0.0410	0.0853	0.41	0.41	7.1	7.7	1
5	0.0410	0.0853	0.41	0.41	7.1	7.9	1
6	0.0410	0.0853	0.41	0.41	7.1	8	1
7	0.0410	0.0853	0.41	0.41	6.8	7.4	1
Colon	0.0026	0.0044	42	42	6.6	7.6	0.1

Body Weight (kg)	Glomerular Filtration Rate (GFR) (mL/min/kg)	Plasma Density (g/mL)	Blood Density (g/mL)	Hematocrit	Liver Weight (g/kg)	MPPGL (mg protein/g liver)	HPGL (million hepatocytes/g liver)
0.25	5.2	1.024	1.05	0.503	40	45	125

Sprague–Dawley Rat: Tissue Related Parameters

Tissue	Volume (mL/kg)	Blood Flow Rates (mL/min/kg)	Tissue Compositions—Fractional Tissue Volume Content (L/kg)			
			Phospholipid	Neutral Lipid	Water	Interstitial Fraction
Adipose	40.0	1.6	0.0021	0.8529	0.12	0.067
Bone	63.2	10.12	0.0005	0.0222	0.35	0.139
Brain	6.8	5.32	0.0532	0.0393	0.75	0.013
Gut	40	52	0.015	0.032	0.7	0.166
Heart	3.2	15.68	0.0141	0.0117	0.77	0.09
Kidney	9.2	36.92	0.0269	0.0334	0.752	0.196
Liver	41.2	80.0	0.025	0.035	0.7	0.169
Lung	4.0	203.2	0.017	0.0199	0.78	0.241
Muscle	487.6	30	0.0103	0.0087	0.743	0.119
Pancreas	5.2	4	0.0188	0.0723	0.66	0.119
Skin	160	20	0.0155	0.0205	0.7	0.253
Spleen	2.4	5	0.0136	0.0077	0.77	0.15
Stomach	4.4	8.2	0.0182	0.0338	0.794	0.216
Testes	10	1.8	0	0	0.859	0.158
Arterial blood	22.4	10.8	0.0039	0.0012	0.63	0
Venous blood	45.2		0.0008	0.0014	0.939	1

Beagle Dog

Radius of Small Intestine (cm)	Length of Jejunum (cm)	Flux of Water From Mucosal to Serosal Side (J_{ms}) (mL/min/cm)
0.5	154	0.33

Intestinal Compartment	Intestinal Transit Rate Constant (k_t) (/min)		Volume of Intestinal Compartments (mL)		Intestinal pH		Permeability Multipliers
	Fed	Fasted	Fed	Fasted	Fed	Fasted	
Stomach	0.0055	0.08	1000	14.54	2.1	1.5	0.0001
1	0.063	0.063	30.54	30.54	6	6	1
2	0.063	0.063	32	32	6	6	1
3	0.063	0.063	32	32	6	6	1
4	0.063	0.063	20.1	20.1	6.2	6.2	1
5	0.063	0.063	20.1	20.1	6.2	6.2	1
6	0.063	0.063	20.1	20.1	6.2	6.2	1
7	0.063	0.063	20.1	20.1	6.4	7.4	1
Colon	0.00177	0.00177	290.9	290.9	6.5	6.5	0.1

Body Weight (kg)	Glomerular Filtration Rate (GFR) (mL/min/kg)	Plasma Density (g/mL)	Blood Density (g/mL)	Hematocrit	Liver Weight (g/kg)	MPPGL (mg protein/g liver)	HPGL (million hepatocytes/g liver)
12	6.13	1.024	1.05	0.503	32	43	120

Beagle Dog: Tissue Related Parameters

Tissue	Volume (mL/kg)	Blood Flow Rates (mL/min/kg)
Adipose	130	6.1
Bone	126	3.7
Brain	7.2	4.5
Gut	48	20.3
Heart	12	15.68
Kidney	6	21.6
Liver	48	33
Lung	12	121.4
Muscle	553	25
Pancreas	5.2	1.5
Skin	120	10
Spleen	3.6	2.4
Stomach	4.4	3.6
Testes	10	1.8
Arterial blood	22.4	
Venous blood	45.2	

Cynomolgus Monkey

Radius of Small Intestine (cm)	Length of Jejunum (cm)	Flux of Water From Mucosal to Serosal Side (J_{ms}) (mL/min/cm)
0.6	115.38	0.02

Intestinal Compartment	Intestinal Transit Rate Constant (k_t) (/min) Fed	Fasted	Volume of Intestinal Compartments (mL) Fed	Fasted	Intestinal pH Fed	Fasted	Permeability Multipliers
Stomach	0.0093	0.04	100	7	2.5	2.5	0.0001
1	0.0430	0.04	15	15	5.6	5.6	1
2	0.0430	0.04	16	16	5.6	5.6	1
3	0.0430	0.04	16	16	5.6	5.6	1
4	0.0430	0.04	10	10	5.8	5.8	1
5	0.0430	0.04	10	10	5.8	5.8	1
6	0.0430	0.04	10	10	5.8	5.8	1
7	0.0430	0.04	10	10	6	6	1
Colon	0.0007	0.0007	146	146	5.1	5.1	0.1

Body Weight (kg)	Glomerular Filtration Rate (GFR) (mL/min/kg)	Plasma Density (g/mL)	Blood Density (g/mL)	Hematocrit	Liver Weight (g/kg)	MPPGL (mg protein/g liver)	HPGL (million hepatocytes/g liver)
3	2.1	1.024	1.05	0.41	32	45	120

Cynomolgus Monkey: Tissue Related Parameters

Tissue	Volume (mL/kg)	Blood Flow Rates (mL/min/kg)
Adipose	130.4	4
Bone	63.2	10.2
Brain	19	14.4
Gut	46	24
Heart	3.4	12
Kidney	6	27.6
Liver	27	43.4
Lung	7.7	135.6
Muscle	500	18
Pancreas	5.2	2.04
Skin	100	10.8
Spleen	1.7	4.2
Stomach	4.4	3
Testes	10	1.8
Arterial blood	22.4	
Venous blood	45.2	

Albumin Levels in Different Species[1,2]

Species	Strain	Concentration of Albumin[a] (μmol/L)
Mouse	CD1	455–515
Rat	Sprague–Dawley	576–667
Rat	Wistar	439–500
Guinea pig	Hartley	364–409
Rabbit	NZW	455–515
Monkey	Rhesus	652–657
Dog	Beagle	439–515
Minipig	Gottingen	439–470

[a]Knowing the molecular weight of albumin, (around 67 kDa) concentration in g/L can be evaluated.

APPENDIX B HUMAN PHYSIOLOGICAL PARAMETERS

Human

Radius of Small Intestine (cm)	Length of Jejunum (cm)	Flux of Water From Mucosal to Serosal Side (J_{ms}) (mL/min/cm)	Gall Bladder Emptying Rate (mL/min) Fed	Gall Bladder Emptying Rate (mL/min) Fasted
1	250	0.02	2.5	0.5–1

Intestinal Compartment	Intestinal Transit Rate Constant (k_t) (/min) Fed	Intestinal Transit Rate Constant (k_t) (/min) Fasted	Volume of Intestinal Compartments (mL) Fed	Volume of Intestinal Compartments (mL) Fasted	Intestinal pH Fed	Intestinal pH Fasted	Permeability Multipliers
Stomach	0.017	0.066	1000	50	5	2	0.0001
1	0.0095	0.035	105	105	6	6	1
2	0.0095	0.035	110	110	6.2	6.2	1
3	0.0095	0.035	110	110	6.6	6.6	1
4	0.0095	0.035	69	69	6.8	6.8	1
5	0.0095	0.035	69	69	7	7	1
6	0.0095	0.035	69	69	7.2	7.2	1
7	0.0095	0.035	69	69	7.4	7.4	1
Colon	0.0014	0.0014	7000	1000	7	7	0.1

Enterocyte volume in small intestine: 0.52 L; Enterocyte volume in colon: 0.007L.

Body Weight (kg)	Glomerular Filtration Rate (GFR) (mL/min/kg)	Plasma Density (g/mL)	Blood Density (g/mL)	Hematocrit	Liver Weight (g/kg)	MPPGL (mg protein/g liver)	HPGL (million hepatocytes/g liver)
70	1.79	1.024	1.05	0.45	24	29	95

Human: Tissue Related Parameters

| Tissue | Volume (mL/kg) | Blood Flow Rates (mL/min/kg) | Tissue Compositions—Fractional Tissue Volume Content (L/kg) ||| Interstitial Fraction |
			Phospholipid	Neutral Lipid	Water	
Adipose	143	3.7	0.002	0.79	0.18	0.067
Bone	124	3.6	0.0005	0.074	0.439	0.139
Brain	20.7	10	0.0565	0.051	0.77	0.013
Gut	23.6	13	0.0163	0.0487	0.718	0.166
Heart	3.8	2.14	0.0166	0.0115	0.758	0.09
Kidney	4.4	15.7	0.0162	0.0207	0.783	0.196
Liver	24.1	21	0.0252	0.0348	0.751	0.169
Lung	16.7	71	0.009	0.003	0.811	0.241
Muscle	429	10.7	0.0072	0.0238	0.76	0.119
Pancreas	1.2	1.9	0.0188	0.0723	0.66	0.119
Skin	111	4.3	0.0111	0.0284	0.718	0.253
Spleen	2.7	1.1	0.0198	0.0201	0.788	0.15
Stomach	2.2	0.56	0.0182	0.0338	0.784	0.216
Testes	0.51	0.04	0	0	0.859	0.158
Arterial blood	25.7		0.0033	0.0022	0.651	0
Venous blood	51.4		0.00225	0.0035	0.945	1

Concentrations of plasma proteins in human are given in Table 3.1, Chapter 3.

Human Physiological Parameters Needed for Biologics[3,a]

	Vascular Volume (mL)	Interstitial Volume (mL)
Blood	5,200.0	
Lung	66.5	199.5
Heart	24.5	66.5
Kidney	105.0	353.5
Liver	350.0	875
Spleen	35.0	70
Bone	5,862.0	12,477.5

[a]Scaled up from murine values[4] proportional to body weight.[5]

Preclinical and Human Data
1. Physiological parameters for human and preclinical species[6]
2. Physiological parameters for rats and humans[7]
3. Hepatocellularity in preclinical (rabbit, rat, mouse) and human: 139+25, 114+20, 117+30, and 135+10 million cells/g liver[8]
4. Rat hepatocyte cell volume; 3.9 µL/million hepatocytes[9]
5. Effect of aging on cardiac output, regional blood flows, and body composition in Fischer 344 mice[10]
6. Absorption physiology, tissue composition, cardiac output, regional blood flows, organ size, and body weight in humans and preclinical (rats, dogs, mice)[11]
7. Extravascular albumin in rats[12]
8. Tissue compositions in rat and human[13–17]
9. Abundances of CYP P450 isoforms along the gut in rats[18]
10. A comprehensive database of anatomical and physiological parameters in preclinical species published by RIVM[19]
11. Interspecies variation in liver weight and hepatic blood flow[20]
12. Physiological and anatomical differences between the gastrointestinal tracts of humans and commonly used laboratory animals[21]
13. Physiological parameters related to drug absorption[22,23]
14. Physiological parameters in commonly used laboratory animals[24]

Allometric and Other Equations for Getting Human Parameters
15. Regression equations for plasma, blood, and tissue volumes and GFR based on age, gender, body weight, and height[25]
16. Height dependence on age; weight dependence on height; CYP abundances in Caucasian and Japanese[26]
17. Cardiac output as a function of age, equation relating standard liver volume to body surface area and liver density[27]

18. Body surface area from height and weight[28]
19. Use of allometry in predicting anatomical and physiological parameters of mammals[29]

Human Data
20. Anatomical and physiological data[30]
21. General reference[31]
22. A book section devoted to physiological parameters and databases for PBPK modeling[32]
23. CYP abundances in human liver[33]
24. Inter-system extrapolation factors (ISEFs) for CYPs[34]
25. Relative activity factors (RAFs) for human CYPs[35,36]
26. Turnover half-lives of human hepatic CYPs[37,38]
27. Contains references for variation of gut physiology with age, gender, race, food, and disease[39,40]
28. Compiled information on morphology (radius, transit time), volume, pH of stomach and small intestine, transit time, concentration of bile in fed–fasted conditions in humans[42]
29. MPPGI (mg microsomal protein per gram intestine) in duodenum, jejunum, ileum from HIM; mg microsomal protein per cm; P-gp, BCRP, and CYP3A4 in duodenum, jejunum, ileum, and colon[41,42]
30. CYP abundances along gut in humans[43]
31. Renal parameters[44]

Population
32. Third National Health and Nutrition Examination Survey (NHANES III)[45]
33. Organ volumes and blood flow rates related to race, gender, age, BMI, body weight and height[40]
34. Parameters (volumes of selected organs and tissues; blood flows for the organs and tissues; and the total cardiac output under resting conditions and average daily inhalation rate) are expressed as records of correlated values for the approximately 30,000 individuals evaluated in the NHANES III survey[46]
35. Anatomical and physiological data[47]
36. EPA-sponsored compilation of physiological parameter values (e.g., alveolar ventilation, blood flow and tissue volumes, glomerular filtration rate) in children, adults, and elderly[48,49]
37. Pregnancy-related physiological parameters[49]
38. The embryonic development of the blood-brain barrier, and changes in pathology are described[50]

39. Developmental changes in body weight, tissue weights, cardiac output, hepatic blood flow, and enzyme activity[51]
40. Developmental changes in CYPs 2C9, 2C19, and 3A4[52,53]
41. Database for healthy and health-impaired elderly[54]
42. Age dependency of cerebral P-gp[55]

REFERENCES

1. Loeb WF, Quimby FW. *Clinical Chemistry of Laboratory Animals*, 2nd ed. Taylor & Francis, Philalelphia 1999.
2. Gabrielsson J, Dolgos H, Gillberg PG, Bredberg U, Benthem B, Duker G. Early integration of pharmacokinetic and dynamic reasoning is essential for optimal development of lead compounds: Strategic considerations. *Drug Discov Today*. 2009;14(7–8):358–372.
3. Davda JP, Jain M, Batra SK, Gwilt PR, Robinson DH. A physiologically-based pharmacokinetic (PBPK) model to characterize and predict the disposition of monoclonal antibody CC49 and its single chain fv constructs. *Int Immunopharmacol*. 2008;8(3):401–413.
4. Baxter LT, Zhu H, Mackensen DG, Jain RK. Physiologically-based pharmacokinetic model for specific and nonspecific monoclonal antibodies and fragments in normal tissues and human tumor xenografts in nude mice. *Cancer Res*. 1994;54(6):1517–1528.
5. Zhu H, Baxter LT, Jain RK. Potential and limitations of radioimmunodetection and radioimmunotherapy with monoclonal antibodies. *J Nucl Med*. 1997;38(5):731–741.
6. Davies B, Morris T. Physiological parameters in laboratory animals and humans. *Pharm Res*. 1993;10(7):1093–1095.
7. Bernareggi A, Rowland M. Physiologic modeling of cyclosporin kinetics in rat and man. *J Pharmacokinet Biopharm*. 1991;19(1):21–50.
8. Sohlenius-Sternbeck AK. Determination of the hepatocellularity number for human, dog, rabbit, rat and mouse livers from protein concentration measurements. *Toxicol in Vitro*. 2006;20(8):1582–1586.
9. Reinoso RF, Telfer BA, Brennan BS, Rowland M. Uptake of teicoplanin by isolated rat hepatocytes: Comparison with in vivo hepatic distribution. *Drug Metab Dispos*. 2001;29(4 Pt 1):453–459.
10. Delp MD, Evans MV, Duan C. Effects of aging on cardiac output, regional blood flow, and body composition in Fischer-344 rats. *J Appl Physiol*. 1998;85(5):1813–1822.
11. Brown RP, Delp MD, Lindstedt SL, Rhomberg LR, Beliles RP. Physiological parameter values for physiologically-based pharmacokinetic models. *Toxicol Ind Health*. 1997;13(4):407–484.
12. Katz J, Bonorris G, Golden S, Sellers AL. Extravascular albumin mass and exchange in rat tissues. *Clin Sci*. 1970;39(6):705–724.

13. Rodgers T, Rowland M. Physiologically-based pharmacokinetic modeling. 2: Predicting the tissue distribution of acids, very weak bases, neutrals and zwitterions. *J Pharm Sci.* 2006;95(6):1238–1257.
14. Rodgers T, Leahy D, Rowland M. Tissue distribution of basic drugs: Accounting for enantiomeric, compound and regional differences amongst beta-blocking drugs in rat. *J Pharm Sci.* 2005;94(6):1237–1248.
15. Rodgers T, Leahy D, Rowland M. Physiologically-based pharmacokinetic modeling. 1: Predicting the tissue distribution of moderate-to-strong bases. *J Pharm Sci.* 2005;94(6):1259–1276.
16. Poulin P, Theil FP. Prediction of pharmacokinetics prior to in vivo studies. 1. Mechanism-based prediction of volume of distribution. *J Pharm Sci.* 2002;91(1):129–156.
17. Poulin P, Theil FP. Prediction of pharmacokinetics prior to in vivo studies. II. Generic physiologically-based pharmacokinetic models of drug disposition. *J Pharm Sci.* 2002;91(5):1358–1370.
18. Mitschke D, Reichel A, Fricker G, Moenning U. Characterization of cytochrome P450 protein expression along the entire length of the intestine of male and female rats. *Drug Metab Dispos.* 2008;36(6):1039–1045.
19. de Zwart LL, Rompelberg CJM, Sips AJAM, Welink J, van Engelen JGM Rompelberg CJM, Sips AJAM, Welink J, van Engelen JGM. Anatomical and physiological differences between various species used in studies on the pharmacokinetics and toxicology of xenobiotics. A review of literature. Report no. 623860010, Bilthoven, The Netherlands: National Institute for Public Health and the Environment http://www.rivm.nl/bibliotheek/rapporten/623860010.pdf, 1999.
20. Boxenbaum H. Interspecies variation in liver weight, hepatic blood flow, and antipyrine intrinsic clearance: Extrapolation of data to benzodiazepines and phenytoin. *J Pharmacokinet Biopharm.* 1980;8(2):165–176.
21. Kararli TT. Comparison of the gastrointestinal anatomy, physiology, and biochemistry of humans and commonly used laboratory animals. *Biopharm Drug Dispos.* 1995;16(5):351–380.
22. Wilson C, Washington N, eds. *Physiological Pharmaceutics. Biological Barriers to Drug Absorption.* Chichester, UK: Ellis Horwood, 1989.
23. Hardy JG, Davis SS, Wilson CG, eds. *Drug Delivery to the Gastrointestinal Tract.* Chichester, UK: Ellis Horwood, 1989.
24. Flecknell P. *Laboratory Animal Anaesthesia.* London: Academic, 2009.
25. Yang F, Tong X, McCarver DG, Hines RN, Beard DA. Population-based analysis of methadone distribution and metabolism using an age-dependent physiologically-based pharmacokinetic model. *J Pharmacokinet Pharmacodyn.* 2006;33(4):485–518.
26. Inoue S, et al. Prediction of in vivo drug clearance from in vitro data. II: Potential inter-ethnic differences. *Xenobiotica.* 2006;36(6):499–513.
27. Howgate EM, Rowland Yeo K, Proctor NJ, Tucker GT, Rostami-Hodjegan A. Prediction of in vivo drug clearance from in vitro data. I: Impact of inter-individual variability. *Xenobiotica.* 2006;36(6):473–497.
28. Du Bois D, Du Bois EF. A formula to estimate the approximate surface area if height and weight be known. 1916. *Nutrition.* 1989;5(5):303–11; discussion 312–313.

29. Lindstedt L, Schaeffer PJ. Use of allometry in predicting anatomical and physiological parameters of mammals. *Lab Anim.* 2002;36(1):1–19.
30. Guyton AC, Hall JE. *Medical Physiology*, 10th ed. New York: WB Saunders, 2000.
31. Rowland M, Tozer TN. *Clinical Pharmacokinetics and Pharmacodynamics Concepts and Applications*, 4th ed. Philadelphia: Lippincott Williams & Wilkins, 2011.
32. Johns DO, Oesterling Owens E, Thompson CM, Hattis D, Krishnan K. *Physiological Parameters and Databases for PBPK Modeling*. Hoboken, NJ, Wiley, 2010, p. 107.
33. Rowland-Yeo K, Rostami-Hodjegan A, Tucker GT. Abundance of cytochromes P450 in human liver: A meta-analysis. *Br J Clin Pharmacol.* 2004;57:687.
34. Proctor NJ, Tucker GT, Rostami-Hodjegan A. Predicting drug clearance from recombinantly expressed CYPs: Intersystem extrapolation factors. *Xenobiotica.* 2004;34(2):151–178.
35. Soars MG, Gelboin HV, Krausz KW, Riley RJ. A comparison of relative abundance, activity factor and inhibitory monoclonal antibody approaches in the characterization of human CYP enzymology. *Br J Clin Pharmacol.* 2003;55(2):175–181.
36. Youdim KA, et al. Application of CYP3A4 in vitro data to predict clinical drug-drug interactions; predictions of compounds as objects of interaction. *Br J Clin Pharmacol.* 2008;65(5):680–692.
37. Ghanbari F, et al. A critical evaluation of the experimental design of studies of mechanism based enzyme inhibition, with implications for in vitro-in vivo extrapolation. *Curr Drug Metab.* 2006;7(3):315–334.
38. Yang J, et al. Cytochrome P450 turnover: Regulation of synthesis and degradation, methods for determining rates, and implications for the prediction of drug interactions. *Curr Drug Metab.* 2008;9(5):384–394.
39. Jamei M, et al. Population-based mechanistic prediction of oral drug absorption. *AAPS J.* 2009;11(2):225–237.
40. Willmann S, et al. Development of a physiology-based whole-body population model for assessing the influence of individual variability on the pharmacokinetics of drugs. *J Pharmacokinet Pharmacodyn.* 2007;34(3):401–431.
41. Bruyere A, et al. Effect of variations in the amounts of P-glycoprotein (ABCB1), BCRP (ABCG2) and CYP3A4 along the human small intestine on PBPK models for predicting intestinal first-pass. *Mol Pharm.* 2010;7(5):1596–1607.
42. Keldenich J. Measurement and prediction of oral absorption. *Chem Biodivers.* 2009;6(11):2000–2013.
43. Paine MF, Hart HL, Ludington SS, Haining RL, Rettie AE, Zeldin DC. The human intestinal cytochrome P450 "pie." *Drug Metab Dispos.* 2006;34(5):880–886.
44. DeWoskin RS, Thompson CM. Renal clearance parameters for PBPK model analysis of early lifestage differences in the disposition of environmental toxicants. *Regul Toxicol Pharmacol.* 2008;51(1):66–86.
45. Center for Disease Control and Prevention. Third national health and nutrition examination survey (NHANES III). Available at: http://www.cdc.gov/nchs/nhanes.htm.
46. Price PS, et al. Modeling interindividual variation in physiological factors used in PBPK models of humans. *Crit Rev Toxicol.* 2003;33(5):469–503.

47. International Commission on Radiological Protection (ICRP). Basic anatomical and physiological data for use in radiological protection: Reference values. ICRP publication 89. Amsterdam: Elsevier Science, 2002.
48. Young JF, Branham WS, Sheehan DM, Baker ME, Wosilait WD, Luecke RH. Physiological "constants" for PBPK models for pregnancy. *J Toxicol Environ Health.* 1997;52(5):385–401.
49. ILSI. Physiological information database. http://www.ilsi.org/ResearchFoundation/Pages/PhysiologicalParametersDatabase.aspx.
50. Abbott NJ, Patabendige AA, Dolman DE, Yusof SR, Begley DJ. Structure and function of the blood-brain barrier. *Neurobiol Dis.* 2010;37(1):13–25.
51. Bouzom F, Walther B. Pharmacokinetic predictions in children by using the physiologically-based pharmacokinetic modelling. *Fundam Clin Pharmacol.* 2008; 22(6):579–587.
52. Koukouritaki SB, et al. Developmental expression of human hepatic CYP2C9 and CYP2C19. *J Pharmacol Exp Ther.* 2004;308(3):965–974.
53. Johnson TN, Tanner MS, Taylor CJ, Tucker GT. Enterocytic CYP3A4 in a paediatric population: Developmental changes and the effect of coeliac disease and cystic fibrosis. *Br J Clin Pharmacol.* 2001;51(5):451–460.
54. Thompson CM, et al. Database for physiologically-based pharmacokinetic (PBPK) modeling: Physiological data for healthy and health-impaired elderly. *J Toxicol Environ Health B Crit Rev.* 2009;12(1):1–24.
55. Bauer M, et al. Age dependency of cerebral P-gp function measured with (R)-[11C] verapamil and PET. *Eur J Clin Pharmacol.* 2009;65(9):941–946.

INDEX

A
ABC efflux transporter, 30–31, 128–129
Absolute neutrophil count (ANC), 353
Absorption, distribution, metabolism, excretion and toxicity (ADMET), 4
Absorption modeling, 76–83
Absorption rate constant, 67, 71, 84, 189, 201, 270
Accumulation, 33, 36, 95–96, 132, 190, 230, 252, 290, 292, 362, 375
ACE inhibitors, 69
Acetylation, 319, 354
Acetylcholinesterase, 271, 354
Acetyl salicylic acid, 319
Acidic phospholipids, 92, 101–103
Active metabolite, 141, 185, 254, 339–340, 344, 375, 391
Adaptation, 234, 301–307, 332, 339, 367, 372
Advanced compartmental absorption and transit model (ACAT), 78
Adverse effect, 301, 313, 357–359, 364
Affinity, 20–21, 29, 33, 38, 96, 122, 129, 186–187, 212, 219, 224–225, 227–229
Agonist, 301–302, 305–306, 308–309
Albumin, 20–21, 92, 95, 100–101, 104, 164, 166–168, 229
Allometric scaling, 239, 242, 252
α_1-acidic glycoprotein (AGP), 20–21
Alpha-2 adrenergic agent, 306
Ambien, 306
Amorphous, 48–49, 64, 83–84
Antagonist, 299, 300, 301, 306, 307, 308, 316, 332, 348, 365, 370, 371, 372, 373
Antibiotics, 59, 69, 184, 349, 354
Antibody-dependent cellular cytoxicity (ADCC), 214, 216–217, 254
Antibody-directed enzyme prodrug therapy (ADEPT), 244

Antidepressant, 21, 306, 353
Antigen
  shed, 231
  soluble, 231
Antisense, 211, 245–254
Antisense oligonucleotides (ASOs), 245, 247, 249, 251
Anxiolytic, 306
A posteriori, 169, 174
Apparent volume of distribution, 95, 114
A priori, 169, 174, 384, 386
Area under the concentration-time curve (AUC), 107
Area under the curve (AUC) ratio, 186, 188–189, 198, 201
Area under the first moment curve (AUMC), 36–37
Arginine vasopressin or antidiuretic hormone (ADH), 212
Aspirin, 319
Autoimmune disease, 338
Autoinhibition, 288, 291–292

B
Basal abundance, 194, 196, 198
Baseline level, 364, 372
Bayesian methods, 171
Bile, 20, 29
Bile acid, 31, 52, 57–58, 128
Bile salt, 57–58, 61, 128, 217, 287
Bile salt export pump (BSEP), 31–33, 128, 163, 187
Binding affinity, 229, 345
Binding site analysis, 6, 10
Bioassay, 229, 254–255
Bioavailability, 4, 9–11, 35
Biological systems modeling, 6, 10
Biologics, 8, 111, 157, 210–211, 239, 368, 416

Physiologically-Based Pharmacokinetic (PBPK) Modeling and Simulations: Principles, Methods, and Applications in the Pharmaceutical Industry, First Edition. Sheila Annie Peters.
© 2012 John Wiley & Sons, Inc. Published 2012 by John Wiley & Sons, Inc.

423

Biomarkers, 247, 311–314, 339, 353, 360, 364, 372, 392–393
Biopharmaceutics classification system (BCS), 60, 64, 83
Biotransformation, 18, 25, 29, 38, 53, 120, 127, 230, 391
Blood–brain barrier, 14, 29, 90, 94, 99, 114, 122, 188, 340, 352
Blood–plasma ratio, 20–23, 96, 98, 104–106, 111–112, 114
Boltzman constant, 47
Bottom-up modeling, 16
Boundary layer thickness, 45, 48
Brain penetration, 93, 105–111, 340
Brain slices, 106, 108–109, 112
Brambell receptor (FcRn), 227
Breast cancer resistance protein (BCRP), 31–33, 59, 62, 74–75, 94, 128, 163, 187

C

Caco-2, 67–71, 74–76, 128, 154, 191, 266, 277–278, 280–281, 283
Canalicular efflux, 123, 128–129, 134, 287
Candidate drug target profile, 344–346
Capacity-limited, 339, 375
Capecitabine, 362
Catabolic degradation, 225, 230
Central nervous system (CNS), 93, 247, 337
Cerebrospinal fluid, 93, 95, 108
Child-Pugh class, 390
Chimeric, 212–214, 227, 251, 255
Chlorthiazide, 188
Cholescystokinin, 57
Chronic dosing, 37, 345, 353
Cimetidine, 52, 188
Cirrhosis, 163, 167–168, 390
Classical receptor theory, 315–320
Clearance, 9, 22–29
Clinical endpoint, 313–314, 339, 372, 375
Clinical trial simulation, 6, 10, 366–367
Coefficient of variation (CV), 171
Collagen, 127, 129, 135, 219
Colloid osmotic pressure, 218
Commercial PBPK software, 158–159
Co-morbidities, 364, 388–393, 395
Compartment, 15
Compartmental absorption and transit model (CAT), 78
Compartmental analysis, 335
Compartmental model, 335
Compartmental PK Modeling, 5–6
Competitive antagonist, 301, 308–310, 334, 372, 374
Competitive inhibition, 194

Complementarity-determining region (CDR), 212–213
Complement cascade, 217
Complement-dependent cell-mediated Cytotoxicity (CDCC), 214, 216–217, 254
Complement-dependent cytotoxicity (CDC), 214, 216–217, 254
Compound-dependent parameters, 81–82, 159, 240, 242, 275
Concentration-effect, 306–315, 320, 335, 339, 357, 361, 373, 375–376
Concentration-response curve, 308–311, 315, 318, 342, 356, 361, 374
Concomitant drug, 184, 393, 395, 402
Conformational changes, 374
Constitutive androstane receptor (CAR), 354
Convective transport, 218–219, 224, 230, 233
Correlated covariates, 171
Co-solvent, 65, 70, 266
Counter regulation, 307, 332–334
Coupling, 14, 305, 320–321, 372, 374
Covariance, 171–172
Critical path initiative, 371, 393
Crystal form, 47
Crystalline nanosuspension, 64, 66
Crystallinity, 45, 65–66, 266
Cyclic adenosine monophosphate (cAMP), 328–330, 375
Cyclic nucleotide phosphodiesterase (PDE), 375
Cyclodextrins, 64, 83
CYP induction, 135, 191–192
CYP reaction phenotyping, 125, 135, 192, 201
Cytochrome P450 (CYP), 8, 19, 53, 121, 124–125, 127, 144, 184–185, 189
Cytomegalovirus (CMV)-induced retinitis, 245

D

Data integration, 9, 263–271, 368, 403
Delayed pharmacological response, 321–332
Desensitization, 305–307, 332, 334, 373, 375–376
Dexamethasone, 129
Diffusion, 29–30, 32, 45–48, 52, 58, 71
Diffusion coefficient, 47, 58, 239
Discontinuous capillaries, 91, 114
Disease model, 345, 349, 364, 369, 371
Dissolution, 19–20, 44–51, 56, 58, 60, 64, 66, 72, 74, 78–79, 167, 264, 270, 278
Distribution
  flow-limited, 157
  permeability-limited, 100, 111, 157–158, 231, 338
Distributional delay, 307, 316, 322, 332–333, 337–338, 340

INDEX

Distribution coefficient, 114
Dosage regimen, 224, 361, 364
Dose estimation, 37–40, 347, 350, 356
Dose proportionality, 47, 132
Dose-response, 307–315, 341–343, 362, 367, 371, 376, 384
Dosing interval, 18, 35–37, 189, 244, 346, 349, 387
Dosing rate, 35–36, 346
Double-stranded RNA, 245, 247–249
Doxorubicin, 66, 390
Drug development, 4–6, 8–9, 11, 34, 48, 108, 280, 301, 315, 360–370, 374, 390, 403
Drug discovery, 4–6, 8–12, 16, 40
Drug-drug interaction (DDI), 8, 29, 33–34, 38, 184–188, 190, 192, 194–195, 197–201, 287–288, 290, 346, 351, 395
Drug effect, 9, 22, 300, 305–306, 308, 315–316, 320, 325, 334, 343–344, 366–367, 373
Drug-food interaction, 59
Druggable, 4, 10
Drug metabolizing enzymes (DMEs), 27, 32, 34, 53–54, 56, 120, 166, 168, 187, 191, 284, 286, 291, 323, 354, 385, 389, 392–393, 401

E

E_{max}, 191–192
EC_{50}, 193–195, 308–310, 313, 319–321, 325, 328, 330–332, 334, 350–352, 355, 359, 373, 376
Efalizumab, 226
Effective permeability, 66, 68, 72–73, 83
Efficacy, 4–6
Effux, 83
Elimination, biliary, 29, 168, 252, 290, 354
Elimination rate constant, 24, 36, 155
Empirical model, 5, 16, 159, 335, 338
Endogenous proteins, 212, 219, 224
Endopeptidase, 223
Endosomal space, 237–238
Endothelial permeability, 219
Enhanced permeability and retention (EPR), 66, 154, 231
Enterocytes, 18–19, 33, 44–46, 51–54, 56, 59, 63, 76, 80, 82–83, 139, 354
Enterohepatic recirculation (EHR), 29, 40, 80, 139, 164, 277–279, 283–285, 294
Environmental factors, 167, 402
Enzyme degradation rate constant, 197
Enzyme induction, 39, 192, 194
Enzyme inhibition, 39, 190, 198, 277
Enzyme-linked immunosorbent assay (ELISA), 255

Enzyme saturation, 26, 36, 54
Enzyme turnover rate, 195, 197–198
Epidermal growth factor receptor (EGFR), 312, 395
Epigenetics, 247, 254, 390
Equilibrium dialysis, 21, 106, 113
Erythropoietin, 211–212, 223–225, 234
Esterases, 53–54, 121
European Medicines Agency (EMA), 186–187, 189, 403
Exclusion volume, 219
Exogenous proteins, 15, 219–220
Exopeptidase, 223
Extended release, 56, 63, 166, 284, 357
Extensive metabolizer, 174, 391
Extravasation, 218–220, 225–227, 230, 232, 240
Extrinsic factors, 393, 395

F

Facilitated diffusion, 29–30, 32, 52
Fasted state small intestinal fluid (FaSSIF), 74, 266, 277–278, 291
Fc-Rn
   protection, 227–230, 232–233, 236–237
Fed state small intestinal fluid (FeSSIF), 74
Fick's first law, 47
First-order, 23–24, 47, 80, 191, 193–194, 197, 230, 236, 322, 324, 334
First-pass metabolism, 20, 53, 163, 340, 390
First time in man, 364
Flip-flop kinetics, 280
Fluorescence activated cell sorting (FACS), 231
5-fluorouracil, 362
Food and Drug Administration (FDA), 4, 184, 186–187, 189, 192, 210, 337, 371, 373–374, 393, 403
Forerunner compound, 344, 346
Formulation, 4, 20, 40
Formulation-dependent factors, 20, 63–66
Fraction absorbed, 44, 56, 58–59, 63–64, 69, 83, 291
Fraction unbound in plasma, 20, 27, 36, 38, 80, 95, 167–168, 277, 280, 346, 348
Fraction unbound in tissue 97
Freudlich–Ostwald equation, 48
Full parametric receptor model, 320–321
Fuzzy simulation, 162
FvFc antibody, 232

G

Gamma amino-butyric acid (GABA) receptor, 301, 303, 305
Gapmer, 251

Gastric emptying, 20, 56–61, 78–79, 164, 166–167, 169, 264, 266, 270–271, 277–278, 280–282, 284, 291, 294, 323, 341–342
Gastric pH, 61, 164, 215, 323
Gastrointestinal tract, 19, 44, 389
Gefitinib, 45, 312, 395
Gene array, 401
Gene therapy, 211
Genetic polymorphism, 163, 165–168, 174, 186, 385, 388, 390, 392
Globulins, 20, 211
Glomerular fitration rate (GFR), 28–29, 40, 124, 141–143, 164, 223, 327, 408, 410, 412, 415
Glucuronidation, 38, 53, 192, 282–284, 354, 364, 392
Glycocholic acid, 57
Glycosaminoglycan, 219
Glycosylation, 224, 227
Glycyrrhizic acid, 364–365
Gonadotropin releasing hormone, 212
Goodness of fit, 275
G-protein coupled receptor (GPCR), 301, 304
Granulocyte colony-stimulating factor (G-CSF), 212, 223–224
Growth factors, 8, 212, 231, 302–303, 312, 375
Guanylyl cyclase receptor, 302
Gut bioavailability, 19, 38, 44–56, 60–62, 83, 166, 189, 201
Gut extraction, 24, 33, 38, 44–45, 54, 56, 62–63, 76–77, 288
Gut microflora, 29, 58, 284

H
Half-life, 18, 22–29
HAMA/HACA/HAHA, 255
Hayduk–Laudie equation, 47
Hematocrit, 23, 167, 323, 390, 408, 410, 412, 415
Henderson–Hasselbalch equation, 49
Hepatic extraction ratio, 27
Hepatic metabolism, 25, 60, 125–127, 189, 193, 225, 265, 293
Hepatic uptake assay, 130–134
Hepato-biliary, 124–137, 144, 153, 287, 290–292
Hepatocytes, 26, 54, 120, 122–123, 125–130
Hepatocytes per gram liver (HPGL), 167–168, 408, 410, 412, 415
Herceptin, 393
Hill equation, 308, 310, 313, 315, 335
Hit identification, 402
HMG-CoA reductase, 95, 187, 303, 337
Homeostasis, 218, 306, 332, 335, 366, 374
Homologous receptor down-regulation, 373

Homology modeling, 6, 10, 125
High-performance liquid chromatography (HPLC), 72
Human ether-à-go-go-related gene (hERG) channel, 10, 354
 modeling, 6, 10
Human gastric fluid (HGF), 74
Human intestinal fluid (HIF), 74
Humanized antibodies, 212, 214, 255
Hydrodynamic volume, 224
Hydrophilic drug, 19, 27–28, 51, 59, 66, 95, 100, 111, 354, 403
Hydrophobicity, 62, 67
Hydroxypropyl-β-cyclodextrin (HP-β-CD), 64
Hypnotic, 306
Hypothesis generation, 9, 274–278, 403
Hypothesis testing, 7, 83, 153, 271, 291, 336–338, 350, 368, 371, 403
Hysteresis, 305, 315–316, 321, 332, 350, 373, 375–376

I
IC_{50}, 190–192
Immunoassay(s), 229, 255
Immunogenicity, 213, 223–224, 227, 255
Immunoglobulin (Ig), 212, 214–215, 217, 229
Indirect response models, 244, 324–332, 334, 355, 364
Inducer, 33, 58, 60, 124, 185, 193–198, 303
Infusion, 18, 34–36, 38, 71, 156, 215, 247, 249, 253, 351–353, 355, 357, 367
Inhibition constant, 190, 193, 196
Inhibitor, 8, 33, 58, 65, 69, 74
Initial estimates, 325, 331
In silico models, 66–67, 98–105, 110, 125
In situ brain penetration, 105
Interfering RNA (RNAi), 247
Interferon, 211–212, 225, 234
Interindividual differences, 15, 56, 341, 364, 392
Interleukins, 212
Internalization, 128, 135, 219, 221–225, 233, 235–238, 240–241, 256, 306, 334, 373
Interspecies scaling, 416
Interstitial fluid, 21, 90, 93, 105–109, 217–219, 228–229
Interstitial space, 93–94, 106–108, 111, 217, 219, 231, 233–234, 237–238
Intestinal loss, 44, 53, 62, 79–80, 82–83, 139, 277–278, 281–284
Intestinal pH, 58, 61, 166, 323, 354
Intestinal secretions, 20, 61
Intramuscular administration (IM), 18, 215
Intravenous administration (IV), 340

INDEX

Intrinsic activity, 308, 323, 373
Intrinsic clearance, 9, 22, 25, 27, 122, 125, 127, 130, 134, 138, 144, 154–155, 167, 192, 194–198, 265, 274–275, 288, 401–403
Intrinsic solubility, 48, 53, 83
Inverse agonists, 301, 373
Investigational new drug (IND) application, 368, 373
In vitro–in vivo correlation, 350
Ionotropic receptors, 302
Iressa, 45, 312, 395
Irinotecan, 391–392
Irreversible inhibitor, 373–374

K
Kinetic solubility, 48, 73, 83

L
Lag time, 169, 197
Latin hypercube method, 171, 174
Lattice energy, 64
L-dopa, 52, 93, 340
Lead generation, 4, 6, 10–11, 48, 83, 263, 268, 344, 346–347, 359, 402
Lead optimization, 6, 8–9, 11, 40, 83, 263, 268, 274, 282, 284, 290, 344, 347, 359, 402
Ligand-receptor complex, 317
Lipid and bile acid transporter (LBAT), 52
Lipinski rule of 5, 76
Lipophilicity, 20, 22, 24, 29, 38, 45, 62, 64, 67, 92
Lipoproteins, 20–21, 92, 100–101, 104, 357–358
Liver slices, 53, 125–127, 134–135, 140
Loading dose, 37, 355
Locked nucleic acid (LNA), 250–251
Log-normal distribution, 165, 169, 174–175
Luminal degradation, 44, 53–56, 78, 82, 282–284
Luminal stability, 67, 74
Lunesta, 306
Luteinizing hormone, 212, 303
Lymphatic system, 56, 217–218
Lymph fluid, 217–218, 237
Lymphocyte marker, 243
Lysergic acid diethylamide, 339
Lysosomal degradation, 221, 224, 229, 234
Lysosomal protease, 223, 228

M
Macrophages, 217, 229, 254, 357–358
Madine–Darby canine kidney MDCK, 67, 70, 74, 105–106, 125, 128, 140, 191
Maintenance dose, 37
Market differentiation, 357, 368, 394–395

Markov chain Monte Carlo methods (MCMC), 162, 171
MATLAB, 157
Matrix-assisted laser desorption/ionization imaging mass spectrometry (MALDI-IMS), 110, 114
Maximum absorbable dose (MAD), 77, 84
Mean residence time (MRT), 36–37
Mechanism-based inhibition (MBI), 184, 197, 201
Mechanism-based PD modeling, 315, 336
Mechanistic PBPK/PD integration, 341–342
Membrane vesicular transporter assay, 129
Metabolic intermediate complex (MIC), 184
Metabolism, 4–6, 8
2′-O-methoxyethyl (MOE), 251
Metabolite, 6, 10, 16, 29–30
Metformin, 69, 95, 188
Michaelis–Menten, 190–191, 194, 308
Micro RNA (miRNAs), 247–248, 255
Midazolam, 390
Migrating motor complex (MMC), 57
Milligram protein per gram liver (MPPGL), 167–168, 408, 410, 412, 415
Minimum inhibitory concentration, 349
Minimum toxic dose (MTD), 174
Miscelle, 46, 58
Model assumptions, 157–158
Model-based drug development, 4, 6, 11,
Monocarboxylic acid transporter (MCT), 52
Monoclonal antibodies (mAb), 8, 211–215, 224–230, 255
Monoprotic acid, 104
Monoprotic base, 102, 104
Monte Carlo simulations, 162
Morpholinos, 251–252
mRNA turnover rate, 197
Multidrug resistance protein (MRP2), 30–33, 59, 62, 74–75, 122, 128, 163
Multidrug and toxin extruder transporters (MATE1 and MATE2K), 30

N
N-acetyl transferase (NAT), 62, 121
Negative feedback, 332
Nephrotoxicity, 190
Neurotransmitters, 302–303
Neutropenia, 224, 392
New chemical entity (NCE), 40, 120, 185
New Drug Application (NDA)/Biologic License Application (BLA), 368, 374
Nicotine, 93, 188, 334, 376
Nitroso group, 184
Noncompartmental analysis, 230

Noncompetitive (unsurmountable) antagonist, 301, 310, 374
Noncompetitive inhibition, 125, 184, 190, 301–302, 310
Nonlinear pharmacokinetics, 34
Normal distribution, 175
Normalized sensitivity coefficient, 172, 269
Noyes–Brunner equation, 45, 47–48, 64
Na$^+$-taurocholate cotransporting polypeptide (NTCP), 31, 128
Nuclear receptor, 190, 197–198, 301–302, 304
Nuclease degradation, 249, 251

O

OATP1B1, 30, 32–33, 38, 95, 128, 163, 167, 187, 190, 290
OATP1B3, 32, 95, 163, 187
OATP2B1, 31–32, 95
OCT, 31, 52, 69
Octanol-water partition coefficient, 66–67, 77, 100
Oligodeoxynucleotides (ODN), 249, 252–253
Oligonucleotides
  2'-O-methyl, 251
  antisense, 245
  first-generation, 249
  morpholino, 251
  phosphodiester, 247, 249
  phosphorothioate second generation, 249
Oligopeptide transporter, 31, 52
Omeprazole, 319, 344
Onset of action, 280, 291
Opsonization, 226, 255
Organic anion transporter (OAT), 31, 127–128
Organic cation transporter (OCT), 31, 287
Osmolarity, 56, 60, 84
Oxytocin, 212

P

Pancreatic secretions, 59–60
Parallel artificial membrane permeability assay (PAMPA), 68
Parameter sensitivity analysis, 9, 270–271
Partial agonist, 301–302, 308, 311, 374
Particle disintegration, 167
Particle size, 20, 47–48, 56, 64, 167, 266, 277
Partition coefficient, 98–105
PBPK/PD modeling, 6, 10–11, 230, 244, 340–341, 357, 362, 367–368
Pegfilgrastim, 224
Pegylation, 224, 255
Peptide nucleic aid (PNA), 250–252
Perfusion, 54, 67–68, 71–72
Perfusion rate, 54, 71–72, 91, 111, 157

Peripheral blood mononuclear cells, 253
Permeability
  carrier-mediated, 51–53
  effective, 66, 68, 72–73, 83
  paracellular, 51–53
  passive, 63, 66–68, 74, 84, 94, 130, 142, 266, 288
  transcellular, 51–53
Perpetrator, 33, 184–186, 188, 190, 194–195, 198, 291
Personalized medicine, 10, 392–395, 403, 405
P-glycoprotein, 128, 187
Pharmacodynamic response half-life, 317, 374
Pharmacodynamics, 221, 227, 299
Pharmacogenetics, 393
Pharmacogenomic biomarkers, 312, 372, 392–393, 395
Pharmacokinetic parameters, 159, 163, 386
Pharmacokinetics, 209–253
Pharmacological target, 219–220, 225, 237–238, 256, 300–302, 357
Pharmacometrics, 371, 374
Pharmacophore modeling, 6, 125
Phase I, II and III clinical trials, 362
Phosphorodiamidate morpholino oligonucleotide (PMO), 251
Phosphorylated-CRKL, 312
Phosphorylated-EGFR, 312
Physiological limit, 329
Physiological model, 13–16, 43–83
Physiological parameters, 33, 56, 62, 82, 102, 111, 136, 159, 162, 166, 169, 171–173
Piwi, 247
PK_a, 22, 49–51, 63, 92, 98, 100, 102
PK/PD, 336
PK/PD correlation, 321, 339–341, 349
PK/PD modeling, 5–6, 10–11, 159, 230–244, 300, 335–336, 340, 353, 357, 362, 364, 367–368, 371
Polar surface area (PSA), 29, 63, 66, 84
Polyadynylation, 245
Polyethylene glycol (PEG), 64, 224, 255
Polymorphic forms, 48, 64, 83–84
Polymorphism, 29, 40
Polymorphic enzymes, 40
Population modeling, 6, 9, 11, 361, 384–386, 388
Positron emission tomography (PET), 110, 231
Potency, 6, 40, 187, 191, 263, 308
Precursor pool depletion, 307, 332
Prediction of PK Properties, 6, 11
Pregnane X receptor (PXR), 190, 197, 354
Pre-systemic metabolism, 72, 218
Primary hepatocytes, 125–126

INDEX **429**

Probenecid, 168
Procainamide, 188
Protein-based biologicals, 247
Protein binding, 18, 20–23, 92, 97, 120, 141, 154, 164, 184, 190, 247, 252, 293, 342, 345–346, 350, 353, 355, 401
Protein modeling, 6, 11
Proteolysis, 215, 223–225, 230
Proteolytic degradation 220, 224
Proteomics, 4, 313
Proximal tubular cells, 190

Q
QSAR and QSPR models, 11
QTc prolongation, 354
Quantitative structure bioavailability relationship (QSBR), 77
Quasi irreversible, 191, 375

R
Rate and extent of tissue distribution, 91–92, 105, 110, 120
Rate of inactivation, 197
Reabsorption, 28, 82, 124, 136, 139, 141–143, 327
Reaction phenotyping, 125, 135, 192, 201
Reactive metabolite prediction, 6, 11
Rebound, 301–307, 332, 339, 346, 352, 357, 359, 372, 374–375
Receptor-mediated endocytosis, 19, 51, 219, 222, 225, 256
Receptor occupancy, 301, 313, 317, 319–320, 322, 342–343, 352, 356, 367, 372, 374–375
Receptor reserve, 320, 375
Receptor tyrosine kinase, 302
Recombinant protein-based therapeutics, 212
Renal
  elimination, 11, 22, 28–29, 32, 34, 155, 168, 223–226, 247, 252, 354
  excretion, 120, 122, 124–136, 139, 141–144, 163, 223, 252
  lymphatic, 56, 154, 217–220, 224–226, 230–231, 239
Renal tubular uptake, 188
Reversible enzyme inhibition, 190
RNAaseH, 245
RNA-Induced Silencing Complex (RISC), 247–248, 255
RNA interference (RNAi), 211, 245–247, 256
Route of drug administration
  buccal, 18, 218, 313
  infusion, 18
  inhalational, 293

intramuscular, 18, 215
intravenous, 18
per oral, 18
rectal, 18, 218
subcutaneous, 18, 215, 247
sublingual, 18
topical, 18
transdermal, 18, 218

S
S9, 76–77, 125–126, 135, 140
Safety, 4–6, 11
Safety margins, 22, 353, 355–356, 359, 367, 369
Sandwich hepatocytes, 129–130
Saturation solubility, 45, 48
Scaffold hopping, 6, 12
Scaling, 22
Scaling factor, 47, 76, 130, 136, 140, 167, 194, 402
Second messenger, 301, 329, 354, 373, 375
Selective serotonin re-uptake inhibitors (SSRI), 306, 353
Semiparametric receptor model, 320
Semisynthetic opioids, 337
Sensitivity analysis
  global, 175, 271
  local, 172
Sensitivity coefficient, 172, 269
Sensitization, 301–307, 332, 339, 375
Serotonin re-uptake inhibitors, 306, 337
Sigmoid E_{max} model, 307–315
Signature discrepancy, 277, 288
Simulated gastric fluid (SGF), 74
Simulated intestinal fluid (SIF), 74
Single chain Fv, 232
Site of action, 111–113, 210, 218
Site of metabolism prediction, 6, 12
SLC antiport transporters, 30
Small interfering RNA (siRNA), 247–248, 251–252, 255
Small intestinal secretions, 20
Software, 78, 157–159, 186, 361, 405
Solubility, 20, 34, 38, 44–51
Solubility product, 51
Somatostatin, 212
Spare receptors, 315, 320, 375
Species differences in physiology, 56–62, 91–98, 341, 353
Spingosine-1-phosphate receptor, 338
Static models, 8, 186, 193–195, 198–201
Steady state pharmacokinetics, 34–37
Steady state volume of distribution (V_{ss}), 37, 133, 227, 230

Steric hindrance, 38, 40, 224
Stokes–Einstein equation, 47
Structure-activity relationship (SAR), 405
Subcutaneous administration, 247
Subpopulation, 163–164
Subpopulation modeling, 388
Sulphation, 354
Surface area, 29, 45, 47, 53, 59, 63–64, 66, 71–72, 77–78, 84, 93, 95, 134, 138, 163–164, 166, 234, 239, 253, 266, 386, 401
Surrogate biomarker or endpoint, 313–314, 375
Systemic availability, 218, 249

T

Tachyphylaxis, 305–306, 375
Tamoxifen, 391
Target density, 229
Target engagement, 313, 350, 352, 359–361, 372, 375, 390, 394
Target-mediated disposition, 256
Target-mediated drug disposition, 219, 221, 225, 235, 337, 339, 375–376
Target patient population, 386–388
Target population, 16, 40, 267–268, 341, 362, 370, 386, 388–389
Target tissue concentration, 113, 231, 300–301, 313, 315, 337–338, 350
Target validation, 359, 376, 402
Taurocholic acid, 57
TGN1412, 230
Therapeutic macromolecules, 211, 230–231, 244
Therapeutic window, 22, 39, 56, 124, 174, 188, 267, 306, 360, 366, 384–385
Thermodynamic or equilibrium solubility, 48, 72–74, 84
Tight junction porosity, 62
Time-dependent inhibition (TDI), 184–185, 191, 194–195, 197–198, 201
Tissue composition, 92, 95, 97–98, 110, 140, 165–167, 339, 356, 386, 409, 415
Tissue partition coefficient, 24, 80, 97–105, 154, 239, 253, 274–275, 337, 350
Tissue uptake, 219, 231–232, 252
Tolerance, 301–305
Top-down modeling, 16
Transcytosis, 19, 51, 94, 217, 219–220, 224, 226, 229–230

Transduction, 10, 307, 312–313, 315–316, 319–320, 322, 324, 332–333, 335, 339, 342, 374, 376
Transfected cell lines, 125, 128, 135
Transit time, 58–62, 84, 166–167, 341, 354, 364
Translational PBPK/PD, 353–356
Transporter-based inhibition, 186
Transporters
  efflux, 20, 30–31, 33, 59, 74–75, 82–83
  uptake, 31, 33, 52, 59, 95, 99, 122, 128, 167, 187–188, 190–191, 195, 200
Trastuzumab, 393–394
Tubular reabsorption, 28, 136
Tubular secretion, 28, 124, 389
Turnover, 129, 184, 317, 324–325, 335
Turnover delay, 340
Turnover systems, 325, 350
Two-pore formalism, 218, 230, 232

U

Ubiquitin-proteasome pathway, 223
Uncertainty, 161–174
Unstirred water layer, 44, 46, 58, 61, 69
Uridine 5'-diphospho-glucuronosyltransferase (UGT), 19, 53–54, 59, 121, 125, 135, 166, 186–187, 192–193, 230, 282, 286, 361, 403
Ussing chamber, 67, 70, 76, 284

V

Value chain, 5, 10, 267–268, 312, 340, 344–345, 401–405
Variability, 161–174
Victim, 33–34, 38, 184–190
Villi, 58–59, 61, 201
Virtual screening, 4, 6, 11–12
Viscosity, 47, 59–60
Voltage-gated ion channels, 303

W

Watson–Crick hybridization, 249
Weibull distribution, 47, 169–170, 175
Well-stirred model, 26, 196
Wettability, 47
Whole-body autoradiography (WBA), 108–110, 114

X

Xenobiotic, 58, 92–98, 120–125, 130